LITTLE ICE AGES

Since *The Little Ice Age* was published in 1988, interest in climatic history has grown rapidly and research in the area has flourished. A vast amount of new data has become available from sources such as ice cores, speleothems and tree rings. The picture that we have of past climates and glacier oscillations has extended further into the past and has become more detailed. However, the knowledge of climate change on the decennial and centennial timescale, to which glacier history can contribute, is scarce and is in demand when attempting to predict future change, especially with regard to global warming.

New chapters and material have been included throughout the book, which tend to confirm and elaborate on the conclusions of the first edition. The glacial evidence has been presented in the context of the oceanographic and icecap studies that have provided such exciting results. *Little Ice Ages* is structured in three parts:

- Part 1 details the evidence for glacier variations in the last thousand years in different parts of the world and the associated climatic fluctuations.
- Part 2 brings together the evidence for the timing of glacier variations in the course of the Holocene.
- Part 3 views the Holocene record in a longer time context, especially as it appears in ice cores, and goes on to consider the likely causes of climatic variability on a Little Ice Age timescale and some of its physical, biological and human consequences.

It becomes apparent in *Little Ice Ages* that the glacier record provides a valuable indication of the nature of climatic fluctuations on the land areas of the globe. The record points to periods of cooling which were more numerous and less continuous than was believed to be the case twenty years ago. There appears to be no single explanation for the variability. Volcanism, solar variability and ocean currents have all played their parts and prediction continues to present many problems. Some authorities have thrown doubt on the existence of the Little Ice Age, but *Little Ice Ages* makes the case for a climatic sequence that can usefully be called the Little Ice Age and which had predecessors occurring at intervals of several centuries throughout much of the last 10,000 years.

The late **Jean M. Grove** was Fellow and Director of Studies in Geography at Girton College, Cambridge.

Routledge Studies in Physical Geography and Environment

This series provides a platform for books which break new ground in the understanding of the physical environment. Individual titles will focus on developments within the main subdisciplines of physical geography and explore the physical characteristics of regions and countries. Titles will also explore the human/environment interface.

1 Environmental Issues in the Mediterranean
J. Wainwright and J. B. Thornes

2 The Environmental History of the World
Humankind's Changing Role in the Community of Life
J. Donald Hughes

3 History and Climate Change
A Eurocentric Perspective
Neville Brown

4 Cities and Climate Change
Harriet Bulkeley and Michelle Betsill

5 Little Ice Ages
Ancient and Modern
Second edition
Jean M. Grove

LITTLE ICE AGES
Ancient and Modern

VOLUME I

Second edition

JEAN M. GROVE

Routledge
Taylor & Francis Group

LONDON AND NEW YORK

First published 1988 as The Little Ice Age
by Methuen & Co. Ltd

Second edition 2004

Published 1990, 2001, 2002, 2003 by Routledge
2 Park Square, Milton Park, Abingdon, Oxon, OX14 4RN
52 Vanderbilt Avenue, New York, NY 10017

First issued in paperback 2018

Routledge is an imprint of the Taylor & Francis Group, an informa business

British Library Cataloguing in Publication Data
A catalogue record for this book is available from the British Library

Library of Congress Cataloging in Publication Data
Grove, Jean M.
Little ice ages: ancient and modern / Jean M. Grove. – 2nd ed.
p. cm. – (Routledge studies in physical geography and environment; 5)
Rev. ed. of: The Little Ice Age. 1988.
Includes bibliographical references and indexes.
1. Climatic changes. 2. Glaciers. I. Grove, Jean M. Little Ice Age. II. Title. III. Series.
QC981.8.I23G76 2004
551.6′09–dc22
2003016884

Typeset in Galliard by Graphicraft Limited, Hong Kong

ISBN 13: 978-1-138-86707-9 (pbk) (vol 1)
ISBN 13: 978-0-415-33422-8 (hbk) (vol 1)
ISBN 13: 978-0-415-09948-6 (set)

CONTENTS

List of plates   vii
List of figures   x
List of tables   xvi
Preface   xviii
Acknowledgements   xx
Notes on the text   xxiii

VOLUME I

1 INTRODUCTION   1

PART 1 THE LITTLE ICE AGE OF THE SECOND MILLENNIUM   11

2 ICELANDIC GLACIERS AND SEA ICE   13

3 SCANDINAVIA   61

4 THE MONT BLANC MASSIF   103

5 THE ÖTZTAL, EASTERN ALPS   125

6 SWITZERLAND   153

7 SOUTHERN EUROPE: THE PYRENEES, MARITIME ALPS AND APENNINES   186

8 ASIA   219

9 NORTH AMERICA   257

10 ARCTIC ISLANDS   294

11 LOW LATITUDES: TROPICAL LATIN AMERICA, EAST AFRICA AND NEW GUINEA 313

12 SOUTHERN HEMISPHERE MID LATITUDES: THE SOUTHERN ANDES AND NEW ZEALAND 330

13 ANTARCTICA AND THE SUB-ANTARCTIC ISLANDS 365

14 LITTLE ICE AGE CLIMATE 371

VOLUME II

PART 2   **THE HOLOCENE** 403

15 GLACIAL HISTORY OF THE HOLOCENE 405

PART 3   **CONTEXT, CAUSES AND CONSEQUENCES** 507

16 THE LITTLE ICE AGE IN CONTEXT: ICE CORE EVIDENCE OF QUATERNARY ENVIRONMENTAL FLUCTUATIONS 509

17 CAUSES OF THE LITTLE ICE AGE AND SIMILAR FLUCTUATIONS 560

18 CONSEQUENCES OF THE LITTLE ICE AGE CLIMATIC FLUCTUATION 591

19 A SUMMING UP 642

Bibliography 644
Index 715

PLATES

Frontispiece Jean and Dick Grove on the Aletsch Glacier, 1961   xxvi

2.1   Landsat image of Vatnajökull taken on 22 September 1979   34
3.1   Tverrbreen, Jostedal, by Johannes Flintøe, 1822–33   80
3.2   Bersetbreen in Krundalen, by J. C. Dahl, 1844   80
3.3   Schematic sketches of glaciers at the head of Krundalen   81
3.4   Lodalsbreen and Stegholtsbreen   81
3.5   Nigardsbreen seen from Elvekroken in 1839, by J. C. Dahl   82
3.6   Nigardsbreen in 1839 by J. C. Dahl   82
3.7   Nigardsbreen by Joachim Frich   83
4.1   The Mer de Glace in 1823, by Samuel Birmann   114
4.2   The shrunken Mer de Glace seen from Montenvers, August 1958   118
4.3   Miage moraines with the bed of the Lac de Combal, August 1958   121
5.1a The advancing Vernagtferner, drawn by a Capuchin friar, Georg Respichler, portraying
      the glacier as it was on 16 May 1678   130
5.1b The original key to Respichler's drawing   131
5.2   Latschferner in August 1772   135
5.3   The Gurglersee in 1772   136
5.4   The Rofnersee in August 1772 as seen from a position on the Zwerchwand   137
5.5   The Vernagt ice in 1772 seen from a distance   138
5.6   Hauslab's map of Vernagtferner in 1817   139
5.7   The end of the Vernagtferner in 1847   139
5.8   The Vernagtferner and Guslarferner in 1888   144
5.9   The Gurglerferner in 1888   145
6.1   The dam formed by the Allalingletscher in the background, with the extended
      Schwarzberggletscher entering the Mattmarksee in the middle distance, by Thalés Fielding   154
6.2a The Rhône Glacier in 1777 by Alexandre-Charles Besson   162
6.2b The Rhône Glacier in 1835 sketched by Thomas Fearnley from a site above the glacier   162
6.2c The Rhône Glacier in 1848 in an aquarelle by Hogard showing the series of moraines
      left during retreat   163
6.2d The Rhône Glacier in 1862 when painted by Eugen Adam   163
6.2e The Rhône Glacier in 1912   164
6.3a The Glacier de Giétroz spilling over into the upper Val de Bagnes   167
6.3b The Giétroz dam in May 1818   167

6.4	Ignace Venetz about 1815, painted in oil by Laurent Ritz	168
6.5a	The Unterer Grindelwald, painted by Joseph Plepp	170
6.5b	The front of the Lower Grindelwald in 1762, painted by Johann Ludwig Aberli	170
6.5c	The Unterer Grindelwald in 1826 by Samuel Birmann	171
6.5d	Grindelwald and the Unterer Grindelwaldgletscher photographed in 1885 by Jules Beck	171
7.1	Monte Perdido, September 1981	192
7.2	Aerial photograph of the Néouvielle Massif in 1983	195
7.3a	Glacier du Pays Baché, 13 August 1963	196
7.3b	Glacier du Pays Baché, 10 September 1989	196
7.4	'Le Glacier de la Maladeta vu de la Pena Blanc'	197
7.5	'Villa Russell' in the 1890s	198
7.6	The SW face of Balaitous	200
7.7a	Glacier d'Ossoue, Vignemale, 19 September 1911	203
7.7b	Glacier d'Ossoue, Vignemale, 12 August 1988	203
7.8a	The glaciers of the Cirque de Gavarnie in 1906	204
7.8b	The glaciers of the Cirque de Gavarnie, 10 August 1989	204
7.9a	Glacier des Oulettes, 3 August 1924	206
7.9b	Glacier des Oulettes, 15 August 1988	206
7.10	Glacier Las Néous, 1905, showing bare ice surface and approximate ice extent in 1989	209
7.11a	Glacier Maledia, Italy, August 1949	214
7.11b	Glacier Maledia, Italy, 30 August 1974	214
7.12a	Calderone Glacier in 1949	217
7.12b	Calderone Glacier in 1990	217
8.1	The Himalaya between Mt Everest and Sikkim (Paiku Hu, 29.0°N, 86.5°E), 6 October 1984	231
8.2	The Sachen Glacier, Nanga Parbat with Sango Sarr Lake, painted by Adolf Schlagintweit in 1856	242
8.3	The Sachen Glacier and its moraines with Sango Sarr in 1958	242
8.4	Northwestern Tibet and the Kun Lun from the Shuttle in October 1984	248
8.5	The growth of lichen on rocks being measured in order to calibrate a lichen growth curve for the Urumqui region	251
8.6	Glacier No. 1, the most extensively studied glacier in China, August 1985	252
8.7	Holocene and Little Ice Age moraines in front of an unnamed glacier, near Glacier No. 1, August 1985	253
9.1	Moraines left by the retreat of the Emmons Glacier on Mt Rainier, July 1970	261
9.2	The Columbia icefield with its Holocene and Little Ice Age moraines, July 1970	267
9.3	William Vaux photographing Illecillewaet Glacier, which was already in retreat, in August 1898 from 'observation rock'	272
9.4	Mary Vaux at the foot of the Illecillewaet Glacier, 17 August 1899	273
9.5	The smooth tongue of the retreating Illecillewaet Glacier in the summer of 1902	273
11.1a	The Kibo glaciers as seen from about 4330 m in the 1880s	326
11.1b	The Kibo glaciers as seen from about 4330 m in 1986	326
12.1	The calving front of the San Rafael Glacier in 1986	335
12.2	The Southern Patagonian Icefield from the west, with Lago O'Higgins, Lago Viedma and Lago Argentino, 10 March 1973	341
12.3	The Southern Patagonian Icefield seen from the east with Lago Argentino, 31 October 1985	342

12.4	View from Mt Cook range to the Tasman and Murchison glaciers, by Julius Von Haast	348
12.5	Watercolour of the Tasman and Murchison glaciers painted by John Gully on the basis of Julius Von Haast's sketch	348
12.6	The moraine-covered surface of the Mueller Glacier, typical of the glaciers east of the watershed	350
12.7	The Franz Josef Glacier as portrayed by Sir William Fox in 1872	356
12.8	The Franz Josef Glacier in about 1905	356
12.9	The Franz Josef Glacier at the end of the 1930s	357
12.10	The Franz Josef Glacier during its advance in 1985	358
12.11	The Franz Josef Glacier in January 2002 as seen from Sentinel Rock	358

FIGURES

1.1   The relationship between changes in climate causing periods of positive and negative
        mass balance leading to glacial expansion and recession and possible methods of dating
        mentioned in the text   5
1.2   The relationship between ^{14}C years and calendar years calibrated from tree rings   5
1.3   Lichenometric dating curves, relating thalli diameter to age, differ according to climate
        and also are affected by the rock type of the substrate   7
1.4   Winter snowfall in Scotland in the Little Ice Age   7
1.5   Tree-ring data collected from the sites shown have been used to construct ring-width
        indices for the period from the mid sixteenth century to the late twentieth centuries   8
1.6   Comparison of independently derived indicators of environmental and climatic change
        covering both the Medieval Warm Period and the Little Ice Age   9
2.1   Iceland: major icecaps, administrative boundaries and some localities mentioned in the text   14
2.2   Surface currents in the seas around Iceland   16
2.3a   Koch's (1945) generalized diagram of the severity of sea ice incidence around Iceland
        from the ninth to the twentieth centuries   17
2.3b   Bergthórsson's (1969) diagram of the decadal incidence of severe sea ice years around
        Iceland from the tenth to the twentieth centuries   17
2.4   Decadal running means of annual mean temperature extended back from 1850 to
        1600 on a basis of sea ice incidence   18
2.5   Incidence of sea ice around Iceland from the seventeenth to the twentieth centuries   19
2.6   The incidence of mild and cold seasons in Iceland and of sea ice around the coast
        from AD 800 to 1598   21
2.7   Decadal sea ice index for Iceland, 1601–1780   22
2.8a   Decadal winter/spring thermal index for Iceland, 1601–1780   23
2.8b   Regional thermal indices for Iceland, 1601–1780   23
2.9   Monthly sea ice positions around Iceland during the periods 1886–92, 1965–71
        and 1987–93   24
2.10   Temperature and sea ice in the Icelandic sea area, 1950–75   25
2.11   Mýrdalsjökull and Eyjafjallajökull   26
2.12   The tongue of Sólheimajökull in 1704–5, 1886 and 1904   29
2.13   Eyjafjallajökull–Mýrdalsjökull in 1795   31
2.14   Cross-section view of Eyjafjallajökull–Mýrdalsjökull in 1795   32
2.15   Frontal fluctuations of Eyjafjallajökull–Mýrdalsjökull glaciers   32
2.16   Vatnajökull in relation to geological boundaries and principal volcanic features   33

2.17	The most important surging lobes of Vatnajökull	36
2.18	Skeidarárjökull, Breidamerkurjökull and the glaciers of Öræfajökull	39
2.19	Heinabergsjökull and its ice-dammed lakes	43
2.20	Skeidarárjökull and its environs in 1794	44
2.21	Vatnajökull in 1794	45
2.22	Twentieth-century fluctuations of Vatnajökull's outlet glaciers	52
2.23	Drangajökull	54
2.24	Fluctuations of Drangajökull's tongues, 1932–96, showing advances in the 1930s and 1940s	57
2.25	The percentage of advancing, retreating and stationary non-surging glaciers in Iceland, 1930–93	58
2.26	Change in average monthly temperature from 1931–60 to 1961–90	59
3.1	Jostedalsbreen and other glaciers mentioned in the text	62
3.2	Jostedalsbreen	63
3.3	Farms in the parishes north of Jostedalsbreen where there were tax reductions in 1340 or where the land was deserted	64
3.4	Late Holocene glacier activity at Sandskardfonna, Jostedalsbreen area, western Norway	65
3.5	Radiocarbon dates relating to early Little Ice Age glacial advances in Scandinavia	66
3.6	Farms in upper Jostedalen resettled in the late sixteenth and early seventeenth centuries, together with the dates at which they reappeared in the tax rolls	69
3.7	Farms in Stryn and Olden Skipreider which suffered serious physical damage during the Little Ice Age as recorded in *avtak* records	72
3.8	Bødal and Nesdal farms in relation to Kjenndalsbreen	74
3.9	Bersetbreen in 1851, showing the tongue in retreat	84
3.10	The forefields of Austerdalsbreen, Tunsbergdalsbreen and Nigardsbreen, all three providing evidence of rapid retreat after 1930	85
3.11	Retreat of Jostedalsbreen tongues from 1750 to 1993	86
3.12	Thinning of Nigardsbreen since the mid eighteenth century	88
3.13	Cumulative net mass balance of Nigardsbreen, 1962–97	89
3.14	The forefield of Storbreen, Jotunheimen with lichenometric dates of moraines	89
3.15	Winter, summer and net balance of Nigardsbreen, 1962–2001, and of Storbreen, 1949–2001	90
3.16	Accumulated mass balances of Scandinavian glaciers	91
3.17	Folgefonni: farms in the vicinity seriously damaged by landslides, avalanching or flooding between 1663 and 1815	93
3.18	The Svartisen area, northern Norway, showing changes of position of the ice margin during the twentieth century	96
3.19	Lichen growth curve for the Svartisen area	97
3.20	The frontal positions of Engabreen and Österdalsisen since the mid eighteenth century	98
3.21	The frontal positions of selected tongues of Folgefonni (Buarbreen and Bondhusbreen) and Svartisen (Engabreen and Fondalsbreen), together with valley glaciers in Jotunheimen (Styggedalsbreen and Storbreen) and in the Kebnekaise massif of northern Sweden (Storglaciären)	99
4.1	Glaciers of the Mont Blanc massif in the mid twentieth century	104
4.2	Sketch of the Brenva in 1742 by J. D. Forbes (1853)	105
4.3	Tongue of the Mer de Glace	108
4.4	Tongues of the Trient, Tour and Argentière glaciers at various times since their maximum Little Ice Age extents	109

4.5	Sketch of the Miage Glacier by J. D. Forbes (1853)	111
4.6	The Triolet rockfall	112
4.7	Advances and retreats of the major glaciers of Mont Blanc since 1820	115
4.8	Frontal positions of the Brenva, 1818–1979	116
4.9	Glacier de Trient, 1845–1990	119
4.10	Changing height of cross-profiles at specified positions indicates the passage of waves down the Mer de Glace since 1900	122
4.11	Frontal positions of L'Allée Blanche (Lex Blanche)	123
4.12	Frontal positions of the Pré de Bar in the twentieth century	123
4.13	The fluctuations of Mont Blanc glaciers are closely paralleled by those of the Grindelwald Glacier	124
5.1	The principal glacier concentrations in the Austrian Alps	126
5.2	The glaciers of the Ötztaler Alpen	127
5.3	The Schlagintweits' map of the mid nineteenth century	140
5.4	The 1843–5 advance of the Vernagtferner	142
5.5	The diminishing extent of Schalfferner, Mutmalferner, Marzellferner and Niederjochferner between the mid nineteenth and mid twentieth centuries	146
5.6	The fluctuations of the Ötztal glaciers since regular measurements began until 1992	147
5.7	Changes in extent of the tongues of the Hintereisferner and Kesselwandferner since 1847	148
5.8	Annual changes in elevation of four glaciers in the Ötztaler Alpen, 1889–1990	149
5.9	Distribution of advancing, stationary and retreating glaciers in Austria amongst those observed for the year 1983/4	149
6.1	Distribution of glaciers in Switzerland	154
6.2	Calendar ages (AD) indicating the initiation of the Little Ice Age in Switzerland	156
6.3	The fluctuations of the Grosser Aletsch Glacier over the last 3200 years and the Gorner Glacier over the last 1200 years	157
6.4	The advance of the Gorner Glacier in the fourteenth century	158
6.5	A comparison of the fluctuations of the Grosser Aletsch and Gorner glaciers during the Little Ice Age	159
6.6	Variation in position of the front of the Rhône Glacier between 1602 and 1914	165
6.7	The Glacier de Giétroz and its setting	166
6.8	The frontal positions of the Unterer Grindelwaldgletscher, 1590–1970	169
6.9	Advances and retreats of Swiss glaciers in various parts of the country since c.1890	172
6.10	The fluctuations of the Rhône, the Unterer Grindelwald, the Fiescher and the average of 23 Swiss glaciers compared with tree-ring density curves	173
6.11	A comparison of the behaviour of glaciers in Austria, Switzerland and France since 1890	174
6.12	Cumulative mass balances of Alpine glaciers	175
6.13	Seasonal weighted thermal and wetness decadal indices for Switzerland, 1550–1820	180
6.14	Movements of the terminal snouts of the Grindelwald Glacier in comparison with 5-year moving averages of weighted thermal and wetness indices for the summer months (June, July and August), number of rainy days in the summer months as departures from 1901–60 averages, and summer temperatures in Basel as departures from 1901–60 averages	183
6.15	Distribution of advancing, retreating and stationary Swiss glaciers amongst those with fronts observed for the year 1974/5	184
7.1	The principal massifs of the Pyrenees	187
7.2	The Cirque de Gavarnie showing areas covered by ice in 1874 and 1906	189

7.3a   Sketch map of the glaciers and Little Ice Age moraine on the Maladeta massif   190
7.3b   Panorama of the Monts-Maudits (Maladeta massif) as seen from the Port de Vénasque
        to the northeast   190
7.4   Sketch map of the Luchonnais glaciers   191
7.5   The Glaciar NE on Monte Perdido   192
7.6   Sketch map of Glacier d'Ossoue   193
7.7a   Sketch map of the glaciers of the Néouvielle massif   194
7.7b   Pays Baché Glacier, 1906   194
7.7c   Pays Baché Glacier in 1856, 1906 and 1989   194
7.8   The glaciers and snowfields of the Balaitous Massif   200
7.9   Sketch map of the Glacier des Oulettes   202
7.10   The glaciers near the Brèche de Roland in 1925 and 1927   207
7.11a   Recent history of advance and retreat of the ice extending from Port d'Oo to the
        Portillon d'Oo   210
7.11b   Cross-section of the Glacier du Taillon to show decrease in thickness during the
        twentieth century   211
7.12   Location of Italian and French glaciers mentioned in the text   212
7.13   Mt Maledia and Mt Clapier and the glaciers in the Mercantour National Park   213
7.14   The Calderone Glacier   215
8.1   Glacierized and snowline elevations in Asia outside China and the Himalaya   220
8.2   Changes in mass balance of representative glaciers in the Polar Urals, Caucasus and
        western Tien Shan from 1958/9 to 1989/90   221
8.3   Accumulated mass balance of the Tuyuksu Glacier as measured from 1957 to 1990, and
        reconstructed from 1937 to 1956   225
8.4   Key radiocarbon dates relating to Little Ice Age advances in the Altai   227
8.5   Retreat of Altai glaciers in the nineteenth and twentieth centuries   228
8.6   The western Himalaya with Nanga Parbat and Karakoram   230
8.7   The Everest area   234
8.8   Key radiocarbon dates relating to Little Ice Age advances in the Everest area   235
8.9   Advances and retreats of Himalayan glaciers since 1820   236
8.10   The Karakoram glaciers   244
8.11   Comparison of the fluctuations of Himalayan glacier termini with those of the
        Trans-Himalaya since the beginning of the nineteenth century   247
8.12   Glacierized areas in China   249
8.13   Key radiocarbon dates relating to Little Ice Age advances in China   253
9.1   Key to glacier regions of North America mentioned in the text   258
9.2   Nisqually Glacier frontal positions   260
9.3   Cumulative mass balance for South Cascade and Thunder Creek glaciers, 1885–1975   262
9.4   Fluctuations of termini of South Cascade, Le Conte, Dana and Chickamin glaciers and
        moraines and retreat stages of the South Cascade Glacier   263
9.5   Location of dated Little Ice Age moraines in the Canadian Rockies and Premier Ranges   266
9.6   Long tree-ring chronologies and glacier fluctuations in the Canadian Rockies   268
9.7   Dated Little Ice Age moraines in the Canadian Rockies and Premier Ranges   269
9.8   Sampling sites and dated trees at Peyto Glacier   270
9.9   Glacier retreat in the Canadian Rockies   275
9.10   Key radiocarbon dates relating to Little Ice Age advances in the coastal range of British
        Columbia   276

9.11	Glacier Bay in Alaska	277
9.12	Key radiocarbon dates relating to advances of Juneau Icefield glaciers	279
9.13	Lemon Creek Glacier retreat stages since the mid eighteenth-century maximum	280
9.14	Lituya Bay: position of glacier fronts as depicted on the map of the Canadian–Alaskan Boundary Commission of 1894 superimposed on La Pérouse's map of 1799	282
9.15	Distribution of values of accumulation area ratio (AAR) over western North America in 1962 in relation to glacier advance and retreat	289
9.16	Cumulative mass balance of selected glaciers in arctic, cordilleran and maritime North America	291
10.1	Location sketch map of the Arctic	295
10.2	Arctic temperature, 1600–1990	296
10.3	Location sketch map of Greenland	298
10.4	Cumulative net balance of Brøggerbreen, 1912–91	307
10.5	Reconstructed running 5-year means of summer temperature at Longyearbyen	308
10.6	Novaya Zemlya showing inferred isotherms	309
11.1	Distribution of glaciers in low latitudes and the southern hemisphere	314
11.2	Frontal variations of glaciers in the Cordillera Blanca	317
11.3	The Quelccaya summit core	319
11.4	Key radiocarbon dates relating to Little Ice Age advances in tropical South America	319
11.5	The changing mass balance of the Chacaltaya Glacier in relation to El Niños and La Niñas	321
11.6	Changing extent of glaciers on Mt Kenya, 1899, 1963 and 1987	322
11.7a	The retreat of the ice on the eastern side of Mt Speke, Ruwenzori, Uganda, 1906–59	324
11.7b	Fluctuations of the termini of the Speke, Savoia and Elena glaciers between June 1958 and January 1967	324
11.7c	The reduction of ice cover on Mt Baker, Ruwenzori during the twentieth century	324
11.8	The shrinkage of the glaciers around the Kibo summit of Kilimanjaro, East Africa	325
11.9	The mountains of New Guinea	327
11.10	Changing extent of the ice on Puncak Jaya (Carstenz), New Guinea	328
12.1	Mount Tronador	332
12.2	Key radiocarbon dates relating to Little Ice Age advances in temperate South America	334
12.3	The Northern Patagonian Icefield	336
12.4	Outlet glaciers of the Southern Patagonian Icefield	339
12.5	Distribution of glaciers and permanent snowfields in South Island, New Zealand	346
12.6	Sketchplans of the tongues of the Tasman, Godley and Classen glaciers made by J. Von Haast in 1862	347
12.7	Broderick's map of the tongues of the Godley and Classen glaciers and their environs made in 1888	349
12.8	Green's map of the Great Tasman Glacier together with its tributaries and part of the Hooker Glacier made in 1884, with additions from Dr von Lendenfeld's map	351
12.9	The Murchison Glacier in 1892	352
12.10	Forefield of the Mueller Glacier, South Island, New Zealand showing age of moraine rock surfaces from lichens and weathering rinds	353
12.11	Key radiocarbon dates relating to glacial advances in New Zealand	354
12.12	Raw $\partial^{18}O$ records from a speleothem at North Westland, New Zealand	354
12.13	Frontal variations of the Franz Josef Glacier, 1894–1971	359
12.14	The position of the front of the Franz Josef Glacier between 1865 and 1965 as seen from Sentinel Rock	361

12.15	The mass balance record and cumulative mass indices of the Tasman Glacier	362
12.16	Variations of the incidence of negative values of the ENSO index, 1930–95	363
13.1	The location of the Antarctic Peninsula and sub-Antarctic Islands in relation to Antarctica and South America	366
14.1	Central England temperatures by seasons and for the year, 1669–1979	373
14.2	The frequency of southwesterly surface winds over England since AD 1340	374
14.3	Weather reported and mean sea level pressure and wind pattern for July 1695	374
14.4	Mean ring widths of over 500 oak samples from the eastern Mediterranean region	376
14.5	The position of Crete in the Aegean	377
14.6	Winter and spring droughts in Crete, 1547/8–1644/5	379
14.7	Severe winters in Crete, 1547/8–1644/5	379
14.8	Circulation patterns in the winter of 1674/5, when westerly winds predominated, and in the winter of 1690/1, when westerly winds were rare	382
14.9	Distribution of tree-ring sample sites in China	384
14.10	Sketch map of China, to show localities mentioned in the text	385
14.11	The location of cultivation of *Boehmeria nivea* in the eighth century, in AD 1102–5 and in 1264	386
14.12	Warm and cold periods shown by the Dunde and Guliya ice cores	387
14.13	Seasonal decadal mean temperature anomalies for north China	388
14.14	Decadal mean temperature anomalies in north and east China	389
14.15	Examples of flood and drought distribution in China	390
14.16	Variation of *Meiyu* rainfall between 1720 and 1800	391
14.17	Reconstructed sea level pressure over China in the coldest decade, the 1650s and in the mild decade, the 1460s	393
14.18	Decadal accumulation in the Quelccaya summit core compared with the Dunde D-1 core for the period 1610–1980	393
14.19	Decadal frequencies of wind direction derived from direction of cloud movement in Paris and decadal dust frequency in China, interpreted as a proxy for winter wind strength	394
14.20	Sketch map to show present-day distribution of winter and summer precipitation regions in southern Africa	395
14.21	Proxy temperature series based on oxygen isotope analyses of a cave speleothem at Cango Caves, foraminifera in diatomaceous sediment cores taken from shelf deposits off Walvis Bay, and tree-ring analyses	396
14.22	Tyson's models of anomalous meridional circulation over southern Africa during extended wet and dry spells	398
14.23	Composite temperature anomaly series for North America, Europe, China and the northern hemisphere	399
14.24	Millennial northern hemisphere temperature, from instrumental data 1900–99 and reconstruction from AD 1000–1900	400
14.25	A comparison of northern hemisphere temperature reconstructions by Mann *et al.* (1999) and by Esper *et al.* (2002)	401

TABLES

2.1	Eruptions of Katla since AD 844	27
2.2	Dates when Vatnajökull's outlet glaciers surged	37
2.3	Summary of documented glacier advances and retreats in the eighteenth and nineteenth centuries	47
2.4	The recession of Breidamerkurjökull since the late nineteenth century according to Sigbjarnarson (1970), Price (1982) and Sigurðsson (1998)	53
4.1	Retreat of the Chamonix glaciers from the 1640 maximum based on a survey c.1730	113
5.1	Mean daily rates of advance of Vernagtferner from 18 June 1844 to 1 June 1845	143
5.2	Dates of filling and emptying of the Vernagt lake between 1599 and 1846	143
5.3	Nineteenth-century retreat of the Vernagt, Mittelberg, Taschach and Gepatsch glaciers, Austria	143
5.4	Annual changes in the surface height at various altitudes of the Hochjoch, Guslar and Vernagt glaciers between c.1893 and 1940	146
6.1	Rating of temperature indices from documentary evidence	159
6.2	*Grosswetterlagen* with a strong influence on glacier nourishment	176
6.3	Indicators used for the construction of thermal indices for individual months	179
6.4	Warm, cold, wet and dry months, 1525–69 and 1570–1600	181
6.5	Differences in mean seasonal temperature between 1755–1859 and 1860–1965	182
7.1	Observations of the level of the névé of the Glacier d'Ossoue in relation to the Grottes de Cerbillona, 1902–9	201
7.2	Distances of markers from fronts of des Oulettes, d'Ossoue and Le Petit-Vignemale glaciers	201
7.3	Recent changes in the areas of Pyrenean glaciers	208
7.4	Mass balance of the Glacier d'Ossoue	208
7.5	Heights of Las Néous Glacier	209
7.6	Loss of area of the Portillon glaciers	210
7.7	Maps of Ghiacciaio del Calderone	216
8.1	Estimated annual net balance, Lednik Instituta Geografii, 1850–1963	221
8.2	Retreat of the Tsentral'nyy Tuyuksu, 1923–61	225
8.3	Retreat of the Rodzevic Glacier, 1850–1974	228
8.4	Distribution of glaciers in China	250
8.5	Number of monitored glaciers in China, retreating, advancing and stationary, by region in the 1960s–1970s	255
9.1	Comparison of glacier area and terminal altitude of Dome Peak glaciers during the Little Ice Age maximum and in 1963/4	264

9.2   Estimated rates of glacial advance across the Robson forefield   268

9.3   Angel Glacier moraines retreat dates   270

9.4   Little Ice Age moraines dated by ice-damaged trees in the Canadian Rockies   271

9.5   Changes in the area and terminal position of Lemon Creek Glacier, Alaska, *c.*1750–1958 281

9.6   Dates of the outermost moraines of the Juneau Icefield outlet glaciers   282

9.7   Dates AD when glaciers extended beyond their present margins on the eastern, maritime
      flank and western, continental flank of the Kenai Mountains, and around Prince
      William Sound, Alaska   285

11.1 Changes in the thickness, width and surface velocity of Lewis Glacier, 1986–90   323

11.2 (a) Changes in the surface areas of Ruwenzori glaciers in km², (b) areas of Ruwenzori
      glaciers as percentages of those of 1906, (c) glacier cover on Mt Stanley, Mt Speke,
      Mt Baker and Central Ruwenzori in km² and as a percentage of the area in 1906   325

11.3 Rate of retreat of Carstenz Glacier, New Guinea, 1850–1974   329

12.1 Recession of the Humo Glacier, 1914–82   331

12.2 Fluctuations of Mt Tronador glaciers, Argentina, 1942–77   333

12.3 Fluctuations of selected large outlet glaciers of the Southern Patagonian Icefield   343

14.1 Classification of data relating to weather anomalies in Crete, 1548–1648   378

14.2 Thirteenth- and twentieth-century temperatures in Guangdong and southern
      Henan provinces   387

14.3 Severity indices for north and south China   388

14.4 Decadal mean autumn temperature anomalies in Beijing   389

PREFACE

This book brings up to date *The Little Ice Age* that my wife wrote over a period of some twenty-five years between 1961 and 1986. Following its publication in 1988 she continued to collect relevant material with a view towards a second edition. She decided to extend the range of the new edition by adding new chapters on southern Europe and on ice cores, by lengthening the chapter on the Holocene, and by including more material on Antarctica.

The volume of information forthcoming about climate change increased rapidly in the 1980s and 1990s with the growing concern about global warming and the introduction of new techniques in palaeoclimatology. It became clear that the first advances of glaciers in the course of the second millennium AD in Europe and in some other parts of the world had taken place around 1300, and that ice fronts had fluctuated markedly over the succeeding five or six centuries until rapid retreat ensued in the late nineteenth and twentieth centuries. At the same time more light was thrown on the sequence of glacial advances and retreats in the course of the Holocene and Late Pleistocene.

Jean retired from the post she had held for forty-one years, as Director of Studies in Geography at Girton College, Cambridge, in September 1994. Her field researches were hampered by arthritis and other health problems but she continued to keep busy. We both attended the 1997 conference in the University of Reykjavik, Iceland on 'The Climatic History of Northern Europe and the North Atlantic Region over the past 1000 years' which had the subtitle 'Was there a "Little Ice Age" in Northern Europe and the North Atlantic Region and, if so, Where and When?' In a paper entitled 'The initiation of the "Little Ice Age" in regions round the North Atlantic' she defended the use of the term, though she suggested that it should be specifically used to refer to the period in the second millennium when glaciers were in more advanced positions than in the twentieth century, rather than being used to indicate a climatic interval. The following year, at the climatic history conference in the University of East Anglia in Norwich, her paper summarized the evidence for the initiation of the Little Ice Age as a global phenomenon (Jones *et al.* 2001a).[1] She became involved in the planning of the first conference to be organized by the Hanse Institute for Advanced Study in Bremen, entitled 'Climate and History in the North Atlantic Realm' which was eventually held in Delmenhorst in October 1999. Jean's contribution was entitled 'Climatic Change in Northern Europe over the last two thousand years and its possible influence on human activity'.

The last lecture she gave was to the Northeast England Meteorological Society at Durham University in November 1999, immediately before we set off for a four months stay with our son and his family in southernmost New Zealand. En route via California, we drove up Owen's Valley which we had first seen in 1970, visiting the Portals at Lone Pine, the Bristlecone Pines, Lake Mono and Death Valley. In New Zealand

---

1 *History and Climate* edited by P. D. Jones, A. E. J. Ogilvie, T. D. Davies and K. R. Briffa (Kluwer Academic/Plenum Publishers) was dedicated to the memory of Jean Grove.

we renewed our aquaintance with Trevor Chinn and Brian Fitzharris as well as revisiting the advancing Fox and Franz Josef glaciers on the west coast, the Wahoe moraine and the glaciers overlooked by Mount Cook. A few days after our return to Cambridge, in March 2000, Jean flew to New York to speak at a small meeting at Lamont-Doherty arranged by Wally Broecker.

I had usually played a part in her fieldwork activities and had assisted her prepare material for publication. When she died she had completed drafts of seventeen out of the eighteen chapters intended for a second edition of *The Little Ice Age* and I had read and commented on them. She had been undecided whether to publish the new and much enlarged book under the original title or to distinguish it by choosing a new one. After her death I rewrote Chapter 18 on the 'Consequences of the Little Ice Age climatic fluctuation' and re-edited the whole volume, including references to some of the more important relevant papers that had recently appeared.

It has become increasingly apparent that the use of the term Little Ice Age can be misleading. The interval from 1300 to 1900 can now be seen as a time when many if not most of the world's glaciers were more extensive than they had been for some 10,000 years. Recent research has confirmed their fluctuating behaviour in the course of the last 700 years and has indicated that the fluctuations were not all synchronous worldwide. It is also evident that global climate has experienced more short-term variability than glacier fluctuations can record. *The Little Ice Age* summarized the information available up to 1986 about the several other 'little ice ages' that occurred in the course of the Holocene, especially in the last 4000 years.[2] Chapter 15 on this period is now the longest in the book. It seems fitting that this volume should be entitled *Little Ice Ages: Ancient and Modern* to distinguish it from its predecessor and at the same time hint at the main differences between the two books.

A. T. (Dick) Grove

---

2   This was long ago called the Little Ice Age and later the Neoglacial. See Chapter 1.

ACKNOWLEDGEMENTS

Frank Debenham, Vaughan Lewis and Gordon Manley aroused Jean's initial interest in the subject. Jean Mitchell was her early mentor in historical matters and she was encouraged and assisted by Alfred and Harriet Steers.

Arthur Battagel played a crucial role and was wholly responsible for the translation of original documents from Norwegian. Rosemary Graham was responsible for translations from Icelandic and also translated some of the Norwegian printed material. Anita Dowsing assisted with the translation of some of the Norwegian printed material and translated nearly all the material in German. Anne Gellatly was responsible for the fieldwork in southern Europe in 1988–91, reported in Chapter 7.

Parts of the manuscript were read and helpful suggestions and criticisms provided by R. P. Ackert, J. B. Benedict, R. S. Bradley, P. E. Calkin, A. E. Corte, J. Dowdeswell, A. F. Gellatly, J. L. Innes, G. Kaser W. Kick, G. D. Osborn, R. Randall, Ren Meier, M. Sharp, H. M. Spufford, Peter Tyson and P. Wardle. Rosemary Graham read and commented on the whole manuscript and A. Battagel on large parts of the first edition.

Helpful information and advice were received from R. P. Ackert Jr, I. Allison, J. T. Andrews, D. Barclay, T. Blunier, J. R. Blyth, R. S. Bradley, R. Brázdil, N. Bhandari, C. Burn, C. J. Burrows, P. E. Calkin, T. Chinn, A. E. Corte, D. N. Collins, A. Conterio, A. Dugmore, G. Farmer, F. Fridriksson, D. Fletcher, A. F. Gellatly, A. S. Goudie, H. Green, N. J. Griffey, R. H. Grove, A. Holmsen, J. M. R. Hughes, J. L. Innes, W. Karlén, W. Kick, H. H. Lamb, O. Liestøl, B. H. Luckman, P. McCormack, G. Manley, B. Menounas, J. Mercer, G. H. Miller, J. A. Matthews, F. Müller, H. Nichols, D. Norton, A. E. J. Ogilvie, Y. Ono, G. Osborn, C. Pfister, D. Phillips, S. C. Porter, F. Röthlisberger, H. Röthlisberger, J. Ryder, J. Sandnes, M. J. Sharp, Shi Yafeng, O. Sigurðsson, J. L. Sollid, O. Solomina, W. Soon, A. Street-Perrott, E. Sulheim, L. G. Thompson, S. Thórarinsson, A. M. Tvede, T. de Waal, H. E. Waldrop, G. Wells, B. W. Whalley, M. Young and H. J. Zumbühl.

The assistance of the archivists in Riksarkivet, Oslo and in Statsarkivet, Bergen has been unstinting. Other archival material has come from the Canton archives, Sion and the Alexander Turnbull Library, Wellington.

The librarians of the British Antarctic Survey, the Scott Polar Research Institute, the Geography Department of the University of Cambridge, Girton College and the Royal Geographical Society have given generously of their time.

The Sulheims at Spiterstulen in Norway provided support and encouragement in the initial stages of the work. A sabbatical term at the Geography Department of UCLA provided an opportunity to learn about the glaciers of the Sierra Nevada and Dr and Mrs Jack Ives housed us in Boulder. Mrs David Norton afforded kind hospitality and a base during an initial visit to South Island, New Zealand. Since then our son Dr William Grove and his wife Lynne, and Andy and Kate Swanson have looked after us on longer stays. Our youngest son Jonathan showed us round southern Iceland in 1997. Generations of undergraduates and research students have provided stimulus. The editorial and production staff at Routledge provided valuable assistance in the preparation of the text, but all responsibility for re-maining errors and omissions rests with the author.

Finally financial support is gratefully acknowledged from the British Council, Girton College, Cambridge, the Smuts Commonwealth Fund, the University Travel Fund of Cambridge University and EC contracts EV4C.0044UK(H) and EVHC.0073UK(H).

The photographs reproduced as Plates 8.2 and 8.3 were kindly given by the late Dr Wilhelm Kick and those reproduced as 6.2a, 6.2c, 6.2d, 6.5a, 6.5b, 6.5c and 6.5d by Dr H. Zumbühl. Plates 11.1b and 12.11 are from photographs by William Grove, 12.1 by Charles Harpum, 7.3a by P. Hollerman and 8.6 by J. Stevenson. Plates 7.3b, 7.7b, 7.8b, 7.9b, 7.12b are from photos by Anne Gellatly and 4.2, 4.3, 8.5, 8.7, 9.1 and 9.2 from photos by A. T. Grove. The author and the publishers would like to thank the following organizations and individuals for allowing the use of plates: Nasjonalgalleriet, Oslo for 3.1, 3.5, 3.6, 3.7 and 6.2b; Billedgalleri Bergen for 3.2; Öffentliche Kunstsammlung Basel for 4.1 and 6.5c; Landsmuseum Innsbruck for 5.1a; E. T. H. Zürich and Dr H. Röthlisberger for 6.2c and 6.2d; Musée de la Marjorie, Sion for 6.4; Denkschriften der Schweizerischen Naturforschenden Gesellschaft for 6.5a, 6.5b, and 6.5d; Peter Hollerman and *Geographica Helvetica* for 7.3a; Muséum d'Histoire Naturelle de Toulouse for 7.4 and 7.5; *Geografia Fisica e Dinamica Quaternaria* and the Italian Glaciological Society for 7.11a, 7.11b and 7.12a; Whyte Museum of the Canadian Rockies, Banff for 9.3, 9.4 and 9.5; NASA for 2.1, 8.1, 8.4, 12.3 and 17.1; Alexander Turnbull Library, Wellington, New Zealand for 12.4, 12.5, 12.6 and 12.7; Mrs D. W. Fletcher and the Macmillan Brown Library, Canterbury University, New Zealand for 12.8 and 12.9; The Scott Polar Research Institute, Cambridge for 11.1a.

Dr Gellatly and Kevin Burkhill of the Department of Geography, University of Birmingham drew the figures for Chapter 7. Dr Gellatly drew Figures 12.6, 12.7, 12.8 and 12.9 from the original maps in New Zealand archives. The majority of the diagrams in the first five chapters, including the splendid copies of old maps and sketches reproduced as Figures 2.13, 2.21, 4.5, 5.3 and 5.4 were drawn by Pam Lucas. Most of the diagrams in Chapters 6 to 12 were drawn by A.

Shelley and the remainder by M. Young, S. Gutteridge, I. Gulley and I. Agnew in the Geography Department, Cambridge University. Other figures were drawn by A. T. Grove.

The author and the publishers would like to thank the following copyright holders for permission to reproduce figures: Acta Naturalia Islandica for 2.18; Aegean Dendrochronological Project, Cornell University for 14.4; *American Scientist* for 18.3 and 18.4; *Annals of Glaciology* and the International Glaciological Society for 10.4, 10.5, and 11.6; *Arctic, Antarctic and Alpine Research* for 1.5, 6.8, 8.7, 8.9, 9.2, 9.4, 12.1, 12.10, 12.15, 15.20, 15.21 and 18.2; A. A. Balkema and Svets & Zeitlinger for 11.6, 11.10 and 18.1; *Bollettino del Comitato Glaciologico Italiano* for 4.11 and 4.12; *Boreas* and Taylor and Francis for 15.9; Cambridge University Press for 8.10; *Climatic Change* and Kluwer Academic Publishers for 2.7, 3.5, 10.2, 14.4, 14.11, 14.20, 14.22, 18.8, 18.9, 18.12, 18.13, 18.14, 18.15, 18.19 and 18.20; Climatic Research Unit, University of East Anglia for 1.1; Colorado Associated Press for 17.4; Elsevier Science for 1.3; *Geografia Fisica e Dinamica Quaternaria* for 4.8, 4.11, 4.12 and 4.13; *Geografiska Annaler* and Blackwell Science Ltd for 2.19, 3.10b, 3.10b, 3.19, 11.7b, 15.10, 15.11, 15.14 and 15.15; *Geographica Helvetica* for 6.10, 15.4 and 15.7; *Geographical Journal* (Blackwell Publishing) for 3.10a and 12.6; IAHS and the World Glacier Monitoring Service for 5.8 and 9.16; *Global and Planetary Change* (Elsevier Science) for 16.25; *Journal of Glaciology* and International Glaciological Society for 8.1, 9.3, 9.10 and 11.7a; *Journal of Quaternary Science* (John Wiley & Sons Limited) for 16.18; *Journal of Volcanology and Geothermal Research* (Elsevier Science) for 17.9 and 17.10; Kluwer for 3.5, 8.4, 8.8, 8.13, 11.4, 12.2, 12.11, 15.16, 18.6b and 18.8; *Meddelelser om Grønland* and the Danish Polar Center for 2.3a and 2.7; Methuen & Co. Ltd for 14.3 and 15.1; *Nature* for 1.2, 2.10, 14.21c, 16.2, 16.3, 16.5, 16.6, 16.9, 16.12, 16.16, 16.17, 16.20, 16.24, 17.2a, 17.3a and b, 17.7, 17.8 and 17.12; *Norsk Geografisk Tidsskrift* for 3.4, 3.10c, 3.14 and 3.19; *Norsk Geologisk Tidsskrift* for 3.12; Norsk Polarinstitutt for 3.20; *Palaeogeography, Palaeoclimatology, Palaeoecology* (Elsevier Science) for 17.12; *Quarterly Journal of the Royal Meteorological Society* for 14.1; *Quaternary Research* for 1.6, 14.21,

15.2, 15.18, 16.2a, and 16.3; Routledge for 14.7, 16.30 and 16.32; *Science* and the American Association for the Advancement of Science for 11.3, 16.8, 16.11, 16.13, 16.15, 16.19, 16.22, 16.23, 16.27, 16.28, 17.6 and 17.11; *Transactions of the Institute of British Geographers* (Blackwell Publishing) for 18.17, 18.18 and 18.21; *Weather* and the Royal Meteorological Society for 18.16; *Zeitschrift für Geomorphologie* for 4.6; *Zeitschrift für Gletscherkunde und Glazialgeologie* for 4.10, 5.7, 5.9, 10.6, 11.2, 11.7c, 15.19 and 15.24; P. Bergthórsson for 2.3b, 2.4 and 18.9; M. Mann for 14.24; A. Ogilvie for 2.6, 2.8a and 2.8b; D. Robinson for 17.2b; F. Röthlisberger for 15.31; and H. Sigtryggsson for 2.5 and 2.25.

Every effort has been made to contact copyright holders for their permission to reprint material in this book. The publishers would be grateful to hear from any copyright holder who is not acknowledged and will undertake to rectify any errors or omissions in future printings of this book.

NOTES ON THE TEXT

SPELLING AND TRANSLITERATION OF FOREIGN NAMES

It has not always been possible to achieve consistency between text and maps. The Icelandic letters thorn (þ) and eth (ð) are represented in the text by 'th' and 'd'.

Both ø and ö are used in Norwegian maps. More than one system of transliteration is commonly used for Chinese and Russian.

GLOSSARY

Accumulation area ratio of a glacier   ratio of the area above the equilibrium line to the total area.

Bisses   irrigation channels in Valais, Switzerland, running almost parallel to the contours and bringing water long distances down mountain sides from glaciers and snowfields to pastures and vineyards.

Boudinage   term applied to random thinning or thickening under pressure of ice layers of different rigidity.

El Niño   sea surface temperatures along the western coast of South America and along the Equator in the central Pacific are usually anomalously cold for their latitude, because of upwelling cold water associated with the Trade Winds. The cold waters stabilize the air above them and the resultant aridity extends into the central Pacific. At intervals of about 2 to 7 years weakening pressure in the South Pacific High accompanied by a decrease in the strength of the Trade Winds causes warming of surface waters, typically in December to March, but sometimes lasting over a year. The resulting atmospheric instability and convectional activity brings rain to the normally arid coastal regions of Peru and Ecuador and the effects are felt much further afield. The phenomenon is known as El Niño, the boy child, a reference to the Christ child and the Christmas season when the warming normally becomes apparent (see Diaz and Markgraf 1992).

ENSO   an abbreviation of El Niño/Southern Oscillation, the oscillating ocean–atmosphere system, most apparent in the tropical south Pacific, associated with complex global readjustments in atmospheric and oceanic circulation patterns. Normally atmospheric pressure is higher in the eastern than in the western Pacific. As a result surface flow in the tropical Pacific is from east to west, piling up water in the western Pacific so that it is about half a metre higher than in the east, while cool water upwells in the eastern Pacific. When atmospheric pressure rises in the west, the easterly Trade Winds weaken and warm waters from the west flow eastwards bringing heavy rain to coastal South America. At the same time Indonesia and Australia experience drought. The system is connected with climatic events elsewhere in the world especially in the tropics.

Equilibrium line altitude   the line or more often a zone running across a glacier where accumulation is balanced by ablation.

Excesis period   interval between ice retreat from a site and its colonization by vegetation.

Gytta organic matter.

hPa atmospheric pressure used to be given in mb (millibars). These are now called hectoPascals, hPa.

Hlaup Icelandic word for a flood of water from beneath a glacier sometimes, but not always, associated with volcanic activity.

Hypsithermal interval a warm phase in the Holocene broadly equivalent to altithermal or climatic optimum.

Jökulhlaup Icelandic word for catastrophic floods from beneath the ice or from ice marginal lakes.

Mass balance when the accumulation of snow and ice during a winter accumulation season exceeds the ice wastage in the following summer, glaciers increase in volume and are said to have positive mass balances. When the wastage is greater than the accumulation, glaciers shrink and their budgets are said to be negative.

Neoglacial the interval when there were small-scale glacier advances after the hypsithermal interval (see Chapter 1, pp. 3–4).

Névé (French) or *firn* (German) snow which has survived summer melting and is not yet hard ice.

Ogives also known as Forbes bands, are alternating, dark and light bands on the surface of some glaciers, arcuate in plan on account of the faster flow in the middle of the glacier than at the sides. The bands occur below icefalls where the ice surface is broken and snow accumulates in the crevasses in winter but fails to do so in summer. The width of a dark plus a light band consequently corresponds with the distance the glacier moves in a year.

Phenology involves recording the dates of occurrence of specific growth stages in plants.

Precession of the equinoxes a wobble of the earth's axis, causing it to describe a circle in space, results from the gravitational forces exerted by the sun, moon and planets on the earth's equatorial bulge. The complete cycle takes about 26,000 years. A second component, which changes the elliptical orbit of the earth around the sun about one focus, has a cycle of about 22,000 years. At present the earth is furthest from the sun about 4 July, in the northern hemisphere summer. The situation was the reverse about

11,000 years ago; then the northern hemisphere received about 10 per cent more radiation at the top of the atmosphere in summer than it does today, and the southern hemisphere received about 10 per cent less.

Proglacial in front of a glacier margin.

Sandur spread of sand and gravel in Iceland, deposited by jökulhlaups and extending downslope from a glacier front.

Seracs (French) or *siracs* (Spanish) pinnacles of ice on icefalls especially on glaciers in low latitudes where the sun's rays are near vertical in the middle of the day.

Southern Oscillation an atmospheric pressure see-saw extending from the southeastern tropical Pacific at one end to the Australian–Indonesian region at the other. When atmospheric pressure at sea level is high over one it is low over the other. The see-saw oscillates on timescales ranging from months to years. The Southern Oscillation Index is a measure of this phenomenon based on the difference in pressure between Tahiti and Darwin. Associated with the Oscillation are variations in sea surface temperature (see El Niño above and Chapter 11, note 8).

Surging glacier a glacier that, from time to time, flows several times faster than normal. This behaviour probably involves two modes of basal sliding, one normal and one fast.

Thalli (singular *thallus*) patches of lichen, roughly circular, which grow outwards from the centre at a regular rate.

Trimline line often visible on the rock above the surface of a glacier marking its former greater height.

Younger Dryas the term Dryas refers to a creeping, perennial flowering plant, *Dryas octupetula*, confined currently to the high mountains or Arctic regions of Europe, Asia and North America. Younger Dryas is the term applied to the more recent and more severe of two very cold intervals at the end of the Pleistocene, between about 11,000 and 10,000 BP, immediately preceding the Holocene period. Peat deposits from this time in high latitudes and at high altitudes in the northern hemisphere often have a high content of Dryas pollen.

ABBREVIATIONS

AAR	accumulation area ratio	GRIP	Greenland Ice-core Project
AMS	accelerator mass spectrometry	IRSL	infrared-stimulated luminescence
a.s.l.	above sea level	LIA	Little Ice Age
DCR	deglaciation climate reversal	mwe	metres of water equivalent
DEP	dielectric profiling	NADW	North Atlantic deep water
$\partial^{18}O$	the proportion of ^{18}O in a sample as compared with the the isotopic composition of a water standard, Standard Mean Ocean Water or SMOW (see Bradley 1999, chapter 5.2)	NAO	North Atlantic Oscillation
		NHAM	Northern Hemisphere Annular Mode
		ODP	Ocean Drilling Project
		Ppbv	parts per billion by volume
		Ppmv	parts per million by volume
DVI	dust veil index	SST	sea surface temperature
ECM	electrical conductivity measurements	VEI	volcanic explosivity index
ELA	equilibrium line altitude	YD	Younger Dryas (see Glossary)
GISP	Greenland Ice Sheet Project		

Frontispiece Jean and Dick Grove on the Aletsch Glacier, 1961

1
INTRODUCTION

Climatic changes come on several different timescales. Between short-term fluctuations lasting a few years such as El Niños, and changes extending over thousands of years such as glacial and interglacial periods, there are variations over a few centuries that may have profound effects on natural phenomena and human affairs. It is variations on this scale, stretching over several generations, with which we are mainly concerned in studying the Little Ice Age. It was a period that may be seen as beginning in the thirteenth and fourteenth centuries (Porter 1986, Grove 2001b) and culminating between the mid sixteenth and the mid nineteenth centuries when glaciers were bigger than they were in the twentieth century or had been in the tenth to twelfth centuries. It was also a period of lower temperature over most of the globe, sufficiently marked to have had important consequences, especially in certain sensitive areas in high latitudes and at high altitudes where conditions for plant growth and agriculture are marginal.

For hundreds of years in the 'High Middle Ages' climatic conditions in Europe had been kind; there were few poor harvests and famines were infrequent. The pack ice in the Arctic lay far to the north and long sea voyages could be made in the small craft then in use. Communications between Scandinavia, Iceland and Greenland were easier than they were to be again until the twentieth century. Icelanders made their first trip to Greenland about AD 982 and later they reached the Canadian Arctic and may even have penetrated the North West Passage. Grain was grown in Iceland and even in Greenland; the northern fisheries flourished and in mainland Europe vineyards were in production 500 km north of their present limits.

The beneficent times came to an end. Sea ice and stormier seas made the passages between Norway, Iceland and Greenland more difficult after AD 1200; the last report of a voyage to Vinland was made in 1347 (Gad 1970). Life in Greenland became harder; the people were cut off from Iceland and eventually disappeared from history towards the end of the fifteenth century. Grain would no longer ripen in Iceland, first in the north and later in the south and east. As the northern winters became colder, fish migrations took different tracks and life became tougher for fishermen as well as for farmers. In mainland Europe, disastrous harvests were experienced in the latter part of the thirteenth and in the early fourteenth centuries, with famines in England in 1272, 1277, 1283, 1292 and 1311. The years between 1314 and 1319 saw harvests fail in nearly every part of Europe (Grove 1996). Extremes of weather were greater, with severe winters and unusually hot or wet summers. In consequence the boundaries of cultivation contracted, though there were of course other forces as well as climate operating at the time.

In these late medieval times and in succeeding centuries, the impact of climatic fluctuations was felt most painfully and persistently in highland areas. Cultivated areas suffered especially in the uplands of maritime regions, such as the Lammermuir hills of southeast Scotland. In the last few centuries glaciers have advanced in the mountains of Alpine Europe and Scandinavia, in the northlands, and indeed in most other moist and cold parts of the world. In the decades between the late sixteenth and late seventeenth centuries European glaciers swelled and their tongues advanced, destroying high farms and damaging mountain villages. Streams fed from glaciers flooded more frequently, sometimes catastrophically. In many areas this kind of hazard was compounded by landslides and avalanches triggered by increased precipitation and the greater glacial activity associated with it.

The relationship between glacial behaviour and meteorological control is delicate and complicated. The researches of H. W. Ahlmann (1949) and his pupils in Scandinavia and of the Innsbruck group led by H. H. Hoinkes (1970) laid firm foundations for understanding it. If the accumulation of snow and ice during a winter accumulation season exceeds the ice wastage in the following summer, glaciers increase in volume and are said to have positive mass balances. If the wastage is greater than the accumulation, glaciers shrink and their budgets are said to be negative. If a series of positive balances occurs, the volume of ice moving downslope increases and eventually the glacier front advances. The conditions most favourable for positive budgets or mass balances are given by plentiful precipitation in winter and short cool summers causing minimum wastage or ablation. There is a lag between the onset of a particular climatic change and the response of a glacier terminus, which depends on the topography of the glacier and its valley and the flow characteristics of the ice mass. Small temperate valley glaciers with a rapid turnover of ice will respond in a few years; valley glaciers of moderate size in Austria have a response time of about seven years; some glaciers in the Cascades of the northwest United States respond over several decades. Large icecaps have much longer lag periods. Details may be found in Paterson's (1981) *Physics of Glaciers*.

Comparison of glacier behaviour and meteorological records over the instrumental period of the last two centuries provides a basis for extrapolating the climatic record into the past. Manley's (1974) long record of mean monthly temperatures for central England between 1659 and 1973, the result of careful integration of data from diverse records, is extremely valuable because it covers a good deal of the Little Ice Age and allows an impression to be gained of the regional temperature changes associated with the comparatively local glacier fluctuations in Europe. The behaviour of glaciers provides us with one of the most useful indicators of past climatic history for places and times lacking instrumental records, so long as the histories of sensitive glaciers with short response times are used.

Parry's (e.g. 1978) studies of the variations in the extent of cultivated land in upland Britain provided a deeper understanding of the economic and social consequences of climatic fluctuations back into medieval times. Hubert Lamb and his successors in the Climatic Research Group at Norwich extended our knowledge of past climatic conditions (Lamb 1972b, 1977, 1982). There is an increasing volume of evidence indicating the importance for agriculture and rural prosperity of long-term change in precipitation and temperature. This has necessitated a reconsideration of the influence of climatic change on the demographic and economic decline in the fourteenth century so long attributed to the Black Death (Lamb 1977: 454–7, Grove 1996).

Historians have long been inclined not only to overlook but deliberately to discount the influence of climatic change on human affairs (Russell 1948, van Bath 1963, Hoskins 1968). Even Le Roy Ladurie (1971), despite his substantial book on the history of glaciation in Europe, *Times of Feast, Times of Famine*, is undecided on whether a difference in secular mean temperature of 1 °C has any substantial influence on agriculture and other human activities. These negative or agnostic attitudes may be explained in part by historians having made use of climatic chronologies that are now known to be incorrect (e.g. Britton 1937, Brooks 1949). Some early climatic historians were less rigorous in their use of sources than is now required and they often made extensive use of compilations and other secondary material, some of which were unreliable (Bell and Ogilvie 1978).

Studies of the Little Ice Age, and also of climatic variations on a similar scale in the more distant past, are very significant for the future. The great importance of the potential for global warming resulting from human activity and particularly the burning of fossil fuels is universally recognized. It is evidently essential to try to distinguish between climatic changes that are essentially natural and those that caused by human activity.

This book brings up to date *The Little Ice Age* by referring to the results of the very numerous studies that have been made since the mid 1980s. As in the first edition, the early chapters, here in Part 1, present a history of the Little Ice Age based on the records of glacier advance and retreat in northwest Europe and the Pyrenees. Information is then brought together about the behaviour of glaciers in this period in other mountainous parts of the world, including parts of Antarctica. In Part 2, the Little Ice Age is set in the longer time perspective of the Holocene, the period since the time of the Last Glaciation during which, in the words of a famous prehistorian, 'man has made himself' (Childe 1936). In Part 3 a completely new chapter presents the climatic results of analysing cores from ice sheets and from ocean floor sediments, especially in the Holocene, setting the current interglacial in the time perspective of the Quaternary Period. A fresh attempt is made to throw light on the possible causes of the Little Ice Age and, finally, an assessment is made of its significance for an understanding of human affairs.

1.1 THE TERM 'LITTLE ICE AGE' AND THE TITLE OF THIS BOOK

The term 'Little Ice Age' is widely used to describe the period lasting a few centuries between the Middle Ages and the warming of the first half of the twentieth century, a period during which glaciers in many parts of the world expanded and fluctuated about more advanced positions than those they occupied in the centuries before or after this generally cooler interval. A number of objections have been made to the employment of this term 'Little Ice Age' and these must be considered before the nature of the phenomena involved is explored.

The main objection is that the term was originally applied to a quite different time period, that it is now used in several different ways by current authors, and that as far as the dominant usage is concerned it is a misnomer. It is also argued that the Little Ice Age was not worldwide, that continental glaciation did not increase, and that 'there was not any sustained low global temperature during the period' (Landsberg 1985, Bradley and Jones 1992c), in short that it was not an Ice Age and was insignificant in scale (Landsberg personal communication 1983).

It is certainly true that Matthes (1939), who introduced the term Little Ice Age into scientific literature, intended it to describe an 'epoch of renewed but moderate glaciation which followed the warmest part of the Holocene'. Matthes was a very close observer who was especially concerned with the glaciers of the Sierra

Nevada of California, which he believed would not have survived the warmth of the climatic optimum of the mid Holocene. These little glaciers had at first been mistaken for snowfields and it was not until 1872 that John Muir was able to demonstrate that they were formed of moving ice. Matthes noted the fresh appearance of their frontal moraines:

> they are made up of many small terminal moraines, laid against and on top of each other, as is clearly shown in instances where individual moraines lie spread out in a series one behind another, with concentrically curving crests. They record for each glacier many repeated advances, all of approximately the same magnitude. How many centuries of glacier oscillation are represented by these moraine accumulations it is difficult to estimate.

Matthes had no means of dating the features he described but he concluded that the glaciers had reformed after the climatic optimum and estimated that they might have reappeared 'as recently as about 2000 BC', although he recognized that larger icestreams such as those on Mt Rainier had probably persisted since the Ice Age.

Matthes (1950) was well aware of the record of repeated glacial advances in Europe during the past 400 years and the extent of the recession after 1850, which he regarded as 'a turning point in the modern glacial history of central Europe; for since then the trend of climate, not only in Europe but throughout the world, has been distinctly milder and the glaciers have been in recession almost everywhere' (p. 155). Interestingly he commented that this general recession did not herald the end of his Little Ice Age but that it was much more likely that it represented merely one of the mild fluctuations which he surmised had occurred repeatedly during the last 4000 years. This surmise has received detailed verification as a result of investigations made in the Alps of Switzerland and Austria, discussed in Chapter 15.

If the concept of mid Holocene conditions so mild that at least in some regions small glaciers disappeared completely is correct, then there is room for a term to cover the subsequent period of several millennia during which glaciers have reappeared and fluctuated in extent. 'Little Ice Age' is still occasionally used in this way (e.g. Benedict 1968) but it has been generally overtaken by 'Neoglaciation', used for instance by Porter and Denton (1967) to describe the interval of rebirth or renewed growth of glaciers after a time of maximum hypsithermal shrinkage during a Holocene warm period lasting, according to them, until about 2600 years BP. The use of this term 'Neoglaciation' is not without difficulty. Porter and Denton noted that 'rather complex low order changes of climate characterize the hypsithermal interval resulting in several early Neoglacial episodes of glacial expansion'.

The concept of Neoglaciation is not recognized as useful by all workers in the European Alps, where the sequence of Holocene glacial events has been worked out in most detail, and the term does not provide a separate label for the period of several centuries which saw the last expansion of glaciers in many parts of the world. Evidence of not one but a whole sequence of such events has accumulated, as Matthes surmised would be the case, and the advent of a whole range of dating techniques in the last few decades has made it possible to distinguish them one from another.

The term 'Little Ice Age' has been widely employed by geographers, geologists, glaciologists and, most significantly, climatologists, to describe the period of glacial advance of the last few centuries or 'the cold Little Ice Age climate of about 1550 to 1800' (Lamb 1977: 104). It is true that the dates assigned to it are not always identical, being influenced by the locational experience of individual workers and the volume and accuracy of the evidence available to them. Synchronous climate change over the globe cannot be assumed without proof, but the widespread indications of rapid twentieth-century glacial retreat and the striking similarity of much data coming from widely separated areas seem to indicate a coherence which justifies a single name.

Alternatives to 'Little Ice Age' have been suggested. Ladurie (1971: 223) noted that German authors, including Kinzl and Mayr, used 'Fernau' but remarked that, as custom makes law, 'Little Ice Age' might well be accepted. More recently some German workers, for example Heine (1983), have begun to use a translation of the English term, '*Kleine Eiszeit*'.

Historical evidence of Little Ice Age events is much more plentiful in Europe than elsewhere but the documentation from other continents, though scantier, is supported by a great volume of field evidence (e.g. Hope *et al.* 1976, Hastenrath 1984) which is presented in Chapters 8, 9, 10, 11 and 12. It emerges that the Little Ice Age was a global phenomenon that began in or around the fourteenth century (Grove 2001a, b, Broecker 2001) and it is shown in Chapter 15 that it was not unique in the Holocene. Involving fluctuations in temperature of 2 °C or less, it was none the less sufficient to cause advance of the Greenland ice edge (Weidick 1968) and to be associated with measurable meteorological, geomorphological and vegetational changes.

It is certainly true that lower temperatures were not sustained throughout the period. The Little Ice Age itself consisted of a series of frequent fluctuations, such as those exhibited in Manley's temperature curves for central England and worked out in great detail for Switzerland by Pfister (Chapter 6). Such fluctuations consist of individual years and clusters of years in which weather conditions depart strongly from longer term means. Average conditions throughout the Little Ice Age were, none the less, such that mountain glaciers advanced to more forward positions than those they had occupied for several centuries, or in some areas even millennia, and fluctuated about those positions until the warming phase in the decades around 1900 brought them back to where they had been in earlier Holocene warm periods. It has been nicely demonstrated that certain Swiss glaciers such as the Ferpècle were of comparable extent in the 1980s and before about 3500 years BP (Röthlisberger *et al.* 1980 and Chapter 15). This finding provides excellent confirmation of Matthes's hunch that the recent recession represents merely 'one of the mild fluctuations that has occurred repeatedly in the last 4000 years'.

In the chapters that follow, the course of events in Europe is considered first, the emphasis being placed on historical evidence which still provides the most complete and accurate record. The material has been deliberately selected so as to illustrate not only a variety of documentary source types but also to give some idea of the reactions of contemporaries. In particular, the somewhat anecdotal style of the earlier part of Chapter 5 has been adopted for this reason. Contemporary records are to be preferred to secondary sources when using historical evidence and great care has to be taken over both reliability and interpretation (Ingram *et al.* 1978). Paintings, sketches and lithographs as well as written documents can yield valuable evidence of glacial extent and character if their dates are known

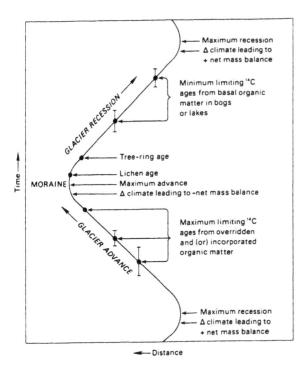

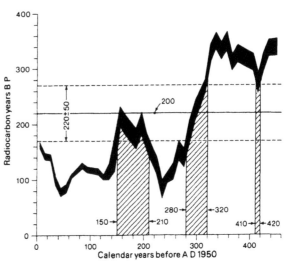

Figure 1.1 The relationship between changes in climate causing periods of positive and negative mass balance leading to glacial expansion and recession and possible methods of dating mentioned in the text (*Source*: from Porter 1979)

Figure 1.2 The relationship between ^{14}C years and calendar years calibrated from tree rings. The width of the curve, marked in black, is twice the standard deviation given by the laboratory. Any one radiocarbon year is equivalent to more than one calendar year. Thus a radiocarbon age of 220 ± 50 years is equivalent to all the calendar dates with the intervals AD 120–210, 280–320 and 410–420 (*Source*: from Stuiver 1978)

and if the artists were intent on accurate representation of nature (Zumbühl 1980).

European historical sources are more plentiful and go back further than those from other continents. Chronologies elsewhere have to depend heavily on the dating of moraines, which generally involves obtaining bracketing dates of greater or lesser accuracy, that is maximum and/or minimum estimates of the time when the debris forming a particular moraine was deposited (Figure 1.1).

1.2 DATING METHODS

It is beyond the scope of this book to discuss critically and in full the variety of dating methods that are available (for a general survey see Porter 1981a and for a critique of many of the methods involved see Bradley 1999). It may however be useful to give a short introductory sketch of the principal methods involved, with their pitfalls and limitations.

Radiocarbon dating of organic material underlain or overlain by moraine nearly always suffers from the defect of time lags of unknown length between the period of moraine construction and the deposition or accumulation of the dated material. In many cases it is only possible to obtain a maximum or a minimum limiting date but not both. A major limitation of ^{14}C dating arises for the Little Ice Age period, quite apart from normal constraints such as sample contamination and variations in methodology and accuracy by both fieldworkers and laboratories. Calibration of ^{14}C values with calendar ages of rings from long-lived trees has revealed non-systematic relationships between them (Klein *et al.* 1982, Stuiver 1986, Stuiver and Reimer 1993). Within the last 500 years, it is not possible to obtain an unambiguous calendar age from a single radiocarbon date (Figure 1.2). It might be possible to eliminate the ambiguity by obtaining ^{14}C dates of several rings of the same log, but this has rarely if ever been done in this context. It must therefore be accepted that while radiocarbon dating may be used to establish that glaciers in a particular region were affected by oscillations in extent or volume during the Little Ice Age, it cannot at present be used for identifying second-order advances and retreats within the last few centuries and more especially should not be used to prove or disprove synchroneity of such fluctuations from region to region.

Radiocarbon provides a most valuable tool for the dating of earlier Holocene periods of glacier advance (Röthlisberger *et al.* 1980) or retreat (Porter and Orombelli 1985) so long as it is used with care and its accuracy is not overrated. Maximum limiting dates can be obtained only from the ^{14}C ages of organic material beneath or within moraine. The most detailed and satisfactory reconstructions are provided if fossil wood in quantity is available (Furrer and Holzhauser 1984), though even with such material care must be taken over interpretation (Ryder and Thomson 1986). The clearest evidence comes from trees sheared off *in situ* by advancing ice, but this is exceptional in most regions (Holzhauser 1984b).

Radiocarbon dating of buried soils presents complications because of the complex nature of soil organic matter (Matthews 1985). Pre-treatment of soil samples to separate soil organic fractions of different ages is employed in order to acquire the closest possible maximum or minimum age estimates. The ^{14}C age of the oldest uncontaminated organic fraction of the buried soil, which in some cases is known to consist of lichen, provides the closest approximation to the time when soil formation began, while the age of the youngest uncontaminated fraction in the buried soil provides the best possible estimate of the time which has elapsed since burial. The two dates serve to bracket the interval between a time when the surface was exposed and a time when it was covered again by ice. Dates obtained from the top and the bottom of a 5-cm thick A-horizon from beneath a moraine in southern Norway gave dates of 880 ± 35 and 3140 ± 55 (Matthews 1980); dating of several thin slices of a single horizon from beneath a moraine in southern Norway gave dates increasing from 485 ± 60 BP at the top to 4020 ± 70 BP at the bottom (Matthews and Dresser 1983). The steep age gradient with depth revealed indicates both that a sample intended to provide a maximum age for the burial of a palaeosol must be taken from as near the top as possible and the considerable error that may be involved should the surface layer be missing. The complications of ^{14}C dating of soils are least pressing when soils are buried at an early stage of their development. Investigation of immature soils within moraine sequences has provided satisfactory evidence on which to build complex Holocene chronologies in the Alps (Röthlisberger *et al.* 1980), the Himalaya and Karakoram (Röthlisberger and Geyh 1985) and New Zealand (Gellatly *et al.* 1985). However, reliability tests of over 300 ^{14}C ages of soils within moraines suggest a maximum resolution of ± 200 years, taking into account the advice of the International Study Group (1982) that the uncertainty should be taken as twice the laboratory value for the standard deviation.

It is sometimes possible to date the culmination of a recent period of glacier expansion directly using dendrochronology at positions where advancing ice has tilted or damaged trees. Much more commonly, a minimum age for a moraine has been obtained by counting the number of rings in the oldest trees growing on it. Several possible sources of error are involved, including the assumption that the oldest tree has been found, and that this is a member of the first generation, the unknown length of time between the stabilization of debris and the establishment of seedlings (the *excesis* period), and the interval required for trees to grow to a height at which it is possible to obtain a core (Lawrence 1950a, Sigafoos and Hendricks 1961, 1972). Dendrochronology, if used with care, regard being paid to variation in the length of time required for establishment of different species in a given area, can nevertheless be a valuable tool in the context of Little Ice Age investigations (e.g. Carrara and McGimsey 1981).

Lichenometry, a method of dating based on the assumption that the largest lichen growing on a given substrate is the oldest individual and that, if the growth rate for a given species is known, the maximum lichen size will provide a minimum age for the substrate, is a dating method of considerable potential (Beschel 1961). It is particularly valuable in treeless areas and those lacking material for ^{14}C dating (Bradley 1999). Successful use of lichenometry demands meticulous care. A growth curve should be set up individually for each region and the nature of the evidence available for this will determine its accuracy and the time depth in which it can be used within reasonable error limits. It has to be accepted that there is a delay between exposure of a surface and colonization and a further delay

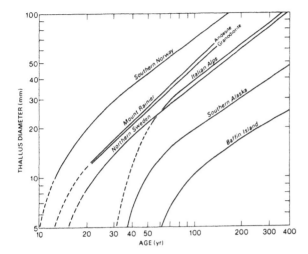

Figure 1.3 Lichenometric dating curves, relating thalli diameter to age, differ according to climate and also are affected by the rock type of the substrate (*Source*: Bradley 1999)

before thalli become visible to the naked eye. Lichen species grow at different rates and must be differentiated from each other in the field; this can be particularly difficult when thalli are very small (Calkin and Ellis 1984). Moreover, both rock surface characteristics and climatic factors affect growth rates (Figure 1.3).

Given favourable circumstances, lichenometry can provide a dating tool accurate to ± 5 years over the last 200 years (Porter 1981b). Dating error increases with increasing age, because lichen growth tends to diminish with time, perhaps at an exponentially decreasing rate (Porter 1981a) and also because of the extrapolation of growth curves to give time depth. Published lichen chronologies differ both in methodology, which is not always even stated, and in reliability. An extremely cautious approach must therefore be adopted as far as correlation from region to region is concerned.

The sophistication of approach to lichenometry has increased markedly in the last decade or two. *Rhizocarpon geographicum* has been used in the majority of lichenometric studies but 29 species of *Rhizocarpon* have been recognized in Europe and 45 in the Arctic. Innes (1982, 1983a), working in Norway, showed that *Rhizocarpon alpicola*, which colonizes deposits later than the *geographicum* group, has a faster growth rate, and that it is necessary to differentiate between the two to avoid important dating errors. This suggests that it may be necessary to re-examine the validity of some of the earlier growth curves, including that of Denton and Karlén (1973a). Techniques of sampling in the field and subsequent analysis are still undergoing detailed review (e.g. Innes 1983b, 1984, 1985b, McCarroll 1994).

The potential importance of lichenometry remains great. In the high Arctic, where other methods are liable to be least applicable, growth rates are low and the time range may conceivably exceed 5000 years. In maritime northern latitudes where growth rates are high, the time range is no more than a few hundred years.

While much of the earlier work on glacier chronology was based on a single method of dating, there has been an increasing realization of the value of a more broadly based approach (Burke and Birkeland 1979, McCarroll 1994). In western North America, the additional evidence presented by tephra layers, themselves radiocarbon dated, has provided a useful framework for Holocene events (Chapter 15.6.3). On Mt Rainier lichenometry has been used together with tree-ring dating to identify periods of glacier expansion during the Little Ice Age (Chapter 9.1 and Burbank 1981). In New Zealand a chronology has been built up using documentary records, lichenometry and radiocarbon dating of soils, supplemented by weathering-rind

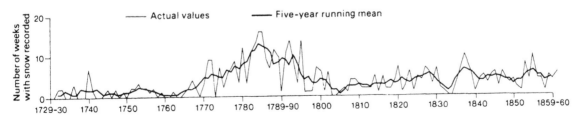

Figure 1.4 Winter snowfall in Scotland in the Little Ice Age

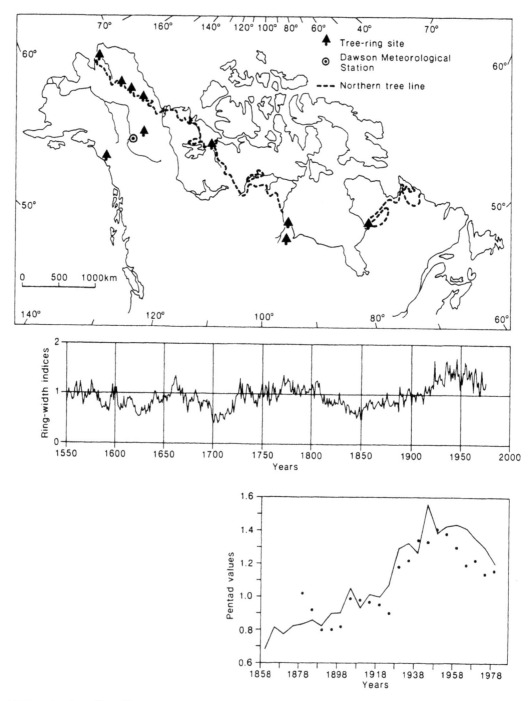

Figure 1.5 Tree-ring data collected from the sites shown have been used to construct ring-width indices for the period from the mid sixteenth to the late twentieth centuries. Pentad values for these indices are plotted against a temperature curve for the northern hemisphere derived independently from meteorological observations (*Source*: from Jacoby and Cook 1981)

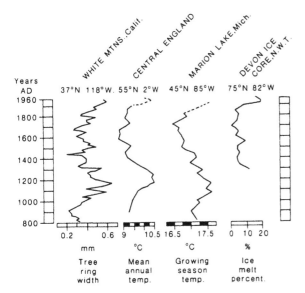

Figure 1.6 Comparison of independently derived indicators of environmental and climatic change covering both the Medieval Warm Period and the Little Ice Age (*Source*: after Bernabo 1981)

analysis and studies of soil and vegetation development (Chapter 12.7 and Birkeland 1981, Gellatly 1985a, b, Gellatly *et al.* 1985). The consistency achieved by using several techniques provides a more convincing result than could otherwise be obtained. There is a need to apply such multi-parameter approaches elsewhere to build up a satisfactory global picture of events in the Little Ice Age and earlier parts of the Holocene.

Little Ice Age chronologies from individual regions may be compared with evidence from other sources, such as documentary evidence of extent of winter snowfall (Figure 1.4) or time series of tree-ring widths which may yield estimates of summer temperature (Figure 1.5) (Jacoby and Cook 1981). A thorough comparison of this sort comes from Pfister's (1981) reconstruction of Little Ice Age climate in Switzerland (Chapter 6.3), which has provided an important line of approach to the problem of the extent of the impact of Little Ice Age climate on the population of mountain Europe. Comparison of the glacial evidence with that of other indicators can also be adopted for longer time periods (Figure 1.6) but an attempt to do this for the whole of the Holocene may be premature at the present time. The glacier chronology is not yet sufficiently securely

based (Chapter 15.2) and complications arise because of the differing sensitivities of the various alternative indicators.

The complexity of geomorphic responses inhibits simple correlation between glacial deposits of the Holocene and those of alpine and sub-alpine lake sediments (Harbor 1985). The conclusion of Davis and Botkin (1985), that fossil pollen deposits are not normally able to resolve climatic changes of the order of 100 to 200 years or to record very brief climatic events, underlines the value of obtaining a soundly based glacier chronology for the Holocene. Cores from ice sheets and ocean floor sediments (Chapter 16) are helping to provide a better idea of the global applicability and time context of a Holocene chronology.

About fifteen ice cores extend back into the last glaciation and some reach the penultimate glaciation (Bradley 1999). Dating by analysing the ^{14}C content of bubbles in the ice is uncertain because of leakage in the course of snow accumulation. Seasonal variations of accumulation, temperature and dustfall allow annual layers to be distinguished and counted. Reference horizons are provided by radioactive fallout from nuclear bomb tests in the 1950s and 1960s, by the dust and acid deposition from major volcanic eruptions, and by 'spikes' in the ^{10}Be record.

1.3 THE CURRENT RELEVANCE OF THE LITTLE ICE AGE AND EARLIER LITTLE ICE AGES

Investigations of the characteristics of the Little Ice Age and the incidence of other such intervals of cooler climate in the past are of practical as well as academic interest. Prediction of future climatic oscillations, dominated as they may be by increasing carbon dioxide (CO_2) in the atmosphere, demands greatly increased knowledge and understanding of the course of events in the past. Ice core evidence appears to show that shifts in climate took place in the last interglacial and early in the Holocene within a few decades, at a speed that was quite unsuspected. Ocean currents, above all the Gulf Stream, are believed to be liable to falter and transform regional climates in mid latitudes even more drastically than the El Niño is known to disturb low latitude weather. Recent advances in identifying variations in the sun's irradiance suggest that solar variability has an important influence on climate change. The short-term effects of volcanism are undoubted (Chapter 17).

The forcing factors, internal and external, which have operated in the past to upset the stability of the climatic system have not disappeared and will continue to influence the course of events (Chapter 17). Humans are still vulnerable to the effects of even minor variations (Chapter 18.3). Semi-arid lands both north and south of the Himalaya depend crucially on glacier water, yet runoff from the ice is affected to a measurable extent even by such changes as have occurred in the last few decades (Collins 1984). Land ice and its changes provide a useful indicator of changes in the energy balance of the earth's surface and it is for this reason that the monitoring which is now regularly undertaken (Haeberli 1985a, b, Haeberli *et al.* 1999) must be accompanied by continued research into past history.

PART I

THE LITTLE ICE AGE OF THE SECOND MILLENNIUM

2

ICELANDIC GLACIERS AND SEA ICE

Meteorological conditions around Iceland have an important bearing on the weather of northwest Europe and the fluctuations of Iceland's climate are symptomatic of those of a much wider area. The timing of glacial changes in this exceptionally sensitive climatic environment is therefore an essential ingredient of a study of the temporal behaviour of European glaciers. The documentary evidence from Iceland in the Middle Ages is unusually specific and life there is highly dependent on climate. It therefore provides an appropriate starting point for a regional study of the Little Ice Age and its impact.

Icecaps dominate the scene. Of the 11,800 km² which are ice-covered, over 11,600 km² are accounted for by the six largest icecaps, Vatnajökull, Langjökull, Hofsjökull, Mýrdalsjökull, Drangajökull and Eyjafjallajökull. In form and size they are intermediate between Alpine icestreams and the ice sheets of Greenland and Pleistocene Europe (see Figure 2.1).

Built by volcanic eruptions during the Cainozoic, Iceland is young both geologically and morphologically. Basalt areas in the east and west are separated by a central depression filled by palagonite, a mixture of volcanic, glacial and aeolian deposits. This zone, running through central Iceland and bounded by local faults running north–south in the north and northeast–southwest in the south, is part of the enormous fault system extending far beyond the confines of the island as the mid Atlantic Ridge, the boundary between the European and American plates. During the Quaternary period, large amounts of basalt were extruded and faulting gave vertical displacements of hundreds of metres. The lava plateaus to the east and west have been faulted into blocks, tilting towards the central depression and with their greatest elevations towards the coast.

Iceland was first settled in the ninth century. It is thought that two-thirds of the area was then vegetated, with birch woodlands at low elevations and dwarf willow shrubs above 300 m. Livestock introduced by the settlers rapidly degraded the vegetation cover. Today less than a quarter of the island is vegetated, with birch woodland occupying only 1 per cent of the total area. Human activity is only part of the story. Environmental conditions are so marginal that any climatic shift has always been of immediate concern to the inhabitants. A fall in summer temperature of 1 °C has been found to result in modern times in a 15 per cent reduction in Icelandic crop yields (Fridriksson 1969), so it is no wonder that long before the deterministic geographical writings of the early twentieth century Icelanders were well aware of the interaction of climate, plant growth and prosperity.

The sensitivity of the economy to climatic fluctuations, together with the extraordinary literary bent of the Icelanders and the emergence of a line of gifted field scientists (Thórarinsson 1960) has resulted in Icelandic records relating to the Little Ice Age being of exceptional length and richness. However, the record of Icelandic glacial fluctuations presents special difficulties of interpretation on account of the geographical setting. In the first place, the climate of Iceland is greatly influenced by the distribution of sea ice and its variation through time. Currently, sea ice is extensively developed in winter to the north of the country; during the Little Ice Age it often extended considerably further south. Second, being situated on the tectonically mobile mid Atlantic Ridge, Iceland has several active

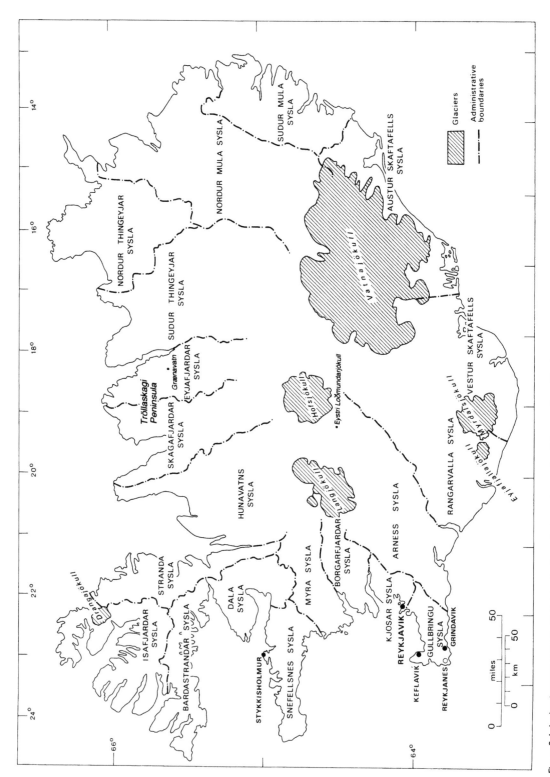

Figure 2.1 Iceland: major icecaps, administrative boundaries and some localities mentioned in the text

volcanoes and the heat flow and tremors associated with them have at times affected the movements and extent of the ice. Third, the disposition of the icecaps themselves presents special problems. Some of them rest on rock surfaces far below the snowline of the present day; if they were destroyed they would not reform under present climatic conditions. Furthermore, some are so large that marginal lobes and tongues could conceivably respond at different times to the same climatic fluctuations.

After a consideration of sea ice and volcanism, attention is directed here to the icecaps in the south and east of the country. The southern margins of Eyjafjallajökull, Mýrdalsjökull and the southern and eastern margins of Vatnajökull are bordered by lowlands that have long been settled. Both Mýrdalsjökull and Vatnajökull overlie volcanoes and several of the lobes of Vatnajökull are known to surge from time to time. Such surges are abnormally rapid advances not directly attributable to climatic events but rather to such features as instability related to the shape of the glacier bed and the presence of water at the base of the ice. Finally, attention is directed to Drangajökull, in the northwest, unaffected by volcanism although its outlet glaciers are now believed to surge (Sigurðsson 1993a).

2.1 VARIATIONS IN THE EXTENT OF SEA ICE AROUND ICELAND

Iceland is situated where warm water from the Atlantic and cold water from the Arctic converge. A branch of the North Atlantic Drift, the Irminger current, is deflected westward by a submarine ridge to flow along the south and west coasts, before sinking beneath the East Greenland current (Figure 2.2). A branch of the East Greenland current sweeps round the north and east coasts of Iceland and, at certain times, brings drift ice close to the land along the north and east and even the south coast. At other times, the sea ice lies further west, drifting down the coast of Greenland and through the Denmark Strait, the ice edge keeping well away from Iceland. The incidence of sea ice near the Icelandic coast has accordingly varied through the historic period. A 'normal' ice year has been defined as one when the ice edge from January to April is about 90 to 100 km away from Straumnes, in northwest Iceland. In a 'mild' ice year the ice edge is about 200 to 240 km away, in a 'severe' ice year it extends along the northern coast, and in an extremely severe year the ice is carried southward along the east coast by the East Greenland current and even reaches the south coast (Eythórsson and Sigtryggsson 1971). For most of the twentieth century the sea ice edge lay to the north of Iceland.

Irish monks are thought to have arrived in Iceland in 795. One of them, Dicuil,[1] was most probably describing events in Iceland when he recorded in his book, *Liber de Mensura Orbis Terrae*, written about 825, that:

> It is now thirty years since clerics who lived on that island from the first day of February to the first day of August told me that . . . those authors are wrong and give wrong information who have written that the sea will be solid about Thule[2]. . . since these men voyaged at the natural time of great cold . . . after one day's sailing from there to the north they found the frozen sea.
> (Translation according to Tierney 1967, paragraphs 11 and 12: 75)

Information for the medieval period is scanty and fragmentary with, for instance, laconic mentions of 'great sea ice around Iceland in 1261', and 'sea ice surrounded nearly all Iceland in 1275' (Jóhannesson 1956).

Thoroddsen, who published the first influential study of sea ice in 1874, included in his book *The Climate*

---

1   Dicuil was not only a monk but also a grammarian and geographer, who went to the Frankish court at the end of the eighth century. His exact dates are uncertain but his known and dated works were written between AD 814 and 825. His book, *De Mensura Orbis Terrae*, was probably completed in 825. It contains what is apparently the earliest mention of the discovery and settlement of Iceland. Irish monks, arriving in Iceland in 795, were struck by the long days of midsummer. Their descriptions appear to confirm that they were describing Iceland rather than Shetland, Orkney or the Faroes, which are too far south to make it at all likely that they that could have reached the sea ice after sailing for one day in the small ships of the eighth century.

2   Thule was the Greek and Roman term for the most northerly lands in the world. Pytheas (*c*.300 BC) had heard of it as an inhabited island, where grain grew sparingly and ripened poorly. As there is no evidence that Iceland was then inhabited or that grain was cultivated there, Pytheas cannot have been referring to Iceland. The name 'Thule' was used by many later writers, and has been identified variously as a part of coastal Norway, Shetland or Iceland.

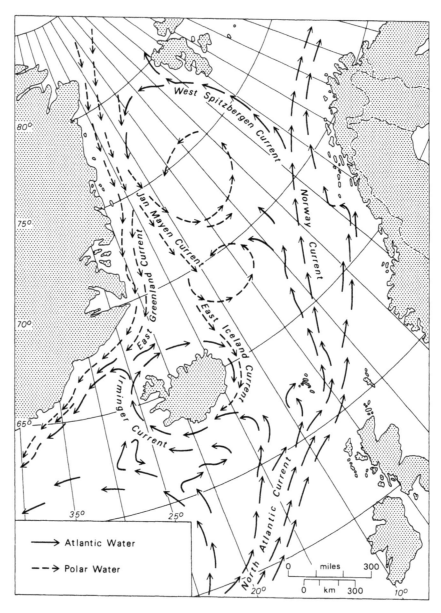

Figure 2.2 Surface currents in the seas around Iceland (*Source:* from Skov 1970)

of Iceland in a Thousand Years (1916–17) a separate monograph in which he drew together a vast amount of information ranging from the *Landnámabók* to Danish nautical and meteorological yearbooks. The more recent literature was the most plentiful and most accurate, the older more fragmentary and incidental in character. He surveyed the sagas and ancient annals

but met a gap in the historical writings between the Nyi Annáll of 1430 and the Reformation of 1540–50. He had access to the few sixteenth- and seventeenth-century annals already printed at that time, that is *Gottskálksannál, Biskupsannáll* and *Skardsárannáll*. For the seventeenth and eighteenth centuries he was able to use nearly all the important annals, but most

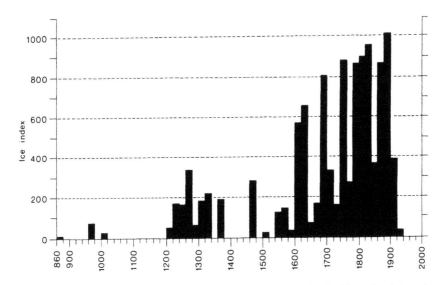

Figure 2.3a Koch's (1945) generalized diagram of the severity of sea ice incidence around Iceland from the ninth to the twentieth centuries. The index is the product of the number of stretches of coast (each about 135 km long) and the number of weeks when sea ice was noted along each of them

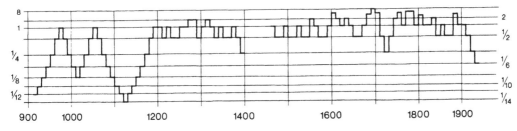

Figure 2.3b Bergthórsson's (1969) diagram of the decadal incidence of severe sea ice years around Iceland from the tenth to the twentieth centuries

of them were still in manuscript and he found them very disorganized and difficult to use. He also searched many other sources, both manuscript and printed, the nineteenth-century material including periodicals and newspapers as well as books, in an effort to gather together all the information he could find concerning climate and weather in Iceland; its analysis he left to others.

Koch (1945), largely on a basis of Thoroddsen's compilation, reconstructed sea ice variations since about AD 800. He divided the coast into stretches, each 135 km long; and devised an ice index which was the product of the number of weeks sea ice was noted and the number of stretches of coast from which it was seen (Figure 2.3a). He was moved to make this reconstruction by the publication of Thórarinsson's (1943) ac-

count of the oscillations of the Icelandic glaciers which prompted him to attempt to throw light on the climatic changes which caused the glaciers to expand in the eighteenth and nineteenth centuries. He was not entirely uncritical of Thoroddsen's methods, pointing out that Thoroddsen sometimes assumed the presence of sea ice on a basis of indirect evidence. Koch himself only made use of records which specifically mentioned sea ice. He concluded that there had been great changes in the climate of Iceland during historic times, but recognized that not all cold periods were necessarily associated with the presence of sea ice.

Bergthórsson's (1969) graph of the severe ice years over the eleven centuries since the Norse settlement bears a considerable resemblance to Koch's (Figure 2.3b). He defined a severe sea ice year as one when sea

DECADAL RUNNING MEANS OF TEMPERATURE

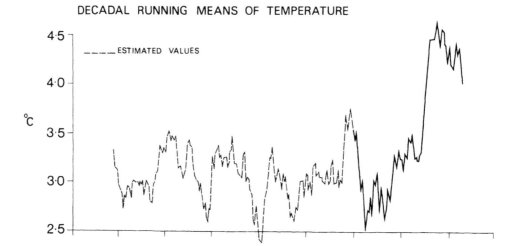

ICE INCIDENCE

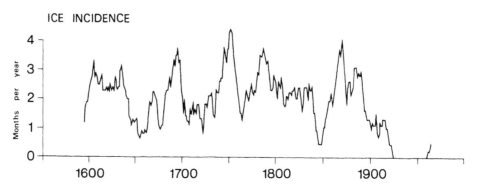

Figure 2.4 Decadal running means of annual mean temperature extended back from 1850 to 1600 on a basis of sea ice incidence (*Source:* from Bergthórsson 1969)

ice is known to have reached the south coast or alternatively as one in which there were deaths from starvation. It could be questioned whether the association between severe sea ice years and famine should be taken to be so close, and the continuity of the graph in periods where information is very sparse could be misleading. Bergthórsson found written history in the early fifteenth century to be so sparse that in his graph he left a gap for this period. Noting the good correlation between temperature and sea ice since the beginning of meteorological observations, he went on to infer past variations in temperature from the sea ice data

(Figure 2.4). A more detailed diagram showing year-to-year variability of sea ice was prepared by Sigtryggsson (1972) (Figure 2.5).

At the time when Thoroddsen was working, methods of dealing with historical sources were less developed than they are today (Bell and Ogilvie 1978, Ingram *et al.* 1981, Ogilvie 1991). Vilmundarson (1972) investigated Thoroddsen's sources for some unusually cold years in the seventeenth century and found that while some were accurate others were misleading. It is possible, for example, that sea ice conditions in 1639 were somewhat less severe than

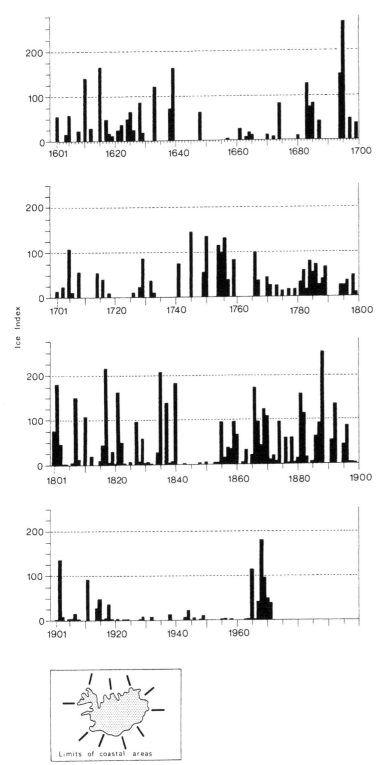

Figure 2.5 Incidence of sea ice around Iceland from the seventeenth to the twentieth centuries (*Source*: from Sigtryggsson 1972)

Thoroddsen supposed. He used the account from the *Skardsárannáll*, 'winter hard from Christmas onward. Sea ice drifted round the country the whole winter. It came along the east coast, then along past Sudernes' (the part of the southern peninsula in the vicinity of Keflavík; see Figure 2.1). 'It was accompanied by severe weather.' The *Skardsárannáll* was a contemporary description of events but, Vilmundarson pointed out, it is not necessarily adequate to determine which annal contained a contemporary account and then use that to determine conditions in a given year. The *Skardsárannáll* was composed by a man who lived far away on the north coast of Iceland and he was writing in general terms rather than with precision. Another account written at Grindavík, south of Keflavík, in 1639 was consulted by Thorlákur Magnússon when he wrote the *Sjávarborgarannáll* between 1727 and 1730, from which it appears that in 1639 the sea ice drifted as far as Reykjanes but no further (Figure 2.1).

> Terrible sea ice arrived three weeks after Easter and moving round the country from the east drifted ashore along the south coast all the way to Reykjanes. It drifted back and forth, east or west along the coast according to wind and current, past the Flitting days [late in May]. No use could be made of the ice either by seal hunting or by driftwood gathering. It cut away all the seaweed on the beaches at Grindavik, not a strand of any kind remaining.

Clearly conditions were severe but Thoroddsen's account, taken up by Koch, gives a somewhat exaggerated idea of the extent of ice round the southern peninsula towards Keflavík.

The section of the *Eyrarannáll* dealing with the period from 1673 to 1703 was considered by Vilmundarson to have been written concurrently with events and to be reliable, at any rate for the earlier years, though the later section must be used cautiously as the author, Magnús Magnússon of Eyri, in Seydisfjördur in northwest Iceland (1630–1704), was by that time suffering from the effects of old age. The picture which the *Eyrarannáll* presents of 1695 is of considerable importance, for it came in a period when glaciers were advancing in Norway as well as Iceland:

> The winter was fairly good with periods of little snow and favourable weather on land, still there were periods of extremely severe frost in between, so that all the fjords were frozen and fishing was hampered . . . sea ice came also to the north and west coasts, into the fjords after the New Year and stayed until summer . . . the drift ice froze together so that one could travel on horseback from one promontory to another across all the fjords also out beyond Flatey in Breidifjördur. No one could remember such ice cover, nor had anyone heard of such ice from older people . . . they also told of such a girth of ice in the sea round this country that ships could hardly reach the shore except in a small area in the south. The same frosts and severe conditions came to most parts of this country; in most places sheep and horses perished in large numbers, and most people had to slaughter half their stock of cattle and sheep, both in order to save hay and for food since fishing could not be conducted because of the extensive ice cover. . . . In Strandasýsla people fished through the ice 1.5 to 2 Danish miles offshore and shark, flounder and other fish were transported on horseback to the shore.

Vilmundarson not only looked critically at Thoroddsen's use of sources but also pointed to some he had overlooked, especially the regular reports on climatic and other conditions made to the Danish Governor by the Sheriffs of the various areas of Iceland, which became customary in the eighteenth century.

Ogilvie (1984, 1986, 1991, 1992, 1994) made new reconstructions of the sea ice and climatic record of Iceland from medieval times to 1800, based on a critical examination of all the available documentary sources and a careful selection of reliable data. She made use, not only of all the previously used sources, but also of other material, unused by the earlier workers such as Koch and Bergthórsson, including letters from Sheriffs and from the Governor of Iceland to the Danish Exchequer. She rejected many of Thoroddsen's sources because of dating errors and also discriminated between alternative versions of the same event, eliminating unreliable and spurious items and incorporating a great deal of new material, much of it from manuscript sources (Figure 2.6).[3]

Evidence for the early medieval period Ogilvie found to be scanty and fragmentary. Indications that the climate was favorable at the time of the Norse settlement and for some time afterwards, though essentially circumstantial, she accepted as pointing to a mild climatic period. The first indications of increasing rigour of the

---

3 Many of the main sources are discussed in detail in Ogilvie (1992).

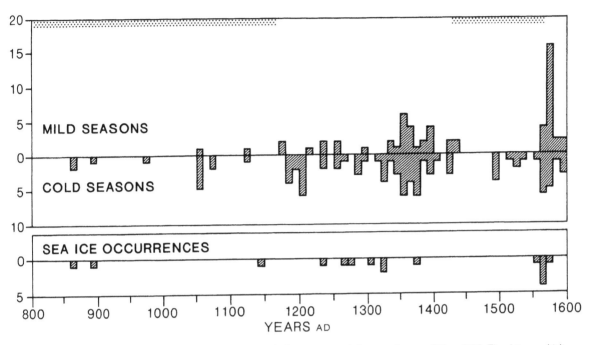

Figure 2.6 The incidence of mild and cold seasons in Iceland and of sea ice around the coast from AD 800 to 1598. The data on which this diagram is based are described in Ogilvie (1991). Stippling at the top of the diagram indicates periods of very poor data (*Source:* after Ogilvie 1991)

climate appear in the 1180s. There were other references to cold years in the next three decades but not from 1212 to 1223. Cold seasons came sporadically over the next thirty years and then the 1280s and 1290s saw more continuously harsh conditions. The first record of very extensive sea ice came in 1261 when 'there was sea ice all round Iceland' (Ogilvie 1991: 241). Ogilvie (1984) cites a statement of 1287 that 'at this time many severe winters came at once and following them people died of hunger'. In 1291 the annals refer to a 'great snow winter' and 'the great ice and livestock death winter' (Ogilvie 1991: 242).[4]

Much more information is available for the fourteenth century, which seems to have been very variable climatically. The first two decades were rather mild, though in 1306 there was 'sea ice all summer in the north'. The early 1320s, in contrast, were harsh, with sea ice close at hand, and the weather in 1320, 1321

and 1323 was very cold. In the next two decades only two winters were severe, but from the 1350s temperatures were lower, with hard winters in 1350, 1351, 1355 and 1362. In 1365 there began the most severe period of the fourteenth century, with particularly hard winters recorded in 1370, 1371, 1374, 1376 and 1379. Though sea ice is mentioned only for 1374, this cannot be taken as proof of its absence in the other years. Around 1350 Arngrímur Brandsson had written that 'on the sea are great quantities of drift ice that fill up the northern harbours'. By the 1360s it seems likely that it had become recognized that ice from the north was disrupting the old sailing routes from Norway. Ogilvie's climatic findings do not agree with Bergthórsson's graph, which shows the second half of the fourteenth century as relatively mild. He had based his temperature curve on the occurrence of sea ice, mentions of which, for this period, are rare.

---

4   Petterson (1914) and Bull (1915) had found evidence of an economic decline in the late Middle Ages associated with severe years and drift ice around Iceland within the period 1291 to 1392, according to Ogilvie and Jónsson (2001).

An independent study cited by Ogilvie is that of Teitsson (1975), based on the number of polar-bear skins mentioned in church inventories. They are held to indicate a period of great cold from 1350 to 1380, when the priests were particularly glad to have a warm fur rug on which to stand in church, and sea ice was near enough to the coast for bears from Greenland to come ashore. Evidence of fourteenth-century glacial advances in Greenland and parts of Scandinavia suggests a strong possibility that glaciers may also have expanded in Iceland at this time.

Unsurprisingly, investigations around Mýrdalsjökull have now revealed that one of the units of the compound moraine fronting Gígjökull was deposited around 700 BP (Kirkbride and Dugmore 2001a) and there is evidence of thirteenth-century advances of Svínafellsjökull in Oræfi (Gudmundsson 1997).

Comparatively mild conditions seem to have returned from about 1380 to 1430, though winters were long and severe in 1424 and 1426. Contemporary sources are rare for the period 1430 to 1560; Bergthórsson's graph indicating mild conditions has little factual basis. However, trade with England at this time, involving cloth rather than grain, suggests that food was not particularly short.

Annual data, mainly from *Gottskálksannáll*, show that while the 1560s were predominantly cold, the 1570s were very mild, sea ice being referred to only in 1572.[5] The last decades of the century, in contrast, were undoubtedly cold. Oddur Einarsson wrote in the late 1580s that

> the Icelanders who have settled on the northern coasts are never safe from this most terrible visitor. The ice is always to be found between Iceland and Greenland although sometimes it is absent from the shores of Iceland for many years at a time . . . sometimes it is scarcely to be seen for a whole decade or longer . . . sometimes it occurs almost every year.
>
> (Cited Ogilvie 1992, translated from Einarsson 1971)

Ogilvie used the wealth of documentary information for 1601 to 1780 to construct decadal sea ice indices. She found the series generally agreed with a winter/spring thermal index series which she constructed for

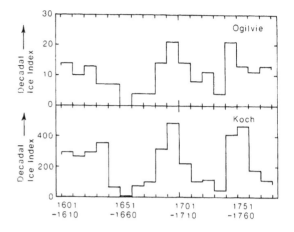

Figure 2.7 Decadal sea ice index for Iceland, 1601–1780. The upper values are from Ogilvie (1984) and are derived from the number of seasons of sea ice occurrence (winter, spring and/or summer) weighted by the number of regions reporting ice (i.e. 1 to 4). The lower values are an equivalent index based on Koch (1945)

Iceland (Figures 2.7 and 2.8). Its main features correspond to those distinguished by Koch: a remarkably mild period from about 1640 to 1670 was followed by one of increasing severity, with the decade 1690 to 1700 extremely cold. The Maunder Minimum sunspot period of 1675–1715 (see Chapter 17), was strikingly variable though not particularly unusual. Both 1705 and 1715 were heavy ice years, but no winters between 1701 and 1715 were as severe overall as those of the 1690s. The 1630s and 1730s, which were cold on land, saw little sea ice; in these decades the response of the sea ice to climatic fluctuation seems to have lagged behind conditions on land.

Ogilvie's documentary sources were detailed enough to allow her to provide five to ten indications of conditions on land for each season, and also thermal indices for the north, west and south of the country as well as for the island as a whole (Figure 2.8). Although there was a good deal of correspondence overall, fluctuations in the severity of winters were not simultaneous throughout the country; the 1630s, for example, were milder in the west than in the north and south; the reverse being true in the 1640s. The cooling trend in

---

5   The Icelandic annals include descriptions of the climate in many mild seasons, which contradicts the common view that historical sources are prone to dwell on extremes, ignoring average conditions.

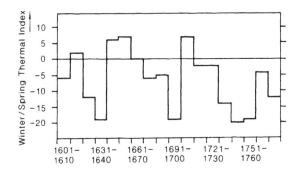

Figure 2.8a Decadal winter/spring thermal index for Iceland, 1601–1780 (*Source*: from Ogilvie 1986)

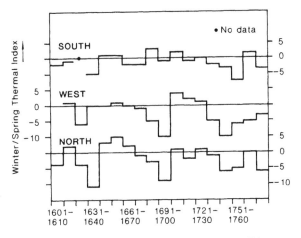

Figure 2.8b Regional thermal indices for Iceland, 1601–1780 (*Source*: from Ogilvie 1986)

the latter part of the seventeenth century, so obvious in the north, is less so in the south, while the cooling of the early eighteenth century is more evident in the south than the north.[6]

Apart from its first decade, the overriding characteristic of the eighteenth century was the great severity of weather events. The cold of the 1740s and 1750s approached that of the 1640s and 1690s. Sea ice was recorded in seven out of ten years in the 1740s and in half the years of the following decade which also saw a catastrophic eruption of Katla. In 1759 'Sea ice

surrounded all the eastern and southern parts of the country up to Reykjanes for two or three weeks during June and July' (Sheriff's letter, quoted by Ogilvie 1992). Milder weather followed in the 1760s, and although sea ice reached the coast in five years that decade, it failed to impinge on the coast of southern Iceland. The next decade was colder, with the incidence of sea ice as great as it had been in the 1690s and 1740s. The coldest decade in the entire period since 1601, and probably since 1501, was the 1780s. Sea ice was present every year of that decade, and throughout every year from 1782 to 1785. The suffering caused by the Laki eruption of 1783 was intensified by severe weather which continued into the 1790s. When at length the cold slackened, sea ice remained extensive off the northern, northwestern and eastern coasts from winter right through to the summer of 1801, not leaving Húnavatns sysla in the northwest until August (Ogilvie 1992).

Sea ice continued to be extensive round Iceland until 1840 (Ogilvie and Jónsson 2001, figure 3a) and it is apparent from Sigtryggsson's diagram (see Figure 2.5) that, except for a spell of fourteen years between 1840 and 1854, it remained so throughout the nineteenth century. The last year in the century when sea ice grounded along the north coast was 1877. This, as it happens, was also the year when the Danish Meteorological Institute began collecting reports on sea ice from around Iceland and the Greenland Sea. Observations were later collected for the whole of the Arctic Ocean from ships, sealing vessels and coastal stations. The work was continued by the Icelandic Meteorological Office and during the Second World War by the British Meteorological Office. Since 1887 there has been no gap in the records, though for 1945 to 1951 they are less full than in some others, partly because seal catching off the east Greenland coast had yet to revert to its pre-war level. Furthermore, the whole subject of sea ice around Iceland was by then a minority interest because it had scarcely affected the country for several decades.

Jón Eythórsson considered this situation unsatisfactory and took a lead in organizing regular drift ice observations, publishing them in the journal *Jökull*

---

6 The extent of regional variations discerned from the archival data is much as might be expected in view of the differences found between various parts of the country during the period of meteorological measurements, especially in the sea ice period 1965–71, shown in Figure 2.9 (Einarsson 1991).

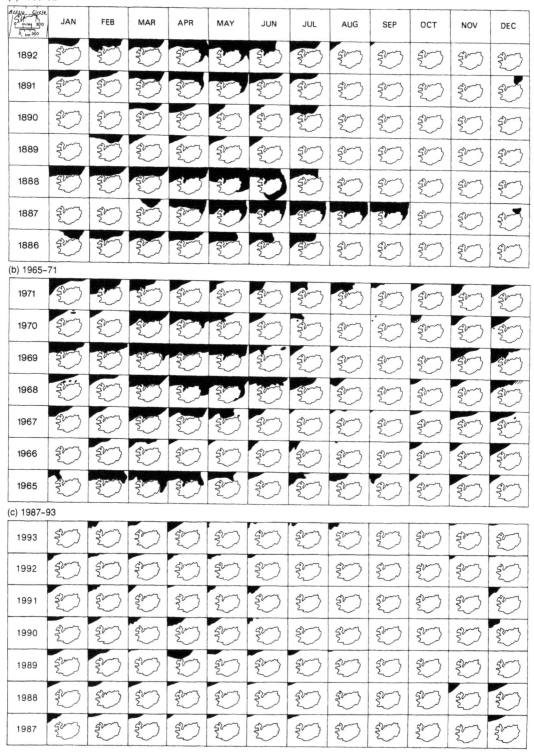

Figure 2.9 Monthly sea ice positions around Iceland during the periods: (a) 1886–92, (b) 1965–71 and (c) 1987–93. Icing was more severe in the October–December quarter of the second period than the first period. No sea ice arrived at the coast in 1987–93. Although Iceland was experiencing a cool period the influence of polar water was lacking (*Source*: after Sygtryggsson 1972, and monthly data from US Navy–NOAA Joint Ice Center)

from 1953 to 1966. Later results were published in the same journal from 1967 to 1971 by Sigtryggsson and Sigurðsson, and after that in an official publication, *Hafís við strendur Islands*, devoted to the detailed presentation of sea ice data. Icelandic interest in the subject reawakened as the ice returned to nearby sea areas (Jónsson 1969). Sources now include daily satellite pictures of the main ice areas, which in the dark winter months include infrared imagery.

Sea ice returned rather abruptly to Icelandic waters in 1965 and subsequently appeared every spring. Ice conditions in 1968 were worse than in any year since 1888; in May it came right up to the jetties at Akureyri and was stranded a few miles further north at Dalvík.[7] Figure 2.9 (a) and (b) allows a comparison to be made between the ice years from 1965 to 1971 and the last set of years of comparable severity from 1886 to 1892. The south coast was not as severely affected as it was in June 1888 and in no year did the drift ice ground on the north coast as it had in 1877. But the greater proximity of the ice to the northern coast in November and December as compared to the nineteenth century is evident.

The increased ice incidence in the 1960s was associated with the development in the 1950s of an anomalous ridge of high atmospheric pressure over Greenland, which strengthened in the early 1960s to reach a peak intensity of 12 mbars above the 1900–39 'normal'. It was coupled with a slight decrease in pressure over the Norwegian and Barents seas, causing a strong pressure gradient promoting northerly airstreams, and a steep decline of mean winter air temperatures over the Norwegian–Greenland seas. The mean annual temperature at Stykkishólmur in western Iceland dropped in 1965 and succeeding years until it was well below the average value since observations began in 1846, for the first time since 1925 (excepting only 1943 and 1949) (Sigfúsdóttir 1969). Mean surface temperatures also dropped over the North Atlantic. As the northerly winds increased in strength, the East Icelandic Current changed from being an arctic current, as it was from 1948 to 1958, and from 1964 to 1971, to become a polar current (Malmberg 1969 and Figure 2.10). An arctic current does not promote the preservation of sea

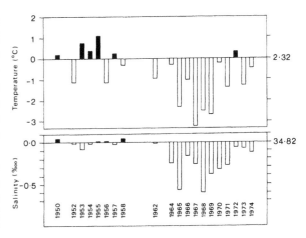

Figure 2.10 Temperature and sea ice in the Icelandic sea area, 1950–75. Arctic water has a slight vertical stratification gradient in the surface layer which prevents surface water cooling to the freezing point before vertical convection starts. Polar water, with a salinity of less than 34.7 per cent, cools to the freezing point of sea water at −1.8 °C before convection starts and so drift ice forms (*Source*: from Dickson et al. 1975)

ice, because a slight vertical stratification gradient in the surface layer prevents the water from cooling to freezing point before vertical convection starts, but a polar current with a salinity of less than 34.7 per cent can cool to the freezing point of sea water, −1.8 °C, before convection starts and so drift ice forms more readily and remains as long as the current retains its surface character. The influence of sea ice and low sea surface temperatures resulted in large negative temperature deviations from the 1936–85 mean, especially in northern and eastern Iceland.

After 1970, the high pressure anomaly over Greenland collapsed dramatically, the mean pressure at sea level in winter falling by 9.6 mbars between 1966–70 and 1970–4, and the polar influence off northern Iceland weakening. Polar water of low salinity failed to reach as far south and Atlantic waters returned to the sea areas north of Iceland. A period with cool summers followed, from 1972 to 1990, but no sea ice arrived at the coast (Figure 2.9c, Einarsson 1991). With reduced ablation losses, more than half the glaciers began to advance (Sigurðsson and Jónsson 1995).

---

7   Personal communication (6 January 1995) from Dick Phillips who witnessed the incident.

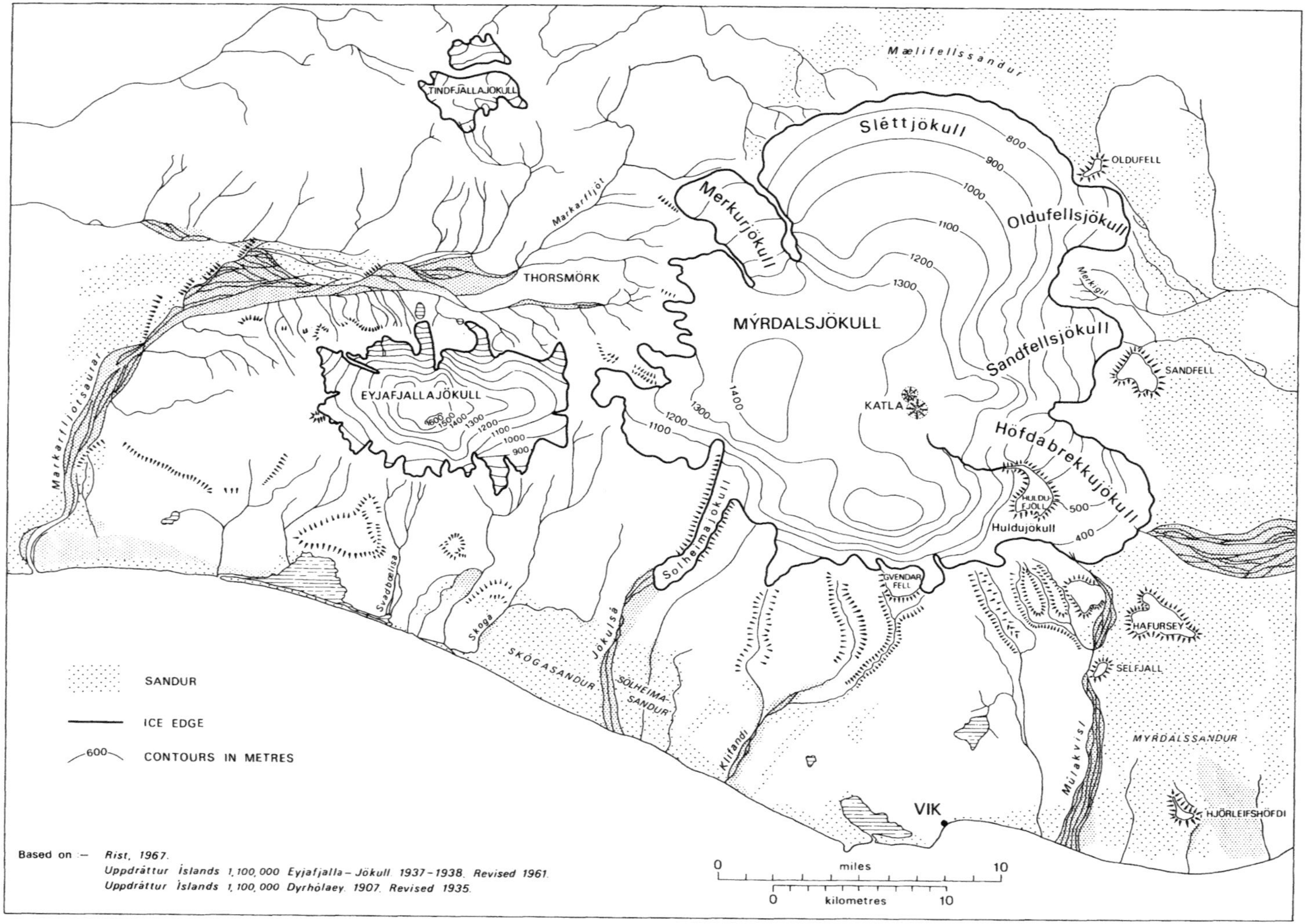

Figure 2.11 Mýrdalsjökull and Eyjafjallajökull (Source: after Rist 1967b)

The economic significance of the variations in the extent of sea ice round the country is considered at a later stage in connection with the associated history of glacier fluctuations in Iceland (see Chapter 18).

2.2 THE GLACIAL HISTORY OF MÝRDALSJÖKULL AND EYJAFJALLAJÖKULL

In the Little Ice Age, Mýrdalsjökull and Eyjafjallajökull together formed a single icecap which separated into a larger eastern and smaller western portion only in the middle of the twentieth century (Figure 2.11). Mýrdalsjökull[8] with an area of 596 km² lies on top of Katla, the second most active volcano in Iceland. Katla's eruptions have been accompanied by catastrophic floods from beneath the ice, known as *jökulhlaups*. Some other jökulhlaups are caused by the emptying of ice marginal lakes[9] (Björnsson 1992).

The area south and east of Mýrdalsjökull was the first part of Iceland to be fully settled. The coastal settlements have always had to struggle to survive volcanic eruptions, floods, ice advances and avalanches. Many farms have had to be abandoned; the rest have concentrated in the safer localities, giving a cluster pattern of settlement unusual in Iceland.

The Mýrdalsjökull ice lens is 200–370 m thick (Rist 1967a) and rises to over 1400 m. Vík, the nearest meteorological station, records the second highest mean annual precipitation in Iceland (2256 mm, 1931–60). On Mýrdalsjökull, accumulation and ablation both reach very high values. The northern and eastern margins of the ice are lobate and relatively gently sloping. On the south side, valley-type outlet glaciers descend over rock steps and display ogives in their lower sections. Smaller, steeper tongues of ice drain towards Thorsmörk in the west. The margins of Eyjafjallajökull are also rather precipitous. The highest point, Eyjafjall, rises to over 1600 m, a height exceeded in Iceland only by Öraefi, the great volcano south of Vatnajökull.

Table 2.1 Eruptions of Katla since AD 844

Eruptions suspected but uncertain		Eruptions verified by eye witness accounts
894		1625
900		1660
934		1721
1000		1755
1245 ⎤	hlaups from eruptions	1823
1262 ⎦	of Sólheimajökull	1860
1311	'Sturluhlaup' – legendary	1918
1332		1955 minor eruption
1416	'Höfdalhlaup' – legendary	
1485	tephra evidence	
1580		

Sources: Thoroddsen 1905–6, Rist 1955, 1967c, Thórarinsson 1959
Note: There is not complete agreement between available authorities on the dates of the eruptions, especially the earlier ones

Katla is situated on a line of fissure eruptions that can be traced running southwest from Vatnajökull and disappearing under Mýrdalsjökull. Its crater, Kötlugjá, is normally beneath the ice but after eruptions it can readily be discerned. Such eruptions have caused a great deal of damage from lava flows, ash falls and *Kötluhlaups*[10] (that is jökulhlaups coming from Kötlugjá), destroying farms and ruining pastures. Early jökulhlaups had carried great loads of sediments from beneath the ice and deposited them to form a gently sloping apron, a *sandur*, running down into the sea. Upon these sandur the first settlements were established, most of them being destroyed between the ninth and eleventh centuries by hlaups from Kötlujökull. After the fourteenth century eruptions became even more frequent. Various accounts of these give a vivid impression of their consequences (Table 2.1).

The Kötluhlaup of November 1660 carried away all the houses and the church of Höfdabrekka; hardly a

---

8   According to the World Glacier Inventory 1989.
9   Jökulhlaups also come from five other geothermal areas in Iceland, and currently from fifteen ice-dammed lakes. Since the 1930s jökulhlaups have been monitored and data are available on their frequency, duration and impact (Björnsson 1992).
10 Kötluhlaups are well characterized by both flood magnitude and duration and by extremely high sediment concentrations. In 1918 the flood waters rose to about 10^5 m³ per second in a few hours and returned to pre-flood levels within 24–36 hours (Maizels 1991). Smaller floods associated with minor volcanic activity occurred in 1955 and 1999.

stone was left on the original site and so much material was carried down to the shore that a dry beach appeared where previously fishing boats had operated in waters 20 fathoms deep. In 1721 masses of ice from the front of the icecap floated out to sea to form a barrier 3 nautical miles offshore. In 1755 ice, clay and solid rock were hurled into the sea for a distance of more than 15 miles; 'in some places where it was formerly forty fathoms deep, the tops of the newly deposited rocks were now seen towering above the water' (Henderson 1819: 248). But the principal damage, so far as the survivors were concerned, was the destruction of pasture.

Thórarinsson (1957) estimated that, even discounting their ice content, the discharge of Kötluhlaups may reach 100,000 m³ per second briefly, a figure which is comparable with the mean discharge rate of the Amazon. Björnsson (1992) considered this an underestimate; according to him, the largest floods from Katla can discharge 100,000 to 300,000 m³ per second for two to five days. He listed 17 eruptions of Katla, the last in 1918. A minor Kötluhlaup in 1955 was preceded by an earth tremor five hours before the flood reached the ice margin. Björnsson suspected that both tremor and hlaup were caused by a subglacial eruption melting the adjacent ice. The water is estimated to have flowed under the ice at a velocity of about 10 m per second on a gradient of about 4°.

The largest and most destructive jökulhlaups in prehistoric times have emerged from the ice lobes east of Katla: Höfdabrekkujökull and Huldujökull. Their peak discharges, caused by eruptions in ice-filled calderas in northern Vatnajökull, notably Bárdarbunga and Kverkfjöll, are estimated to have reached 400,000 m³ per second (Björnsson 1992). The accompanying ice collapses were on such a large scale that these two ice lobes are of little or no value for tracing events of climatic significance. The northern lobes are too remote from settlements for details of their movements to have been observed. The most useful information for our purposes comes from ice-dammed lakes and the floods associated with them at the southern outlets of Mýrdalsjökull and Eyjafjallajökull, though even these glaciers have not been altogether free of jökulhlaups and the effects of volcanism.

2.2.1 Sólheimajökull

Sólheimajökull in southwest Mýrdalsjökull is drained by one of the most dangerous rivers in Iceland, the Jökulsá á Sólheimasandi. Its other name is Fúlilækur, stinking river, from the hydrogen sulphide it emits. Bárdarson (1934: 41) mentions jökulhlaups emerging from Sólheimajökull in AD 930 and 1262 and attributes them to volcanic eruptions. However, he does not mention his sources and his explanation for the hlaups is not necessarily the only one available.

Árni Magnússon described Sólheimajökull in 1704–5 as 'a flat, low outrunner of Mýrdalsjökull, extending in a bend southwards from that glacier and then to the west'.[11] The description is accompanied by a sketch (Figure 2.12) which is the earliest known map of an Icelandic glacier. Magnússon explains that it had grown westwards so as to block a canyon carrying meltwater from Mýrdalsjökull, the water escaping by a tunnel under the tongue of Sólheimajökull.

When the tunnel is blocked up, the water which would otherwise have run out of the canyon mentioned above, forms an enormous deep lake. When the ice blocking up the tunnel can no longer withstand the pressure of the water, the tunnel is opened *cum impeta* and everything inside it is crushed. Thus arise the jökulhlaups in Jökulsá. . . . The jökulhlaups usually occur once a year and are smaller the shorter the intervals between them. . . . Sólheimajökull both lifts and slides over the surrounding land, so that the difference can be seen from one year to the next. People say of the main glacier [Mýrdalsjökull] that it does not grow visibly except in places where it descends into ravines, which are gradually filled up by small glaciers.

From this passage and from the sketch it is clear that Sólheimajökull had blocked the Jökulsá canyon a good many years before 1704. According to Ólafson and

---

11 Árni Magnússon (1663–1730), the founder of the Arnamagnaean collection, travelled through Iceland between 1702 and 1712 and recorded many geographical and topographical details in his 'Chorographica islandica'. The Sólheimajökull material is probably in the handwriting of his secretary and dates from his visits to Vestur-Skaftafells sysla in 1704 and 1705.

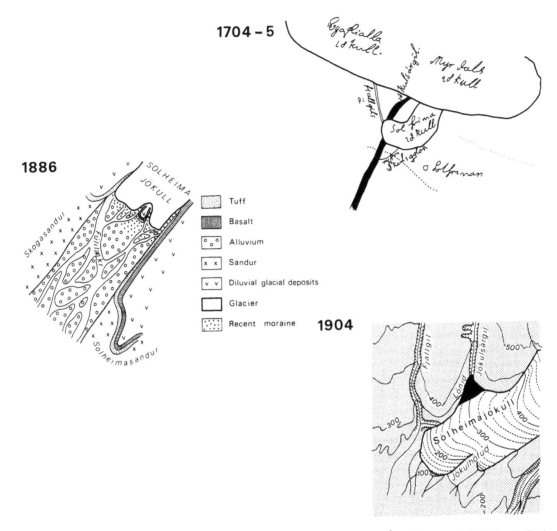

Figure 2.12 The tongue of Sólheimajökull mapped in 1704–5 after Thórarinsson (1960) from Árni Magnússon; in 1886 after Keilhack (1886); and in 1904 after Thórarinsson (1960) from the Danish General Survey map of 1904

Pálsson (1772: 763),[12] it continued to advance after Katla's 1755 eruption, swelling and thickening very greatly; then it retreated. Thoroddsen (1905–6: 183) recorded that by 1783 it had withdrawn and the Jökulsá was flowing freely between the glacier and Skógafjall, which means that Sólheimajökull was considerably

---

12 Eggert Ólafsson, a naturalist and poet, travelled through the country every summer from 1750 to 1757 with Bjarni Pálsson, a physicist, and eventually published a very comprehensive, if not especially original, account of the observations they made. He was too much affected by the abstruse scientific theories then current, but fortunately he recorded the views of the country people, which were often based on detailed practical knowledge, although himself rejecting some of the soundest of them. Thus he gave the local view on the alimentation of glaciers which 'reached high up in the air, where it is much colder than on the flat low-lying land. On them rain will change to snow and ice, and as they always attract rain, clouds and fog, they will maintain their size and grow unless the sun can every year melt as much as is added to them.' While Ólafsson's informants had grasped the basic notion of glacier mass balance, he succeeded in convincing himself that glaciers were probably nourished by penetration of sea water through subterranean passages.

smaller than in 1705. By 1794 the glacier had advanced again, reaching Skógafjall, blocking the Jökulsá and damming back a lake which, according to Sveinn Pálsson (1945),[13] flooded especially often in 1794. Neither his map nor his panoramic view (Figures 2.13 and 2.14) provide further detail but it would seem likely that Sólheimajökull's tongue was then in a similar position to that of 1704. Eythórsson (1931) recorded that a reliable local man, 80 years old, told him that by 1820 the ice covered the rock called Jökulhöfud which, according to Thoroddsen (1905–6: 183), was still covered in 1860. The glacier may then have begun to recede because local people told Eythórsson in 1930 that the ice had not completely covered Jökulhöfud for sixty or seventy years. Certainly by 1883 the greater

part of Jökulhöfud was free of ice, for a map drawn by Keilhack (Figure 2.12) shows the glacier dividing into a wider western and a narrower eastern part either side of Jökulhöfud, with moraine labelled recent and possibly dating from 1860, extending about 100 m in front of the ice. The situation seems to have been much the same in 1893 (Thoroddsen 1905–6: 183).

A Danish map of 1904 (Figure 2.12) shows that the front of Sólheimajökull had retreated to about 100 m above sea level, as compared with 50 m when Keilhack (1886) had mapped it. Jökulhöfud was free of ice on its northern side and the eastern terminus of the glacier had retired about 200 m since 1883, though the western one had remained more or less stationary. The retreat continued, with a mean rate of recession of 30

---

13 Sveinn Pálsson (1762–1840) was the most notable glaciologist of eighteenth-century Iceland. He was the son of a farmer from Skagafjördur in the north. He studied natural history and medicine in Copenhagen from 1787 to 1791 and then, in the succeeding four summers, he travelled widely through Iceland making geographical, geological and botanical observations. In 1795 he married Thórunn, the daughter of Bjarni Pálsson, who had travelled with Eggert Ólafsson. In 1799 he was appointed doctor to the south of Iceland, serving an area stretching from Hellisheidi to Skeidarársandur, as well as the Westmann Islands. He held this appointment till 1824 and for most of the time lived and also farmed at Sydri Vík in Myrdalur. Despite this double occupation in two outstandingly onerous fields he made careful weather records from 1791 to the end of his life and undertook a great deal of research and writing. His most important glaciological work was the treatise which he sent in 1795 to the Natural History Society in Copenhagen, which financed his travels. He wrote this in Danish. Parts of it were published in Den Norske Turistforening's Yearbooks for 1881, 1882 and 1884, but it was not published in full until it was edited and brought out in Icelandic by Jón Eythórsson in 1945. As a result it was entirely overlooked elsewhere in Europe and played no part in the mainstream of glaciological thought. Had it done so, it must surely have ranked as the most important and illuminating work on the subject written during the eighteenth century.

Pálsson had read Strøm's *Physisk og oeconomisk Beskrivelse over Fogderiet Søndmør* published in 1762, which contained Wiingaard's description of the fluctuations of some of the Jostedal tongues (see Chapter 3), and also the very much less outstanding discussion of glacier motion included in Fleischer's *Forsøg til en Natur-Historie*. Through this, but not directly, he learnt of the work of Walcher, Gruner, de Saussure and other Alpine glaciologists. He made such use as he could of existing knowledge gained in this way but treated it critically. Unlike his predecessors he thought it worthwhile to give detailed accounts of individual glaciers. He also drew many maps, that of Vatnajökull being especially impressive (see Figure 2.21).

Pálsson was the first glaciologist to perceive that convection rather than radiation could dominate the ablation process on temperate glaciers, writing that 'I also venture to affirm that the rays of the sun do not play as much a part in melting the firn of the plateau ices, not even in cold weather when they fall directly on them, as misty weather without precipitation does' (quoted by Thórarinsson 1960: 14). This view was eventually to be confirmed by the work of the Swedish-Icelandic investigations of Vatnajökull in the 1930s. It is not surprising that Pálsson was the first to differentiate clearly between the oscillations of the Vatnajökull margins caused by climatic changes and those caused by volcanic activity. His work is a major source for any study of Icelandic glaciers in the eighteenth century. Thórarinsson (1960) summed up his contribution most effectively: 'His treatise on glaciers constitutes a last phase and a culmination of a glaciology that may be called Icelandic in the sense that it was principally based on knowledge of glaciers in Iceland.' That knowledge was to a large extent common to the country people who lived along the southern margin of Vatnajökull, in close contact with its advancing outlet glaciers and its glacier rivers and sandur. It was a knowledge which had gradually accumulated during nine centuries, because this people was in large measure endowed with 'Man's nature to wish to see and experience the things that they had heard about and thus to learn whether the facts are as told or not' (Larson 1917). What is more remarkable is that Pálsson should have been able to distil, interpret and build upon this knowledge whilst living for forty years in extreme poverty, working as a doctor, fisherman and farmer, caring for his people, his land and his large family.

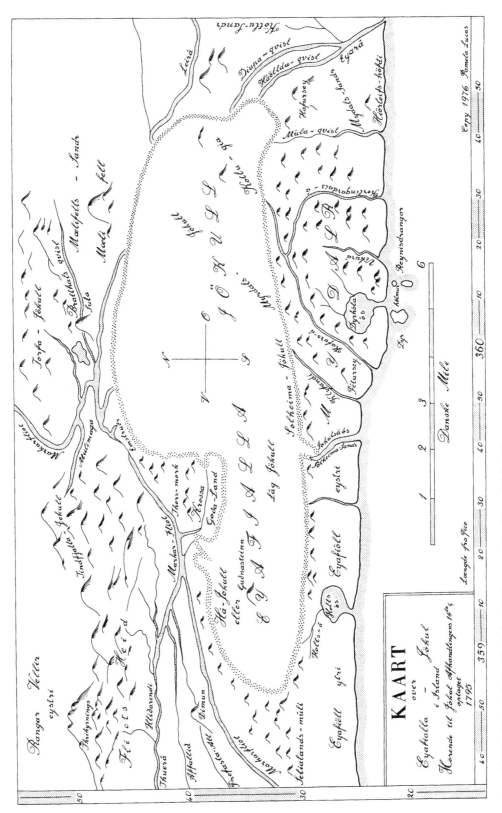

Figure 2.13 Eyjafjallajökull–Mýrdalsjökull in 1795 as mapped by Pálsson (1945 edn)

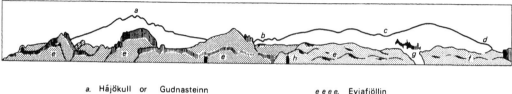

a. Hájökull or Gudnasteinn	*e e e e.* Eyjafjöllin
b. Lágjökull	*f.* Mýdalsfjöllin
c. Sólheimajökull	*g.* Jökull, which Fúlilækur flows under
d. Mýrdalsjökull	*h.* Skógafoss

Figure 2.14 Cross-section view of Eyjafjallajökull–Mýrdalsjökull in 1795 by Pálsson (1945 edn)

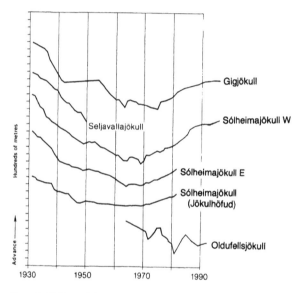

Figure 2.15 Frontal fluctuations of Eyjafjallajökull–Mýrdalsjökull glaciers (*Sources:* from figures published in *Jökull* by Eythórsson, Rist and Sigurðsson)

and was advancing or stagnant until about 1860. Since then it has, on the whole, been in recession. . . . Its maximum extent in the first half of the 18th century – probably approximately the same as in the middle of the 19th century – was the largest in historical times.

(1939: 236)

No major jökulhlaups occurred over the period of the Sólheimajökull record, though volcanic activity was certainly associated with its expansion about 1755. The timing of the advances and retreats is believed to be of climatic significance and may now be compared with those of neighbouring glaciers.

2.2.2 Gígjökull

Gígjökull, the glacier east of Jökultungur, and another near it, came down onto the plain in 1756 and were described by Ólafsson (Bárdarson 1934: 44–5) as alternating and advancing every year. But he also commented that Eyjafjallajökull was smaller than it had been earlier and that two hills which had not been seen for a long time were sticking out of the ice. He thought that this thinning was due to the eruption of Katla in 1755, but Katla is a considerable distance away. Gígjökull and its neighbour still extended down to the plain when Sveinn Pálsson saw them in 1794. For the nineteenth century we have no information. When it was mapped in 1907, the front of Gígjökull was 200 m from the outermost moraines of historic times (Bárdarson 1934: 44). Local people reported that it had been advancing somewhat for a few years before 1930, when a permanent marker was erected. Subsequently, in the 1930s and late 1950s, retreat was rapid, but between 1960 and 1990 Gígjökull regained 231 m of the 675 m lost between 1930 and 1960.

Einar Einarsson (1966), a farmer of Skammadalshóll in Myradalur, was told by his father and grandfather that

to 40 m per year between 1930 and 1937, and the ice rapidly thinning. After 1930 the positions of the front of Sólheimajökull and other glaciers were measured practically every year (Sigurðsson and Jónsson 1995). As Figure 2.15 shows, the retreat was maintained except for a slight advance in 1964–5, until Sólheimajökull reached its minimum extension in 1969, after retreating a kilometre in the previous thirty-nine years. Advances since then brought the terminus halfway back to its position in 1930 (Figure 2.15).

Summing up the oscillations of Sólheimajökull, Thórarinsson (1939) decided that in 1705 it was of

about the same extent as in 1904. In 1783 it was considerably smaller than in 1904. In 1794 it had resumed its advance; at about 1820 it had reached its 1705 position

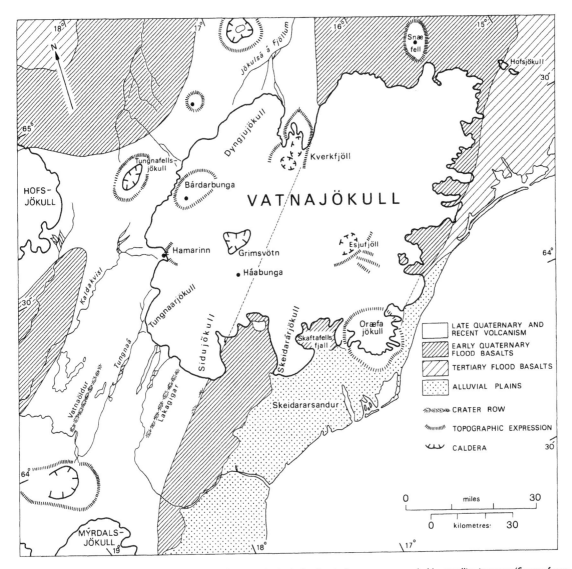

Figure 2.16 Vatnajökull in relation to geological boundaries and principal volcanic features, as revealed by satellite imagery (*Source:* from Thórarinsson *et al.* 1973)

the glaciers of southern Mýrdalsjökull advanced between 1870 and 1890 but afterwards retreat was general until 1966. Local people living on the south side of Eyjafallajökull told Eythórsson (1931) that Seljavalla- jökull retreated between 1907 and 1930 and that the glaciers had become much thinner. All of this informa- tion corresponds to the data for Sólheimajökull.[14]

2.3 VATNAJÖKULL

Vatnajökull, which was also known as Klofajökull, is the largest icecap in Europe, and the world's largest temperate icecap (Figure 2.16). It exhibits a closer jux- taposition and a more intimate interplay of vulcanism and glaciation than is to be found anywhere else on

---

14   The only exception seems to have been the 'glacier east of Jökultungur which advanced in the period 1907–30' (Eythórsson 1931). Unfortunately no details are available.

Plate 2.1 Landsat image of Vatnajökull taken on 22 September 1979

such a large scale. It rests on a group of active vol-canoes centred on Grímsvötn. To the east of the Grímsvötn caldera is an extinct caldera in the little-known Esjufjöll area; to the northwest a volcano underlies Bárdarbunga. Jökulhlaups in the Jökulsá á Fjöllum led Thórarinsson (1957) to suggest that there might be an eruption centre in Dyngjujökull in the northwest. Satellite imagery shows two elliptical fea-tures, presumed to be calderas, in the Kverkfjöll area a little further east, and also a volcano-tectonic line running from these caldera SSW across Vatnajökull, through Grímsvötn, to Sídujökull. To the west of this line the volcanoes are active; to the east they appear to be extinct with the very important exception of

Öræfajökull. Volcanic activity since Iceland was settled has been associated mainly with Grímsvötn and Bárdarbunga, together with their associated rift systems. Integrating the features of the subglacial topography, earthquake activity and jökulhlaups has provided much information about the volcanic systems underlying Vatnajökull (Björnsson and Einarsson 1990).

The continuing difficulty of forecasting the timing, volume and location of jöklhlaups emanating from volcanic activity under large icecaps was demonstrated by the eruption in the Bárdarbunga system in 1996, and particularly by the unexpected delay before the flood emerged.[15] Jökulhlaups are relatively common, many occurring elsewhere in Iceland and playing an

---

15 Strong tremors were recorded in the northern part of the Bárdarbunga system on 29 September 1996 and the associated eruption caused major fissuring of the overlying ice on 2 October. Two days later the eruption was emitting an ash cloud 8 km long and 5–6 km high. Meltwater running through the fissure system lifted the ice cover over Grímsvötn and water level in the caldera rose 90 m during the first five days of the eruption. Fine ash fell over northern and eastern Iceland on 3 October, but as it contained only one-tenth of the fluorine in ash from Hekla eruptions it did not endanger grazing

important role in landscape formation. Five of the 24 glacial rivers currently monitored by the Icelandic Energy Authority experience hlaups from time to time (Sigurðsson *et al.* 1992).

An eruption of Öræfajökull in 1362[16] was the largest explosion in Europe since Vesuvius in AD 79. It erupted again in 1727. The most disastrous outburst of all was that of the Laki craters immediately west of Vatnajökull, in the fault zone southwest of Síðujökull, in 1783. It lasted seven months and emitted 15 km³ of lava that covered 450 km² between Vatnajökull and the sea. Even more destructive to life and property was the ash deposited over most of Iceland, poisoning the pasture lands and killing half the cattle and three-quarters of the ponies and sheep.[17] More than 9000 people died of famine, one-fifth of Iceland's population. The total abandonment of Iceland was seriously considered.

2.3.1 The outlet glaciers of Vatnajökull

The first map to represent correctly the whole periphery of Vatnajökull (except Skaftárjökull) was compiled by Ahlmann and Thórarinsson in 1937 (Ahlmann and Thórarinsson 1937: 200, and Figure 2.17). The ice was estimated at this time to cover 8800 km², about 8 per cent of the country; Thórarinsson (1958a) gave a value of 8538 km² and the World Glacier Inventory

(1989) gives 8300 km². Though the icecap has been shrinking during the present century these figures cannot be taken as representing progressive change in the surface area, as they are not equally accurate or strictly comparable. The surface of the icecap is made up of several flattish domes rising to about 1600 m with a broad shallow depression running NNE to SSW from Brúarjökull to Breidamerkurjökull, about 20 km east of the volcano-tectonic line. The margins of the ice consist of great lobes except in the southeast where, from Breidamerkurjökull to Eyjabakkajökull, several valley glaciers flow down towards the coast. The glaciers descending from Öræfajökull, 2119 m, form a largely independent system in the southeast.

Several of the outlet glaciers of Vatnajökull, especially those on the north and west sides, are surging glaciers (numbered 1 to 6 on Figure 2.17). The anomalous behaviour of Brúarjökull and Eyjabakkajökull attracted the attention of Nielsen (1937) and Thórarinsson (1938). At first glaciologists were inclined to explain their sudden advances in terms of subglacial volcanism or seismic activity (Bárdarson 1934, Thórarinsson 1943). By 1960 enough information had been collected to discount such explanations and to show that none of the Vatnajökull surges corresponded in time to volcanic activity and earthquakes, except for those of Dyngjujökull and Síðujökull in 1934 when there was an eruption in the Grímsvötn caldera (Thórarinsson 1964b, 1969).[18]

---

stock. The eruption continued unabated with meltwater continuing to accumulate within the ice and on 7 October a jökulhlaup appeared imminent. Seismic activity was gradually diminished but the water level in Grímsvötn continued to rise; by the 15th it was already at 1504 m. Though a major jökulhlaup to the south was expected, it was envisaged that some of the meltwater might flow out to the north harmlessly into the Jokulsá á Fjöllum (Figure 2.16). By 28 October the water level in Grímsvötn had risen a further 5 m but still the flood was delayed. Eventually the water escaped southwards, emerging unexpectedly on 5 November from the western part of Skeidarárjökull (Figure 2.16) and discharging to the sea over Skeidarársandur. The flood peaked after 11 hours at 45,000 m³ per second and subsided the following day. It had destroyed the road over Skeidarársandur, part of Iceland's only ring road, together with bridges, powerlines and a fibre optic telecommunications cable (see Gudmundsson *et al.* 1997).

16 Before 1362 Öræfajökull was called Knappafell or Knappafellsjökull.

17 Climate had been harsh even before the eruption, except in the south (Ogilvie 1986). Many contemporary descriptions witness to the terrible conditions after the eruption. 'Then the eruption occurred with dust and fumes and the grass, which had been green, was made quite yellow and white by sulphurous rain. After that the grass withered from the roots' according to the Sheriff of Rangárvallasysla (cited Ogilvie 1986).

18 Grímsvötn has been the most active of the volcanoes situated beneath Vatnajökull for some hundred years; 36 eruptions are known to have occurred in the historic period (Björnsson and Einarsson 1990), though as many as 40 or 50 have been estimated as occurring in the last 1100 years. Eruptions definitely took place in 1922, 1934 and 1983. Since 1983 conditions at the eruption site have been observed annually by the Icelandic Glaciological Society (Gudmundsson and Björnsson 1991).

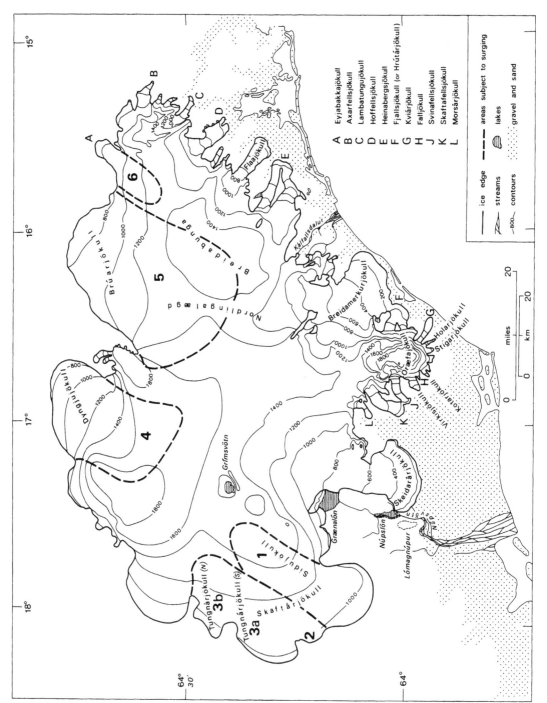

Figure 2.17 The most important surging lobes of Vatnajökull, numbered 1–6 (*Source:* from Thórarinsson 1969)

Table 2.2 Dates when Vatnajökull's outlet glaciers surged

Year	Glacier	Area affected (km²)	Max. advance (km)
1994–5	Tungnaárjökull		(max. rate 10 m per day)
1990–4	Tungnaárjökull		
1991–3	Sídujökull		about 1.16
1984–6	Skeidarárjökull		
1978–9	Dynjujökull		
1963–4	Sídujökull	480	about 0.5
1963	Brúarjökull	1400	about 8.0
1951	Dynjujökull	(700)ᵃ	about 0.3
1945	Skaftárjökull	450	about 0.6
1945	Tungnaárjökull		about 1.0
1934	Sídujökull	480	about 0.6
1934	Dyngjujökull	(700)ᵃ	about 0.3
1890	Brúarjökull	1400	about 10.0
1890	Eyjabakkajökull	110	about 0.6
1810	Brúarjökull	1400?	
late 1720s	Brúarjökull		
1625	Brúarjökull		

Sources: Thórarinsson 1969, Sigurdsson 1988, 1993c, 1998
Note: a values in brackets are rough estimates

2.3.1.1 *Brúarjökull*

The Vatnajökull glaciers that have surged are shown in Table 2.2. Brúarjökull figures prominently; early surges of this glacier are pinpointed by records left by Ólafsson (1772: 792) and Thoroddsen (1914b: 297). Others have followed at intervals of between seventy-three and ninety-five years. In 1964, when rates of movement of the Brúarjökull front were measured for the first time, it was remarked that the noise made by the surging glacier could be heard 2 or 3 km away, but during the winter of 1889/90 the rumbling advance had been audible at a distance of 50 to 60 km. The scale of the phenomenon is very great, with as much as 700 km³ of ice being involved, spread over an area of 1400 km² (Thórarinsson 1969).

Brúarjökull is only the extreme example of surging glaciers emanating from Vatnajökull. Most of the other glaciers around the periphery, except those in the south-east, are known to have surged during the last hundred years, and this has been predominantly a period of glacial recession; in periods of expansion surging may well have been more frequent. The history of Brúarjökull is not easy to establish because the inland margins of Vatnajökull are so remote. In any case there is not much evidence left on the ground as moraines formed by surging glaciers here are generally small (Thórarinsson 1964a, b), though according to Todtmann (1960) those of 1890 are quite big. It is only with the availability of remote sensing data that it has been possible to begin to assemble a general picture of the activity in recent years (see, for instance, Hall *et al.* 1995).

Some of the surges of Vatnajökull could have been climatically induced; their low surface gradients make the outlet glaciers very sensitive to changes in temperature and precipitation. Thórarinsson (1964) argued that their morphology may result in stress increasing until a critical point is reached in the balance between accumulation and ablation, when flow suddenly accelerates.[19] By confining our attention mainly to the

---

19 It is not intended here to provide a resumé of the development of thought on the genesis of surging. However it may be noted that Sigurdsson and Jónsson (1995) consider that although the records of surging and mixed-type glaciers show variations unrelated to climate, the maximum extension of surge-type glaciers at the end of surges, and the minimum extension just before a surge begins, appear to be influenced by long-term climatic change.

glaciers of the southern and eastern margins, for which it happens that most historical information exists, we can avoid some of the uncertainties attributable to surging, and concentrate on climatically controlled glacier oscillations.

Bárdarson (1934) made the first attempt to survey the recent history of Icelandic glaciers in a critical manner. He used data first brought together by Thoroddsen, together with the results of field studies made by Eythórsson and Eiríksson, and various cartographic data. He did not compare the oscillations of the various glaciers, nor did he attempt to relate the fluctuations to changes in climate. Thórarinsson's (1943) synthesis made use of material resulting from his own and Ahlmann's investigations of Vatnajökull in the 1930s, plus information from such early sources as Ísleifur Einarsson's Land Registers and descriptions of Austur-Skaftafells sysla from 1708–9 and 1712[20] (Figure 2.1). Since 1950, the journal of the Icelandic Glaciological Society, *Jökull*, has published the results of current glaciological research.

Although detailed information about the extent of the glaciers before 1700 is sparse, it is clear that at the time of the settlement and for long afterwards the southern and eastern tongues of Vatnajökull were smaller than in the eighteenth and nineteenth centuries. Farms built between Vatnajökull and the coast by the first generations of settlers were damaged by later ice advances.

2.3.1.2 Skeidarárjökull and Núpslón

Skeidarárjökull is further west than the other glaciers to be discussed and its movements are affected by the eruptions of Grímsvötn. Nevertheless it is worth noting some features of its history, which is relatively well known because the glacier front was commonly observed and remarked upon by travellers on their way to the districts of Myrar and Lón and farms near Öræfajökull, such as Skaftafell (Figure 2.18).

Skeidarárjökull still holds back a lake, Grænalón, at 635 metres on its west side as it emerges from Vatnajökull (Figure 2.17). When the ice front is in an advanced position it touches the ridge of Lómagnúpur

and dams up the Súla to form another lake, 10 km further south, known as Núpslón. Grænalón is the largest glacier lake in Iceland, with a volume of 1500–2000 million m^3; its area during recent decades has varied between 10 and 15 km^2 (Björnsson and Pálsson 1989). From time to time it causes jökulhlaups, as it did in 1898, 1935, 1939 and 1986. Núpslón has never been known to have an area greater than 5 km^2, which, with a depth of about 20 m, gives a maximum water volume of about 100 million m^3. Both Thoroddsen (e.g. 1892: 111) and Sveinn Pálsson (1882: 53) thought that a flood in 1201, described in an MS edition of *Biskupasögur* written between 1212 and 1220, and in *Sturlungasaga*, also thirteenth century, was due to a hlaup from Núpslón. Thoroddsen considered this one of the most important proofs that Icelandic glaciers in saga days were as large as in his own. This argument was rejected by Thórarinsson (1939) on the grounds that there are no moraines or shorelines round Skeidarárjökull to indicate that it was larger in the thirteenth century than at its greatest extension in the nineteenth century. Moreover a lake of Núpslón's size can fill in four to eight days and probably run off early in the summer. It would not reform in the same year because the large runoff from Skeidarárjökull would keep the tunnel open. But the description in the sagas indicated that the 1201 hlaup lasted some days and that it must have occurred in mid August, for the bishop, whose journey was impeded by it, reached Lón on St Bartholomew's Day, that is 24 August. Thórarinsson took the view that it is far more likely that this hlaup, the earliest for which there are records, was due to the emptying of Grænalón rather than Núpslón. He argues that a thinner Skeidarárjökull would have allowed Grænalón to empty subglacially, even without the intervention of a volcanogenic hlaup. There is strong evidence that volcanism was dormant in the Grímsvötn area until the fourteenth century. There are no records of eruptions of Öræfi before 1362 or from the Skeidarárjökull area before 1389 (Thoroddsen 1925: 83). The whole history of settlement in Skaftafells sysla indicates volcanic inactivity in saga times. In particular the Eyrarhorn farm (see Figure 2.18), one of

---

20   Unfortunately the section of Magnússon's and Vídalín's *Jardabók* (1702–12) for Skaftafells sysla and Múla sysla is lost, probably burnt in the great Copenhagen fire of 1728.

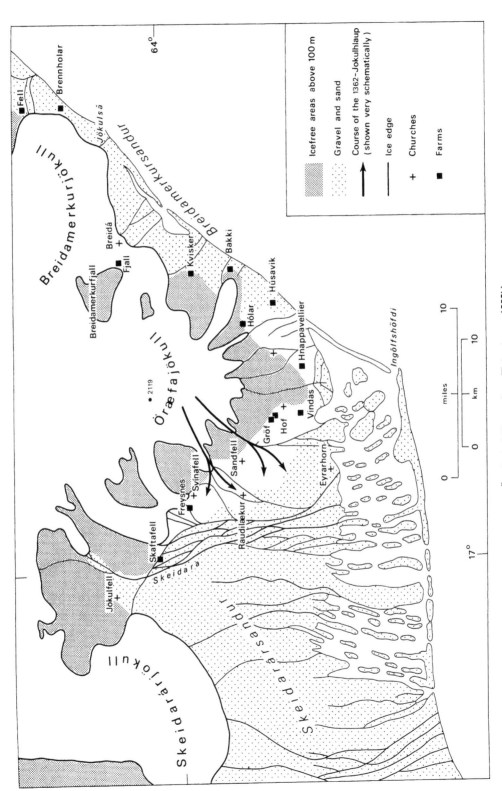

Figure 2.18 Skeidarárjökull, Breidamerkurjökull and the glaciers of Öræfajökull (*Source:* from Thórarinsson 1958b)

considerable size situated on land now covered by the Skeidara lagoons, was being worked up until the late fifteenth century, and in Thórarinsson's view, since this area has been completely flooded by all the known volcanogenic jökulhlaups from Skeidarárjökull, no farm could possibly have existed there if Grímsvötn had been active. This being the case, he concluded that the 1201 hlaup provides strong evidence that 'the purely climatologically conditioned outlet glaciers of Vatnajökull must have been considerably less extensive in the saga period than they are now' (1939: 231).

Taking this lack of activity in the earlier centuries into account, Thórarinsson advanced an interesting supportive argument for climatic deterioration since the end of the fourteenth century. He noted that 28 jökulhlaups were recorded from Skeidarársandur between the end of the fourteenth century and 1938. A hlaup which occurred in 1934, although not reckoned amongst the largest, was estimated to have involved some 15 km³ of water. Thórarinsson took 10 km³ of water as a safe value for the average total discharge of a hlaup, considering that this was probably a very substantial underestimate in fact. This gives an aggregate water loss for 28 hlaups of 280 km³ of water or about 300 km³ of ice. Now this is of the order of twice the mid twentieth-century volume of Skeidarárjökull below the firnline. Thórarinsson concluded that, as Skeidarárjökull did not recede despite these enormous losses, the climate must have been more favourable to glacier growth after volcanic activity began in the fourteenth century than it had been in saga times. The general drift of these arguments is substantiated by positive evidence from the pre-Little Ice Age period from Breidamerkurjökull.

2.3.1.3 Breidamerkurjökull

Land adjacent to and now covered by the ice of Breidamerkurjökull was settled in Landnám times (AD 870–930) and is said to have been occupied by many farms and to have been well covered by grass until 1100 or later (Ólafsson and Pálsson 1772: 786–8). There was certainly a large area of scrub

woodland, 'Breidamörk', on the western slopes of Breidamerkurjökull, and two farms, Fjall and Breidá, in the area which was eventually to be first devastated and then covered by ice from Breidamerkurjökull, and its neighbour, Hrútárjökull (now called Fjallsjökull). According to the *Landnámebók*, Fjall, at the southern end of Breidamerkurfjall, was first cultivated in about 900, by Thórdur Illugi who had taken in the land. In 1179, when the pastures were rather good, the church of Raudilækur, in Hérad,[21] had grazing for 160 wethers on Fjall lands (*Diplomatarium Islandicum*, I: 148, cited Thórarinsson 1943). Fjall was mentioned again in 1387 and was still being cultivated (*Diplomatarium Islandicum*, III: 401). Indeed, agricultural activity probably had another two centuries to run before trouble came. As late as 1660 Brennhólar to the east of Breidamerkurjökull is mentioned as being leasehold of Fjall (Bárdarson 1934: 27). The other farm, Breidá, seems to have been a large and prosperous farm in saga times, the home of chieftains of noble birth (Eythórsson 1952). It was adopted by the author of *Burnt Njál's Saga* as the home of Kári, one of the heroes of that sad tale. Whilst the saga is to be considered as fictional, the choice of Breidá as Kári's home would scarcely have been made if it had not been recognized as an important and prosperous place. However that may be, there is no doubt of the existence of Breidá, and that it was situated at the southern end of Breidamerkurfjall. Breidá was still prosperous in 1343 when, according to documentary evidence, the church at Breidá owned all the farmlands, including a wood and two other farms at Hólar and Hellir Eystri as well as other property (Bárdarson 1934: 28).

The great eruption of Öræfi in 1362 (see Figure 2.18) had consequences that were both widespread and dramatic. According to an approximately contemporaneous account of the year 1362,

Volcanic eruption in three places in the South and kept burning from flitting days [i.e. end of May] until the autumn with such monstrous fury as to lay waste the whole of Litlahérad as well as a great deal of Hornafjördur and Lónshverfi districts, causing desolation for a distance of some 100 miles. At the same time there was a glacier burst

21   This district was known as Hérad or Hérad milli Sanda as long as Raudilækur existed, but later came to be known as Litlahérad (Thórarinsson 1958a).

from Knappafellsjökull [i.e. Öræfajökull] into the sea carrying such quantities of rocks, gravel and mud as to form a sandur plain where there had previously been thirty fathoms of water. Two parishes, those of Hof and Raudilækur, were entirely wiped out. On even ground one sank in the sand up to the middle of the leg, and winds swept it into such drifts that buildings were almost obliterated. Ash was carried over the northern country to such a degree that footprints became visible on it. As an accompaniment to this pumice might be seen floating off the west coast in such masses that ships would hardly make their way through.

(*Islandske Annaler*: 226, cited
Thórarinsson 1958b: 26)

Thórarinsson (1958b) made a very detailed study of the 1362 eruption which was based not only on careful assessment of documentary evidence but especially upon measurements of the very extensive tephra layer associated with it. Volcanogenic ice advances and hlaups from Fallsjökull and Róturfjallsjökull destroyed some farms in Hérad, but despite the long-established tradition that they were the chief cause of devastation, Thórarinsson showed that the hlaups probably swept west, not directly south to the sea, and that the major cause of destruction was the tephra fall. He estimated that the volume of tephra was at least 10 km^3, corresponding to about 2 km^3 of solid rhyolitic rock. Hérad was ruined permanently and the farms of the districts of Sudursveit, Myrar and Nes, east of Öræfajökull, were abandoned for some years.

We have no direct evidence as to whether Breidá and Fjall were abandoned after the eruptions, but in 1387 Breidá no longer had any cattle and the church there had lost its ornaments (Eythórsson 1952). This certainly reflects the results of the eruption, but clearly shows that Breidá survived it. Documents dated 1525 indicate that the farm was still in cultivation then, although no mention of a church is to be found (*Diplomatarium Islandicum*, IX: 158, 247). Farming is again mentioned in documents of 1587 and 1697 (*Blanda*, I: 49) but 1697 may have been the last year during which anyone managed to cultivate this land.

Evidence of the existence of farms near Breidamerkurjökull and Hrútárjökull (Fjallsjökull) is

sufficiently concrete and reliable to substantiate the general proposition that the Vatnajökull tongues were substantially smaller at the time of the settlement of Iceland and for several centuries thereafter than they have been for the last three centuries or so. Nothing very much is to be gleaned, however, about the time at which the swelling leading to the Little Ice Age maxima may have begun.[22] The occurrence of events such as the 1362 tephra fall, causing widespread environmental deterioration, would in any case make the identification of the onset of the climatic change involved more difficult, except perhaps as far as changes in the distribution of sea ice were concerned.

2.3.2 Early and mid eighteenth-century advances of Vatnajökull

By the end of the seventeenth century the Vatnajökull glaciers were already much enlarged. The Fjall farm seems to have been abandoned by about 1695 at the latest. In the land register of 1708–9 we read that '14 years ago tún [meadows near the house] and ruined buildings were still to be seen, but all that is now in the ice'. Again in 1712 it appears in the register of abandoned farms in Öræfi: 'Fjall was the name of a farm west of Breidamörk. It is now surrounded by ice. Twelve years ago ruins could still be seen.' The date of final abandonment is not known; there was still some activity at Breidá in 1697, but it was completely deserted by 1698/9 (Bárdarson 1934: 29).

According to a document of 1700–9 (Jónsson 1914: 44, cited Bárdarson 1934: 29), Breidamerkurjökull and Hrútárjökull had united in front of Breidamerkurfjall at the end of the seventeenth century. In 1709 there was another reference to Breidá being deserted but it was said to have had the use of wood at Breidamerkurmúli, the part of Breidamerkurfjall farthest to the north, which was surrounded that year by the glacier (Ísleifur Einarsson 1709, cited Bárdarson 1934: 29). Einarsson's cadastral register of 1712 mentions that 'the ruins of the farm buildings can still be seen', though the ice was evidently then close to them.

---

22 A long sequence of tephra horizons within the ice of Vatnajökull includes one identified as that recording the Öræfajökull eruption of 1362. This is being investigated with a view to using it in studies of the history of mass balance, though it might be expected that difficulties will arise because the icecap is temperate (Larsen *et al.* 1995).

There was said to have been a stone slab in the ruins, supposedly on the tomb of Kári Sölmundarson, the original settler, who had gone there about 1017 (*Blanda*, I: 49–50), but this had disappeared under the ice by 1712.

An early map of Vatnajökull, made by Knopf who surveyed Skaftafells sysla in 1732, shows the ice margins about 6 km from the shore at Breidá. A verbal account of Iceland by Knopf, written down by a Swede in Norway and dated 14 August 1741, contains a description of Breidamerkurjökull:

> On the south side of the country, in Skattefialls Syzel, a quarter of a mile from the sea, there is a glacier called Breide Märker Giökel or Giakel, which moves as much as 30 ells forward once or more times every year, and then as much back again after a short time. The movement is so much the more obvious as this glacier, as it were, rests on the sandur, and in advancing pushes the sandur in front of it like high walls or breastworks.
> (Kålund 1916, cited Thórarinsson 1943: 25)

The question now arises as to whether the behaviour of Breidamerkurjökull and Hrútárjökull during the seventeenth and early eighteenth centuries was typical of the Vatnajökull outlets as a group.

A grassy area of Hafrafell between Skaftafellsjökull and Svínafellsjökull, on the west side of Öræfi, traditionally used for grazing goats, became more and more difficult to reach as the two glaciers advanced towards each other (Figure 2.18). By the early eighteenth century, according to Ísleifur Einarsson's Land Register of 1708/9, 'The Farm has the right of summer grazing on the part of Freysnes Farm called Hafrafell, but this right cannot now be executed as everything is covered by ice.' Árni Magnússon's 'Chorographica' is rather more explicit. 'Between Svínafell and Skaftafell lies Hafrafell, a large grassy hill, formerly accessible by paths, and with grazing for sheep in summertime; now this hill is so surrounded by glacier tongues that it can only be reached on foot and with great difficulty' (quoted Thórarinsson 1943: 32). The position was described in similar terms in 1746, 'Further east lies Hafrafell, surrounded by glaciers' (Stefánsson 1746, cited Thórarinsson 1943: 32). The ice had clearly not retreated to any marked extent at this time.

When Ólafsson and Pálsson visited the Stígárjökull in 1756, it descended through a gorge to the pastures near the tún of Knappavellir, and the people at the farm told the travellers that the ice had been advancing in recent decades. Thórarinsson (1956a, 1943) concludes that Stígárjökull, like the other Öræfi tongues, was larger in the mid eighteenth century than in the twentieth.

Heinabergsjökull, to the northeast of Breidamerkurjökull, was also in an advanced position in 1756, for Ólafsson reported that 'Heinabergsjökull has destroyed the central part of Hornafjördur, above which it is situated, one arm having reached right down on the plain to the farm Heinaberg, which nevertheless is still inhabited' (Ólafsson and Pálsson 1772: 790) (see Figure 2.19). A manuscript description of Austur-Skaftafells sysla dated 1746 by Sigurdur Stefánsson (quoted by Thórarinsson 1943: 22) mentions that 'In Heinabergsjökull there is a hill called Hafrafell, with some grass and great precipices'. This account confirms that the snout had come right down to the plain, and states that Heinabergsvötn (or Dalvatn), the ice-dammed lake formed when this tongue is in a forward position, had 'spoilt many adjacent farms, devastated many of them completely, washing away grassland, leaving only shingle, clay and sand'. This shows that the ice had already dammed the lake during the first half of the eighteenth century and was near its maximum extension in the middle of the century (Figure 2.19).

Svínafellsjökull in Lón was also enlarged in the mid-eighteenth century, for according to Sigurdur Stefánsson, who lived in the neighbourhood, 'A glacier comes down on both sides of Svínafell, which is a low but precipitous hill, with some grass, ling and forest vegetation' (Thórarinsson 1943: 19).

Skeidarárjökull, like the other Vatnajökull glaciers, advanced during the first half of the eighteenth century. When Knopf mapped the southern margin of Vatnajökull in 1732, he wrote that Skeidarárjökull 'slid out 600 years ago. The same glacier moves back and forth.' The first statement is interesting but unsubstantiated; the second is certainly to be taken seriously. Öræfi had erupted in 1727 (Thorláksson 1740, cited Henderson 1819: 208–12).

> In 1727, when both the *Öræfi* and *Northern Skeiderá* volcanic Yökuls were in activity, this low Yökul began to rock, to the great danger and consternation of some people who happened to be travelling on the sand before it. According to the account they afterwards gave, it moved backwards and forwards, undulating at the same time like the waves of the sea, and spouting from its foundations innu-

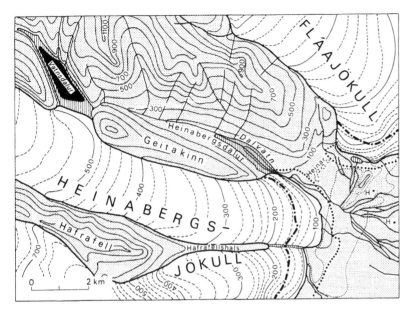

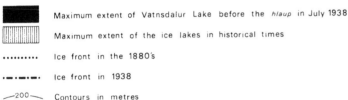

Figure 2.19 Heinabergsjökull and its
ice-dammed lakes. H = Hólar, a farm
(*Source*: from Thórarinsson 1939)

merable rivers, which appeared and vanished again almost instantaneously, in proportion to the agitation of the Yökul. As the progress it made was inconsiderable, the spectators saved themselves on a sand-bank, but the suddenness and unexpectedness with which the rivers continued to rush forth, rendered it impossible to travel any more that way the whole summer.

(Ólafsson and Pálsson 1772: 780, cited
Henderson 1819: 215)

By 1756 things had settled down again, with the glacier in a forward position, for when Ólafsson and Pálsson crossed the Skeidarársandur they found that 'the mountain slope north of the Núper is beautiful and fertile and lush grass and birches are growing there, which is all the more astonishing as the glacier is so close to it and the cold Núpsvötn glacier river, rushing out of the glacier crevasses, floods the bottom of the said slope' (Ólafsson and Pálsson 1772: 777).

The obvious interpretation of this passage is that Skeidarárjökull was sufficiently enlarged to have reached Lómagnúpur and so dammed up Núpslón. This most probably reflects climatic conditions and not only volcanic activity.

It may be concluded that the Vatnajökull glaciers advanced rapidly between 1690 and 1710 and were more extensive than at any time earlier in the historic period. For the next few decades their fronts were stationary or fluctuated within fairly narrow limits. Towards 1750 to 1760 marked advances or readvances took place. Thórarinsson (1943: 47) considered that most of them reached their maximum Little Ice Age positions at that time, though it seems difficult to prove on the evidence so far available that the mid eighteenth-century advances were much greater than those around 1710.

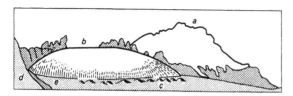

a. Öræfajökull

b. Skeidarárjökull has advanced between Færiness to the east and Súlutinda to the west

c. Skeidarársandur and some heaps of pebbles at the terminus of the glacier

d. The lower part of Lómagnúpur

e. Núpsvötn

a. Skeidarárjökull	f. Lómagnúpur
b. Skeidarársandur	g. Skeidará
c. Sea	h. Morsá
d. Eyjafjallajökull	i. Jökulfell
e. Fljótshverfisfjöllin	k. Midfell

Figure 2.20 Skeidarárjökull and its environs in 1794, showing the convex shape of the tongue and the existence of the lake Núpsvötn or Núpslón at that time (*Source*: from Pálsson's 1945 edn sketches)

2.3.3 The mid eighteenth- to late nineteenth-century period of enlarged glaciers

In the second half of the nineteenth century, the southern margins of Vatnajökull were visited by several foreign travellers and scientists. In 1857 the Swedish geologist Otto Torell visited the southeast coast, studying the glaciers and fluvio-glacial deposits. He was followed by C. W. Paijkull, whose account (1866) of his observations in 1865 aroused interest in jökulhlaups outside Iceland. Although there were no systematic measurements of any of the Icelandic glaciers, useful measurements at the margins of a number of Vatnajökull outlets were made by Helland in 1881.

The Skeidarárjökull lobe was bulging markedly in 1782 and the following year (Figure 2.20):

> I was told that before the last hlaup occurred in 1784, Skeidarárjökull was so high that the foremost point of Lómagnúpur could be seen only as a small cliff above the glacier from the highish mountain near Skaftafell sæter, since the glacier reached right up to the above-mentioned moraines lying below its front. Now [i.e. ten years later], on the contrary, almost the whole of the upper half of the same [Lómagnúpur] can be seen above the glacier from the farm itself.
>
> (Pálsson 1882: 51)

The flood in 1783 probably caused thinning, which would account for the retreat of the ice observed by Pálsson when he was mapping Vatnajökull in 1793–4 (Figure 2.21). Pálsson (1882: 47) found that

up at the glacier, large oblong bands of gravel follow the glacier margins at a distance of a couple of hundred fathoms almost all the way. These have obviously been tumbled or pushed up by an advancing glacier and indicate how far it extended before it began to recede.

Although by 1794 Núpslón had disappeared because of the recession, Sveinn Pálsson made the first specific mention of the lake: 'the southwest corner of Skeidarárjökull, from which the Súla comes, sometimes glides forwards till it collides with Lómagnúpur in front: Súla and Núpsá are then completely blocked and dammed up into a lake which finally breaks out, causing great floods' (1882: 33). He showed part of the lake in his diagram (Figure 2.20).

Thórarinsson (1939: 228) thought Núpslón probably came into existence after 1700, and judged from the indistinct shorelines and moraines formed by the mid eighteenth-century advance that it did not last for very long. These moraines, which appear on the General Staff map of 1904, were in part swept away by later hlaups, and although some, such as the ridge known as Sangígar on the western sandur, remain, reconstructing past events is now difficult.

Henderson, travelling across Skeidarársandur in 1815, had to traverse several acres of dead ice 'the surface of which has the same appearance as the rest of the sand'. He described the ice as:

the remains of the projection which took place in 1787 . . . . This region may be about three quarters of a mile

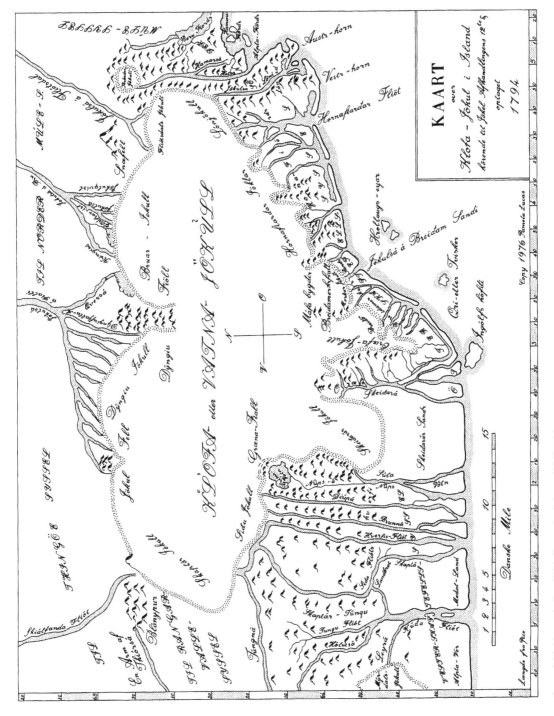

Figure 2.21 Vatnajökull in 1794. Drawn by Pálsson (1945 edn)

from the present margin of the Yökul; and near the middle of the intervening space are a number of inferior heights
which have been left on the regress of the Yökul in 1812,
the last time it was observed to be in motion.

(Henderson 1819: 215–16)

So it would appear that there was an advance in 1787
before the retreat in the 1790s observed by Pálsson,
and that the subsequent recession was followed or interrupted by a stationary period from 1812 to 1815.

The Revd P. M. Thorarensen's description of the
parishes of Öræfi, dated 6 July 1839, indicates that by
1830 Skeidarárjökull was advancing again. Helland gives
the distance between the southwest corner of the glacier and the mountainside west of Núpsvötn as 113 m
in 1857, although it is not clear from where he got this
information (Helland 1882b: 208). Advance was not
long sustained after that, however, and the formation
of the lake in the mid nineteenth century was probably
temporary if, as is probable, it occurred. A description of Kálfafell parish, dated 12 July 1859 (cited
Thórarinsson 1939: 228) by Jón Sigurðsson of Kálfafell,
includes a statement that

> Many years ago this glacier advanced rapidly westwards
> towards the mountains on the inner side of the so-called
> Lambhagi cave, E. of and inside Lómagnúpur, but it gradu-
> ally receded again. A few years ago it again advanced west-
> wards as far as Núpsvötn, but has now almost receded,
> and since last year is still moving in the same direction.

The Háalda moraine,[23] rising 40 m above the sandur
surface and running half a kilometre from (and within)
the eighteenth-century moraine, was probably formed
by the mid nineteenth-century advance, which presumably ceased about 1850. Thórarinsson (1943: 35) reckoned that in 1850 the glacier front was about 800 m
forward of the position shown in the 1904 map.

According to Helland (1882a), Skeidarárjökull
retreated 450 m between 1857 and 1880, and 88 m in
the year 1880/1. He estimated that the ice had thinned
some 62 m along the line between Sandfell in Öræfi
and Súlutindar, so that there was now a clear view across
the glacier in that direction, whilst in the summer of
1880 only the highest peaks of Súlutindar could be
seen. The distance between the southwest corner of

the glacier and the mountainside west of Núpsvötn was
now 750 m, so the ice had retreated 637 m between
1857 and 1881; the edge of Skeidarárjökull was now
60 m above sea level at the lowest.

Between 1892 and 1894, when Thoroddsen visited
the area, the tongue thinned again. There was a major
jökulhlaup in 1892, and the thinning of the tongue in
1893 may well have been due to this (Thoroddsen
1914b: 243).

It has now to be seen whether the general picture of
mid eighteenth-century advances and then retreat at
the end of the century, followed by mid nineteenth-
century advances to less forward positions again followed by retreat, was reflected by other glaciers in the
region and so is to be considered as climatic in origin.
The observations of Skeidarárjökull were made by particularly reliable observers, but it is essential that any
ideas of the sequence of glacier fluctuations gained from
its behaviour should be substantiated by ample evidence
from elsewhere before being accepted as significant
climatically.

Information about Svínafellsjökull and Skaftafellsjökull in the eighteenth and nineteenth centuries is
much sparser than for Skeidarárjökull, but in general
they appear to have behaved in the same way. In the
early 1790s Sveinn Pálsson found that

> East of Svínafell farm there are two large glaciers which
> are said to be always growing. Coming down two separate
> clefts, they unite, and descend to a fair distance out on
> to the flat land without the slightest melting in the
> summer. . . . The southern arm comes close to Svínafell
> farm . . . but the northern close to Skaftafell, the most
> northerly farm in Öræfa parish.
>
> (1882: 41)

Thus these glaciers were still enlarging in the last
decade of the eighteenth century and, according to
Thórarinsson's estimate, were probably about the same
size as in 1904.

According to the parish descriptions of the 1830s
by Thórarensen, the Öræfi outlet glaciers, like their
larger neighbour, were advancing in about 1830 and
by 1865 the ice was only five minutes' walk from
Svínafell farm; the position was much the same in 1904
(Paijkull, cited Thórarinsson 1943: 33). In 1881 the

---

23   The Háalda moraine, shown on sheet 87, Öræfi SW, of the 1904 General Staff Map, was swept away by a hlaup in 1922.

Table 2.3 Summary of documented glacier advances and retreats in the eighteenth and nineteenth centuries

Date	Source	Glacier	Observation
1704–5	Árni Magnússon	Sólheimajökull	already blocked Jökulsá canyon
1772	Pálsson	..	advancing
1783	Thoroddsen	..	smaller than in 1705
1794	Pálsson	..	advance had blocked Jökulsá again
1820	Eythórsson	..	Jökulhöfud covered by ice
1883	Keilhack	..	Jökulhöfud free of ice
1905	Thórarinsson	..	had retreated to its 1705 position
1201	Sveinn Pálsson	Skeidarárjökull	hlaup from Núpslón
900–1400	[see p. 40]	Breidamerkurjökull	Fjall and Breidá farms still cultivated
1695	[see p. 41]	..	Fjall farm abandoned by this date
1697	[see p. 41]	..	Breidá last cultivated
1712	[see p. 41]	..	Fjall farm buildings covered by ice
1708–9	Land Register	Hafrafell	surrounded by Skaftafellsjökull and Svínafellsjökull ice
1746	Sigurdur Stefánsson	Heinabergsjökull	had dammed back Dalvatn lake
1756	Ólafsson	..	reached the farm Heinaberg
1756	..	Breidamerkurjökull	still advancing
1756	Ólafsson and Pálsson	Stígárjökull	had been advancing in recent decades
1756	..	Skeidarárjökull	was damming back Núpslón
1780s	Pálsson	Svínafellsjökull	still enlarging
1782	Pálsson	Skeidarárjökull	bulging markedly
1784	..	..	last hlaup from Núpslón
1787	Henderson	..	glacier advanced sharply
1794	Pálsson	..	retreating – Núpslón disappears
1794	..	Breidamerkurjökull	reached a maximum then retreated
1812–15	Henderson	Skeidarárjökull	recession interrupted by a stationary period
1820	Thienemann	Breidamerkurjökull	advance of *c.*1000 m, still 2 km from sea
1830	P. M. Thorarensen	..	advancing again
1836	Gaimard	..	only 400 m from the sea
1836	..	Skaftafellsjökull	close to Heinabergsjökull (and for *c.*50 y)
1839	Gunnlaugsson	Breidamerkurjökull	2200 m from the sea at Jökúlsa
1869	Paijkull	..	reached down to beach
1857–81	Helland	..	retreat of about 600 m
1880	Thórarinsson	Heinabergsjökull	advanced until recession began in 1887
1892	Thoroddsen	Breidamerkurjökull	*c.*250–1000 m from sea

margin of Svínafellsjökull, according to Helland, was 98 m above sea level.

We have little or no eighteenth-century evidence, but the Öræfi outlets were evidently generally affected by advances in the 1830s. P. M. Thórarensen was probably referring in fact to Kótárjökull or Virkisjökull when, in 1839, he wrote that Sandfellsjökull was advancing to the WSW, though not as rapidly as Skeidarárjökull, and that the Öræfi glaciers in general were advancing strongly. Helland (1882b: 206) found a 137-foot-high moraine in front of Fallsjökull–Virkisjökull in 1881, the snout of which was 110 m above sea level (Figure 2.17). The ice seems to have been thinning immedi-ately before 1881 and stagnant from 1881 to 1884. By 1904 the ice front was about 600 m behind the moraine, but that ice was dead; the moving ice was 1000 m from the moraine. Helland evidently saw the first stages of real recession here. These glaciers were not affected by advances in the late nineteenth century.

When the Hrútárjökull (Fallsjökull) was investigated by Sveinn Pálsson in 1794, he recorded that it covered the entire Fjall estate and a few years earlier had joined Breidamerkurjökull. The two glaciers had been united earlier in the century (p. 41), so there must have been a recession sometime between 1709 and 1793/4. Hrútárjökull was still confluent with its neighbour in 1894, when Thoroddsen saw it and found it to be advancing, unlike the other Öræfi glaciers.

Eggert Ólafsson wrote of his visit to the Breidá area in 1756 that 'As we were riding back, the guide took us to the western edge of the glacier to the place where the church had been, but we saw no other traces of the farm but a couple of green cultivated meadows and a heap of stones pushed together' (Ólafsson and Pálsson 1772: 784–8, 840–4). It is clear from his description that Breidamerkurjökull must have been advancing in 1756.

Sveinn Pálsson travelled along the front of Breidamerkurjökull in both 1793 and 1794. In 1793 he found that the

> Jökulsá divides it not externally but internally, into two parts. The part lying to the west of the river is steeper and has a lower, narrower and more uneven margin. It is more crevassed and is covered with a quantity of grit and has some considerable mounds of grit in front of it. The part to the east, on the other hand, is quite compact and without crevasses. It has a steep margin, 16 to 20 fathoms in height, no grit on its surfaces except such loose sand as has been deposited on it by strong winds, and this gives this part a greyish-white colour. There are no, or very few, grit mounds in front of it which shows that the western part has been and still is in motion, though no withdrawal has so far been observed, as with Skeidarárjökull, but rather a gentle advance. The eastern part, however, so far, shows no sign of movement. . . . Such was this glacier in the summer of 1793 when I visited it for the first time, but next summer, or in 1794, things had changed. The part to the east of the river was clearly seen to have advanced over 200 fathoms from its position of the previous year. The smooth, unbroken margin described by Ólafsson and Pálsson 40 years ago was now quite unrecognisable. Here were large crevasses and sharp-pointed pyramids; there it was excavated like some filigree work, great pieces of ice thrust out, and where the margin still seemed more or less smooth and unchanged, it had greatly increased in height and had bulged out in its centre like a wall of greensward at the point of bursting, or slit down its length on account of the water which had accumulated behind it. There was too a constant rumbling from the entrails of the glacier. Here and there were small streams jetted from the fissures, accompanied by an exceedingly nasty, damp mist from cavities in the glacier, and a very penetrating cold. The eastern part of the glacier at Vedurá had advanced furthest . . . but now appears to have started to withdraw, as mounds of grit were clearly to be seen in front of its

margin. The advance is said to have begun suddenly, so to speak, and to a large extent without any appreciable discharge of water, at Whitsuntide in the said year 1794. Since that time people in the neighbourhood of Hornafjördur have complained of the continual mist, cold and sleet which is said to have come from this uneasy glacier.

> (Pálsson 1882: 33–5)[24]

It was very fortunate that Pálsson visited Breidamerkurjökull at exactly the right time to observe this late eighteenth-century advance, which evidently took place later here than in other Vatnajökull outlets. From his description it must be suspected that when he saw it in 1793, the western part of the lobe had already begun to retreat from its maximum, although it was still more advanced than the eastern part which was to advance so decisively the following year.

The 1794 withdrawal of the eastern part of Breidamerkurjökull was only temporary. Henderson travelled across the sandur in 1815 and reported renewed evidence of advance:

> Of its progress towards the sea, I was furnished with the amplest proof on passing along the margin. About the distance of a quarter of a mile from the south-east corner of the Yökul, I was surprised to find it traversing the track made in the sand by those who had travelled this way the preceding year; and, before reaching the point, I again discovered a track, which had been made only eight days previous to my arrival, lost and swallowed up in the ice. The same fact is confirmed by a comparison of the present length of the river, with what it was about fifty years ago. Olafsen and Povelsen, describing it as the shortest river in Iceland, state it to have been scarcely a Danish mile, or about five British miles, from its egress to its junction with the sea at the time they passed it; whereas it does not now appear to exceed a British mile in length.

> (Henderson 1819: 196)

Henderson had some odd ideas on the origin of Breidamerkurjökull which are without scientific merit, but his statement that 'it is only in summer it advances, after a strong thaw on the snow-mountains; at which time, also, the river which it discharges, is poured forth, now at one place, and now at another' (1819: 197) rings true. He expected the advance to continue:

> if this field of ice be not entirely carried away by some awful convulsion in the mountains behind it, the progress

---

24   Henderson's (1819: 189–9) translation of Pálsson is somewhat inaccurate.

it is making will soon bring it to the sea; and, in the course of a few years, all communication between the southern and eastern districts by this route will be cut off.

This very rapid advance was still continuing in 1820, when, according to Thienemann (cited Bárdarson 1934: 25), central Breidamerkurjökull advanced about 1000 m and its movement was so fast that on some days it advanced 4 to 8 m. It is not completely clear whether this was a surge or perhaps the result of a hlaup. Thórarinsson (1969) thought it was probably a surge. When Thienemann crossed the sandur in 1821 the glacier was retreating, leaving frontal moraines 10 to 13 m high, and the distance from the margin to the sea was about $\frac{1}{4}$ Meile, or 2 km (Thienemann 1824: 311–13, cited Bárdarson 1934). From all this it is clear that the Breidamerkurjökull fluctuations in the 1790s and early nineteenth century were dominated by rapid advances, interspersed with recessions which were probably slower, and it is possible that at least one surge occurred.

Gunnlaugsson visited Öræfi in 1835 and surveyed Austur-Skaftafells sysla in 1839. On his map he showed Breidamerkurjökull 2200 m from the sea at Jökúlsa, and the distance rather less to the east of the river. In the west the margin is shown 800 m in front of Breidamerkurfjall. Thus the ice was probably further from the sea at Jökúlsa in 1835 than it had been in 1794.[25]

In 1836 a Frenchman, Gaimard, travelling in Iceland, reported that the shortest distance from Breidamerkurjökull to the sea was about 400 m (Gaimard 1838, II: 237). When Thorarensen wrote his description of Hof and Sandfell parish in 1839, the glacier was enlarged, for he wrote, 'nobody crosses these glaciers except for driving sheep to Hafrafell and Breidamerkurfjall, both of which are surrounded by glaciers'.

Alternate advance and retreat near the maximum extension seem to have continued for the rest of the nineteenth century. Paijkull (1866: 122–5, quoted Thórarinsson 1943) saw the tongue in 1865 after the glacier had advanced in 1861, forming a moraine and

then receding slightly once more. In 1869 the margin expanded rapidly so that in June and July it was encroaching on the beach and the moraines on its eastern edge. Great floods which devastated the farm Fjall were associated with the advance (Figure 2.18). Kålund found the ice barely 1000 ells, or 600 metres, from the sea (Kålund 1872–82, II: 278).

According to Helland (1882b: 208–25), Breidamerkurjökull retreated after 1875 and was still retreating in 1881, by which time the ice edge was about 20 m above sea level at Jökulsá and about 11 to 12 m above sea level further east. Thoroddsen, a decade later, found the centre of the lobe 256 m from the sea, although the ice was 1000 m from the sea at Jökulsá. Breidamerkurjökull was then at, or very close to, its greatest extension during the 1890s, and although its margins had fluctuated a number of times during the preceding century, there is no evidence of really substantial retreat between the rather delayed advances of the 1790s and those of the 1890s. Despite the unusually rapid advances of this glacier, especially around 1820, and the possibility that not all the intermediate fluctuations between 1790 and 1890 have been identified, the main outlines of the oscillations of Breidamerkurjökull may be identified with reasonable certainty.[26]

Most of the rest of the information about Vatnajökull's eastern outlets is concerned with the fluctuations of Heinabergsjökull, Fláajökull and Hoffellsjökull. Heinabergsjökull (see Figure 2.19) was apparently enlarged throughout the period between the mid eighteenth century and the late 1880s, when it reached its maximum extension in historic times (Thórarinsson 1943: 23). The lakes dammed up by this glacier in both Vatnsdalur and Heinabergsdalur have been notorious for their hlaups. Thórarinsson reckoned that settlement in the western part of Myrar (Figure 2.17) would not have been possible if Heinabergsjökull had been large enough to cause hlaups. He could find no evidence of interruption of settlement in Myrar before 1700; but the glacier

---

25 Gunnlaugsson probably had access to the data of Scheele and Frisak, who had surveyed there in 1813, and that of Aschlund from 1817 (Thórarinsson 1943: 27).

26 Many of the figures quoted for the distance between Breidamerkurjökull and the sea at various dates are merely estimates. Moreover this sandur coastline is not static and so statements of distance of the ice from from the water must be treated with caution.

evidently became sufficiently enlarged to dam up lakes sometime between 1708 and 1783. It is probably significant that a list of abandoned farms in Austur-Skaftafells sysla made by Jón Helgason of Hoffell in 1783 lists three farms in Myrar which had been occupied in 1708, according to Einarsson's Land Register (Thórarinsson 1939: 223–4).

In 1839 both Fláajökull and Heinabergsjökull were advancing and had been doing so for some years. The Revd Jón Bjarnason wrote in his description of the parish of Einholt, dated 24 December 1839 (quoted Thórarinsson 1943: 21):

> these glaciers are growing and gradually sliding out over the low-lying land is evident; great is their unrest, and loud crashes can be heard far away, when the lakes are drained off underneath them, rushing forth with terrible speed and doing great damage to the soil on the plains, meadows and grazing grounds. I presume they are impassable almost everywhere, especially near the settlements . . . when the rivers are dammed, which often happens, especially in Heinabergsdalur, the water collects in a kind of lake.

Gaimard, travelling through Skaftafells sysla three years earlier, had described the two glaciers and their junctions: 'Nous côtoyâmes encore, pendant deux à trois heures, deux grands glaciers qui, après être descendus des Jökulls Skálafels et Heinabergs, se réunissent en un seul. Semblables aux autres par leurs aiguilles, ils offrent cependant cela de remarquable, que leur réunion où leur point de contact est indiquée par une ligne ou crête de moraine à la surface de la glace' (Gaimard 1838, II: 240).

Fláajökull almost reached Heinabergsjökull in 1857 when Torell visited the area. According to local tradition the two glaciers were nearly united in the 1860s. This fits in with Paijkull's description (1866: 63). During his visit in 1865 he was told of a place 'where the jökulhlaup is due to two glaciers coming from different valleys blocking up a third valley with no glacier but running water. The water collecting there will eventually break through the glacier and the jökulhlaup will begin.' The lake mentioned in the 'third valley' was probably Dalvatn. Thórarinsson (1939: 217–24) collected oral evidence about the lakes from the local farmers and accepted their account that in the 1860s and 1870s Heinabergsjökull was large and thick and its front close to the outwash moraines on the sandur. The

southeast corner of the tongue nearly met Fláajökull. Vatnsdalur was full of water up to the col at 464 m, and it had been full as long as the oldest inhabitants could remember. Its outlet was over the col to Heinabergsdalur, which in its turn was dammed by Heinabergsjökull, forming the Dalvatn lake. This emptied almost every year, usually in the early summer, the flood rarely lasting more than one day.

Heinabergsjökull was nearly stationary in the 1870s and 1880s, tending to advance rather than recede, and attained its maximum late in the 1880s. Thórarinsson found that the moraines from which the ice began to recede in the 1880s were everywhere the outermost on the sandur. The ice finally began to recede in 1887. As the tongue thinned at the end of the nineteenth century, so Dalvatn began to shrink. By 1897 the Vatnsdalur lake was too low to drain into Heinabergsdalur. By 1889 Heinabergsjökull was thin enough for the lake to drain out under it. Between 1898 and 1938, when Thórarinsson visited the area, the Vatnsdalur hlaup was an annual event, occurring earlier in the summer as the glacier tongue thinned.

We have no information about pre-nineteenth-century Fláajökull, but it is known to have oscillated between 1882 and 1894, advancing and retreating three times in those twelve years. Thoroddsen (1914b: 221) found it extended to its moraines in 1894, but it is uncertain which moraines he meant. It is clear that advance until well into the 1890s, with a strong tendency to fluctuate near the position of maximum extension, was not a peculiarity of the Breidamerkurjökull lobe.

There is little data to be had about Hoffellsjökull (Figure 2.17) before the eighteenth century. This glacier is divided into two by a hill, and so is double-tongued. In the description of Austur-Skaftafells sysla by Sigurdur Stefánsson of Holt in Hornafjördur, dated July 1746 (quoted Thórarinsson 1943: 19), he wrote that 'then follows the small hill, Svínafell, which incidentally was formerly called Gölltur [The Boar]. A glacier comes down to the plain on both sides of Svínafell, which is a low but precipitous hill with some grass, ling, and forest vegetation.' Thórarinsson concluded that Hoffellsjökull was probably about the same size in 1746, or slightly larger, than it was in 1903 when it was first mapped. There are no moraines that might suggest that it was larger in the eighteenth century than it was round about 1890.

A parish account by the Revd Th. Erlendsson (quoted Thórarinsson 1943: 20) mentions that hlaups from ice-dammed lakes on the east side of Hoffelsjökull–Svínafellsjökull made the river impassable when they drained all the way down the main road from Bjarnanes, where Erlendsson lived, to Holt. This indication of thick ice is supported by the statement that 'it comes close to the north side of Svínafell and even extends up its slope'. Erlendsson recorded that the outlet glaciers from the eastern part of Vatnajökull 'go slightly forward and back, but more forwards'. This further substantiates the impression that the glaciers of Vatnajökull were tending to advance in the decades before 1840 and were thick in the mid nineteenth century. Svínafellsjökull was still damming up lakes and causing hlaups in 1873, when one of its branches reached the plains east of Svínafell (parish account by Bergur Jónsson, 1873, cited Thórarinsson 1943: 20) but was said to have 'looked plane', so it may have been thinning or receding a little then. According to local residents, it advanced slightly in the 1880s and reached its maximum extension in 1890 or soon after that date.

Summing up the evidence from the mid eighteenth- to late nineteenth-century period, it is fair to say the information available from the outlets of southern and eastern Vatnajökull gives a consistent picture. All the glaciers were in an enlarged state in the 1750s and were subject to minor fluctuations throughout the next 150 years, although they remained generally enlarged. The mid eighteenth-century advances of Breidamerkurjökull were perhaps a little delayed, while Skeidarárjökull was already retreating in the latter part of the nineteenth century, having reached its maximum extent in the mid eighteenth and not in the late nineteenth century like some of the other tongues. The consistency of the pattern which emerges is pleasing and not a little surprising in view of the possible complications.

2.3.4 Twentieth-century retreat

Reliable mapping of the Icelandic glaciers began in 1902, and the Danish General Staff map of 1904 provides a very useful basis for comparisons of the extensions of Vatnajökull ice at different times. Calculation indicates that the volume of Vatnajökull has decreased by 180 km^3 during the twentieth century (Einarsson *et al.* 1994). Regular measurement of the frontal positions of glaciers in Iceland was initiated by Eythórsson[27] in 1931, continued under his guidance till 1967, when responsibility passed first to Sigurjón Rist, and then in 1987 to Oddur Sigurðsson. The important work of the Swedish–Icelandic expeditions to Vatnajökull of 1936–8, reported in *Geografiska Annaler*, was the forerunner of much scientific investigation of this icecap which has taken place since and has been reported in large part in *Jökull*. Air photography, which began in 1937/8, made possible the extension of mapping to the whole of Vatnajökull, and in the 1970s satellite images added the possibility of monitoring the ice front at regular intervals.

In general, the Vatnajökull outlet glaciers melted back rapidly in the 1890s. Skeidarárjökull's margin rose about 15 m in altitude between Helland's measurements of 1881 and the 1904 map. Of the Öræfi glaciers, the Virkisjökull front rose 48–58 m between 1881 and 1904, involving a horizontal recession of at least 600 m, and Svínafellsjökull's margin rose 22–30 m. Breidamerkurjökull receded 250 m east of Jökulsá, about 500 m at Jökulsá and about 300 m in the southwestern part of the lobe. Hrútarjökull was exceptional in that it appears in much the same position on the 1904 map as it was when Thoroddsen saw it in 1894 and found it advancing. It is possible that a late nineteenth-century advance here, followed by retreat, may have brought the margin back in the early twentieth

---

27 Jón Eythórsson (1895–1968) first studied natural history in Copenhagen and then meteorology in Oslo in the Bjerknes school. He worked with Ahlmann in Jotunheimen, helping to set up the first high-altitude meteorological station in the northlands. Returning to Iceland he became a forecaster with the meteorological office, an association that lasted practically all his life. He was not only a full-time meteorologist but also wrote and translated a large number of books and was responsible for the publication of such important works as Pálsson's treatise (1945). He made his greatest impact as a glaciologist. He initiated systematic observations of the margins of the major Icelandic glaciers in 1932 and, with the help of local volunteers, carried these on for nearly forty years. He published regular reports of the drift ice situation from 1953 to 1966, founded the Icelandic Glaciological Society, and was editor of its journal *Jökull*.

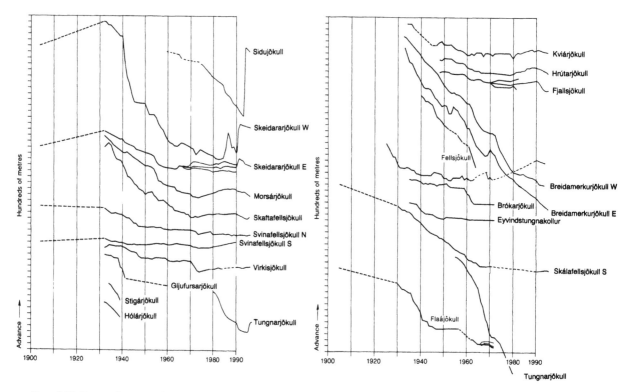

Figure 2.22 Twentieth-century fluctuations of Vatnajökull's outlet glaciers (*Sources*: from figures published in *Jökull* by Eythórsson, Rist and Sigurðsson)

century to roughly the 1894 position. In the period *c*.1890–1903 Svínafellsjökull retreated 420 m and Hoffellsjökull also withdrew. Fláajökull retreated 300 m between its maximum positions in 1894 and 1903, and Heinabergsjökull 420 m between 1897 and 1903. (Thórarinsson 1943). Figure 2.22 shows the course of recession in the twentieth century; clearly there was catastrophic recession and thinning of Vatnajökull, bringing the fronts of its outlet glaciers back to positions similar to those they occupied at the time of the Norse settlement.

Twentieth-century recession is best documented for Breidamerkurjökull, though this glacier may not be entirely representative because it spreads out into such a broad terminal lobe. Maps on a scale of 1:100,000 by the Danish Geodetic Survey are available for 1904 and on a scale of 1:50,000 by the US Army Map Service for 1945/6. Durham University Exploration Society made a plane-table map of the tongue on a scale of 1:25,000 in 1951; the Department of Geography, Glasgow

University, produced a 1:30,000 map in 1965 from 1:15,000 air photographs; Iceland's Survey Department, Landmælingar Íslands, photographed the area in 1980/1. Using the maps and supplementing the information they provided with aneroid barometer measurements of the heights of trimlines marking former glacier margins, Sigbjarnarson (1970) made estimates of the changes in Breidamerkurjökull's surface area and volume from 1894 to 1968 as shown in Table 2.4. Price (1982) re-examined the evidence and, making use of the latest air photograph cover, found rates of retreat somewhat faster than those given by Sigbjarnarson persisting to 1980. Precision in these matters where the glacier front is lengthy and uneven cannot be expected, but detailed and frequent surveys at west Breidá by groups from the University of East Anglia tend to support Price's figures.

Breidamerkurjökull, which constitutes about 14 per cent of the total area of Vatnajökull (Thórarinsson 1958a), seems to have diminished in volume

Table 2.4 The recession of Breidamerkurjökull since the late nineteenth century according to (a) Sigbjarnarson (1970), (b) Price (1982) and (c) Sigurðsson (1998)

Period	Frontal retreat (m)	m/y	Loss in area (km²)	km²/y	Loss in volume (km³)	10^6 m³/y
(a)						
1894–1904	120	12	3.0	0.3	2.8	280
1904–45	1130	28	25.5	0.6	21.2	520
1945–68	1046	45	25.3	1.1	25.0	1120
1894–1968	2296	31	53.8	0.7	49.0	660
(b)						
1903–45		30–40				
1945–65		53–62				
1965–80		48–70				
(c)						
1980–94		20–30				

between 1894 and 1968 by about 49 km³. On this basis Sigbjarnarson calculated that the shrinkage of Vatnajökull has been somewhere between 268 and 350 km³, that is some 8 to 10 per cent of its entire mass. Should this continue it would take 600 years for Vatnajökull to waste away entirely. Whether melting at a similar rate over a period as long as six centuries would ever be likely to occur is, of course, another matter. Even the precipitate retreat of Breidamerkurjökull since detailed measurements began in the 1930s has not been unbroken, though the tendency for standstill or even slight advance since 1940 has been less marked than with many other Vatnajökull tongues. Breidamerkurjökull (Nygrædnabakki) retreated 1271 m between 1966 and 1992 (Sigurðsson 1993c). The slight advances which occurred (see Figure 2.22) in the 1960s and thereafter were not by any manner of means restricted to Vatnajökull, and are therefore of all the greater significance (Chapter 14).

2.4 DRANGAJÖKULL

Drangajökull is a relatively small icecap in the far northwest of Iceland. It had an area of only 160 km² in 1988. While it has attracted rather little attention compared with Vatnajökull, the behaviour of its outlet glaciers is of special significance climatically because no volcanic eruptions have occurred either under it or near it in historic times. Eythórsson (1935) took the view that it was particularly worthy of investigation because 'its variations must therefore chiefly be due to climatic changes'. But Thórarinsson (1969) pointed to at least two advances in the 1930s which are difficult to explain climatologically. Since 1993 all three of Drangajökull's main outlets, Kaldalónjökull, Reykjafjardarjökull and Leirrufjardarjökull, have been officially rated as surging (Sigurðsson 1993a).

From the long axis of Drangajökull, which runs northwest to southeast, glaciers radiate towards the sea which, except to the south, is only a short distance away (see Figure 2.23). These outlet glaciers enter valleys that were settled until quite recently: Kaldalón to the southwest, Leirufjördur to the northwest and Reykjafjördur and Tharalátursfjördur to the northeast. Data relating to these valleys were assembled by Bárdarson (1934: 51–8). Eythórsson made some detailed investigations in 1931, fixing markers and instituting regular measurements of the main outlet glaciers and these have since continued.

Drangajökull was visited by many of the earlier Icelandic glaciologists. Ólafsson and Pálsson travelled in Vestfirdir in 1753, 1754 and 1757, and Olavius in 1775. Sveinn Pálsson unfortunately never went there and depended upon information from Ólafsson and Pálsson; it was probably reliable because Eggert Ólafsson, like Olavius, was a Vestfirdir man. Gunnlaugsson published a map showing Drangajökull in 1844, but he was

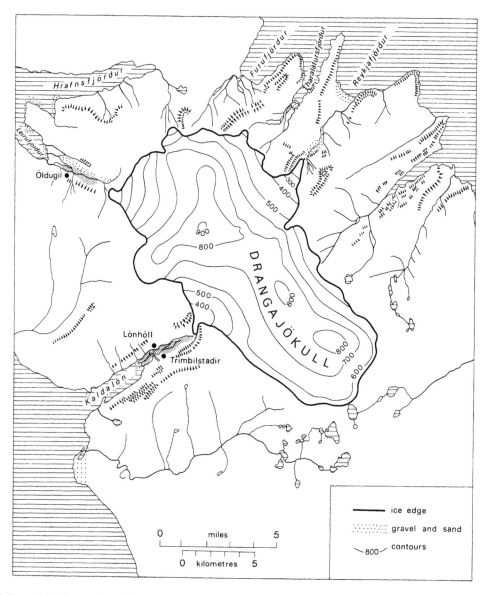

Figure 2.23 Drangajökull (*Source:* from Eythórsson 1931)

obliged to base that part of it dealing with the interior, including Drangajökull, on some earlier nineteenth-century geodetic observations made by others, on older maps and on local information. He did not himself visit Drangajökull and so his map cannot be used for comparative purposes or to elucidate changes in the area of the icecap (Thórarinsson 1943: 13). Thoroddsen investigated Drangajökull in the summer of 1886 and

1887, and it was mapped by the Danish General Staff in 1913/14.

2.4.1 The main seventeenth- and eighteenth-century advance phase

The narrow valleys of Kaldalón, Leirufjördur, Reykjafjördur and Tharalátursfjördur all contain some lowland that was farmed for centuries, and in each case there is evidence of early Little Ice Age devastation of cultivated farms and farmhouses. The outermost moraines in the Kaldalón valley are certainly pre-Little Ice Age (Eythórsson 1935, John and Sugden 1962). Two farms, Lónhóll and Trimbilsstadir, inside these moraines, were ruined before 1710 according to Árni Magnússon. John and Sugden recorded that 'local people believe that Trimbilsstadir was destroyed by the glacier around 1600'. This would suggest that Drangajökull's expansion was much swifter than that of Vatnajökull, but the evidence is too vague to be certain. John and Sugden (1962: 352) query whether the farms were actually destroyed, as 'even if they were inhabited in 1700–56, a glacier snout less than half a mile away and inevitable flooding of the valley floor periodically, would have rendered all agriculture impossible and would surely have forced the inhabitants to move'. They may indeed have moved away earlier, for although Olavius records that Lónhóll was devastated by a great 'glacier-burst' in 1741 it may have been untenanted at the time.

Ólafsson and Pálsson visited the valley in 1754 and 'all the local people reported that the ice now covers areas where there was green grassy ground twenty years ago. . . . The inhabitants also assured me that the ice had retreated from time to time' (1772: 516). So the front must have been more advanced in 1754 than in the 1730s, and probably more than in 1710, although expansion was not continuous in the first half of the eighteenth century.

In Leirufjördur there are said to be two main terminal moraines, with remains of smaller ridges on the flat below them (Eythórsson 1935: 54).[28] The outermost of these crosses the valley opposite Öldugil, where a small tributary stream enters the valley from the south. There was once a farm near this stream, also called Öldugil. In Magnússon and Vídalín's *Jardabók* of 1710, Öldugil is said to have been deserted between 1400 and 1500, 'and glacier-bursts and floods destroyed this dwelling according to common opinion, so that what is to be seen of the ruins is now [1710] situated quite close to the margin of the glacier; according to state-

ments of people still alive, the glacier has overwhelmed the whole area of the former farm' (Magnússon, cited Eythórsson 1935: 128). Olavius mentioned several other deserted farms near Öldugil, of which one, Svidningsstadir, was specifically said to have been destroyed by the glacier (Olavius, cited Eythórsson 1935: 129). It is uncertain from all this whether Öldugil was ruined by floods or by ice, though it is more likely that it was flooded. Very substantial advance of the ice had certainly taken place by 1710.

Deserted farm sites are also to be found in the Reykjafjördur valley, where again a series of old frontal moraines cross the floor. The outermost, not more than 300 m long, on the north side of the river, was found by Eythórsson just below the narrowing of the valley at Stórahorn. This moraine formed the limits of the meadowland when Eythórsson saw it; upstream were only gravel flats with scanty vegetation. There was one farm, Kirkjuból, in the grassy valley below this moraine. It was said to have been deserted finally about 1740, although recorded as untenanted in 1525 (Eythórsson 1935: 130). Árni Magnússon describes an old untenanted farm, Knittilstadir, close to the glacier, but Eythórsson (1935: 130) was informed locally that Knittilstadir was in the outer part of the valley. Olavius mentions two other farms, Nedra-Horn and Fremra-Horn, leaseholds of Kirkjuból, which were devastated in the seventeenth century or earlier. At Fremra-Horn the damage was definitely said to have been caused by the glacier (Olavius 1780: 63).

Clearly, the rather wide valley of Reykjafjördur was once well occupied and farmed despite its remoteness. The desertions which took place in the sixteenth century may not have had anything to do with deterioration of climate or glacial advance; there is no evidence either way. But farms in the upper part of this valley, and specifically Fremra-Horn, were apparently destroyed by glacial advances and perhaps associated floods in the seventeenth century or possibly earlier.

The glacier in Tharalátursfördur is described by Árni Magnússon as having destroyed the inhabited land in the valley before living memory (Magnússon and Vídalín 1710) while Olavius (1964–5: 63) says that

---

28 According to Bárdarson there are four moraines here, according to Thoroddsen three. Presumably the less developed moraines mentioned by Eythórsson account for the discrepancy.

Drangajökull reached to the head of the fjord. Evidently the glacier was in a very advanced position in the mid-eighteenth century, but there is no evidence as to when the advance began. Eythórsson estimated that the front was of the order of 2000 m in advance of its 1913 position in 1770.

The situation in the mid eighteenth century was nicely summed up by Ólafsson:

> In view of the size of this ice mountain which is twelve [Danish] miles long by six miles broad, and of its situation so close to the settlements and the sea on all sides, it is not astonishing for it to cause snow and fog, wind, cold and unsettled weather. The waxing and waning of the glacier are also remarkable. All the residents agree in saying that the glacier covers ground which was green and fertile twenty years ago. The constant winds, which for some years at a time may be either easterly or northeasterly from the ice mountain, or westerly or southwesterly from the sea, are presumably the prime cause of this.
>
> (Ólafsson and Pálsson 1772: 516–17)

There is no doubt that the Drangajökull outlets had advanced over previously farmed land by the end of the seventeenth century, and by the mid eighteenth century were more enlarged than they had been since the surrounding valleys were settled.

2.4.2 The mid eighteenth- and late nineteenth-century condition of Drangajökull

Unfortunately there is very little information about Drangajökull between the visit of Olavius in 1775 and Thoroddsen's in 1887. Thoroddsen thought that the Kaldalón ice had retreated 500 or 600 m in total between the mid eighteenth century and 1887 (Thoroddsen 1905–6: 175). He wrote that:

> From the head of the ford to the glacier there is a distance of about half a Danish mile; the ground is perfectly smooth except for three moraines which stretch in half circles between the hillsides. The moraine nearest to the sea is grass-grown and was formerly covered with brushwood; the second moraine is bare of vegetation and such is naturally the case with the innermost moraine. Twenty to thirty years ago the glacier extended as far as the innermost moraine, but it has since retreated so that now there are 2–3,000 fathoms between the moraine and the glacier tongue.
>
> (Thoroddsen 1911: 19, cited Eythórsson 1935: 125)

The glacier front was swollen and 130–170 m high when Thoroddsen saw it, so it was probably advancing again at that time.

Local people told Thoroddsen that the Leirufjördur glacier formed the outermost moraine in the valley in the decade 1837–47, that is, the ice reached nearly as far as it had in 1710 (Thoroddsen 1905–6: 36). He cut a cross in a basalt block, then 90 m downstream from the front, in the hope of providing the opportunity to measure future oscillations.

When he visited Reykjafjördur in 1887 Thoroddsen was told that the glacier there had advanced very noticeably in about 1840, and had reached its greatest extension in 1846, pushing a moraine ahead of it (Thoroddsen 1914b, II: 74). It probably did not begin to retreat till about 1850, but by 1886 had retreated about 1500 m from its maximum. Thoroddsen collected rather less information about the Tharalátursfördur glacier, but he did note that it retreated several hundred fathoms between 1860 and 1880, and that its horseshoe-shaped moraines might date from 1840 or 1860–70.

The evidence available is certainly insufficient to permit identification of all the fluctuations occurring between the mid eighteenth and late nineteenth centuries. It seems likely, however, that the marked advance which occurred around 1840 may have been preceded by a slight recession, while there is no doubt that the mid nineteenth century advance was succeeded by a retreat which was both rapid and substantial.

2.4.3 The great ice recession of the late nineteenth and twentieth centuries

When Eythórsson worked on the Drangajökull in 1931 he took the opportunity to compare the positions and characteristics of the outlet glaciers in that year with those recorded by Thoroddsen, at the same time as initiating measurement of frontal oscillations.

By 1931 the Kaldalón glacier had retreated about 340 m since 1887 and its thickness had diminished even more drastically. Where Thoroddsen had described it as precipitous, Eythórsson found the tongue thin and not very steep. The Leirufjördur and Reykjafjördur glaciers had retreated much more in the same period. The Leirufjördur tongue had become 'short and insignificant' and hardly extended 200 m beyond the main

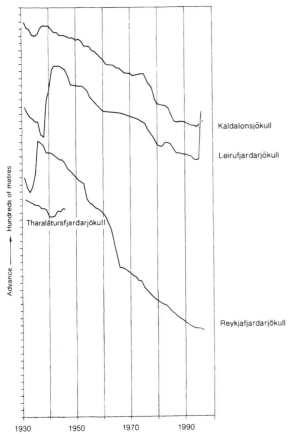

Kaldalonsjökull

Leirufjardarjökull

Tharalátursfjardarjökull

Reykjafjardarjökull

Advance ———⟶ Hundreds of metres

1930   1950   1970   1990

Figure 2.24 Fluctuations of Drangajökull's tongues, 1932–96, showing advances in the 1930s and 1940s (*Sources*: from measurements published in *Jökull* by Eythórsson, Rist and Sigurðsson, and from figures kindly provided by Sigurðsson)

thinning rapidly in 1931 but, judging from the map, had retreated only a little since 1913/14. Its tongue still reached to 201 m above sea level.

Thoroddsen and Eythórsson thus witnessed two stages in the main retreat from the Little Ice Age extensions of the Drangajökull outlets. This retreat continued after 1931, although not without interruptions; the results of the measurements made since then are shown on Figure 2.24. No evidence has been forthcoming from Drangajökull of any substantial advances in the late nineteenth century, but between 1939 and 1942 the Leirufjördur ice suddenly advanced 1 km of which 540 m took place in the winter of 1938/9, while Reykjafjardarjökull advanced 750 m during 1934–6. It is difficult to explain these advances except as surges. However, after slackening their retreat in the decade around 1970, the glaciers here resumed their withdrawal, though Kaldalonsjökull and Leirufjardarjökull again showed signs of stabilization in the 1980s. Then, in the early 1990s, both Kaldalonsjökull and Leirufjardarjökull rapidly advanced, confirming earlier suspicions that their aberrant behaviour is due to surging (Sigurðsson 1998). It may be noticed here that these surges took place during a period in which the proportion of advancing Icelandic glaciers had risen (see Figure 2.25).

2.5 VARIATIONS IN GLACIERS, CLIMATE AND SEA ICE

Although data on sea ice and glacial conditions in Iceland in early medieval times are very fragmentary, one gains a general impression from them that in settlement times from AD 870 to 930, and saga times from AD 930 to 1030, the climate was milder than it has been since. This impression is strengthened when one turns one's attention to Greenland. Farming settlements established there when the settlement of Iceland was proceeding, throve for a time and acted as bases for the Vinland voyages (Gad 1970 and see Chapter 18). In Iceland the population expanded for a century or two and a new society emerged with its own culture and institutions, notably the Althing, the 'grandmother of parliaments'. All the evidence points to the environmental circumstances in which the Norse settlement of Iceland took place and prospered as having been favourable. The sea ice lay far from sight, glaciers were

margin of Drangajökull when Eythórsson saw it. He reckoned that the front had retired about 3 km between 1837–47 and 1931, with the most rapid retreat of about 85 m per year during the period 1898–1913. Retreat was accelerated at this time, as several hundred metres of the lower glacier became isolated. Reykjafjardarjökull reached almost to the valley floor in 1931. Eythórsson judged from the 1913/14 map that this tongue must have retreated about 100 m between 1886 and 1913. Thoroddsen gave the distance from the 1840 moraines to the 1886 position as 1500 m, but the distance between these positions on the 1914 map is also shown as 1500 m. Thoroddsen's figure in this case was probably guesswork. The Tharalátursfördur glacier was

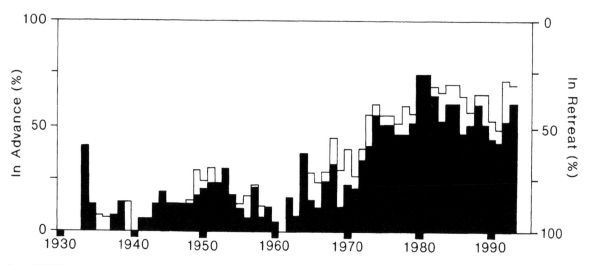

Figure 2.25 The percentage of advancing (shown in black), retreating and stationary non-surging glaciers in Iceland, 1930–93 (Source: from Sigurðsson and Jónsson 1995)

withdrawn, grass generally grew well and sea voyages in open boats in the North Atlantic and adjacent Arctic seas were not too difficult.

Evidence of climatic deterioration about the thirteenth century is forthcoming from a number of different sources. The incidence of sea ice was increasing. By the end of the twelfth century cereals were no longer being cultivated in northern and eastern Iceland. Thórarinsson (1944, 1956b: 16) quotes from a document of 1350 to the effect that 'grain, but only barley, is grown in a few places in the south country', and showed, using both documentary evidence and pollen analysis, that grain-growing persisted in a few places along the southern shore of Faxaflói until the end of the sixteenth century. Meanwhile, Greenland lost contact with the other Norse lands. There is evidence that the ground was frozen more severely, bodies buried in the graveyard at Herjolfsnes between 1350 and 1500 being better preserved than those buried earlier (Gad 1970: 154). The Norse settlements in Greenland were finally extinguished in mysterious circumstances towards the end of the fifteenth century (see Chapter 18).

Confirmation of the authenticity of the thirteenth-century climatic deterioration and some indication of its cause was provided by oxygen isotope analysis of the upper parts of a deep ice core from the Greenland ice sheet (Dansgaard et al. 1969).[29] Oxygen isotope analysis of such cores provides a record of mean annual surface air temperature (Robin 1976, Bradley 1999: 130–9) and in the core from Camp Century a strong cooling around 1200 followed by a further cooling around 1300 was indicated. Analyses of cores from central Greenland have since revealed more detailed results of general importance for the North Atlantic region (see Chapter 16).

Very little information is forthcoming about glaciers or sea ice in the fourteenth and fifteenth centuries. It seems that harsh conditions were experienced in the late sixteenth century. After a remarkably mild interval from 1640 to 1660 the cold greatly intensified and reached a maximum in the last decades of the century. The glaciers Vatnajökull and Drangajökull were already much enlarged and they readvanced about 1750 to 1760 at a time when the sea ice index was also high. A fluctuating retreat of the ice was followed by advances about 1840 to 1850, which happens to have been the only decade in the nineteenth century when the sea ice index was low. At the end of the century, when sea ice was more in evidence than ever before on record, some of the Vatnajökull glaciers reached their historic maxima

---

29   More recent and detailed investigations of ice cores in Greenland are discussed in Chapter 16.

whilst those of Drangajökull continued to retreat. A rapid warming in the 1920s[30] led to rapid retreat of most glaciers in the 1930s. But in the 1930s some of the Drangajökull glaciers staged a dramatic advance while those of Vatnajökull continued their retreat.

By the 1960s nearly all the Icelandic glaciers had shrunk to occupy smaller areas than at any time since the seventeenth century, or even earlier, and the sea ice was far removed from the north coast. There was a sudden return of sea ice in the years around 1970; some of the glaciers ceased to retreat and several readvanced. By the early 1980s, retreat seemed to have been resumed, though advances of some glaciers were to follow in the next decade. In 1992, out of thirty-four glaciers being regularly monitored, eleven were retreating, two were stationary and twenty-one were advancing. Ablation was exceptionally low in the cool summer of 1992, following the eruption of Pinatubo. A very different situation emerged in the summer of 1996, after a winter with extraordinarily little snow and with glacier surfaces unwontedly bare and vulnerable to melting.

The relationship between glacier variations and climatic change can be traced most closely since 1930 when glacier monitoring began[31] (see Figure 2.26) and meteorological measurements were already long established (Sigurðsson and Jónsson 1995). Very cold decades at the beginning of the twentieth century[32] (Einarsson 1993) were followed by strong warming in the 1920s, and an unusually warm period from 1926 to 1946. Within ten years glaciers were in retreat, and the whole period from 1930 to 1970 was one of predominant withdrawal. By 1962 all the monitored glaciers had retreated from their 1930 positions.

Summer temperatures had fallen gradually after 1940, and yearly mean temperatures followed in the mid 1960s, and remained below the levels of 1931–60. Average monthly temperature for 1961–90 was lower than the corresponding average for 1931–60 in

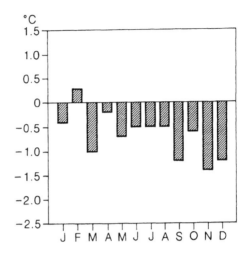

Figure 2.26 Change in average monthly temperature from 1931–60 to 1961–90, based on an average of 14 stations (*Source*: from Jónsson 1991)

all months except February (Jónsson 1991 and Figure 2.26). The rate of recession tended to level off around 1960[33] and after 1970 many glaciers started to advance, though a few, especially in the southeast, have remained stationary. Advances were swiftest in southern, central and northern Iceland, where ice has reoccupied about half of the area from which it had retreated since 1930, as in the case of Sólheimajökull (Figure 2.15). Increases have been least in the west, where a maximum of only about a quarter of the lost ground has been recovered. Temperature has evidently been the predominant control of the glacier fluctuations; there has been no apparent trend in the precipitation values recorded (Jónsson 1991).

2.6 THE LITTLE ICE AGE IN ICELAND

The documentary evidence makes it plain that most Icelandic glaciers reached Little Ice Age maxima in the mid eighteenth century. Dating of moraines by

---

30 See Ogilvie and Jónsson (2001, figure 4).
31 Monitoring of twenty-seven glacier snouts began in 1930, and was extended to eleven more during 1948–72 (Sigurðsson and Jónsson 1995).
32 Nine of the fourteen coldest years in Iceland during the twentieth century were clustered in the first two decades (Einarsson 1993).
33 Nonetheless sediment loads in the Jökulsá á Sólheimasandi basin decreased by 48 per cent over 1973–92, though the rate of decrease was slowing by the end of the period (Lawler and Wright 1996).

lichenometry has suggested a more general nineteenth-century maximum but only a few glaciers, notably Heinabergsjökull, have been observed to have advanced as far in the early and mid nineteenth century as in the preceding one.[34]

In recent years the characterization and dating of Icelandic tephra falls has presented a valuable opportunity to refine the chronology of ice advances and retreats and to resolve the conflict of evidence. The tephra investigations do not support the findings from lichenometry (Larsen *et al.* 1998; see also Chapter 15). Kirkbride and Dugmore (2001a) have used tephra of known ages to date Little Ice Age moraine sequences of Steinholtsjökull and Gigjökull (northern outlets from the Eyjafjallajökull icecap) previously dated by lichenometry. They show that dates from lichens are unreliable for deposits more than 160 years old. Deposits of small outlet glaciers which they dated by tephra to the mid eighteenth and early nineteenth centuries had been given lichenometric dates which spuriously cluster in the mid to late nineteenth century. Kirkbride and Dugmore (2001b) accept that Heinbergsjökull and one or two other glaciers may have reached maxima in the late nineteenth century but argue that most glaciers of small or moderate size were largest in the mid to late eighteenth century, during the period of great cold distinguished by Lamb (1979) and Ogilvie (1992), when conditions in Iceland were most favorable for glacier expansion.

The lives of Icelanders have everywhere been directly affected by the movements of ice on sea and land and by variations in the severity of the climate. Their sagas and historical records provide an unusually long and detailed record of climate fluctuations. Fluctuations in the fronts of many of the glaciers have been associated with surging not simply attributable to climatic conditions in the immediately preceding years. Volcanoes affected the activity of many glaciers in the early centuries of the settlement and more recently. Direct information about ice advances and retreats is lacking until the seventeenth century when it is evident that Drangojökull ice had advanced over previously farmed land and by the mid eighteenth century was more extensive than it had been since the settlement of surrounding valleys several hundreds of years earlier.

Evidence of cooling comes from the third quarter of the thirteenth century and especially from the 1320s, 1350s and 1370s. The last decades of the sixteenth and seventeenth centuries were especially cold and so were the 1740s, 1750s and 1780s. Most glaciers reached their greatest extents in the eighteenth century and then retreated only to advance again in the nineteenth century though generally not so far as previously. A more prolonged period of ice retreat, beginning in the nineteenth century and continuing into the early years of the twentieth century, was interrupted for two decades after 1970 but has since been resumed.

---

34 Ogilvie and Jónsson (2001) quote Thoroddsen (1916–17) to the effect that climatic conditions in the nineteenth century were considerably better than in the seventeenth and eighteenth centuries.

3

SCANDINAVIA

The Scandinavian glaciers are scattered over several mountainous areas between 60° and 71° north, mainly in Norway, and all within 180 km of the west coast (Figure 3.1). They are nourished by snow brought by cyclonic depressions from September to April, the weather conditions of the autumn and spring months being critical for the glacier budgets. A large proportion of the total area under ice, which exceeds that of the Alps, is made up of Jostedalsbreen, 487 km², the largest icecap of mainland Europe (Østrem et al. 1988). Many of the glaciers are in the far north, in remote, sparsely populated areas, and it is fortunate, for our purpose, that Jostedalsbreen is well to the south where the adjoining valleys have been cultivated for centuries and records detail critical features of their history.

3.1 JOSTEDALSBREEN

3.1.1 The question of ice advances in the fourteenth century

Jostedalsbreen outlet glaciers overlook the cultivated fields of Oldendalen, a valley in the north tributary to inner Nordfjord, and also finger down into Jostedalen itself in the east, and into the heads of Veitestrandsvatnet and Fjærlandsfjorden (Figure 3.2). All these glaciers respond to changes in the mass balance of the icecap feeding them, some more sensitively than others.

The farming economy of the valleys was based, as it still is, on pastoralism, with reliance being placed on summer pastures (saeter) on the higher ground, plus hay and cereals near the farmsteads at lower levels. Settlements were established in the coastal areas of Nordfjord at an early stage and tracks to them crossing over Jostedalsbreen are known to have been in use in the twelfth century. Most of the farms with which we shall be concerned in Oldendalen and Lodalen were in existence in the early fourteenth century and a church had been built in Jostedal by 1332. It has generally been supposed that the climate at this time was relatively benign (Øyen 1907, Hoel and Werenskiold 1962). Nevertheless, life for most of the people would have been hard, with natural hazard never far away.

Some kind of disaster evidently afflicted the area in the first half of the fourteenth century, judging from a letter of 1340 in the Norwegian archives (Diplomatarium Norvegicum, IX, no. 120)[1] which lists farms that had been deserted and rent reductions that had been ordered. The farms deserted or in part abandoned, listed as audnir, that is wasteland, may have been out of cultivation for many years as a result of long-term difficulties of some kind (Figure 3.3). The

---

1 Diplomatarium Norvegicum, IX, no. 120 gives in full a letter of 19 January 1340 concerned with proceedings held at the church in Olden on the day before the feast of Fabian and Sebastian in the twenty-first year of the reign of King Magnus Eriksson (1319–50). Deserted lands were listed and so were farms for which tax reductions were ordered. The reductions totalled in all 81¾ månadsmatsleigar, one månadsmatsleiga equalling one laup of butter or its equivalent (see p. 72). The exact date and nature of the event or events causing reductions were not specified. Lands designated as audnir were recognized as having gone out of permanent cultivation.

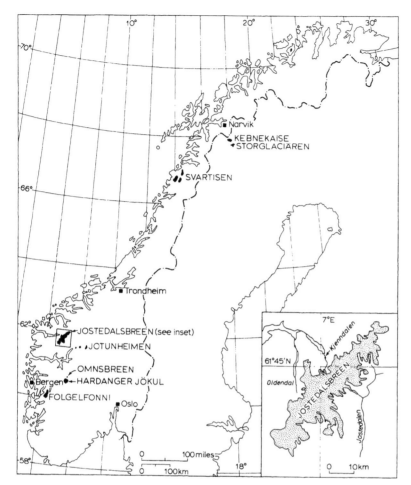

Figure 3.1 Jostedalsbreen and other glaciers mentioned in the text

rent reductions would only have been made in response to some sudden event in the years immediately preceding 1340 causing serious environmental disruption. Exactly what happened is unknown but it is significant that the farms that suffered most, namely Bødal and the two Nesdals, were situated at the head of Lovatnet; Tyva and Seta were at the lower end of the lake; and others somewhat less seriously affected were along the lake's northeast shore. This suggests the possibility that an enormous rockfall at the precipitous southern end may have created a great wave that swept down the lake from one end to the other. Damage on a lesser scale in Oldendalen and on Innviksfjorden could conceivably have occurred at the same time. A rockfall into the head of Lovatnet in 1905 caused a wave that destroyed Bødal farm, which was then rebuilt on a safer

site. It was finally destroyed by an even larger rockfall in 1936.

The Black Death, which caused a catastrophic decline in population throughout Norway, is commonly seen as having been responsible for many farms being abandoned in marginal areas in the fourteenth century. However, the abandonment of farmland in the Nordfjord area and the damage mentioned above took place several years before the Black Death. These events could be explained by climatic deterioration and associated mass movements of the kind which, it will be shown, affected the same sites in the Little Ice Age of the late seventeenth and eighteenth centuries.

There is some stratigraphical evidence indicating fourteenth-century ice advance, but this is not plentiful, possibly because it has been concealed or swept

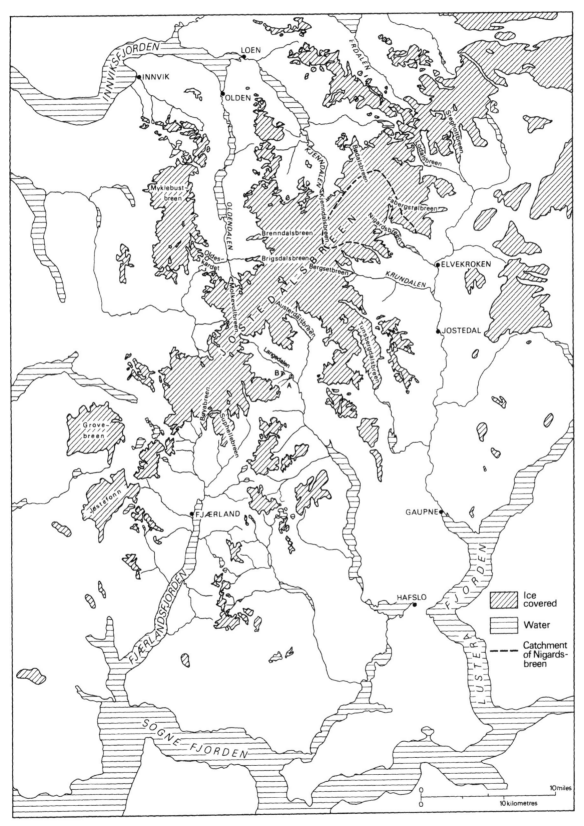

Figure 3.2 Jostedalsbreen (*Source:* from Norges Geografiske Oppmåling 1:50,000 maps 1418.III, 1418.IV and 1318.II)

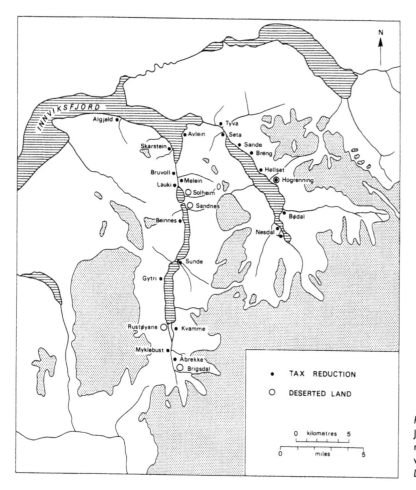

Figure 3.3 Farms in the parishes north of Jostedalsbreen where there were tax reductions in 1340 or where the land was deserted (*Source:* data from *Diplomatarium Norvegicum*, IX, no. 120)

away by more extensive and widespread advances later in the Little Ice Age, especially during the eighteenth century. Persuasive dating of the initial expansion comes from a section in the distal part of a glaciofluvial fan deposited by Sandskardfonna, a cirque glacier north-west of Jostedalsbreen (Figure 3.4) (Nesje and Rye 1993). The presence in the youngest palaeosols of stumps more than 10 cm in diameter and probably less than a hundred years old when buried indicate periods of little or no meltwater supply and low glacial activity around 890 ± 60 BP (T-5813) and 620 ± 70 BP (T-10028), that is around cal AD 1040–1230 and 1295–1410 (according to the calibration of Stuiver and Reimer 1986).[2] Unfortunately the logs were not *in situ*, and it was not possible to count their rings, which would have allowed them to be placed much more precisely by comparison with long ring sequences. There seems no doubt that during warm periods the soil surfaces became stable and vegetation, including alder and birch trees, was able to grow up. Dates from the logs provide maximum ages for the initiation of the Little Ice Age, indicating that a short, introductory cold phase in the thirteenth century, followed by a warm interlude, led into a prolonged Little Ice Age, culminating in the

---

2   Other periods with little glacial activity were similarly identified as occurring in cal AD 670–790, 770–900 and 880–900.

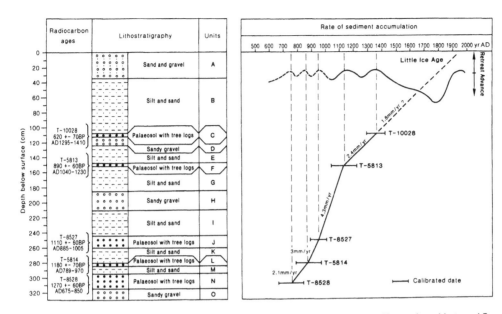

Figure 3.4 Late Holocene glacier activity at Sandskardfonna, Jostedalsbreen area, western Norway (*Source:* from Nesje and Rye 1993)

eighteenth century. The radiocarbon evidence alone does not show decisively whether the Little Ice Age began here in the thirteenth or fourteenth century. The Sandskardfonna section also reveals several earlier episodes of glacial activity, between AD 600 and about 1000, which are considered not as part of the Little Ice Age, but the equivalent of an earlier cold period identified in the eastern Alps as Göschener II (see Chapter 15).

In the 1970s, thinning of the central part of Omnesbreen, a small glacier north of Hardangervidda (Figure 3.1), revealed an extensive area of fresh till (ground moraine) overlying humus and plant remains dated to 550 ± 110 BP (T-1485), 440 ± 120 BP (T-1578) and 430 ± 100 BP (T-1479), cal AD 1300–1440, 1410–1640 and 1410–1630 (Elven 1978, see Figure 3.5). The undisturbed nature of the plant remains points to rapid climatic deterioration, causing the plants to become embedded in increasingly permanent snowbeds. The species involved grow at somewhat lower altitudes today and Elven concluded that the change in climate had been too swift for vegetation to respond before being overwhelmed by the ice. Having calibrated the ^{14}C ages of the plant fragments, accord-

ing to the procedure of Ralph *et al.* (1973), Elven concluded that the cooling had taken place in the fourteenth century.

Similar data have come from several glaciers in the Jotunheimen, to the east of Jostedalsbreen (Matthews 1991). A sample of moss (CAR-839, 270 ± 60 BP) from beneath the outermost moraine of Storbreen (calibrated using 95 per cent confidence limits, according to Stuiver and Reimer 1986) gives possible age ranges between 1450–1680 and 1735–1806. Though Storbreen evidently began to advance some time after 1450 the dating is too uncertain for it to be attributed to the early part of the Little Ice Age. Radiocarbon dating of mosses extracted from sediments in front of the outermost frontal moraine of Sagabreen gave an age of 910 ± 55 BP (CAR-750). Calibration suggests that the Sagabreen front was already expanding at some time between AD 1000 and 1250, probably during a cold phase documented at Sandskardfonna (see Figure 3.5). The outermost moraine, in the light of further ^{14}C data, appears unlikely to have been deposited before 1520. The earliest Little Ice Age advance of Bøverbreen formed a moraine damming a lake before 1440. Here intact mosses found immediately below lacustrine silts gave a

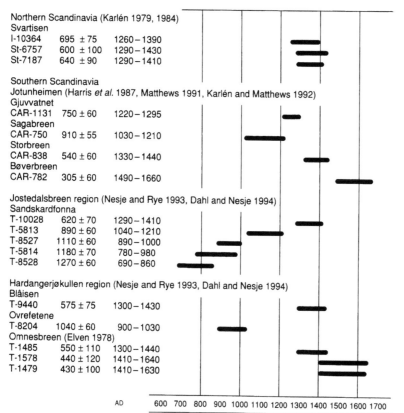

Northern Scandinavia (Karlén 1979, 1984)
Svartisen
I-10364   695 ± 75   1260–1390
St-6757   600 ± 100   1290–1430
St-7187   640 ± 90   1290–1410

Southern Scandinavia
Jotunheimen (Harris et al. 1987, Matthews 1991, Karlén and Matthews 1992)
Gjuvvatnet
CAR-1131   750 ± 60   1220–1295
Sagabreen
CAR-750   910 ± 55   1030–1210
Storbreen
CAR-838   540 ± 60   1330–1440
Bøverbreen
CAR-782   305 ± 60   1490–1660

Jostedalsbreen region (Nesje and Rye 1993, Dahl and Nesje 1994)
Sandskardfonna
T-10028   620 ± 70   1290–1410
T-5813   890 ± 60   1040–1210
T-8527   1110 ± 60   890–1000
T-5814   1180 ± 70   780–980
T-8528   1270 ± 60   690–860

Hardangerjøkullen region (Nesje and Rye 1993, Dahl and Nesje 1994)
Blåisen
T-9440   575 ± 75   1300–1430
Ovrefetene
T-8204   1040 ± 60   900–1030
Omnesbreen (Elven 1978)
T-1485   550 ± 110   1300–1440
T-1578   440 ± 120   1410–1640
T-1479   430 ± 100   1410–1630

AD   600 700 800 900 1000 1100 1200 1300 1400 1500 1600 1700

Figure 3.5 Radiocarbon dates relating to early Little Ice Age glacial advances in Scandinavia

Notes: The numbers in the margin give the laboratory identification number, the ^{14}C age and the calibrated age (AD) for 68% probability. Horizontal lines represent the calibrated ages

date of 305 ± 60 BP (CAR-782), yielding age ranges of 1440–1670, 1749–60 and 1773–93 (Stuiver and Becker 1986). Matthews (1991) investigated the forefield of Haugabreen, a southern outlet of the Mylebustbreen icecap, to the west of Jostedalsbreen. He regarded the results as suggestive rather than conclusive, despite relating directly to moraine age.

Evans et al. (1994) attempted to date moraines about 15 km south of Sandane in western Norway by geomorphological mapping and lichenometry and concluded that some of them were formed in the thirteenth or fourteenth century. After a critical examination their dating was rejected by Matthews et al. (1995).

Initial Little Ice Age expansion northwest of Jostedalsbreen has been dated to later than 890 ± 60 BP, that is since cal AD 1030–1220 (Nesje et al. 1991) from laminated sediments indicative of meltwater transport, deposited in Vanndalsvatnet by a stream from the small icecap Vanndalsbreen.

Many other local studies appear to show that glaciers began to advance in the late fifteenth and sixteenth centuries. Andersen and Sollid (1971) described peat protruding from beneath the front of Tverrbreen (now called Tuftebreen), one of Jostedalsbreen's outlet glaciers, and obtained a date from it of 410 ± 60 BP (T-779). Calibration according to Damon et al. (1974) indicates a date for the peat of about AD 1540. This was assumed to be the date of initial moraine formation even though, as it was noted, this is incompatible with the historical records of ice advance. From beneath the moraine fronting Storbreen in Jotunheimen, moss radiocarbon-dated to 664 ± 45 (SRR-1083) and 532 ± 40 (SRR-1084) and, again calibrated according to Damon et al. (1974), was also originally taken to show glacial expansion in southern Norway in the second half of the fifteenth or the first half of the sixteenth century (Griffey and Matthews 1978). However, as the upper of the two moss layers sampled had the older date, the authors concluded that 'no importance can be attached to the difference between the two dates'.

Temperatures reconstructed from tree rings show similar long-term trends on both sides of the Scandes, but differ in the magnitude and timing of the extremes. Climate reconstructions for northern Sweden show summers from 1200 to 1360 as generally either near or below the long-term mean (Briffa *et al.* 1992) and at Forfjorddalen on the west coast, summer temperatures were predominantly low from 1375 to 1440 (Kirchhefer 2001). At both inland and coastal sites temperatures were generally high from 1475 until a sharp deterioration affected the coast around 1540 and the interior between 1570 and 1580. Summers in the interior remained generally cold until the middle of the eighteenth century whereas on the coast minima at about 1605, 1645 and 1680 correspond with those of cold years over much of Europe. The authors recognize that their standardization methods, though intended to capture long-timescale variability, need further exploration and refinement, that their reconstructions refer only to summer temperatures, and that correlation between temperature fluctuations in northern Fennoscandia and southern Scandinavia may be low. If more logs, preferably *in situ*, could be found in the south and their rings correlated with long sequences to reveal the dates at which they were killed, as has been possible elsewhere (Luckman 1995, 1996a, b; see Chapter 9, p. 271), this could supplement the historical data in a more precise manner.

The great majority of field studies of the Little Ice Age in southern Norway depend on radiocarbon dating. Despite improved calibration, most of the radiocarbon dates from peats and other organic materials have errors which can be as much as three times the laboratory errors stated (Stuiver 1978, International Study Group 1982 and see Figure 1.2). The time interval between the peats and the glacial sediments overlying them is uncertain. Evidently, studies such as those discussed above cannot reveal unambiguously the century in which Little Ice Age advances began (Matthews 1982, Grove 1985).

We are left dependent on historical sources for more precise information. In view of the widespread indications from these that the Little Ice Age had begun by the fourteenth century, if not earlier, coming from other regions and discussed in later chapters, it seems reasonable to assume that the difficulties which triggered tax reductions and caused abandonment of lands in Oldendal and Lodalen before 1340 were associated with early Little Ice Age climatic deterioration. The sparsity of evidence relating to the early phases of the period is attributable to ice advances in later centuries, especially the eighteenth, having been so much more extensive as to conceal the effects of earlier glacial expansion.

3.1.2 The expansion of Jostedalsbreen 1680–1750

According to Schøning (1761) there were crop failures in Norway in 1600, 1601 and 1602 and again in 1632 and 1634.[3] Gerhard Schøning's compilation has not yet received full critical analysis but, according to Sandnes (1971: 336), he used contemporary seventeenth-century records. There were certainly crop failures in the mid seventeenth century: a petition made in June 1644 by people in the Jølster area, immediately west of Jostedalsbreen, explained 'there is such great misery here that many in this little skipreide are starving and will soon die of hunger, some having to eat bark mixed with chaff instead of bread' (*Samlinger til det norske Folks Sprog og Historie*, V: 493). In 1648, King Christian IV addressed an open letter to the farmers of Norway, in which he granted 'the same relief granted them in previous years'. This was because of 'crop failure, poor fisheries and cattle pestilence' (*Norske Rigsregistranter*, VIII: 602). A letter in almost identical terms was sent the following year (*Norske Rigsregistranter*, IX: 278). By this time it seems reasonable to infer that Jostedalsbreen was enlarging, for in the 1680s ice began to creep down into the surrounding valleys.

The earliest reliable evidence of direct damage to farmland by advancing ice in Scandinavia comes from Jostedal in a brief account, dated 1684, of the arraignment before the local court or *ting* of two farmers,

---

3 Schøning (1761) lists a number of cold years from 1294 to about 1340 and several years of acute harvest failure in Norway from 1315 onwards, but this part of his compilation has yet to receive critical analysis.

Knut Grov and (illegible) Berset, for non-payment of *landskyld* by the proprietor of their farms, a widow called Brigitte Munthe (*Tingbok for Indre Sogns Fogderi*, no. 14, 1684). They pleaded that they could not pay because their high pastures had been covered by advancing ice. The exact position of these *saeter* grasslands and the huts which stood on them is unknown but Grov and Berset farms were, as they still are today, in Krundalen, a right-hand tributary of Jostedalen. There is not a great deal of space in upper Krundalen and the implication is that if the land here was suitable as pasture it must have been ice-free for a considerable period. It is significant that Brigitte Munthe submitted that the two farmers could use pastures at Kriken and Espe, also her properties, as there was sufficient grass there. Kriken and Espe lay on the eastern side of Jostedalen, the side of the valley away from Jostedalsbreen. The advance of the ice after 1684 was swift. According to the records of an inquiry held in August 1742, Knut Grov's son Ole, born about 1678, and a neighbour Ole Bierch, born about 1672, 'men of good repute', explained that

> they remembered that in their youth, the said glacier (Tverrbreen) had been only high up on the mountain in the narrowest neck of Tufteskar, but it had forced its way through the gap, and down onto the flat fields towards the river, and, according to Rasmus Cronen's explanation, it had advanced 100 fathoms [*c.*200 m] in only 10 years.

They made no mention of temporary retreats or halts between 1684 and 1742, or of earlier advances of the glacier. This is hardly surprising in view of the history of settlement in the valley. However, the assessors did look to the future and measured the distance from the glacier to a cairn erected near the river 'so that it may be noted how much the glacier may advance in future years' (*Kongelige Resolusjoner, Bergens Stift*, 1740–3).

The effects of the Black Death in Norway, though severe, were uneven (Hovstad 1971). Many upland areas were left with little or no population for prolonged periods. After the plague of 1349–50 Jostedalen was deserted. There was still one farm occupied in Krundalen in 1374, but eventually this too lay empty. Survivors migrated to richer lowland areas or to coastal situations. It was not until the population had risen substantially in the sixteenth century that remote valleys like Jostedalen were reoccupied. The resettlement of Jostedalen can be traced by examination of the tax

rolls (Laberg 1944). When a farm was taken in from the wild a few years' grace was allowed before it was assessed for land rent, taxes or tithes. It may be assumed that farms were reoccupied during the decades preceding the dates at which they first appear in the tax rolls. The results of Laberg's investigations are shown on Figure 3.6. Jostedalen was not farmed again after its desertion until the years immediately preceding 1585, when several immigrants came in and settled down. The Krundalen farms, Grov and Berset, were taken in just at the time, as will appear later, when the glaciers around Mont Blanc are known to have been expanding to cover or damage farms and high villages in the Arve valley (Grove 1966, Ladurie 1971 and Chapter 4). The grasslands at the head of Krundalen must not only have been clear of ice in the late sixteenth century but also for a long period before that, as they were available for summer grazing. Conditions in Jostedalen were still sufficiently attractive for further newcomers to arrive in the early part of the seventeenth century; several farms, including Kriken and Espe, appear in the tax rolls in 1611.

When Jostedalsbreen eventually expanded in the late seventeenth century, the damage it caused to farmland led to on-the-spot investigations (*avtaksforretninger*) and courts of inquiry and hence to an accumulation of documents. From these it is clear that the advance of Tverrbreen was by no means exceptional in its extent or timing. Foss, vicar of Jostedal, gave an account of his parish in 1744 in which he noted that ice had covered six fields (*jorder*) near the head of Mjølverdalen and explained how the ice there and in Krundalen, as it pushed forward, had widened as the valleys opened out.

> Its colour is sky blue and it is as hard as the hardest stone ever could be with big crevasses and deep hollows and gaps all over and right down to the bottom. Nobody can tell its depth although they have tried to measure it. When at times it pushes forward a great sound is heard, like that of an organ and it pushes in front of it unmeasurable masses of soil, grit and rocks bigger than any house could be, which it then crushes small like sand. In summer there is an awful cold wind blowing off it. The snow which falls on it in winter vanishes in summer but the ice glacier grows bigger and bigger.

The summer cold damaged the crops and so chilled the people working in the fields that they had to go clad for winter even in the summer weather. 'The vol-

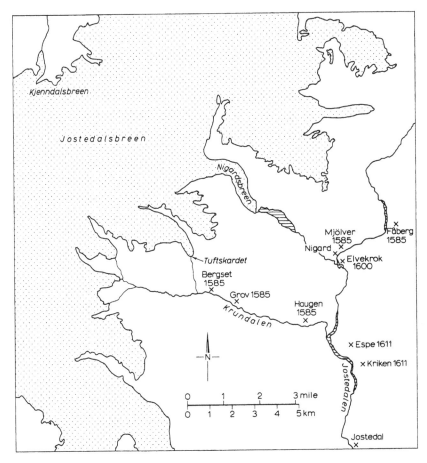

Figure 3.6 Farms in upper Jostedalen resettled in the late sixteenth and early seventeenth centuries, together with the dates at which they reappeared in the tax rolls (*Source:* data from Laberg 1944)

umes of water coming from the two valleys make an enormous river that not only damages what poor roads there are, but also undermines the soil itself, damaging meadowland and removing large trees' (Foss in Laberg 1948: 207).

An inquiry into the tax assessment of Mjølver farm on 16 November 1735, shortly before the farm buildings of the highest *brak*[4] disappeared beneath Nigardsbreen, provides more detail.

Guttorm Johanssen Mielvær of Jostedal Skipreide appeared before the court and, with the consent of the court, re-

quested the common people to give a true account of his farm in Jostedal, of its situation and of the extraordinary damage which the well-known Jostedal snow or ice glacier inflicts each year on Mielvær farm on which he lived. To which the people replied that they could in all God's truth, state and testify that the well-known Jostedal glacier, which extends over seven parishes, has grown so much, year after year, that it has carried away not only the greater part of the farmed meadowland but also that the cold given off by the glacier prevents any corn growing or ripening, so that the poor man, Guttorm Johanssen who lives there, and his predecessors have each year had to beg for fodder and seedcorn. Nor for some years has he been able to pay his landlord his dues which have been in arrears for many

---

4 These farms were compound; each family occupied separate buildings in the farm cluster and farmed a separate piece of land known as a *bruk*.

years and he has nothing whatever with which to pay them. His landlord. Hr Christopher Munthe, has had nothing from this farm, except relief of taxes which this poor man does his best to pay, mostly by begging. They also explained that providing the above-mentioned glacier continued to grow as it had been doing for some years and was now within a stone's throw of the house, the farm would be completely carried away within a few years and would never again be habitable. These were the true facts according to the Jostedal people present.

(Eide 1955: 18)

Thus although Nigardsbreen had taken longer to reach the grasslands in Mjølverdalen, between 1710 and 1735 the front had in fact moved forward some 2800 m.

The forecast made in 1735 proved to be accurate. By 22/23 August 1742, when Mjølver farm was again examined for tax relief purposes, two of its three *bruk* were derelict. Only the one worked by Ole Ottesen was still being farmed, probably because 'this part lies behind a large hill which in some measure shields it from the strong cold winds which the glacier exhales'. On Anna Aasen's *bruk*, corn could no longer be ripened 'because of the cold given out by the glacier'. The officials found that

the glacier has advanced to within 20 alen [12.15 m] of the cabins or derelict farm buildings onto Rev. Chr. Munthe's part which was abandoned by their people four years ago. The said glacier comes down from the great ice-mountain and covers the whole valley from one mountainside to the other and is some thousand alen wide at its margin where, immediately in front of the farm, it is over 140 alen [88 m] in height measured by instruments and reckoning. . . . This glacier has, as we saw and had explained to us, covered all the farm's hayfields, grazing lands and summer pastures over the distance of a quarter of a mile [equivalent to 2.7 km] in which the grazing land, as well as the upper part of the Mielvær Farm has been carried away and the little remaining of the derelict pasture and arable land belonging to the Rev. Chr. Munthe has also become derelict by reason of the cold that rises from the glacier and is so ruined that it could not provide forage for one beast.

(*Kongelige Resolusjoner, Bergens Stift*, 1740–3)

Nigardsbreen continued to advance and Foss tells us that by 1743

the glacier had carried away buildings; pushing them over and tumbling them in front of it with a great mass of soil, grit and great rocks from the bed and had crushed the buildings to very small pieces which are still to be seen,

and the man who lived there has had to leave his farm in haste with his people and possessions and seek shelter where he could.

Foss went on to note that the glacier 'had approached another farm called Bjerkehougen and carried away its fields and meadows, only the buildings remaining, and it is thus uninhabitable' (Foss 1750: 17, 18). Erikstad (1990) states that the advance culminated in 1748, since when its snout has retreated 4.5 km.

In Krundalen, at about the same time, Bersetbreen

filled the whole of the valley from side to side and from the top down to the stream. . . . Further, the glacier called Vetledalsbreen has, from the south, forced itself between the mountains into the valley in front of the mountains called Høye Nipen, and it is by these glaciers (which are all joined to the great glacier or mountain of ice covering all the mountains, which we were told, were said to be one and a half [Norwegian] miles [16 or 17 km] wide . . .), that part of Berset farm's hayfields and grazing lands such as their cattle enclosures and summer pastures, has been almost completely taken away.

The farm's arable land and meadows were next examined:

as regards the arable land, all was still quite green, part in ear and part just come into ear, so that there was little or nothing in them, as it advances more than formerly down the valley. . . . The freeholder, Anders Berset, then stated that during these years of crop failure he had got no seedcorn but the meagre crop he had won had been of a kind of light corn or chaff which, with straw and pine bark, he had ground to flour and mixed in order to keep alive.

(*Kongelige Resolusjoner, Bergens Stift*, 1740–3)

Farms in the main valley of Jostedalen were affected by summer cold, avalanches and floods in the eighteenth century. Ormberg, wedged into a narrow flat on the floor of the valley, was badly damaged by flooding in 1741. The farmer told a court of inquiry of many years with poor harvests; in some he had got back what he had sown, in others a little more, so that like his neighbours he had to mix pine and elm bark with his corn. Much of the arable land which lay along the river was eroded away in 1741 so that a strip 1334 alen long by 210 alen wide lost 84 alen of its breadth; further up the river a strip 336 alen long by 63 wide was also lost (1 alen equalled 0.63 m) and much of the remaining land was covered with sand (Laberg 1944).

Elvekrok was damaged by the river in 1742 and the court of inquiry reported that

it was apparent to us that it was the nearness of the glacier which is the cause of crop failure on this farm, for in the fields where the crops were now in ear, the ears on the side towards the glacier from the west were quite brown, and on the other side green, though some ears were not even in bud because the cold and the strong cold winds which the glacier exhales freezes it away in one night's still weather. The damage to this farm by flooding by the stream which comes from the glacier could not be measured, as we could not cross it, as it runs in several branches, but as far as it was possible to see and carefully observe, more than the half part of the farm's pastures had been washed away and removed, so that it can never recover and be improved, nor, because of the cold, can there be any corn harvested for food let alone seed.

Taxes here were reduced by half (*Kongelige Resolusjoner, Bergens Stift*, 1740–3).

In sum, evidence from the farm histories of Jostedalen dates the descent of the ice from the high tributary valleys to the 1680s and the subsequent encroachment onto permanent farm sites to the period between the 1680s and 1745, a period of increased winter snowfall and short cold summers.

Witnesses of the glaciers' behaviour, it must be emphasized, were not lacking between 1585 and 1684 and they would have left records of any sizeable expansion affecting farm and grazing land. We have records of other natural events from the seventeenth century. For instance it is known that Kruna or Kronen, one of the largest farms in Krundalen, passed from one Sjur to his son Klaus in 1653 after the son's farm had been destroyed by flooding (Laberg 1944). But there is no hint of damage caused by glaciers in documents written before the late seventeenth century. The grasslands overwhelmed by Tverrbreen and Nigardsbreen in the late seventeenth and early eighteenth centuries, it may be concluded, had been free of ice since the resettlement of Jostedalen.

The likelihood that Jostedalsbreen advanced earlier than the late seventeenth century diminishes further in the light of historical evidence from Oldendalen on the northern edge of the icecap. This valley was not completely deserted after the Black Death, a number of farms remaining occupied without interruption. People began to move back earlier than into other valleys around Jostedalsbreen, especially between 1530 and 1620 (Aaland 1973). By 1563 there were twenty-three farms within Olden parish and a further sixteen had been added by 1667. Not only were old farms resettled but new farms were taken in and the valley fully occupied. Farming settlements or *gårder*, consisting of a group of farmhouses and their associated buildings, more like hamlets than unit farms, were increasing in size. Each of the families in a *gård* occupied a *bruk*, a portion of land; Kvamme *gård*, for example, was composed of three *bruk* in 1602, five in 1606, and six in the 1640s.

The most significant histories from the point of view of the present inquiry are those of two farms, Åbrekke and Tungøen (Tungøyane), which both held grazing land up in Brenndalen, a hanging valley tributary to Oldendalen.[5] These farms, established before 1563, huddled beneath the rock step that terminates Brenndalen at its western end on either side of the stream from Brenndalsbreen, which joins the river from the Brigsdal and Melkevoll glaciers in upper Oldendalen (Figure 3.7). The neat farm buildings of Åbrekke can still be seen today at the foot of the step where the stream descends from the hanging valley of Brenndalen. According to tradition, after the settlement of the farms the ice was just visible on the skyline and did not reach the floor of Brenndalen. Both Åbrekke and Tungøen certainly used Brenndalen for pasturing their cattle. Documents from the period 1563 to 1670 do not reflect any anxiety about the future of the farms and no damage to their lands is recorded. In the tax register (*matrikkel*) of 1667[6] Tungøen is listed as a farm of

---

5   Rekstad (1901) published long quotations from documents concerned with the history of Tungøen and Åbrekke, but was not conversant with the complete histories of the two farms. Aaland (1932: 34 ff.) gives a more complete but summary account without references. Eide's (1955) version was the fullest but unfortunately he also gave no references.

6   The assumption of absolute monarchy by the Dano-Norwegian Crown in 1660 brought about a measure of centralization in government and some uniformity to a complex of ancient taxes and dues. The whole of Norway was reassessed for land rent (*landskyld*) between 1665 and 1670. The reassessment of Nordfjord in 1667 provides a very useful index against which to measure tax reliefs granted to many farms during the subsequent period of climatic deterioration.

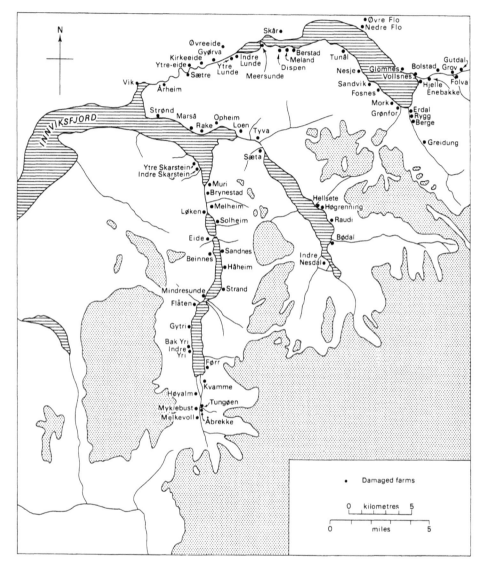

Figure 3.7 Farms in Stryn and Olden Skipreider which suffered serious physical damage during the Little Ice Age as recorded in *avtak* records

three *bruk* and with the taxable value, in butter as was commonly the case, of 2 *laup* and 1 *pund*: 24 *merker* of butter were equivalent to 1 *pund*, and there were 3 *pund* to 1 *laup*. A *laup* was about equal to 18 kg, val-ued for tax purposes at 2 *riksdaler* in coin, although the market value of 18 kg of butter must, by the seven-teenth century, have been higher.[7] So the taxable value of Tungøen was about 42 kg of butter. The farm was

---

7   In medieval times taxes and rents were paid in kind. Most of the goods were sent to Bergen, where they were weighed and measured at Bergenhus. Some of the goods were used to provision the fortress, some may have been sent to Copenhagen, but the main bulk was sold to private merchants in Bergen. The transition to payment in cash appears to have taken place in the seventeenth century. The *foged* or sheriff now had to forward cash to Copenhagen. Because many of the peasants

described as consisting of cultivated fields and meadows. It supported thirty-eight cattle and three horses and produced 29 *tønner* of corn from 9 *tønner* of seed.[8] Farms like this were almost entirely self-supporting. The cattle were raised for their milk and sold for slaughter. Any surplus corn and dairy produce was also sold. Farming families with sheep clothed themselves from the wool. They built their own houses and outbuildings. But the standard of living must have been low at the best of times and near penury when crops failed or disaster struck, as happened at Tungøen in the late seventeenth century.

In the 1680s the ice was spilling over into Brenndalen and accumulating below the rockstep at its head. The glacier stream was enlarged and by 1685 Tungøen had suffered great damage from flooding (Aaland 1932: 433). A court of inquiry held in 1693 found serious damage to both arable land and pastures. Rockfalls continued and, though the farm buildings had been moved, it was recorded that 'great rocks continue to fall towards the buildings, so the occupants are in danger of their lives'. The court recommended a reduction of the taxable value by 2 *pund* and 8 *merker*, leaving an assessment of 4 *pund* 16 *merker*. It had much other business to do in the Olden and Stryn area that year (*Rentekammeret, Ordningsavdeling, Affældnings Forretninger, Bergens Stiftamt, Pakke* 3, 1702–84); next door at Åbrekke more than half the arable land and pasture had been carried away by

two terrifying rivers which come from two glaciers . . . in addition, great damage is being done by rocks and grit which frequently fall from the mountain above their arable and pasture and because of this the farm's tenants,

one by one, have become impoverished and have had to leave the farm.

The damage to Bødal in upper Lodalen in 1693 was explicitly attributed to glacier advance:

this farm, Boedahl, has suffered much damage this year to its hayfields outside the fenced lands by a landslide, called the glacier landslide, and last year . . . another landslide called Espe. . . . Immediately inside the fenced land the river has broken out and made a new course across the farm's best meadows and fields, damaging up to five mæler of seed ground and the closer the river approaches the great freshwater lake, the more it spreads out on each side, causing great damage to their meadows and ploughland. A valley called Tiørdal, half of which they formerly owned with the Indre Næsdal farmers, has now been wholly lost and completely ruined by the destructive glacier and by the great river running from it, so that where they formerly had five enclosures and six barns they and their contents have been completely carried away so have neither the benefit of their hay and grass nor can they henceforward obtain any because of the great mass of rocks and boulders that can never be cleared away.

(*Rentekammerarkivet, Fogederegnskaper, Sunnfjord–Nordfjord*, 1702)

'Tiørdal' was Kjenndalen and the damage was caused by Kjenndalsbreen (Figure 3.8). The six barns and other property would have been included in the 1667 assessment as being in no danger and it follows that the advance of the glacier had taken place rapidly between 1667 and 1692/3.

The documentary evidence shows that the advance of the glaciers was associated with increased damage from landslides, rockfalls, avalanches and flooding (Grove 1972, and see Chapter 15). Not only was farming made more difficult but fisheries were also

---

could not pay taxes in cash, the sheriff had to trade the payments he received in kind for cash. The course of the transition from payment in kind to payment in cash has not been investigated in detail for the Nordfjord area. (Private communication from Dr Håkon Hovstad.)

The *mark*, the basic unit for butter, was fixed at the equivalent of 249.875 g and the *bismerpund* at the equivalent 5.997 kg in May 1683. By a decree of 4 March 1684, tenants were given the chance of paying in cash or kind, but if they chose cash then the amount paid was to be an official evaluation of the *landskyld* in kind. A decree of 5 February 1684 urged payment in cash and this, incorporated in a law of Christian V, remained in force until 1965 (*Norsk historisk Leksikon* 1974: 193). Equivalents were generally recognized between units of butter and other goods. Thus in the Nordfjord area, in the eighteenth century, one *laup* of butter equalled two *vog* of fish, equalled two hides, equalled one *skippund* of flour. Because of fluctuations of the *landskyld* over time, a reduction of *landskyld* or tax on one *laup* in the seventeenth and eighteenth centuries indicates more damage than a reduction of one *månadsmatsleiga* in 1340. (Private communication from Professor A. Holmsen.)

8   In Nordfjord a *tønne* was a measure of about 162 litres in the 1660s. A *tønne* varied in volume from district to district and from one time to another.

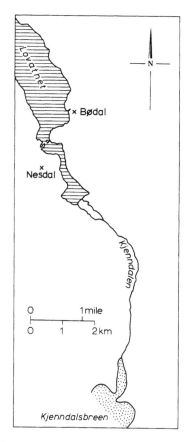

Figure 3.8 Bødal and Nesdal farms in relation to Kjenndalsbreen

were not approved by the higher authorities. As conditions worsened the farmers became desperate and Hugleik Tungøen, a man of unusual initiative and energy, journeyed to Copenhagen twice to seek the King's authority for a further inquiry to be held into the damaging of thirty-six farms in Olden and Stryn. He was successful and delivered the writ authorizing the inquiry to the Provincial Governor in Bergen in July 1702. Then, tragically, he lost his life, freezing to death in a storm on his way home (Eide 1955). Memories are long in Oldendalen and a commemorative pillar was erected in the valley during the last century.

The reports of the 1702 commissioners and of the further courts of inquiry held over the next half century show that conditions at Tungøen, Åbrekke and Bødal were by no means solitary instances and that the climatic deterioration and advance of the ice after 1667, probably as late as the 1680s, had been dramatic and swift. Events at Yri (Figure 3.7), which had been prosperous in the early seventeenth century, with its *landskyld* being increased as late as 1674, showed that Myklebustbreen, an outlying fragment of Jostedalsbreen overlooking the farm, expanded in the last decades of the century. Yri's dues were reduced on account of landslides and glacier encroachment in 1687 (Aaland 1932: 473–80). The 1693 commissioners found that

> the farm Indre-Yri has time and again been caused even more damage, always by the same means, and in particular, last winter when a glacier (Blaabræde) lying above the farm and its adjacent lands (which is of snow as hard as ice and which has lain there summer and winter as long as any man can remember) gradually comes further and further down and finally falls and causes damage, which has now happened here, bringing with it a great mass of rocks, grit and sand from the mountain which has taken away almost all the arable land here and partly covering and ruining it, so it seems impossible ever to be able to improve it or ever bring it into any use. Down by the farm buildings some of the said snow or glacier was still lying, not having been thawed during the summer by sun or rain, and the occupants sorrowfully deplore and confirm the same on oath.

Near by at Bak-Yri 'the said glacier and landslides have not only covered and ruined this farm's arable land with grit and rocks, but also carried away two mill houses with their gear and broken the mills to pieces, item a smithy and two boathouses down by the lake' (*Rentekammeret, Ordningsavdeling, Affældnings*

damaged. The 1693 assessors examining Muri near the entrance to the fjord recorded that

> this farm in times past had again and again had increases made to its *landskyld* by His Majesty's sheriffs, because it had been blessed by God with fisheries, for their land reached them, and they and many others living in the valley were able to supply themselves, but many years have now passed since the fish came to the land, and they now have only the meagre growth of the soil and the produce of their cattle by which to meet the royal taxes and dues and with which to support their wives and children.

After 1700 the ice advanced rapidly down Brenndalen towards the rock step under which Tungøen and Åbrekke stood. Rekstad (1901: 26) records the tradition that on one occasion pieces of ice came down through the smokehole at Åbrekke. But the tax and *landskyd* reductions recommended by the court of 1693

Forretninger, Bergens Stiftamt, Pakke 3, 1702–84). Avalanches compounded the trials of the local people, who eventually had to leave their homes and take shelter under nearby cliffs. Sunde in lower Oldendalen, a large farm at the narrowing of Oldevatnet, was destroyed by a fragment toppling from the edge of the icecap overlooking it (Eide 1955: 27). On the coast at Loen, where the 1667 assessment had taken the value of the fisheries into account, their yield diminished and by 1702 it was recorded that 'a new river course had carried with it a great quantity of grit and earth, thereby spoiling and ruining the best herring-ground'. Expansion of the ice continued into the early decades of the eighteenth century. An inquiry at Stryn in 1726, to see whether the tax relief granted earlier should be rescinded, reported that conditions had not improved anywhere and that most of the farms had suffered further damage. In addition,

> there is another farm in the Skipreide, called Erdal, which for some years now has suffered irreparable damage to pasture and arable from two large rivers and a glacier which is breaking down onto it, so the 5 poor occupants can never remain there permanently unless a most gracious tax relief can be granted.
> (*Tingbok for Nordfjord Sorenskriveri*, no. 8, 1726–8)

The ice in Brenndalen had continued to advance. In July 1722 three men were sent in accordance with Treasury instructions to inspect farms granted tax relief in 1702. They reported absolutely no improvement in the condition of any of them, noting in particular that Tungøen 'has clearly suffered great damage from the rivers, rock falls and the glacier to both *Inn* and *Utmark*. Further, this farm and its buildings are always in danger from the advancing great glacier.' Åbrekke, it was recorded, 'has suffered greater damage than the other farms from the encroaching river and the glacier. Some of the buildings have been moved and the occupants are now running around the parish begging their bread, so that no taxes have been paid for the greater part of the farm for the past two years' (*Tingbok for Nordfjord Sorenskriveri*, no. 6, 1721–3).

The following year a general tax commission examined all the farms and in their report more detail is given:

at Åbrekke they have been keeping themselves alive for several years largely by begging and on the parish. Three years ago the farmed arable land and pastures were completely ruined and carried away by an immoderate burst of water which burst out from beneath the advancing glacier, causing a great trench across the land, a good 30 to 40 alen in length so that only a very small part of their land now remains. The two poor occupants had to move their buildings hurriedly but one of them is still in the greatest danger of his life, remaining there although his house was quite ruined but because of his great poverty he has not been able to leave it and so remains there in an extremely dangerous situation.
> (*Matrikkel over Nordfjord Fogderi*, 1723, vol. 146, fos. 126–39)

After Hugleik Tungøen's death, Rasmus Andersen came to Tungøen, together with his neighbour Rasmus Olsen, he petitioned the Provincial Governor for a further investigation and reduction of his assessment, the substantial relief recommended by the 1723 commissioners not having been implemented, and Andersen explained how the farm had

> in our forefathers' time and in our own suffered great damage both to arable land and meadows by reason of a dangerous and destructive glacier above it and below by the cruel river running through the length of Oldendal. Thus death daily stares us in the face, most particularly from the great and horrid glacier, so that henceforth we are not able, without great risk to our lives, to venture to take up our abode there, and far less are we able to support ourselves and our poor women and children by the miserable remnants of land left to us.
> (Eide 1955: 10–11)

Before the inquiry was held on 12 October 1728, 'These poor and needy people, for fear of the danger and disaster hanging over them in the form of the terrible glacier mentioned, now in 1728 have had to move their buildings and dwellings to a lower place which, with God's support and protection they think to be safer' (Rekstad 1901). The tax was reduced from 2 *laup* to 1 *laup*.

Six years later there was need for yet another inquiry. In the previous year, at midsummer, Tungøen's fields and meadows had been flooded and the glacier now overhung the farm and had destroyed most of its remaining arable land. Only two *mæler* each of barley and oats[9] could now be sown, instead of two *tønner*, and only six cows kept instead of eighteen.

---

9 A *mæle* was a measure of corn which varied in size over time and also differed from one area to another.

The glacier has time after time advanced so far that it is to be feared that in a few years it will come right down to the mainstream and if that should happen (which God in his mercy will surely avert) they can expect nothing but the total ruin and destruction of the whole neighbourhood. In the meantime, the mainstream has also broken out into another river course so that in high summer it goes over the whole field and right up to their poor little buildings which were moved to the place where they now stand in 1728. The other river, which previously came out under the glacier is now completely blocked at that place, but has dug itself another course right down over the arable land and meadows which were intact until 1733. But now that both these rivers, together with numerous lumps of ice from the glacier, mixed with gravel and large stones have joined and broken out over most of the best part of the farm's arable fields and meadows, these poor people with their wives and children both in the years 1733 and in the following year have had to beg for food in order to stay alive and, therefore, have to an even lesser extent been in a position to pay or to guarantee payment of their taxes for that same exceptional year. There was nothing but wretchedness and misery to be seen here, for where there were previously great level fields, both arable and meadow, there is now nothing to be seen but terrible large, wide and deep pits and trenches, which no one can cross but has to take another long way round, either high up on the mountains where there is continual fear of the same glacier, or down on the stones and grit which have been discharged by the river. . . . Things have finally reached such a pass that total ruin and destruction seems imminent.

(Rekstad 1901)

Complete exemption from taxes for 1733 and 1734 was recommended and the assessment for the future reduced to half a *pund*.

The fears of the court were justified. The years around 1740 were cold, with heavy winter snows and short cold summers. The glacier in Brenndalen, until this time fed by ice falling from the edge of Jostedalsbreen, now thickened so that the ice flowed directly from the icecap 4 kilometres down Brenndalen. On 1 December 1743 a great mass of water, ice, stones and gravel swept down and obliterated the new houses of Tungøen, killing all but two of the occupants, a soldier and a small boy. We learn from the military inquiry that followed, on account of the soldier losing his musket and sword, that

the soldier, Anders Pederssen Møchleoen, was then working on the farm and at the same time had his musket and sword with him for cleaning and no one on the farm was

saved but for the soldier mentioned and a little boy, but all the other people, with everything that was in the buildings, were destroyed and covered by stones and grit so that there are no means of recovering any of it.

(Eide 1955: 13)

All that was left to the heirs were two injured cows, two waistcoats, a pair of socks, a blanket and an old skirt (Rekstad 1901). Tungøen was removed from the tax list and the remnants of its fields were taken into the neighbouring farm of Aaberg. Its site is now marked only by a large mound of moraine lying under the Brenndalen step.

In December 1743 there were eleven days of very heavy rain which left a train of devastation up the west coast of Norway from around Stavanger to north of Nordfjord and inland as far as Jotunheimen. Over a hundred successful requests for tax reductions followed (Øvrebø 1970, Battagel 1981).

The tongue of Brenndalsbreen was not the only one to collapse around this time. There is also a record from Vetlefjordbreen, a tongue of Jostefonni at the head of the Vetlefjorden branch of Fjærlandstjorden. Because of the great damage caused by a 'terrifying flow of water from the celebrated glacier' on 14 August 1741, a detailed inspection of six farms in the valley above Vetlefjorden was carried out starting in October 1741 and, because of bad weather ('the rains had so swollen the river that we could not properly measure the damage suffered'), continued in September 1742. The assessors

all went out into the field and went northwards, right up to the well-known and celebrated glacier. From the front of its descending tongue a piece had broken off down between the two great rock-walls, being – as well as we could estimate – 40 fathoms [78 m] wide and 100 fathoms [190 m] long and of enormous thickness.

The collapse of the snout may have taken place at the time of the water burst, but when the assessors went up the valley in October 1741 they made no report of a detached piece, so it may well have occurred between then and September 1742 (*Kongelige Resolusjoner, Bergens Stiftskontor*, 1740–5).

Landslides and other gravitational movements, as well as floods, continued to destroy farmland in the areas surrounding Jostedalsbreen and consequent tax reductions were registered. In 1755 seven farms in Olden Skipreide[10] made a joint petition for relief, explaining that 'the unusually heavy snowfall of last winter caused heavy damage and devastation to our lands' Helleseter [Helect] on Lovatnet was destroyed. At a time

> when the freeholder was out fishing in the fjord, a great avalanche broke out from the mountain and not only swept all the farm buildings and cattle into the lake but also his wife, children and servants, eleven people in all and all of them being killed. The avalanche had also devastated and torn away all the land. . . . Nevertheless the freeholder has started to build again on the outer edge of the farm where he thinks the buildings will be somewhat safer.
>
> (*Rentekammeret, Affældnings Forretninger, Nordfjord Fogderi, Pakke* 5, 1734–61)

All the farms in Olden and Stryn Skipreider which are known to have suffered serious physical damage and were accordingly allowed relief on either *lendskyld* or taxes, or both, are shown on Figure 3.7. The similarity of the distribution to that shown on Figure 3.3 of farms damaged in the fourteenth century is evident.

There are no tax documents mentioning glaciers advancing after 1743, the year of Tungøen's final destruction.

3.1.3 The period of ice withdrawal, 1750 to the present

Foss, writing in 1750, dates the beginning of ice retreat in Jostedalen to 1748: 'one has experienced how, from 1748 it has retired, only very slowly but still noticeably' (Laberg 1948: 209). He was probably basing his account on the behaviour of the large ice tongues near Jostedal vicarage, especially Bersetbreen and Nigardsbreen. A decade later, Wiingaard (1762) wrote of the ice retreating from the mounds of stones left behind by the glacial advances in Krundalen. So the glaciers of Jostedalsbreen seem to have reached their most advanced positions in the Little Ice Age between about 1743 and 1750. From that time until the present the glaciers of Jostedalsbreen and the surrounding areas have retreated, with minor interruptions.

The precision of the record of glacier retreat increases towards the present day. Between 1750 and 1890, Jostedalsbreen was visited by many travellers, climbers and field scientists, whose observations, including for example those of Forbes, are both vivid and reliable. Artists ventured into this part of Norway and their paintings can be compared with later photographs. Regular surveys of the positions of the glacier fronts began at the end of the nineteenth century, organized by Øyen and Rekstad. Since the early 1960s, mass balance studies have been made, with annual volumetric changes being measured for several glaciers, including a number of the Jostedalsbreen tongues.

The retreat from the moraines formed in about 1740–50 was rather slow and seems to have occasioned little comment at the time. Conditions around Jostedalsbreen, and Scandinavia generally, seem to have been cool in the 1770s and 1780s. Slåstad's (1957) index of tree growth in Gudbrandsdalen, converted into a temperature index by Matthews (1976), shows cool summers in the 1770s. (It has to be noted that Slåstad's data are of unknown reliability; the methods used in their collection and analysis were less meticulous than would be employed now.) Sommerfeldt (1972) quotes a contemporary account of the Lensmann (sheriff) of Vågå to the effect that large masses of snow and ice accumulated in the mountains and the glaciers grew markedly, but we have no precise record of any glaciers readvancing as a result. Winters were on the whole dry, and spring came late and lasted for a long time. The records of the duration of ice in the Danish Sound, which are reasonably complete for the 1770s and 1780s, show that the means for these decades of the numbers of days with ice were very high and have never been exceeded.

The winter of 1788/9 was particularly severe, the hardest in Norway for many years, with intense cold before Christmas continuing into April and freezing the soil to a great depth. In the Sound there were 134 days with ice, a value only once exceeded since then (Lamb 1977: 589). Sea ice persisted near the Icelandic coast from May to early June and in the open waters between the Faroes and southeast Iceland surface

---

10   A *skipreide* was an administrative district approximately equal to an English hundred.

water temperatures in the summer were still low, from 1.0° to 1.6 °C below the average for 1921–38. However, to the southwest of Iceland and also at the entrance to the Baltic, temperatures rose above normal and a warm summer in southern Norway culminated in 'almost tropical heat of July' (Jarrman, 25 July 1789, quoted by Sommerfeldt 1972). Then came a great flood in Gudbrandsdalen as a result of unusually heavy rains over a very large area and probably of snowmelt in the mountains. It left a trail of devastation from landslides, rockfalls and flooding over the eastern and central parts of the country (Blyth 1982). Successful *avtak* (tax or rent relief) was registered for 953 farms damaged on 22 July 1789 (Øvrebø 1970). No ice was recorded in the Danish Sound in the following winter, nor in the one after that, and the decadal average for the 1790s was only half that of the preceding decade (Lamb 1977: 589).

It seems likely that glacial retreat was resumed after the eventful year of 1789 at a greater pace. In 1812 a young botanist called Christian Smith, who two years later was to become Professor of Botany at Christiania (Oslo), visited Jostedalsbreen and described the glaciers as being in retreat (1817). Bohr, in 1818 or 1819, wrote of Nigardsbreen that

> The mighty accumulation of moraine which this very glacier at Nigard had formerly pushed before it is now about 1726 [Norwegian] feet [541 m] below its margin whilst the bare sides of the mountain show its depth, now more than 200 feet less than it had once been. . . . The crops at Elvekroken this year were very good, while nothing but the moraine stood between the glacier and the ripe corn.
> (Bohr 1820: 257)

At about the same time he went on to visit Lodalsbreen and Stegholtbreen, the other two glaciers at the head of Jostedal, and recorded that 'the moraine showed clearly that these glaciers too had formerly descended about 1700 feet [538 m] further down; while the dark naked sides of the mountains, as if the surface had been shorn off, showed that they had been almost 200 feet deeper' (Bohr 1820: 259). By 1821, according to Naumann (1824: 201), Lodalsbreen was 1853 feet (586 m) from the 1740s moraine. The following year he tells us that in Mundal, near the southern extremity of Jostedalsbreen, the two tongues of Suphellebreen, which he called Veslebreen and Storebreen, had retreated respectively 1500 and 750 paces in the course of the preceding 100 years. Close by, Bøyabreen had retreated 900 paces over the same period, which Rekstad (1904: 10) interprets as a retreat of 500 m. Kraft (1830: 808) recorded that Mundalsbreen and Vetlefjordbreen had withdrawn considerably during the previous century and that it was evident from the moraines in front of Vetlefjordbreen that it had once reached 2000 paces further down the valley. Vetlefjordbreen had again caused a great deal of damage in 1820 when a large volume of water burst from it, devastating farms downvalley. This water may have been dammed up by the glacier in Svartvassdalen until the ice withdrew sufficiently to release it. (Vetlefjordbreen has since completely disappeared.)

The descriptions of the glacier fronts of Bøyabreen and Nigardsbreen given by Bohr and Neumann[11] suggest the possibility that the glaciers advanced a little in the 1820s. The tongue of Nigardsbreen was 'cut off at the lower edge at a height of 20–30 feet' and this edge was 'almost black from soil, dust and gravel', while the edges of Bøyabreen formed an 'ice wall of about 50 alen in perpendicular height' (Neumann 1923: 540–2). The painting of the terminus of Tverrbreen by Johannes Flintøe, made between 1822 and 1835, shows a smooth but swollen tongue which could be advancing (Plate 3.1). Rekstad (1902: 13, and 1904: 37) states that in Krundalen a moraine was deposited in 1830, 50 m inside the 1750 moraine of Bersetbreen (Plate 3.2). However, the first clear account of renewed advance since the mid eighteenth century is given by a botanist, Lindblom, who visited Nigardsbreen in 1839. He saw the glacier only vaguely through the rain, but recorded that the people living close by had noticed that in recent years it had begun to advance slowly, although it was at a standstill in 1839 (Plates 3.5 and 3.6). The advance can only have been a modest affair because in 1845 Durocher found the distance from the snout of the glacier to the outermost moraine had increased from 541 m in about 1819 to 700 m (Durocher 1847: 104).

---

11 Bishop Neumann travelled in the Sogn area in 1823 and recorded the state of this part of his diocese in considerable detail, but his figures for time and distance are so rounded that they have to be discounted.

Durocher also visited Krundalen and drew a picture of its glaciers (Plate 3.3). While Bersetbreen at the head of the valley had now retreated more than 600 m, Tverrbreen entering from the north, had retired only 350–400 m. Lodalsbreen, according to his measurements, had continued to retreat since Bohr had seen it and was, by this time, 600–700 m from its outer moraine (Plate 3.4).

In 1851 Forbes,[12] following his Alpine expeditions (see Chapter 4), made an excursion to Norway, where the ogives of the outlet glaciers of Jostedalsbreen particularly attracted his attention. He measured the intervals between the ogives on Bersetbreen and found their breadth to be somewhat unequal but an average of 167.7 feet, which 'represents, I have no doubt, very nearly the average annual movement of the glacier'. He counted twenty ogives beneath the icefall, which makes the length of the tongue of the glacier on the flattish valley floor about 1 km in 1851 (see Figure 3.9). The terminus was, Forbes found, 900 yards from the moraine, 'a great moraine evidently modern. Its limits may be at once traced all round for no birch woods grow within them. Beyond question, it is of the same date with the great extent of the Nygaard glacier . . . of which the date is unknown. The Tvaer Brae has a corresponding moraine.' He was also impressed by the evidence of the recent retreat and downwasting of Nigardsbreen, of which he made and published a drawing (Forbes 1853: 169). He knew of the advances currently taking place in the western Alps but makes no mention of any such occurrence in Jostedalen. Prominent moraines below some of the largest glaciers flowing from Jostedalsbreen have been dated to 1848–50 by lichenometry (see Figure 3.10). However, suspicion that the accuracy of the methods involved was insufficient to allow such precise results has

been confirmed by unusually thorough dating of the Nigardsbreen moraines (discussed later in this chapter) from which error limits of ±20 years for the oldest moraines were calculated (Bickerton and Matthews 1992).[13]

The glaciers retreated through the 1860s as Doughty (1865: 143) reported for Nigardsbreen, Blytt (1869: 37) for Nigardsbreen and Bøyabreen, and de Seue (1870: 15) for Bersetbreen. However, de Seue found Bøyabreen and Lodalsbreen advancing in 1869 and his wording suggests that Nigardsbreen may also have been advancing. A large moraine had formed in front of Nigardsbreen by 1873 (Rekstad 1902: 15) when Larson said the ice was advancing (Larson 1875: 11). Rekstad noted (1901: 13) that Brigsdalsbreen was also advancing between 1869 and 1872, though it failed to produce a moraine; he recognized that variations between the characteristics of different glaciers would result in a lack of uniformity in the morainic record. This lack of uniformity is exemplified by Bøyabreen, which advanced from 1880 to 1890 (Rekstad 1904), at a time when there are no reports of other Jostedalsbreen glaciers advancing. Thus the glaciers were almost continuously retreating for a century and a half, without substantial readvances of the kind that have been recorded from the Alps (see Chapters 4, 5 and 6). Small icecaps and isolated glaciers bordering Jostedalsbreen were also shrinking. Bing (1899) wrote that Rauddalsbreen, which eighty years previously 'went out over the valley to the other side', had now retreated and thinned so that it scarcely reached the valley bottom.

Access to the Jostedalsbreen icecap at the present day is difficult for cattle but throughout much of the latter part of the eighteenth century and until well into the nineteenth century ice projected down into the valley heads and provided convenient routeways from

---

12 This was Forbes's last scientific expedition; illness which began in the autumn of 1851 put an end to his active fieldwork (Shairp *et al.* 1873). Forbes had made pioneer studies of the alternate light and dark bands formed beneath the icefall of the Mer de Glace in France and the relationship between such bands or ogives and the speed of the flow of the ice. When he visited Norway he had already observed the expansion of Alpine glaciers towards 1850, e.g. 'I found the Rhône glacier much enlarged since I last visited it' he wrote in his journal for 1846 (Shairp *et al.* 1873: 329).

13 Dating methods were not the same in the three cases illustrated. Quite apart from the techniques involved we may notice that the outer moraine of Nigardsbreen here was dated to 1850, although Foss (Laberg 1948) recorded that the tongue started to retreat in 1748. The only pre-1900 dated surface used by Mottishead and White (1972) at Tunsbergdalsbreen was for the outermost moraine 'dated from historical evidence by Rekstad (1901) presumably by analogy with Åbrekkebreen and Nigardsbreen'. Bickerton and Matthews (1992) produced a table comparing their results with those of some of the earlier workers.

Plate 3.1 (p. 80, top) Tverrbreen, Jostedal, by Johannes Flintøe, 1822–33. The smooth but swollen state of the tongue suggests that Tverrbreen may have been advancing when Flintøe saw it; however, the abandoned left moraine shows that it had retreated from its maximum extent (*Source*: Nasjonalgalleriet, Oslo, Inv. No. 1085, cat. 1968, no. 649)

Plate 3.2 (p. 80, bottom) Bersetbreen in Krundalen, by J. C. Dahl, 1844. The front is steep and possibly advancing but a moraine on the left of the picture has been abandoned (*Source*: Rasmus Meyers Collection, Cat. No. 92, Bergen Billedgalleri)

Plate 3.3 Schematic sketches of glaciers at the head of Krundalen. AA = Tverrbreen; BB = Bersetbreen; CC = Vetlebreen. The portrayal of Tverrbreen suggests that the tongue was substantially withdrawn from its most extended position (*Source*: from Durocher 1847, figure 5)

Plate 3.4 Lodalsbreen to the left and Stegholtsbreen (T). Durocher found Lodalsbreen 600–700 m from its outer moraine. K = Lodalskåpa; N = skyline of Jostedalsbreen; M = medial moraine; L = well-marked lateral moraines. The whole of the medial moraine shown here is now free of ice (*Source*: from Durocher 1847)

Plate 3.5 (p. 82, top) Nigardsbreen seen from Elvekroken in 1839, by J. C. Dahl. The original sketch has the artist's notes written on it (*Source*: Nasjonalgalleriet, Oslo)

Plate 3.6 (p. 82, bottom) Nigardsbreen in 1839 by J. C. Dahl, dated 1847 but based on the sketch of 1839 shown as Plate 3.5. The *sæter* huts must have been erected after the ice retreated as Elvekroken was damaged by the ice. Dahl visited and sketched Nigardsbreen in 1839, the same year as it was visited by Lindblom. The front of Nigardsbreen in that year was said to be at a standstill following a slow advance (p. 78) (*Source*: Nasjonalgalleriet, Oslo, Inv. No. 1477, cat. 1968, no. 367)

Plate 3.7 Nigardsbreen by Joachim Frich, from an illustration in *Norge fremstillet I Tegninger*, Christiania, 1855. It shows two terminal moraines, the outer one with small trees growing on it. The tongue appears to be in retreat

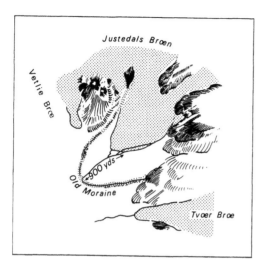

Figure 3.9 Bersetbreen in 1851, sketched by J. D. Forbes (1853), showing the tongue in retreat

Jostedal to the lowlands around Nordfjord. The people, wrote Foss (in Laberg 1948: 209),

> used to go year after year, arranging beforehand so that many could go together. They take with them wooden things made in the winter, such as troughs, baskets, bowls, ladles, wooden spoons and so on, which they sell very easily for a good price. They do not bring much back with them, apart from some oats, herrings and corn. . . . Also drovers go there to buy cattle which they bring back with them and also horses except for those who live in Valders in Christiania Stift who, when they go this way over the glacier and trade in Nordfjord with big droves of cattle and horses do not usually go back this way but by a more convenient route.

Foss also described several ways over the icecap including one from

> Berset to Aalden [Olden], where the first farm you reach is called Qvame. . . . Here the glacier is so big that the clouds blow against the top of it and the highest mountains lie far below as in an abyss so that you can hardly see them, even less arrange your route on them, so, in order not to get lost, wooden posts have been put into the glacier to mark the route.

Norvik (1962) enumerates no fewer then ten main routes over the ice, pointing out that the alternative to

a 30 km trek to the farms in Oldendalen was a 340 km journey down to the fjord, round to Vadheim and then overland.

Slingsby (1904: 249) noted that at the beginning of the nineteenth century many passes were used by traders crossing between Sogn, Jølster and Nordfjord, and suggested that their disuse was due to the shrinkage of the glaciers, though by then other routes were easier and new means of transport were available.

Systematic measurement of the ice fronts of Jostedalsbreen was begun by P. A. Øyen and J. Rekstad. From 1900 to 1940 annual reports appeared in *Naturen*, published by the Bergen Museum. Rekstad handed over responsibility to K. Fægri, of the Bergen Botanical Museum, in 1932. Fægri tabulated the consolidated results from 1900 to 1947 (1948: 301–2). Since then data have appeared in the yearbooks of the *Norsk Polarinstitutt*, and in *Fluctuations of Glaciers* published by the International Commission on Snow and Ice (e.g. Müller 1977). A complete set of Rekstad's glacier photographs is held by the Geology Museum of the University of Bergen.

The main features of the fluctuations after 1890 were two readvances of the glaciers of Jostedalsbreen which reached their most forward positions in 1910 and 1930 (Figure 3.11). The retreat that had continued without major interruptions since 1750 was halted or greatly slowed and then, after 1930, was resumed at a greater rate than ever before only to be halted once more at various dates after 1950, according to the characteristics of individual glaciers. Briksdalsbreen[14] was the first to halt; others followed in due course.

It has long been recognized in Norway that glaciers respond to climatic events at different rates (Rekstad 1902: 14). In the early 1920s the shorter, steeper glaciers of Jostedalsbreen started to advance about three years earlier than the longer ones, such shifts in the positions of the termini reflecting changes in the total volume of the ice, caused by variations in the relationship between accumulation and ablation, and between snowfall and snowmelt.

Olaf Rogstad (1941), of the Norsk Vassdrags og Elektrisitetsvesen (Norwegian Water Power and Electricity Board), made a study in 1940 in which he

---

14 Briksdalsbreen is also known as Brigsdalsbreen.

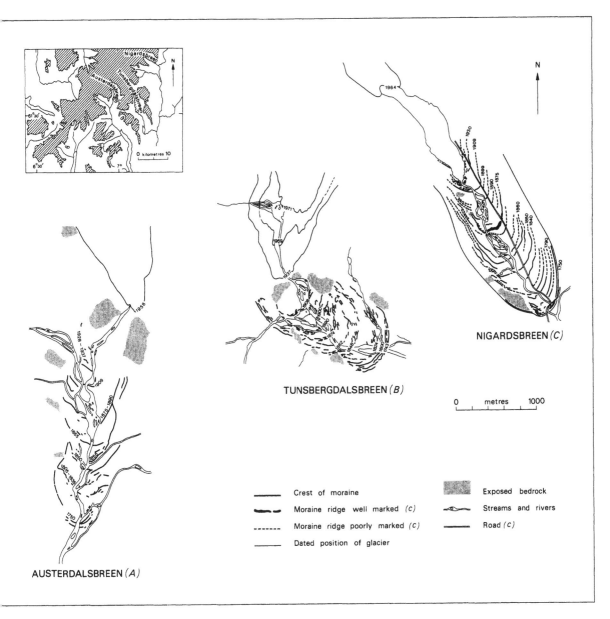

NIGARDSBREEN *(C)*

TUNSBERGDALSBREEN *(B)*

AUSTERDALSBREEN *(A)*

——— Crest of moraine	Exposed bedrock
➤ ⚫ ⚫ Moraine ridge well marked *(c)*	➤⚫ Streams and rivers
------ Moraine ridge poorly marked *(c)*	——— Road *(c)*
——— Dated position of glacier	

0   metres   1000

Figure 3.10 The forefields of (a) Austerdalsbreen, (b) Tunsbergdalsbreen and (c) Nigardsbreen, all three providing evidence of rapid retreat after 1930 (*Sources*: (a) after King 1959, (b) after Mottishead and White 1972, (c) after Andersen and Sollid 1971)

compared the discharge of the glacial rivers emerging from Oldendalen and Kjenndalen with the runoff from the nearby valley of Hornindal, which is free of ice. By this means he obtained some indication of the mass of water being stored in glacier ice or released from the ice reservoir as water in individual years between 1900 and 1940. His results showed that the advances and retreats of the ice in the two valleys lagged four years behind changes in their volumes. The response time is brief because the glaciers involved, including Briksdalsbreen and Melkevollbreen, are relatively short and steep. Rogstad went on to show that advance of

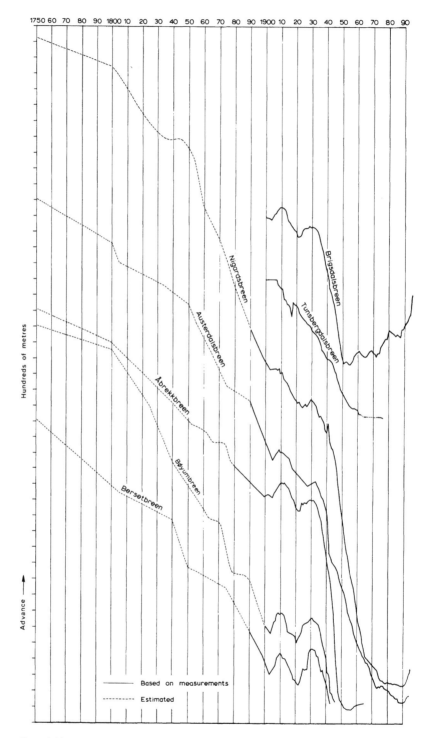

Figure 3.11 Retreat of Jostedalsbreen tongues from 1750 to 1993 (*Sources*: based on Fægri 1933, Liestøl 1963, 1976, 1977a, b, 1978, 1979a, 1980, 1982a, b, Kasser 1967, 1973, Müller 1977)

these tongues was associated with the ice masses increasing as a result of low summer temperatures and higher winter precipitation. Rogstad's conclusions were later substantiated by more sophisticated calculations based on longer records (Nesje 1989, Nesje *et al.* 1995), which also fitted the conclusion of Bickerton and Matthews (1993), based on lichenometric evidence, that since the late eighteenth century terminal moraines around Jostedalsbreen have been formed about five years after temperature minima.

Waning after the maximum extension of the ice was at first slow and broken by halts or minor readvances. Documents record some of them, but the number and complexity of minor fluctuations are revealed more clearly by moraine assemblages on glacier forefields, although not all fluctuations result in moraine formation and some moraines are removed by erosion or concealed by later advances.

Erikstad and Sollid (1986) concluded from lichenometric dating of 101 moraine ridges along N–S and E–W transects in south Norway that the glaciers retired hesitantly from their eighteenth-century maximum, halting and readvancing a short distance on several occasions, with deglaciation being concentrated around 1800, 1820, 1840, 1860, 1880, 1900, 1905 and 1930.[15]

Historic evidence of the timing of moraine formation by Nigardsbreen is sufficient to allow construction of lichen dating curves, as well as assisting in testing the accuracy of the resulting age determinations. Bickerton and Matthews (1992) carefully tested the reliability of the mean predicted ages of Little Ice Age moraines and concluded that their median-predicted dates had an accuracy of *c.*10 per cent, errors being of the order of ± 20 years for the older moraines and ± 5 years for the youngest.[16]

The Nigardsbreen curves were employed to investigate the moraine sequences of seven other outlets of Jostadalsbreen (Bickerton and Matthews 1993). Initial findings were tested for accuracy; the criteria used including whether or not dates were in agreement on both sides of the foreland, and whether predicted dates

using section *Rhizocarpon* were in agreement with those using the *Rhizocarpon* subgenus as a whole. The ages of a few moraines on each foreland, known from historical accounts or measurements, were used to check the lichonometrically derived dates. Only the most firmly dated were used to compare sequences from foreland to foreland. The accuracy achieved was ± 5.5 years for most moraines deposited during the twentieth century, and ± 9.4 years overall.

It emerged that despite large differences in the size of the glaciers, and the times of maximum extension having varied by some decades (Fåbergstølsbreen was found to have reached its maximum extension as early as 1705, for instance), frontal variations have been remarkably synchronous (see figure 11 of Bickerton and Matthews 1993). Dates of moraine formation turned out to cluster in a significant manner, suggesting that the sequences contain a sensitive record of the high frequency climatic variation since the early eighteenth century.

Between 1886, when regular meteorological observations in Norway began, and 1940, mean annual temperatures rose markedly, especially in the north of the country. The rise in temperature was 0.4 °C in Bergen, 1 °C in Finnmark, and 2 °C in Spitsbergen (Hesselberg and Birkeland 1944, 1956). This was enough to cause mean annual isotherms to be elevated 110 m, or to shift 300 km northwards. The warming, which was most marked in winter, was associated with a reduction in atmospheric pressure in northern Europe and an increased frequency of westerly winds. While this goes a long way towards explaining why the glaciers retreated, understanding of the details of their variations involves more than generalized meteorological information and recording of tongue positions. Fortunately, regular mass balance observations on the Norwegian glaciers, begun in the 1940s, have provided much of the data required.

Glaciers act as reservoirs by storing winter snow and releasing meltwater in the summer when it is needed for agriculture. Glacier water has been used for irrigating valleys around Jostedalsbreen for several centuries

15 Excellent geomorphological maps of the glacier and forefield of Memurubreene (1:6500) and of 14 other glaciers and their forefields accompany their paper.
16 Conditions were unusually favourable for lichenometry; results of this degree of precision cannot necessarily be expected in other parts of Scandinavia.

and the notion of the icecap as a reservoir appears in the literature as early as 1758: 'One could consider it as one of nature's peculiar reservoirs which the valleys profit from whilst they are short of rains' (Jostedalsbreden, etc. 1758). Now water is required for generating electricity and, with a view to more thorough exploitation of the energy resources provided by the ice, studies of glacier mass balance have been made in various parts of the country since 1962. The methods used are described in Pytte (1969), Østrem and Stanley (1969) and Nesje and Dahl (2000).

It was at just about the time when mass balance observations began on a selection of glaciers in both maritime and continental environments that a noteworthy fluctuation in climate took place. Records from stations examined by Hesselberg and Birkeland (1956) show that the mean annual temperature in the period 1941–50 had been as much as 0.4 °C lower than in the previous decade. The cooler conditions, accompanied by northerly winds and an increase of annual precipitation near the west coast of about 15 per cent, were to continue for several years. Mean summer temperatures at Bergen (the most convenient station for evaluation of temperature variations on the frontal fluctuations of Briksdalsbreen) between 1901 and 1930 were below the 1961–90 mean (12.5 °C). Then came the 1930s and 1940s, the warmest decades of the century, and in

1948 Briksdalsbreen reached its maximum retreat rate of 84 m per year (Nesje *et al.* 1995). Cooler summers in the late 1940s resulted in positive mass balances and stabilization of the front by the early 1950s. High winter precipitation gave such high positive net balances in 1988/9 and 1989/90 that the front of Briksdalsbreen advanced 75 m in 1992/3 and 80 m in 1993/4, the largest annual advances recorded in the twentieth century. In 1997–9 the ice 300 m behind the glacier front was measured as moving forward at between 17 and 36 cm/day (Kjøllmoen 1999). Altogether, between 1987/8 and 1995/6, Briksdalsbreen advanced 350 m, its snout reaching the outlet of Lake Briksdalsvatnet.

One of the glaciers that has received most attention is Nigardsbreen, which constitutes almost 10 per cent of the total area of Jostedalsbreen. Østrem *et al.* (1976) reconstituted the 1748 longitudinal profile of the glacier by mapping the valley slope trimlines that had attracted the attention of Forbes and Bohr, together with the terminal moraine and other moraines on nunataks. Figure 3.12 makes it plain that the lower glacier thinned more between 1937 and 1974 than between 1748 and 1937. Between 1936 and 1968 the front withdrew 1200 m and a proglacial lake formed. In the 1960s, shrinkage of the glacier diminished and between 1962 and 1975 the total volume of Nigardsbreen increased

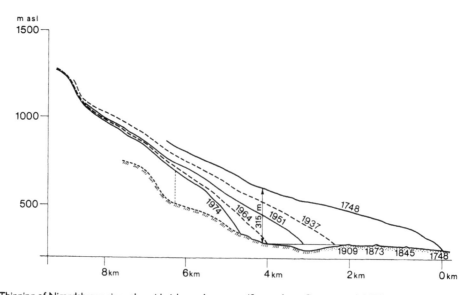

Figure 3.12 Thinning of Nigardsbreen since the mid eighteenth century (*Source:* from Østrem *et al.* 1976)

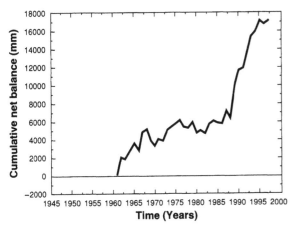

Figure 3.13 Cumulative net mass balance of Nigardsbreen, 1962–97 (*Sources:* from Roland and Haakensen 1985, Kjøllmoen 1999)

by a mean annual thickness of 0.4 m of water equivalent, with positive mass balances amounting to about 2 m for both 1961/2 and 1966/7 (Figure 3.13).

Up to 1975 the front of the glacier continued to retreat at a rate of about 40 m annually; it is a long glacier with a long response time. Between 1976 and 1981 the total volume of Nigardsbreen decreased by a mean annual thickness of 0.16 m of water equivalent, with a negative balance of 1.22 m in the single year 1979–80. But just at the time when the curve for cumulative mass balance for the glacier was tending to diminish, the front of the glacier ceased to retreat and even advanced slightly. (Steeper tongues like Briksdalsbreen and Abrekkebreen had stopped retreating in the early 1950s, as had Tunsbergdalsbreen by 1960.) In the late 1980s and early 1990s, positive mass balances became both frequent and substantial, as with Briksdalsbreen (Figure 3.13). The upper part of Nigardsbreen gained more than 5 m water equivalent in the winter of 1990; the average winter balance was 3.52 m and ablation the following summer was less than average (Elvehøy and Haakensen 1992). Again in 1993, the winter balance was 131 per cent of the average of the previous 32 years values, while an exceptionally cold summer resulted in summer ablation being only 66 per cent of the average (Haakensen 1995). The cumulative mass balance of Nigardsbreen has consequently risen sharply since the late 1980s (Haeberli *et al.* 1994), in strong contrast with southern Norwegian glaciers such as Gräsubreen and Hellstugubreen

(Figure 3.16). Its front advanced 240 m between 1988 and 1999 to regain its 1971 position (Kjøllmoen 1999).

3.2 COMPARISON OF THE LITTLE ICE AGE RECORD FROM JOSTEDALSBREEN AND GLACIER FLUCTUATIONS ELSEWHERE IN SCANDINAVIA

3.2.1 Jotunheimen

Historical documents relating to the fluctuations of ice masses in Scandinavia, other than those already men-

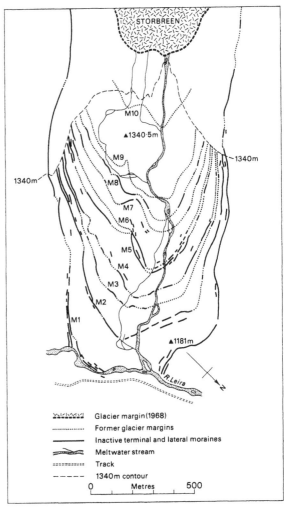

Figure 3.14 The forefield of Storbreen, Jotunheimen. Lichenometric dates of moraines: MI 1750, M2 1807/12, M3 1833, M4 1858, M5 1871 (*Source:* after Matthews 1974)

tioned, are scanty. Many of the valley and small cirque glaciers in Jotunheimen discharge into the Bøvra river, which is known to have caused severe flood damage in 1708 and 1743 (Rekstad 1900, Kleiven 1915). Floods at these times could well have been associated with glacial advances of the kind taking place around Jostedalsbreen about that time. However, there is need for caution in ascribing flooding to glacial advances unless there is specific evidence to that effect. The Bøvra also flooded severely in 1760 and 1763. The greatest flood of all, Store Offsen in 1789, which caused widespread devastation over much of southeastern Norway, was primarily caused by heavy rains that began in July and lasted for a fortnight, though, as we have seen, it can be argued that sluicing of snow from the glaciers may have augmented the flow of rivers draining glacial valleys. The Lensmann of Vågå may have been familiar with Jotunheimen when he stated that the glaciers grew markedly in the years before 1789 (Sommerfeldt 1972).

The few accounts of the Jotunheim glaciers available after 1820 were summarized by Øyen (1894). He himself worked in Jotunheimen between 1892 and 1909, but the measurement of frontal positions of glaciers which he initiated then lapsed until Werenskiold took it up again from 1933 until 1948. Such measure-

ments as are available suggest a general accordance of behaviour of the glaciers of Jotunheimen and Jostedalsbreen. This view is supported by the dates obtained lichenometrically for the moraine sequences of Storbreen by Liestøl (1967) and Matthews (1974) (Figure 3.14) and for those of Svellnosbreen by Green (1981).

Mass balance studies of Storbreen in Jotunheimen, starting in 1948, provide one of the longest such records available for any glacier (Liestøl 1967, 1973 and 1978) (Figures 3.15 and 3.16). A comparison of the records of Storbreen and Nigardsbreen shows that for every one of the years 1962 to 1974 the signs of the mass balances of these glaciers were the same, being either both negative or both positive. But the cumulative mass balances of the two glaciers over the period differed, with Nigardsbreen thickening over the period while Storbreen thinned. Between 1948/9 and 1974/5 the surface of Storbreen was lowered by 7.37 m and every year between 1959 and 1975 its front retreated (Kasser 1967 and 1973, Müller 1977). Loss of mass ceased around 1980. Over the next two decades Storbreen maintained its volume. Its front retreated more than a kilometre in the course of the twentieth century, but in the last decade it was holding its ground (Kjøllmoen 1999).

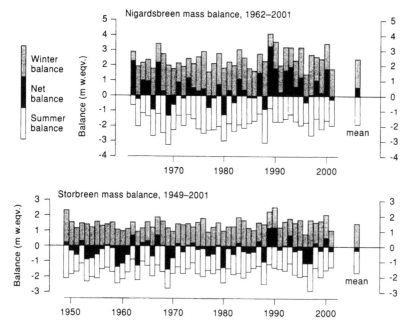

Figure 3.15 Winter, summer and net balance of Nigardsbreen, 1962–2001, and of Storbreen, 1949–2001 (*Source*: from Kjøllmoen 2003)

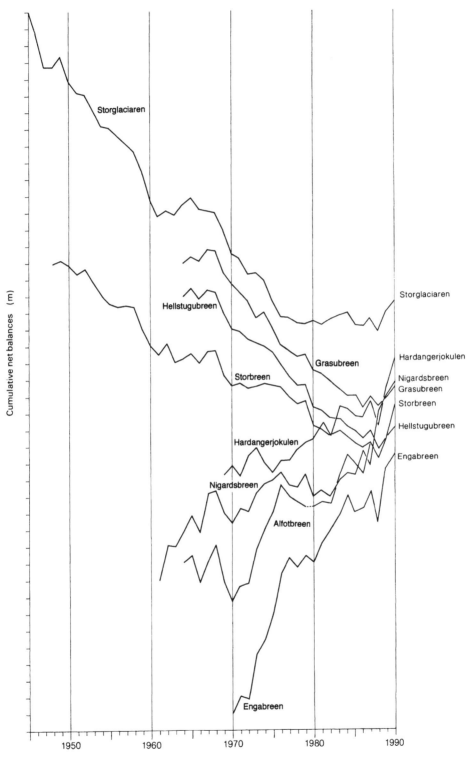

Cumulative net balances (m)

Storglaciaren

Hellstugubreen

Storbreen

Grasubreen

Hardangerjokulen

Nigardsbreen

Alfotbreen

Engabreen

Storglaciaren

Hardangerjokulen

Nigardsbreen
Grasubreen

Storbreen

Hellstugubreen

Engabreen

1950   1960   1970   1980   1990

Figure 3.16 Accumulated mass balances of Scandinavian glaciers. The glaciers in the interior, Storglaciären in northern Sweden and Gråsubreen, Hellstugubreen and Storbreen in Jotunheimen, have been diminishing in volume, while the more maritime glaciers, Hardangerjøkulen, Nigardsbreen and Ålfotbreen in southern Norway and Engabreen in northern Norway, greatly increased their volumes in the early 1990s (*Sources*: from Kasser 1967, 1973, Müller 1977, Haeberli 1985, Kjøllmoen 1999)

The differing cumulative mass balances of Nigardsbreen and Storbreen between 1960 and 1987 typify the contrasting behaviour and economies of the maritime glaciers of Norway on the one hand and the more continental ice masses on the other (Figures 3.15 and 3.16) (Haakensen 1982, 1984, 1995). Between 1973–4 and 1992–3 the mass balance signs of Storbreen and Nigardsbreen differed in 9 of the 19 years. After 1987 the volumes of both glaciers increased, with winter accumulation above average, and summer ablation below average. It appears that the contrasting states of the maritime and continental glaciers of southern Norway are attributable to differences in precipitation near to and further from the coast.

3.2.2 Folgefonni

Somewhat more historical information comes from Folgefonni, the third largest sheet of ice in Norway, on the peninsula between Åkrafjorden, Kvinnheradsfjorden and Sørfjorden (Figure 3.1). Two glaciers, Buarbreen in the east and Bondhusbreen in the west, flow from it down towards the coast and populated valleys. Most information therefore relates to these two tongues.

Hoel and Werenskiold (1962), in the acknowledgement section of their book, refer to a 1677 document dealing with an advance of Buarbreen 'in the period immediately preceding that year', but it has not proved possible to locate such a manuscript in Statsarkivet in Bergen. A court of inquiry held in 1677 found serious damage to Buar farm: 'a cruel mountain had fallen on the pasture . . . and the best part of it taken away, besides which the river had broken out of its course and carried away all the arable' (*Tingbok for Hardanger Fogderi*, 1677, fos. 14b and 15). The report is reconstituted but the missing parts are small and the original entry evidently quite brief. There is no reference to Buarbreen, nor is there any other reference to Buar in 1677 to be found in the *Tingbok* (A. Battagel, personal communication).

All the *avtak* reports from the period 1665 to 1815 for the area surrounding Folgefonni have been examined (*Rentekammeret, Affældnings Forretninger, Hardanger og Sunnhords Fogderier*, 1702–84). Many of the farms in the surrounding zone were damaged in the late seventeenth and the first half of the eighteenth centuries, as were those around Jostedalsbreen (see

Figure 3.17). It is surprising to find that there is no record of Buarbreen having caused direct damage to Buar farm, although a court was held at Buar in 1744 because fields had been destroyed there by the great storms of December in the preceding year.

Øyen (1899) collected together many references to Folgefonni in the literature. He cited Hertzberg, a famous dean in Hardanger, to the effect that

> an old man in Strandebarm parish whose farm is so situated that Folgefonni can be seen from it, rising above a lower mountain has told me that his father said and he himself had noticed, that the glacier had increased in height for in his youth only the upper part of the glacier could be seen from the farm's buildings but now the same part of the glacier could be seen from a good way below the buildings.
> (Hertzberg 1817–18: 720, quoted Øyen 1899: 167)

The reference here was probably to Bondhusbreen and the farm Bondhus farm (see Figure 3.17). It is unclear whether 'in his youth', refers to the old man or to his father, but the period of expansion in question would not seem to have been later than 1750 to 1760.

Bondhusbreen was advancing again in the early nineteenth century, for Hertzberg reported that 'when I was up there in 1807 and took a look at this glacier, I saw clearly that a mass of ice had advanced, for a bank or collection of large and small rocks lay below the glacier's lowest edge, it being driven forward by the ice' (Hertzberg 1817–18: 720, quoted Øyen 1899: 170). Like Nigardsbreen, Bondhusbreen was retreating before the middle of the nineteenth century, for in 1845 Konow noted a 'significant retreat' (Øyen 1899: 176) and Bing reported that

> Folgefonni has thawed significantly during the last 25 to 30 years. . . . This applies particularly to the southern part above Mauranger and Odda and also to the central part, where rock now appears which formerly was always covered by the snowfield . . . whether the glacier tongues have retreated or advanced during the same period, the people here cannot say definitely.
> (Bing 1895, quoted Øyen 1899: 183)

The Folgefonni tongues advanced in the later part of the nineteenth century. According to Yngvar Nielsen, Buarbreen 'advanced more than 80 m in 1870, and 4 m in one week in 1871, but has now retreated 30 to 40 m'. In 1872 it advanced '70 yards' (Wilson 1872), while in 1874 it 'lay above the greensward and cows

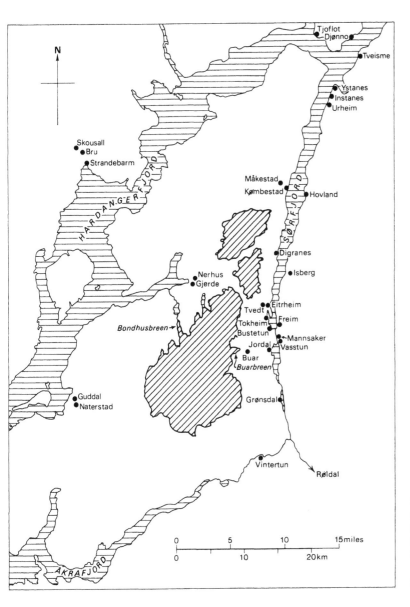

Figure 3.17 Folgefonni: farms in the vicinity seriously damaged by landslides, avalanching or flooding between 1663 and 1815 (*Sources*: drawn from Norges Geografiske Oppmåling 1315.III and 1314.IV. Farm data from *avtakforretninger* reports)

were grazing close beneath it' (Nielsen, quoted Øyen 1899: 188). By 1877, Sexe reported that 'since 1832 Buarbreen has advanced significantly some 2000 paces [*c.*1700 m] and has covered a not inconsiderable stretch of cattle and sheep pastures' (Sexe, quoted Øyen 1899: 189). The front was still advancing in 1878 when Holmstrøm 'measured the distance from the front to Buar farm and found it to be about 947 m'. (The 1974 NGO 1:50,000 map, sheet no. 1315/III shows this

distance to be 1350 m.) The following year, 1879, Nielsen visited the glacier again for the fourth time, twenty years after his first visit and five years after his third.

I thus had a good opportunity to judge its advance, which had never been so apparent as now. The glacier had ploughed up the fresh turf and had rolled it in front of it up onto a small moraine which it continues to push forward. Turf, rock and grit and uprooted trees formed a moraine wall which was certainly remarkable while also being most unpleasant in a way, showing as it did the destructive elements of natural forces. Never had I seen Buarbreen like this. In 1874 it lay above greensward and cows were grazing close beneath it. It did not then appear to be moving, and it was not then easy clearly to imagine its rapid advance. Now, in 1879, the first glance was enough to show that powerful forces were at work here, and the first question I had to ask myself was, how long it would take before the ice tumbled into Buar farm's Innmark, the buildings lying only ten minutes' walk from the present terminal moraine. The glacier would soon reach a projecting rock-spur, which would presumably offer some resistance. It would either have to surmount it or take a course down through the narrow river bed before spreading out again over Buar's hayfields. Assuming it maintains its present rate of advance, this might reasonably take three years, a period which will be very interesting.

(Quoted Øyen 1899: 90)

Bondhusbreen was also advancing in 1879, though Nielsen wrote that 'this advance appears to be correspondingly slight as compared with the rate of Buarbreen'. He also got the impression that the icecap itself was simultaneously thinning, because new areas of exposed rock were appearing, and in 1897 he was told that the route across Folgefonni from Sørfjorden down to Gjerde in Mauranger was as good as over unbroken rock, as it still was in the 1980s.

Of Mysevatn, Yngvar Nielsen reported in 1879, 'the glacier is said once to have gone down into the lake, but has now retreated further from its bank', while Bing in 1896 recorded that Mysevatnbreen was by then 100 m from the lake, while fifty or sixty years earlier it had stretched so far into the water that it was not evident where the end of the lake was. In view of this it is not entirely surprising that Bondhusbreen was retreating in 1892 and had been in overall retreat since 1882. Buarbreen was still advancing in 1892–3 but retreated steadily in 1894, 'perhaps most in the last two years; it has retreated c.50 m and its height is much less. The same in Mauranger, Gjerde and Bondhusdalen' (Øyen 1899: 199). Richter visited the area in 1895 and formed the view that 'there cannot have been a large-scale advance for centuries as there is old vegetation to be found immediately behind the new moraine'. When he saw it, Bondhusbreen 'had a new terminal moraine 5 or 6 m in height not more than 50 m from the front'. In 1897 the glacier was still not as small as some of the older people remembered it to be. Bondhusbreen

continued to retreat slowly until 1899 (Øyen 1899: 217–18).

The late nineteenth-century advances of the Folgefonni tongues appear to have brought them close to the position of maximum Little Ice Age extension or even to have equalled it. Those of the 1870s were apparently in phase with those of the Jostedalsbreen tongues, and measurements of the positions of Buarbreen and Bondhusbreen in the twentieth century (e.g. Fægri 1948: 303 and Liestøl 1963: 188) show fluctuations paralleling those of the Jostedalsbreen tongues, with small advances in 1905 and 1910 and more marked advances in the 1920s, followed by much swifter retreat. This slowed down and was replaced by small advances at times until 1957.

The fluctuations of Blomsterskardbreen, a remote outlet of the southern part of Folgefonni, merit particular attention, although nothing is known about this glacier before the time when it was visited and photographed by Rekstad in August 1904. A photograph taken from the same vantage point in August 1971 showed it 200 to 250 m forward of its 1904 position (Tvede and Liestøl 1977). Examination of available maps, sketches and photographs reveals that the advance took place between 1920 and 1940, and probably mostly within the 1930s. Air photographs taken in 1959 and 1976 show only insignificant changes, with retreat of less than 50 m.

The southern part of Folgefonni receives some of the heaviest precipitation in Norway (Tvede and Liestøl 1977). The mean annual value is calculated to be between 5000 and 5500 mm on the higher parts of Blomsterskardbreen and between 3500 and 4000 mm on the northern part of the icecap feeding Buarbreen and Bondhusbreen. Not only is the precipitation greater on the southern névéfield but Blomsterskardbreen is longer and less steep than the northern outlets. Tvede and Liestøl argue that it is to be expected that Blomsterskardbreen would have a longer response time. It is therefore reasonable to categorize the advanced position reached in 1940 as a response to the strong positive net balances around 1920 which, as we have seen, caused noticeable advances of some of the Jostedalsbreen tongues. However, Blomsterskardbreen, according to Tvede and Liestøl (1977: 321), was 'the only glacier in Scandinavia where a net advance has been documented within the last 70 years'.

3.2.3 Northern Scandinavia

Glaciers and icecaps in northern Norway and Sweden are generally remote and far from farming settlements, and consequently very few records are to be found concerning their positions at times earlier than the late nineteenth century. Evidence for earlier glacial events depends almost entirely on radiocarbon and lichenometric dating (Karlén 1982).

Svartisen (Figure 3.1), the second largest ice sheet in the country, lies on the Arctic Circle and is divided by Vesterdalen,[17] running north and south, into two sections, Vestre Svartisen covering about 220 km² with forty outlet glaciers, and Østre Svartisen, with an area of about 145 km² and twenty outlet tongues.

Karlén (1979) extracted a sample consisting of 70 per cent wood fragments from the top 2 cm of a peat deposit beneath the outermost moraine of Fingerbreen, one of the eastern tongues (Figure 3.18). This gave a date of 695 ± 75 (I-10364) BP, which suggests that the moraine was deposited soon after AD 1260–1390 (see Figure 3.5). Another date of 600 ± 100 (St-6757) was obtained from the upper 2 cm of a peat layer under the forest bed of a delta formed in a lake dammed by the expansion of one of the tongues of the western section of Svartisen into Glomdalen. This sample, calibrating to AD 1290–1430, was considered to predate the glacier expansion closely. However, as Karlén (1979) pointed out, peats may be deposited over a long period of time. He also showed that considerable disparities may occur between dates within the top few centimetres of a single peat layer, and noted that the uppermost section of a peat layer could be missing. In view of these points, maximum radiocarbon dates obtained from peat layers cannot be expected to indicate closely the age of the ice advance succeeding them. His two ¹⁴C dates from Svartisen are compatible with thirteenth- or fourteenth-century advances, but they cannot be used as proof of such advances.

Karlén (1973, 1979) and Karlén and Denton (1976) mapped and dated numerous sets of moraines in the Sarek and Kebnekaise mountains of Sweden between 67° and 68°N, as well as those in the Svartisen area of Norway, and dated them by radiocarbon analysis of soils and by lichenometry. Measurements made on both *Rhizocarpon geographicum* and *Rhizocarpon alpicola* thalli were consistent, with moraines known to be older on a basis of geomorphological mapping bearing larger thalli. Maximum lichen diameters on control surfaces of known age in the Sarek area agreed closely with those of similar age in Kebnekaise. Thus in both regions surfaces dating from 1900 to 1916 bore lichens 21 to 27 mm in diameter, while lichens on surfaces from the seventeenth century gave maximum values of 66 to 85 mm. Only a few surfaces of known age were available in Sarek National Park and so, in view of the close agreement found with Kebnekaise, a lichen growth curve was constructed using twenty-one control points drawn from both Kebnekaise and Sarek. Twelve of these datable surfaces were twentieth century, mostly exposed at times determined from travellers' accounts and photographs. The older control points were provided by copper and silver mine tips from the seventeenth and eighteenth centuries. Most of them were datable only to within a certain span of years and two of them were ambiguous. One of the copper tips was formed between 1884 and 1902 and another either between 1745 and 1751 or between 1699 and 1702.

The growth curve used to date the moraines around Svartisen and the little glaciers of the Okstindan and Salttjellet areas was based on the ages of fourteen surfaces, the great majority late nineteenth-century or early twentieth-century mining tips (Figure 3.19). Only one pre-nineteenth-century control point could be found: a silver mine tip at Nasa, in operation in the seventeenth century. Karlén (1979) gathered lichenometric data from about 125 moraines and concluded that the glaciers in northern Norway had reached advanced positions in the early 1300s and had subsequently retreated. He identified further periods of glacier expansion towards the end of the sixteenth and in the mid and late seventeenth centuries. 'These results also permit relatively good dating of several of the youngest general periods of glacial maxima. These occurred at AD 1780, *c.*1800, 1810–1820, 1860, *c.*1880, 1900–1910, and then *c.*1930.' These results compare closely with those from Kebnekaise and Sarek, where individual Little Ice Age advances culminated about 1590–1620,

---

17   Vesterdalen is also known as Glomdalen.

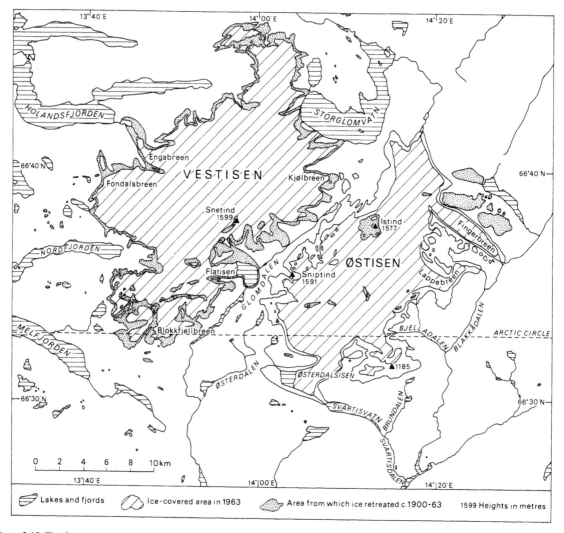

Figure 3.18 The Svartisen area, northern Norway, showing changes of position of the ice margin during the twentieth century
(*Source:* after Theakstone 1988)

1650, 1680, 1700–20, 1780, 1800–10, 1850–60, 1880–90 and 1916–20 (Karlén and Denton 1976).

Karlén's field investigations certainly demonstrated the complex nature of the Little Ice Age in northern Scandinavia. Early Little Ice Age advances were indicated, but the earlier part of the dating curve must be considered the less precise. The dates of many of the minor episodes of glacial expansion identified correspond to those in the Alps (e.g. Ladurie 1971, Messerli *et al.* 1978, Bray 1982). However, lichenometric dating is insufficiently precise to differentiate minor epi-

sodes in the last few centuries unless the lichen growth curves employed have very narrow error limits, and it does not seem that the error limits for the curves for northern Sweden and Norway are sufficiently narrow. Innes (1982, 1983) has pointed to differences in the growth rates of *Rhizocarpon alpicola* and species within the *geographicum* group which result in dating errors if these groups are not treated separately. His work indicated the need for verification of the curves for northern Scandinavia. Innes (1985) subsequently derived an independent growth curve for the area immediately

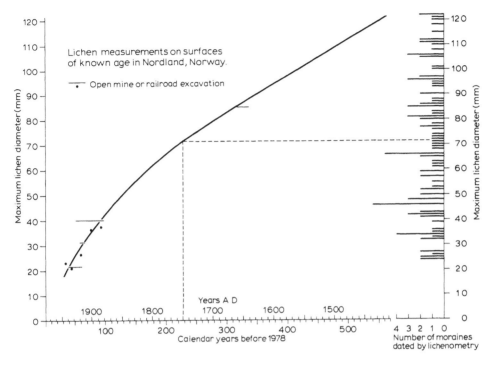

Figure 3.19 Lichen growth curve for the Svartisen area (*Source*: from Karlén 1979)

south of the Svartisen icecap and applied it to some of the moraines of Blockfjellbreen (a southern outlet of the western Svartisen icecap) which had also been dated by Karlén. This curve was primarily based on measurements of *Rhizocarpon* section *Rhizocarpon* lichens on acidic igneous and basic igneous gravestones in two cemeteries in Dunderdalen (between 45 m and 68 m above sea level) and upon a railway embankment built in 1934 at 130 m above sea level. It indicates that the lichens grew faster than Karlén had supposed; this is confirmed by photographs taken in 1910 which show Blockfjellbreen extending into an area which, according to Karlén, had been occupied by ice as long ago as 1780–1830. Innes attributes the discrepancies between the curves to differences in sampling methods and in particular Karlén's reliance on mining tips and ^{14}C dated surfaces. The tips are often the products of intermittent mining activity and therefore difficult to date precisely. Furthermore, as Innes shows, lichen growth rates differ from mica schist to granite-gneiss and may be affected by the metal in ores. The activity of carbon in soils is known to vary with depth (Matthews 1980,

1981, 1984), uppermost horizons are liable to be destroyed, and laboratory error ranges are quite wide. Consequently it seems that Karlén's dating of moraine sequences, though a considerable advance on what had been done before in northern Scandinavia, was not sufficiently accurate to allow it to be used as a basis for comparing Little Ice Age sequences there with other regions.

Of all the glaciers in northern Norway, most information comes from Svartisen. It is clear from the 1743 Matrikkel that Engabreen was by then already far advanced. According to Petter Dass (1647–1707), the glacier had reached the shore fifty years earlier (cited Liestøl 1979b). The 1723 draft Matrikkel lists three small farms in the Svartisen area, Funøren, Fundahlen and Storsteenøren, two of which had their taxes reduced. Storsteenøren (Figure 3.20) was described as being damaged day by day by the river and the glacier and had been deserted, the farm buildings having been carried away by the ice. Fundahlen, with frostbound soil, 'is damaged year by year by the glacier'. According to Øyen (1899), there was still a tale amongst the

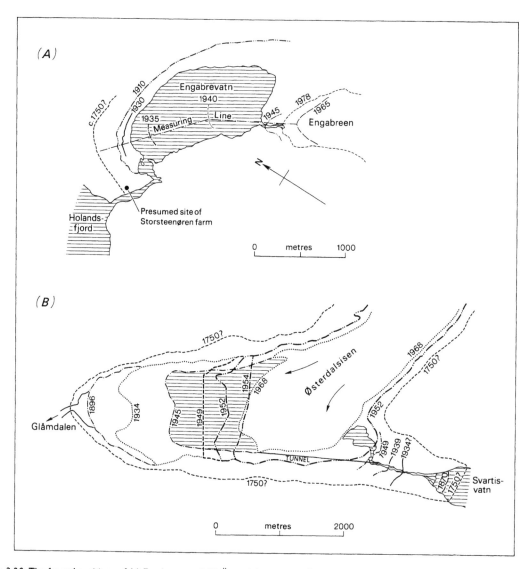

Figure 3.20 The frontal positions of (a) Engabreen and (b) Österdalsisen since the mid eighteenth century (*Source*: from Liestøl 1979b)

people at Rødøya and Meløya that 'when the corn was about to be cut the ice came so close to the buildings that the inhabitants had to leave them and shortly afterwards the glacier buried everything'. The *landskyld* which had amounted to 31 *laup*, 1 *pund* for seventeen farms in the district in 1667 was reduced by 6 *laup*, 2 *pund* and 16 *merker* in total on the 1723 draft Matrikkel, presumably reflecting the deterioration of climate that had taken place. The rolls of 1740 and 1745 listed Storsteenøren as still derelict.

Theakstone (1965, 1990), summarizing available data on the Svartisen glaciers and their fluctuations, found little evidence for the period 1723 to 1800. Subsequent changes he documented in a glacier atlas (Theakstone 1988) from which the following data have been taken. In 1800 Engabreen was so close to the sea that it was reached by the water at flood tides (Rekstad 1893). By 1800 slow retreat was under way and the ice was a hundred feet from the outermost moraine. In 1865 Geikie found it ending in a small lake with a plain of shingle and alluvium in front of it. When Østerdalsisen, the southernmost of the Svartisen outlets, was visited by de Seue in 1873 it had advanced over the previous three years, but Rabot in 1881 found it had retreated to its 1870 position again, while Engabreen was now a kilometre from the fjord. In 1891 Østerdalsisen still reached the edge of Svartisvatnet, as it had in 1873, but no longer carved into it as it had then, and the glacier surface was 15 m lower than it had been twenty years earlier. Fingerbreen, on the east of Østisen, was also in slow retreat. Rekstad examined much of Svartisen in 1890 and 1891 and found widespread evidence of recent retreat (Rekstad 1891–2). This diminution continued through the next decade. Kaiser Wilhelm II in 1898–9 measured a retreat of Engabreen of 60–80 m; Rabot was told by a local farmer in 1898 that the glacier had retreated continuously for the last fifteen years.

Some small advances occurred in the early twentieth century. Engabreen began to advance in 1903. In 1909 Rabot found Engabreen had advanced 100 m and Fonndalsbreen about 60 m between 1907 and 1909. The expansion was short-lived. When Marstrander made extensive observations of the Svartisen glaciers in 1910, Østerdalsisen was thinning rapidly and local people told him that it had retreated 300 to 400 m in the previous twenty years. Flatisen had also retreated considerably since Rekstad's visit in 1890 and Fingerbreen's tongue was only half as wide as it had been when noted by Rabot in 1882. Marstrander reported signs of a new advance in a number of small, steeply sloping glaciers but found all the major trunk glaciers unaffected by it.

Annual measurements of the fluctuations of Engabreen and Fondalsbreen were made from 1909 to 1943 and published in the *Bergens Museums Årbok* (see Figure 3.21). Although minor advances occurred in the period up to 1930, retreat predominated and after 1930 accelerated as it did in the Jostedalsbreen tongues. By 1950 the lower part of Fondalsbreen was detached and Engabreen had retreated 400 m since it was last measured in 1943 and 2 km since 1909. Mean annual surface lowering of Østerdalsisen from 1960 to 1963 was about 7 m and this glacier, of all the Svartisen tongues, was suffering the most serious diminution. By 1965 Vestisen and Østisen were surrounded by a number of stagnating ice masses which had separated

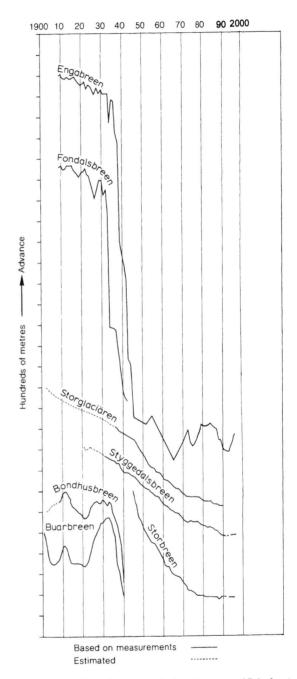

Figure 3.21 The frontal positions of selected tongues of Folgefonni (Buarbreen and Bondhusbreen) and Svartisen (Engabreen and Fondalsbreen), together with valley glaciers in Jotunheimen (Styggedalsbreen and Storbreen) and in the Kebnekaise massif of northern Sweden (Storglaciären) (*Sources*: data from Fægri 1933, Liestøl 1963, 1976, 1977a, b, 1978, 1979a, 1980, 1982a, b, Kasser 1967, 1973, Müller 1977)

off from the two icecaps. The period of rapid retreat was broken in the 1960s in maritime northern Norway, as it was further south near the coast. In 1965 Engabreen began to advance once more and by 1984 it was within 180 m of Engabrevatnet. Positive mass balances were recorded in all but five of the years between 1970 and 1998 (Haakensen 1995, Kjøllmoen 1999). Elvehøy and Haakenson (1992) note that 'Engabreen made a short advance around 1980' and 'the current retreat began in 1986'. Their map (p. 70) shows a retreat of about 100 m between 1980 and 1991. However Haakensen (1995) reported that in 1992 the snout was advancing and was only 30 m above sea level. The forward movement continued until 1998/9 when the glacier pushed up a moraine ridge indicating that it was more advanced than observed for a very long time (Kjøllmoen 1999). The effects on the glacier system of diverting the subglacial river in 1993 for generating electricity have yet to be determined.[18]

The contrast between the increases in volume registered for Engabreen and the contemporary decreases registered until the late 1980s for Storsteinsfjellbreen further east in Norway and Storglaciären in Sweden echoes that between maritime and continental glaciers in southern Norway (Figure 3.16). The mass balance history of Engabreen since observations began in 1969, though not identical with that of Nigardsbreen, is basically similar. The eastern glaciers of Svartisen receive less snowfall and have higher mean equilibrium line altitudes than those on the western side; consequently negative annual net balances occur more frequently and glacier histories have more in common with those in northern Sweden than those in western Svartisen.

The first compilation dealing with Swedish glaciers, by Hamberg (1910), included information on frontal positions, velocity measurements, maps and photographs. Hamberg himself made observations till 1918. He was concerned particularly with Mikkaglaciären, where he made ablation and velocity measurements. In the 1920s and 1930s Hans Wilson Ahlmann (1953) investigated the climatic causes of the contemporary

thinning of glaciers in northern Scandinavia and adjacent regions. His student, Valter Schytt, initiated mass balance observations on Storglaciären, a small valley glacier in the Kebnekaise Mountains, as early as 1945–6.[19] They still continue and form the longest series recorded anywhere in the world (Holmlund *et al.* 1996). Similar observations are made on five other Swedish glaciers of various degrees of continentality, and contrasting in type, exposure and size. Holmlund (1993, 1995) has reported on twentieth-century glacier fluctuations in Sweden and the history of their investigation.

The 'glaciation level', the minimum altitude required to maintain a glacier, has a very marked east–west gradient, being at about 1400 m at the Norwegian–Swedish border and about 2000 m in the continental climate of the eastern Sarek mountains (Enquist 1916). Many of the Swedish glaciers are flat and relatively cold, with their fronts frozen to the ground and inactive. The time they take to respond to any change in climate change is much longer than for glaciers in the Alps or maritime Norway. Most reached a position close to their Holocene maximum around 1900 and then retreated in the twentieth century in response to an increase in summer temperature after 1910 of 1 °C or more. It was as much as two decades after warming began before some of the most continental glaciers, such as Rabots Glaciär, began to retreat. Observations of the frontal position of 20 glaciers were initiated during the 1960s. Some, including Vartasjekna, retreated between 1910 and 1965 at an average of 10–20 m per year. In spite of mild winters in the period 1989–94, positive net balances were recorded on most, but not all, Swedish glaciers on account of positive precipitation departures from the mean. These were greater in the west than the east. In the early 1990s withdrawal slowed down: over half the tongues observed were stationary and a small minority began to advance; for example Isafallsglaciären in the Kebnekaise massif. With a more maritime climate than Storglaciären and a steep ablation area, its response time is only 5–10 years. Most of

---

18 A rock tunnel beneath Engabreen, a by-product of the diversion, gives access to the glacier bed via the Svartisen Subglacial Observatory.

19 See 1:10,000 map of Tarfala basin, edited by Per Holmlund, 1992, Department of Physical Geography, University of Stockholm, S-106 9-1.

the other glaciers in the area have stable fronts or are retreating very slowly. A few have recession rates which have increased for reasons other than climate change, including the effects of calving into lakes, or thinning causing frontal sections to be cut off from the moving ice.

The rise in summer temperature after 1910 is generally thought to have been the dominant factor causing continental glaciers such as Storglaciären in the north and Gråsubreen and Hellstugubreen in the south to retreat this century (Holmlund 1987). In more maritime areas, although variations in temperature have been almost the same as in the interior, increased accumulation has played the decisive role. This view has been disputed by Raper and others (1996). Using a simple geometric model, calibrated with the long tree-ring-based temperature reconstruction for northern Fennoscandia (Briffa *et al.* 1992), and measured mass balance data (1946–92), they reconstructed the volume of Storglaciären back to AD 500 and concluded that the prime cause of the decrease in its volume, at least between 1946 and 1992, was reduced accumulation rather than a rise in temperature. Their reconstructed volume changes agreed well with geomorphological evidence, and they noted that 'where differences exist, deductions can be made about past accumulation'. The reconstructed volume reached a maximum in the cold period from 1580 to 1740.

The meteorological reasons for the negative mass balances before 1980 and more positive ones in the 1980s and early 1990s have been considered by Pohjola and Rogers (1997). They point out that the mass balance variations of Storglaciären and Storbreen in the latter half of the twentieth century correlate well with the strength of maritime airflows. Stronger westerlies in winter and cooler summers in the 1980s and early 1990s accompanied a prolonged highly positive North Atlantic Oscillation Index. The increase in the Index was associated with storm tracks in winter taking a more northerly course, giving greater snowfall over western and northern Scandinavia. In summer, variations in mass balances seem to have been somewhat more closely associated with a Norwegian Sea Index, a measure of the difference in pressure between Valentia in SW Ireland and Vardø in NW Norway. With lower pressure over the Barents Sea, northern Scandinavia in summer was more frequently under the influence of cool, westerly air from the Atlantic, giving cloudy skies and reduced ablation losses.

Studies made on Storsteinfjellbreen, 45 km WNW of Storglaciären, between 1964 and 1968, and by Kjøllmoen and Østrem (1997) between 1991 and 1995, show that the mass balance records of the two glaciers are similar. Their positive and negative years were synchronous except for 1968 when heavy summer snow fell on Storsteinfjellbreen but not on Storglaciären. Between 1960 and 1993, while the surface of the Norwegian glacier dropped more than 60 m at the tongue, its thickness above the equilibrium line increased by up to 20 m. Similarly, the accumulation area of Storglaciären thickened by 5–10 m between 1980 and 1993, largely because of increased winter accumulation, especially between 1989 and 1993. In 1995/6 the North Atlantic Oscillation Index weakened. Accumulation on Storglaciären was below average, and both winter and summer were warmer than average, with the result that the net balance for the year was negative. The winter of the following year was snowy but the greater accumulation was offset by a warm summer (Haeberli *et al.* 1999).

Laumann and Reeh (1993), using a simple degree-day model, calibrated by measured mass balances from Ålfotbreen, Nigardsbreen and Hellstugubreen, and thirty-year normals of temperature and precipitation from nearby meteorological stations, concluded that low-lying maritime glaciers near the Atlantic coast experiencing high precipitation are more sensitive to changes in both temperature and precipitation than glaciers at high elevations further from the coast. All the glaciers they studied would lose mass in a warmer climate unless precipitation were to increase dramatically by something like 25–40 per cent for each degree rise in temperature.

The reversal in the past few decades of the climatic amelioration that characterized the early decades of the twentieth century is suspected to have resulted in greater climatic stress for birch and pine which had extended their ranges uphill, notably, according to Kullman (1997), in the Scandes Mountains in Sweden. More snow has been accumulating, and short-term extreme events such as severe frosts are said to have become more frequent.

3.3 A SUMMARY OF THE LITTLE ICE AGE IN SCANDINAVIA

Though little direct evidence of moraine formation in the thirteenth and fourteenth centuries has been found in Scandinavia, the documentary evidence from the parishes north of Jostadalsbreen is difficult to explain except by climatic deterioration, probably associated with glacier expansion. However, no direct evidence has yet been found of expansion of Scandinavian glaciers before the seventeenth century. The enlarged state of glaciers and ice sheets in the eighteenth century was common to both north and south, as was the great recession of the twentieth century. The great majority of the Scandinavian glaciers about which we have information reached their maximum size in the mid eighteenth century. Matthews (1976) suggested that the equilibrium line altitude of Jostedalsbreen was then 140–5 m lower than now, in response to a temperature depression of the order of 1 °C. (Temperature depression of this order is in accord with the view of Jones (1990) that mean global temperature has varied no more than 1 °C during the last millennium.) Modern and Little Ice Age equilibrium line altitudes for twenty of the Jostedalsbreen outlets have been calculated by Torsnes *et al.* (1993). The mean depression found was 70 m, with values ranging between 25 m at Brigsdalsbreen and 30 m at Nigardsbreen, to 265 m at Skålbreen, depending on the relative sizes of accumulation and ablation areas, themselves closely related to topography.[20] The importance of topographic control on equilibrium altitude depression was also revealed by estimates made for outlets of Hardangerjøkulen (Nesje and Dahl 1991) and Grovabreen (Aa 1996).

Small frontal oscillations were observed in the late nineteenth century. Rekstad recorded an advance of Brigsdalsbreen between 1869 and 1872, while de Seue found Bøyabreen and Lodalsbreen advancing in 1869. When he visited northern Norway in 1873 he found Østerdalsisen larger than it had been in 1870. The small advance and subsequent stationary state of Engabreen in the first three decades of the twentieth century co-incided with the time when many of the Jostedalsbreen outlets advanced or halted. Superimposed on the generally coherent pattern are variations caused by differing degrees of exposure to oceanic influences. Mass balance studies carried out in recent decades have shown that the extent of winter accumulation has dominated the total change in volume of ice masses such as Folgefonni, while summer temperature has usually been identified as the dominant control for those situated further inland (Haakensen and Wold 1981). It is significant that the most deviant behaviour of any of the glaciers has been that of Folgefonni, with some of the most oceanic of the Norwegian glaciers, notably Buarbreen and Bondhusbreen advancing almost as far in the late nineteenth as in the mid eighteenth century. Blomsterskardbreen is apparently the only known glacier in either Iceland or Norway to have reached its maximum extent in the twentieth century. Figure 3.16 shows the cumulative mass balances of a selection of glaciers and demonstrates very clearly the contrast between the decreasing volumes of the more continental ice masses in both Norway and Sweden and the increasing volumes of those in more oceanic situations near the Norwegian coast up to the 1980s. The situation changed in the late 1980s and early 1990s as mass balances of many eastern glaciers increased in Sweden as well as in continental Norway. This increase was caused by heavier winter precipitation, possibly in association with the combined effects of a powerful El Niño and the eruption of Pinatubo.

The accordance of major features of the history of Scandinavian and Icelandic glaciers is apparent. It is particularly noteworthy that the eastern tongues of Vatnajökull in Iceland not affected by surging reached advanced positions, even maxima, in the late nineteenth century at the same time as the glaciers of Folgefonni in Norway: both depend on very heavy precipitation. Blomsterskardbreen in 1980 was something of an anomaly, being near its maximum position; its 1930s expansion, it might be noted, overlapped with an anomalous, extraordinarily rapid advance of the Leirufjördur tongue of Drangajökull in northwest Iceland that has recently been attributed to surging.

---

20   Two approaches were employed. The first depended on accumulation area ratios of the glaciers (see Chapter 8, p. 225), the second method (Kuhle 1988) depended on median elevations and toe-to-headwall altitude ratios. It was concluded that the accumulation area ratio method appeared to give the most reliable results.

4

THE MONT BLANC MASSIF

The snowcap of Mont Blanc is the highest point in Europe west of the Caucasus, the summit of a mountain massif that extends 50 km from Martigny in the northeast to St Gervais in the southwest, separating France and Italy and protruding into the Swiss canton of Valais (Figure 4.1). A part of the Hercynian system, it consists of sedimentary rocks metamorphosed by granitic magmas. The central granites give the near vertical slopes of the Dru, Grandes Jorasses and Aiguille du Géant and are capped at the highest levels by crystalline schists which outcrop in the Grands Mulets and near the summit of Mont Blanc itself.

The massif is delimited by fault-guided valleys. On the French side, the upper Arve flowing down the deep trough of Chamonix is overlooked from the southeast by the Tour, Argentière, Mer de Glace and Les Bossons glaciers. From the main watershed on the Italian side of the massif, precipices on a Himalayan scale drop down to Val Ferret and Val Veni. Meltwater streams of the Brenva, Miage, L'Allée Blanche and a series of smaller glaciers join the Dorea Baltea which escapes between Entrèves and Courmayeur to the Val d'Aosta and the Po.

Although evidence of the Little Ice Age fluctuations of the Mont Blanc glaciers is unusually varied and abundant, cartographic coverage was poor until the nineteenth century. As late as 1863, Adams-Reilly (1864) found it 'strange that the chain of Mont Blanc should be the most visited and at the same time the worst mapped portion of the Alps'. Since then, a wealth of good maps has appeared.[1]

The glaciers have been the subject of over a hundred papers as well as several books (Bordier 1773, Bourrit 1773, 1776a, b, 1787, Viollet-le-Duc 1876) and monographs (Marengo 1881, Mougin and Bernard 1922, Vallot 1900). Some give valuable systematic accounts of the oscillations of particular glaciers for various periods (Rabot 1902, Mougin 1925, Bouverot 1958) or even of all the glaciers of a particular national sector for the whole of the Little Ice Age (Sacco 1918, Mougin and Bernard 1922, Vallot 1900). By far the most complete account of the behaviour of the Mont Blanc glaciers as they are portrayed in original archival sources is by Le Roy Ladurie (1971). Vivian's (1975) résumé of what was known about the glaciers in the French sector includes an extremely detailed discussion of changes of volume, area, length and flow over

---

1 Martel (1744) gave a sketch of the Arve area on which the glaciers of Chamonix are at least represented. That the map is rudimentary is hardly surprising as it was the product of three hours spent after dinner at Montenvers 'which time I was employed in making a plan of the glaciers'. Favre (1867, vol. 1) lists seventy-one maps earlier than that of Martel, showing the Chamonix valley if not necessarily its glaciers. Sacco (1918) provided a useful sketch of the Italian cartography of the area, while Forbes presented a more detailed and critical account of that of the whole massif (1900 edition). In 1922 Vallot published a paper on the evolution of the cartography of Savoie and Mont Blanc. A good coverage of the Mont Blanc area is provided by three 1:50,000 French sheets (Carte de France en 50,000, file XXXVI, 30, 31, 32) and the Institut Géographique National has produced a superb set of maps of the whole chain, based on air photography, on the 1:10,000 scale.

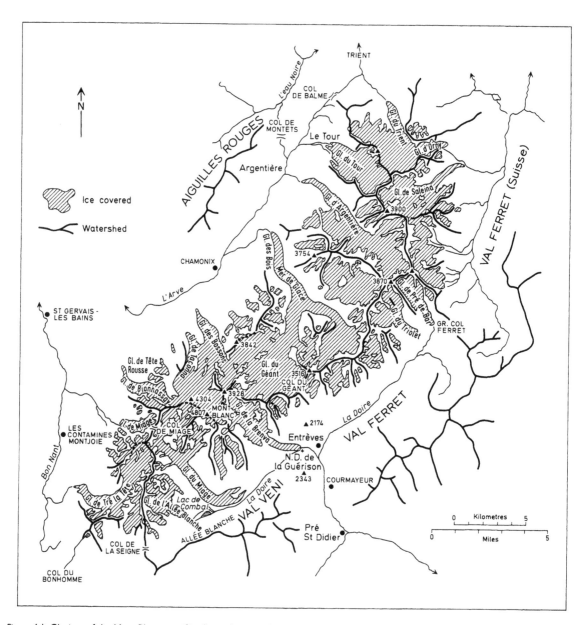

Figure 4.1 Glaciers of the Mont Blanc massif in the mid twentieth century

the period of scientific measurement since the late nineteenth century. Documentary evidence for the Italian versant is scantier. The fluctuations reconstructed in most detail are those of the Brenva by Orombelli and Porter (1982) who made use of maps, photographs and more particularly paintings and other illustrations, as well as documents and instrumental surveys.

4.1 THE GLACIERS OF MONT BLANC FROM THE FOURTEENTH TO THE SIXTEENTH CENTURIES

Some notion of the state of the glaciers during late medieval times may be gathered from local tradition and myth. For example, the story goes that the Brenva

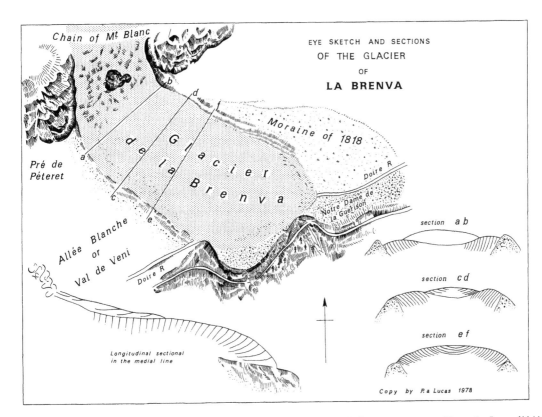

Figure 4.2 Sketch of the Brenva in 1742 by J. D. Forbes (1853). The village of St Jean de Pertuis was presumably on the floor of Val Veni, somewhere within the area covered by the lower part of the tongue, below the chapel of Notre Dame de la Guérison (*Source:* copied from Forbes 1900 edn)

did not always occupy the bottom of Val Veni as it does today; instead there were cultivated fields and meadows. But on 15 July of some year unknown, when the hay was dry and the weather fine, the villagers of St Jean de Pertuis failed to observe the Feast of St Margherita. Next day the glacier came down, engulfing the village and all its inhabitants (Forbes 1843: 207) (Figure 4.2). A tale mentioned by Virgilio (1883) gives another version of the catastrophe, according to which some old chalets 'de Pertus', on the slopes of Mont Noir de Péteret were carried away by the glacier.

Apparently there was once a village named after the martyr St Jean, who was killed by the Gauls in 'Pertu' during the reign of Emperor Maximian at the end of the third or beginning of the fourth century (Viollet-le-Duc 1915, cited Orombelli and Porter 1982: 17). Pertu was probably near the site of modern Pertud. Dollfus-Ausset (1867) records the existence of a manu-

script dating from 1300 which documents the existence of St Jean de Pertuis on the south side of Mont Blanc, on the valley bottom, below the modern chapel of Notre Dame de la Guérison. According to Dollfus-Ausset, this village was destroyed by a landslide or rockfall.

Orombelli and Porter, who had found that glacial advances in the main period of the Little Ice Age were associated with large rockfalls (Porter and Orombelli 1980, 1981, see Chapter 12), were prepared to accept that the Pertu event was associated with an advance of the Brenva into the Val Veni. Ladurie (1971: 221, 327) was unable to find the village in any Val d'Aosta tax list of the late Middle Ages, so the early fourteenth century would seem to be a likely date for its disappearance from the scene. There seems to be less justification for ascribing the village's destruction to the sixteenth century (Sacco 1918) or to around 1600 (Matthes 1942).

Radiocarbon dates for wood from the uppermost parts of the massive lateral moraines of the Brenva do not help to clarify matters very much. Orombelli and Porter (1982) pointed out that a sample with an age of 285 ± 60 (UW-464), when corrected for atmospheric variation in ^{14}C could indicate an advance between about AD 1520 and 1670. It is worth stressing that Orombelli and Porter also found a log only 6 m below the crest of the lateral moraine with a radiocarbon age of 1170 ± 55 (UW-465) which, when corrected according to Stuiver and Reimer (1993) at one standard deviation, gives a calendar age of AD 780 to 980. Thus the complex moraine system of the Brenva, which protrudes right across Val Veni at right angles, cannot be attributed to the oscillations of the terminus over the last few centuries but represents a long-term accumulation, part of which may predate the Holocene.

The moraines of the Miage glacier form a barrier across Val Veni upstream of the Brenva. From time to time a lake has formed behind them, the Lac de Combal. Rabot (1902) suggested that the waters of Combal overflowed at the same time as those of Rutour[2] and, uniting with them at Pré-St-Didier, caused the floods which destroyed the fortified houses of Rubbily and Rovary at Margex sometime before 1340. These events suggest the possibility of late medieval advances of the Miage and Brenva. On the other hand, no evidence has been found, as yet, of ice advances at this time on the other side of Mont Blanc, in the valley of the Arve.

Ladurie (1971: 155) provides indirect evidence that the glaciers of the Arve valley were advancing in the second half of the sixteenth century. In the Chamonix area, tithes of villages and hamlets well away from the glaciers, or protected from them, rose during the sixteenth and early seventeenth centuries. But the nominal value of the tithes of the settlements that were later destroyed or damaged by glacial advances remained generally stationary in the sixteenth century and especially in the latter part of it. This stagnation during a period of inflation suggests a decrease in real productivity. For example, at La Rosière, which first appeared in the accounts in 1390 and was eventually destroyed by the Argentière glacier, the tithes fell from 50 florins in 1577–1600 to 32 florins in 1622, whereas at Vallorcine at a much lower altitude the tithes were 92 florins in 1520, 160 in 1578–90, 300 in 1600 and 400 florins in 1622–5. In southern France generally, populations in the seventeenth century were higher than in the fourteenth and fifteenth centuries; in the Chamonix area between 1458 and 1680 they decreased by a half or more. At Le Châtelard, the number of taxpayers fell from forty in 1458 to eighteen by 1559. Environmental decline is also reflected in reductions in dues paid in respect of mountain pasture. The pasture of Blaitière, under the glacier of Blaitière, for instance, was paying 5 to 7 lbs of cheese to the priory of Chamonix between 1540 and 1580 but only 3 lbs between 1580 and 1602. Deliveries of cheese to the priory fell from 52 lbs in 1540, to 45–7 lbs in 1550–70 and to 39 lbs in 1622.

Not only were the Brenva and other large glaciers safely tucked away in their valleys in the early sixteenth century, but the passes were said to have been much easier. There was even a local tradition that the Col du Géant was once passable, or at least much more practicable than in 1788 when de Saussure made his famous observations there (1779–96, vol. IV).

Windham who, with his companions, was the first to describe a journey to the Chamonix valley in English (1744), wrote:

> Our guides assured us in 1741 that in the time of their fathers the Glacier [Mer de Glace] was but small, and that there was even a Passage through these Valleys, by which they could go into the Val d'Aosta in six hours, but that the Passage was then quite stopped up, and that it went on increasing every year.

This tradition was first recorded by a tax official called Arnod (1691) who tried vainly to force a way over to Chamonix in 1689, accompanied by three hunters 'avec des grappins aux pieds, des hachons et des crocs de fer à la main'. Arnod was making his attempt at least a hundred years too late, a circumstance which was not fully appreciated by Montagnier in his scholarly examination of the Col Major legend (1920–1, 1921–2).[3]

---

2   As in 1284 (H. Aeschlimann 1983, quoted Pelfini 1999).
3   Montagnier's exploration of the cartography and documentary sources of the Col Major legend led him to conclude that the tradition rested partly upon the identification of the 'Col Major' with the Col du Géant. Montagnier traced the cartographic vicissitudes of the Col Major in great detail and established that, despite the contrary opinion of the great

These stories certainly cannot be accepted at face value, especially as in the course of time they lost nothing in the telling; it was even sometimes said that the Chamoniards went over to Courmayeur to attend mass there! Why they should ever have done this when there was a priory in Chamonix is not clear. But the tales are symptomatic, and their essential basis is confirmed by an examination of seventeenth-century documents. A good start has been made (e.g. Röthlisberger 1974) but much research remains to be done on the early history of Alpine passes and cols, and the results would be of glaciological and climatic, as well as historical, interest.

The earliest useful account of the Mont Blanc glaciers occurs in a description of the valley of Chamonix by Bernard Combet, Archdeacon of Tarantais, who visited it in 1580 as arbiter in a tax dispute. He had, remarkably, sufficient interest in the local topography and scenery to record what he saw with some exactness:

> To the right, as you approach from the south, these mountains are white with lofty glaciers, which even spread through rifts in the mountains themselves, and descend almost to the said plain [of the Arve] in at least three places. One thing is clear: those rifts which people call moraines have sometimes caused unavoidable floods, both in the regions through which the waters descend and in the middle of the valley, where they swell the stream said to rise in the Alpages du Tour and which then forms into quite a large river [the Arve].
>
> (Translated from the original by Ladurie 1971: 135)

Despite its apparent simplicity, there has been some disagreement over the interpretation of this text. Rabot (1915) identified Combet's three glaciers as Bossons, Argentière and Tour, and concluded that the advances of the Little Ice Age had not yet begun in 1580, if these were the only glacier tongues which could be seen from the plain. Blanchard (1913) thought the Tour much too far from the other two glaciers to have been

intended, and also dismissed the Tacconaz because its tongue is always higher than that of the Bossons, its alimentation being more restricted. He therefore identified the third tongue as that of the Glacier des Bois, the terminal part of the Mer de Glace. Ladurie (1971) reached the same conclusion, pointing out that Combet specified his exact route, and that nowhere on it could he have seen the Tour glacier. He must therefore have been referring to the Argentière, Bossons and Mer de Glace. These days, the Glacier des Bois is concealed from view in a deep ravine, and if Combet saw it reaching 'almost to the plain' this would imply that the Mer de Glace was more extensive than at any time in the twentieth century and that the advance of the ice was already well under way by 1580.

4.2 LITTLE ICE AGE ADVANCES 1580–1645

Within twenty years of Combet's visit the ice invaded important tracts of cultivated land and forest in the valley of the Arve and additional areas were ruined by glacier torrents. Flooding seems to have increased in direct proportion to the advance of the ice, the plentiful meltstreams from the enlarged tongues being reinforced from time to time by bursting water pockets and outflows from glacial dams. Crops failed and yields became meagre with the deterioration of the climate. When the local people made a series of supplications for tax relief, the local conditions were officially investigated by both ecclesiastical and civil authorities. As a result, precise documentary evidence remains in the archives of Chamonix and Haute Savoie, from which the main pulsations of the advance can be dated, at least on the French side of the mountains. Both the extent and destructive character of the invasion, and the nature of the evidence upon which our knowledge of it is based, may be illustrated from documents concerning the Mer de Glace.

Alpine historian Coolidge (1908: 202), it could not be regarded as an exact geographical term synonymous with the Col du Géant. Montagnier concluded that the Col du Géant could not have been crossed within 75 or 100 years of the date of Arnod's attempt to reopen it. He thought that above the snowline much the same conditions must have been offered to travellers in the Middle Ages as in the twentieth century, and saw little reason for assuming that the glaciers had undergone any great change within the last thousand or more years at these higher levels. Montagnier took too little account of the effects of glacier thinning and thickening, and of the fluctuations of the snowline associated with retreat and advance. The Little Ice Age in the Alps was well advanced by the time of Arnod and conditions were certainly very different then from those a century or more earlier.

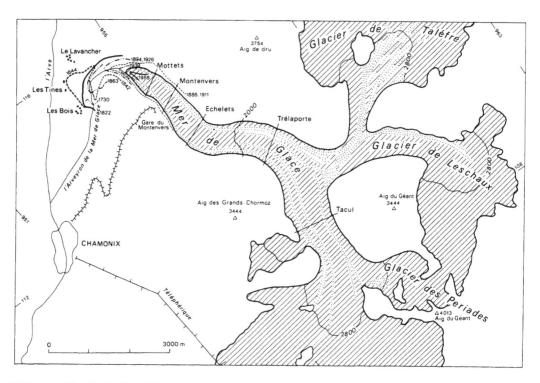

Figure 4.3 Tongue of the Mer de Glace. The rises and falls of the ice surface between 1890 and 1981 at the cross sections shown on this map are illustrated on Figure 4.10 (*Sources*: after Reynaud 1977. Details of the frontal positions from 1644 onwards are after Lliboutry 1965: 726 and Ladurie 1971: 145)

In 1605 the 'Chambre des Comptes de Savoye' made an inquiry into the justice of various requests for tax reduction, and found that by the time of the tillage reform, that is by 1600, the glaciers had spoilt 195 'jornaulx' of land in various parts of the parish of Chamonix, of which ninety belonged to the village of Le Châtelard (Figure 4.3) where twelve houses had been destroyed (Blanchard 1913). Le Châtelard, situated midway between Les Tines and Les Bois, was assessed for tithes, commonly paid in wheat, between 1384 and 1640. In 1564–5 Chamoniards of the valley had bought houses there, and a widow called Perrette purchased some land for her children (Letonnelier 1913). So Le Châtelard had until very recently been a flourishing place and local people had had no doubts for the future. The document of 1605 described the glaciers as still advancing ('the said glaciers, whose ravages continue and progress from one day to the next') and mentioned that Les Bois had had to be abandoned because of danger from the ice, although the greatest advances seem already to have taken place in 1599/1600. In 1610 one Nicolas de Crans (Letonnelier 1913) reported again to the ecclesiastic authorities. By now the Glacier des Bois, 'which is terrible and frightening to look at', had almost completed the destruction of Le Châtelard and a good part of its land. It had also damaged another little settlement called Bonnenuict,[4] but trouble was not limited to that caused directly by

---

4 Ladurie (1971), following and extending the work of Rabot (1920), investigated the positions of the villages destroyed by the Mer de Glace and concluded that Châtelard was midway between the present villages of Les Tines and Les Bois, while Bonnenuict or Bonanay was immediately north of the slope of Le Piget (Figure 4.3). The ruins of Le Châtelard were still to be seen in 1920.

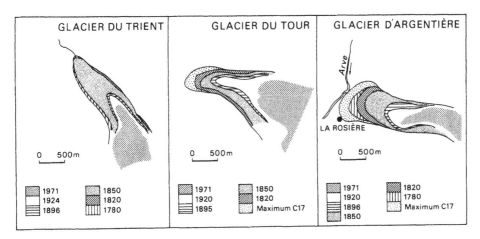

Figure 4.4 Tongues of the Trient, Tour and Argentière glaciers at various times since their maximum Little Ice Age extents. Note the greater extent of the ice in 1780 than 1820 and in 1820 than 1850 (*Source: after Bless 1984*)

the ice. Meltwater streams carried away whole houses and barns and ruined valuable land out on the plain. Ladurie (1971: 144) cites another document from 1610 which quantifies the types of damage. 'The streams have spoiled seventeen journaux of land since the tillage reform (1600). . . . The Arve has spoiled eight . . . and the glaciers two hundred and four and a half.'

There is no evidence from the documents of any recession between 1605 and 1610; the ice fronts may well have remained more or less stationary during the interval. Undoubtedly there was another advance in 1610. The Argentière glacier had already damaged the hamlet of La Rosière as well as the village of Argentière, covering seven houses by 1600 according to the 1605 inquiry (Figure 4.4). But 'it happened that on the twenty second of June 1610 by the overflowing of the glacier of La Rosière, eight houses and forty-five journaux of land were completely destroyed' (cited Ladurie 1971: 148).[5] In 1616 Nicolas de Crans

> went to the village of Châtelard where there are still about six houses, all uninhabited save two, in which live some wretched women and children, although the houses belong to others. Above and adjoining the village, there is a great and horrible glacier of great and incalculable volume which can promise nothing but the destruction of the houses and lands which still remain.

Crans visited the Argentière tongue and found

> The great glacier of La Rosière every now and then goes bounding and thrashing or descending; for the last five or six years . . . it has been impossible to get any crops from the places it has covered. . . . Behind the village of Les Rousier, by the impetuosity of a great horrible glacier which is above and just adjoining the few houses that remain, there have been destroyed forty three journaux [of land] with nothing but stones and little woods of small value, and also eight houses, seven barns, and five little granges have been entirely ruined and destroyed.
>
> (Cited Ladurie 1971: 148–9)

Evidently both the Mer de Glace and the Argentière had moved forward again; their steep fronts in 1616 were about a kilometre outside their positions at the end of the second millennium. The two glaciers were in phase with their neighbours. Ladurie (1971: 151) cites texts to show that not only was the Bossons glacier similarly enlarged but so also were the glaciers on the Italian side of Mont Blanc. On 6 April 1600, a notary living in Aosta was visited by one Jacques Cochet of Les Bois who inquired whether it was true that the parishioners of Courmayeur had sent to Rome to request the Pope to pray that the glaciers might withdraw, and whether it was true that the Italian glaciers, and particularly the Brenva, had actually retreated.

---

5 The extent of a 'journal' at this time is not known.

He was told that the people of Courmayeur had not appealed to Rome, that the glaciers had not retreated, and that they were as threatening as ever.

Thirty years later, in the period 1641–3, there came another advance almost as great as before. The Tour, Argentière and Bossons glaciers were all menacing established settlements in the valley and the Mer de Glace had advanced in the previous two years, 'contre la territoire du village des Bois, des Prés et du Chastellard ença d'une mousquetade' (Blanchard 1913) (150 to 200 m). But the tongue only threatened and did not reach Les Bois, although it had been so nearly reached by ice in 1605 that its inhabitants had to abandon it. It follows that the glacier must have receded somewhat in the intervening years, perhaps between 1610 and 1628. In 1628 there was a great flood of the Arve 'caused by the glaciers' and a text of 1640, cited by Ladurie (1971: 340), refers to the 'third of good and cultivable land lost in about the last ten years through avalanches, falls of snow and glaciers'. It is improbable that we have evidence of all the individual incidents that occurred during the great seventeenth-century advances of the ice. Indeed Jean Duffong, administrator of the priory of Chamonix, described in 1643 how the glaciers had advanced 'again and again' (Blanchard 1913). Duffong related that he had led various processions to exorcize the local glaciers, the most recent having taken place in May 1643. The situation of the Chamoniards was certainly one of considerable distress; another passage of the same date refers to 'les personnes y sont si mal nourries qu'ils sont noirs et affreux, et ne semblent que languissants' (Blanchard 1913). The burden of refugees from the more directly exposed settlements must have been heavy on their neighbours in the valley, struggling with diminished crop yields and recurring floods. In 1644 the Glacier des Bois was so much extended that it was feared that the Arve itself would be dammed, and Charles August de Sales, Bishop of Geneva (nephew of St Francis de Sales), decided to exorcize the glacier again. He did so in June, visiting all four of the glacier tongues and the villages they

threatened. We learn from a document of 1663 that his efforts were successful, for the glaciers retired little by little (Blanchard 1913).

The Courmayeur archives have unfortunately twice been burnt down, so we have little documentary information about the behaviour of the Italian glaciers between 1580 and 1645. Most is known about the Lac de Combal, which was formed by the Miage glacier whenever it sprawled across Val Veni and blocked the Doire (Figures 4.1 and 4.5). Outbursts from Combal are recorded from 1594, 1595, 1629/30, 1640[6] and 1646; these dates fit in well with the glacier evidence from the Arve valley (Baretti 1880, de Tillier 1968). Apart from these Combal records, there is a seventeenth-century tradition of a route through Val Ferret being cut by the Toula glacier (Vaccarone 1884).

4.3 ENLARGEMENT OF THE GLACIERS FROM THE MID SEVENTEENTH TO MID EIGHTEENTH CENTURIES

The early seventeenth century advance was followed by a period of relative quiescence, during which the glaciers remained much larger than before 1600 but were by no means stable. Slow withdrawals, such as that which set in after the visit of the Bishop of Geneva in 1644, were interrupted from time to time by fresh advances, which were less damaging than those which had come earlier, as the ice was now moving over ground already spoilt, but the glacier fronts remained uncomfortably close to villages such as Les Bois and Argentière.

Between 1678 and 1680 there were further outbreaks from the Lac de Combal.[7] Ten years later, the people of Chamonix were in a state of such apprehension that they begged the Bishop of Geneva, then Jean d'Arenthon, to come to see them again, their faith in his powers of exorcism being sufficient for them to offer to pay his expenses. This was clearly not his first visit; according to Ladurie he had already exorcized the glaciers in 1664. Perhaps this was the occasion after which 'the glaciers had retired more than eighty

---

6 A tree stump buried in the moraine records an ice advance of around this date of the same magnitude as the maximum Holocene expansion (H. Aeschlimann 1983, quoted Pelfini 1999).

7 Corbel and Ladurie (1963) obtained a radiocarbon age from wood within the Tacconaz moraines but in view of the ambiguity of radiocarbon dates of the last few centuries their result cannot be taken as evidence of an advance around 1680.

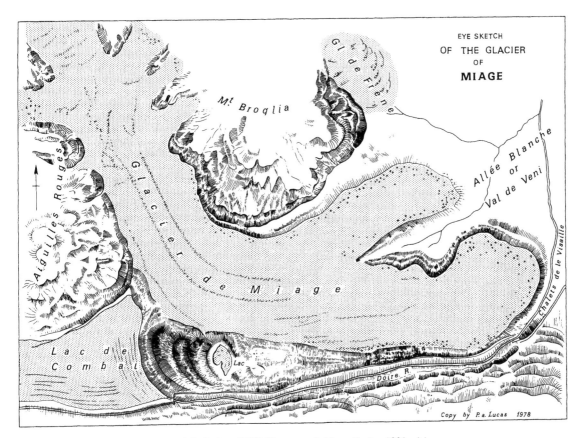

EYE SKETCH
OF THE GLACIER
OF
MIAGE

Figure 4.5 Sketch of the Miage Glacier by J. D. Forbes (1853) (*Source:* copied from Forbes 1900 edn)

paces'. But we learn that after his final visit in 1690 'the glaciers have withdrawn an eighth of a league (about 500 m) from where they were before and they have ceased to cause the havoc that they used to do' (Le Masson 1697: 147). The Brenva was far advanced across Val Veni in 1691 when it was seen by Arnod, a judge from Val d'Aosta, for there was then only a 'very narrow pass' between it and the opposing hillside, near the site of the modern Chapel of Notre Dame de la Guérison.[8] But the balance of evidence is that the Mont Blanc glaciers were retreating slowly during the last decade of the seventeenth century.

The Mont Blanc glaciers do not appear to have enlarged in any very pronounced way in the early eighteenth century. The Chamoniards, now subjects of the King of Sardinia, had not ceased to dispute their taxes and supported their demands for reduction with vivid details of the meteorological and other handicaps under which they were labouring. The syndics of Chamonix made such a supplication in 1716, mentioning that:

doten que leur paroisse devins toujour plus inculte à cause des glacier qui advancent seur leur terre, et qui inonde partie en faisent des grand débordement d'eau, en vuyden

---

8   According to Drygalski and Machatschek (1942: 214) the Brenva covered the floor of Val Veni from 1691 to 1694. But these two dates merely relate to the manuscript of Arnod, which was dated April 1691 and had notes added in 1694. This manuscript, in the archives in Turin (and which was published in 1968) was not available to Drygalski and Machatschek who were quoting it via Vaccarone (1881).

leur lac, et même il i a pleusieur village qui sont en grand danger de périr.

(Letonnelier 1913)

The very general terms of this document contrast with the more precise evidence embodied in the seventeenth-century sources. The syndics were surely sufficiently wise to quote details had any been available. But there is no impression of any sudden change in the situation. Records of an ice and rock fall from the Triolet glacier are dramatic and more detailed:

the highest mountain of the aforesaid Trioly, along with rocks and ice, suddenly collapsed in the night of the 12th September 1717. Boulders, water and ice, all mixed together, rushed with great force over the aforesaid mountains or alps, so that there were covered in the depths all moveable chattels, one hundred and twenty oxen or cows, cheeses and men to the number of seven who perished instantly.

(Translation from Latin citation by Sacco 1918)

The slopes overlooking Val Veni are so steep that even minor advances are very likely to cause rock or ice falls, so this occurrence could well have marked a small forward pulse. There is no clear evidence on this point. De Tillier (1968, cited Porter and Orombelli 1980) recorded that the debris which originated from the fall of a high ice-covered rock onto the Triolet rushed violently downslope, rose against the valley side and then travelled on for a league. When de Saussure saw the glacier in 1781 it was still covered with granitic rock debris. A massive deposit of large angular boulders, extending some 2 kilometres downvalley from moraines dated lichenometrically by Porter and Orombelli to the eighteenth century, was considered by Sacco (1918) to be moraine from the sixteenth to nineteenth centuries. This deposit was the subject of careful investigation by Orombelli and Porter. They concluded that it was formed early in the eighteenth century and has characteristics that permit it to be distinguished satisfactorily from moraine. They inferred that it resulted from the rockfall that wiped out the settlement of Ameiron and Triolet in 1717 (Figure 4.6).[9]

The frontal positions of all the main glaciers in the Arve valley are marked on a detailed plan of landholdings

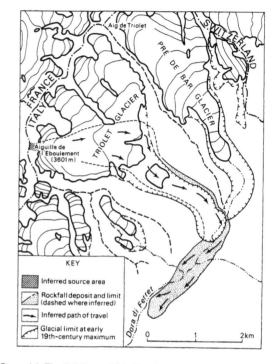

Figure 4.6 The Triolet rockfall. The debris of the giant rockfall from the Aiguille de l'Eboulement which swept the Triolet glacier in 1717 is estimated to have had a volume of 16–20 million cubic metres and to have descended 1860 m over a distance of 7 km in a few minutes. The resulting deposit was thought by Sacco (1918) to be an eighteenth-century moraine and by Mayr (1969) to be an early sub-Atlantic moraine. Porter and Orombelli's (1980) rejection of its status as a moraine depends not only on lichenometric dating but also on characteristics such as lithology, surface gradient, thickness and anomalous extent. The Pré de Bar and probably the Triolet Glacier extended slightly further downvalley in the mid eighteenth century than in the early nineteenth century (*Source:* from Porter and Orombelli 1980)

in the Chamonix area made between 1728 and 1732. Although the general state of cartography still left much to be desired, this cadastral survey was unusually accurate and is of great value as a primary source as the positions of glacier fronts and recent moraines were marked (Guichonnet 1955). The glacier tongues, though still much larger than they are now, had retreated so far that they were no longer immediately

---

9 De Saussure (1779–96, vol. 4: 18) thought that the catastrophe could have been caused by an earth tremor. Forbes (1843: 245) mistakenly recorded the event as having taken place in 1828, and as having been caused by an avalanche or sudden descent of the whole glacier.

Table 4.1 Retreat of the Chamonix glaciers from the 1640 maximum based on a survey *c.*1730

Retreating glaciers	c.1640–1730	1730–1911
Le Tour	414 m	700 m
Argentière	257 m	675 m
Glacier des Bois (Mer de Glace)	250 m	1330 m

Source: from Guichonnet 1955

threatening the villages of Le Tour, Argentière and Les Bois (see Table 4.1).

A document dated by Letonnelier as written after 1730 gives no indication of further advances but makes it clear that conditions had not otherwise much improved. In a further plea for tax relief to the King of Sardinia, much is made of the exposure of the land to rock, ice and snow falls, and the danger of floods and erosion by the Arve and its tributaries. The strong winds and 'cet air glacial cause une certaine aridité et sterilité . . . et malgré tous les soins d'un vigilant labeur, le terrain ne produira que d'avoine et fort peu' (Letonnelier 1913). The villagers, whose fathers had paid their tithes in wheat, were reduced to growing oats and, even so, were harassed by low yields.

The glaciers of Mont Blanc evidently fluctuated from the mid seventeenth to the mid eighteenth centuries, their fronts moving backwards and forwards over the outer fringes of the zone which had been occupied during the great advances of the late sixteenth and early seventeenth centuries. For the local people the time of great disasters was over only to be succeeded by a climate that was still cold and hard, with poor crops and occasional floods.

4.4 THE ADVANCES OF THE MID EIGHTEENTH TO THE MID NINETEENTH CENTURIES

When Windham (1744) visited Savoy in 1741, his guides told him that the Mer de Glace 'went on increasing every year'. However, his description of his descent to the ice over a morainic slope which was 'exceedingly steep, and all of a dry crumbling earth, mixed with gravel and little loose stones, which afforded us no firm footing' makes it clear that the glacier had not

regained its maximum thickness. Soon afterwards Martel (1744) reckoned that the glaciers must once have been 'eighty feet higher than they are now', though the Mer de Glace still spread out below the bar of Mottets and at its tongue was the famous grotto of Arveyron.

Windham's visit was definitely in the pre-tourist age. His party was 'assured on all hands that we shall scarcely find any of the Necessities of Life in those Parts', and so took horses, loaded with provisions, and a tent 'which was of some use to us'. They found the terrible description of the country which they had been given much exaggerated, although they noticed some flood damage; the good stone bridge near Bonneville had 'suffered in the late inundations of the Arve'. Windham and Martel were followed by a gradually increasing number of tourists and by some of the most distinguished of the early glaciologists and field scientists, such as H.-B. de Saussure and J. D. Forbes, both of whom had a passion for accurate measurement and a particular interest in the Mont Blanc area. Writings about the Mont Blanc glaciers during the period accordingly proliferated.

The three last large advances of the Little Ice Age culminated between 1770 and 1780, around 1818 to 1820 and about 1850, the maxima for different glaciers ranging between 1835 and 1855 in this final expansion. We have some clues as to when the glaciers began to swell again, although the advance of the Mer de Glace that had begun in 1741 may have been merely a minor oscillation. De Saussure visited the Tacconaz in 1760 and again in 1778, and found that it had augmented greatly in the interval (1779–96, vol. 4: 432, 463–4). About the same time, the Glacier de l'Allée Blanche advanced onto the plains of Combal, blocking the Doire river, and by 1765 the new lake formed on part of the bed of the old Lac de Combal was already known as the Lac de l'Allée Blanche. It was a matter of importance locally, as it meant that shepherds had to cross the ice to reach their pastures on the slopes to the east of the glacier tongue (Virgilio 1883, Sacco 1918).

The Mer de Glace was advancing in 1778 and continued to do so for a year or two. Engravings by Charles Hackert dated 1780, show the Argentière tongue reaching almost to the river, and the source of the Arveyron to be at the snout of the Glacier des Bois in a position similar to that of 1820 (Forel 1901). The representation of other detail on these pictures is so accurate that full dependence may be placed upon them as a source of evidence. The Italian glaciers were also much enlarged. The Triolet was still increasing in 1781 and the shepherd at Pré de Bar told de Saussure that it had been advancing for the previous eight years (de Saussure 1779–96, vol. 2: 293).

The Brenva was much swollen and by 1767 had overtopped its moraines and drawn close to the fields of Entrèves (de Saussure 1779–96, vol. 2: 293). By 1776 it seems to have reached the southern slopes of Val Veni, for the Doire flowed under the glacier and emerged from a 'beautiful arch of ice' (Bourrit 1776b). In the final decades of the eighteenth century, the Brenva's tongue may have ceased to advance, but a drawing of about 1795 by Jean-Antoine Linck, an artist from Geneva, shows it still occupying the main valley floor, apparently terminating quite close to its position in 2000. But the ice was much thicker than

now, for only two small outcrops of bedrock appear in the middle of the icefall suggesting that the tongue was ready to advance again (Orombelli and Porter 1982, Figure 7).

In 1784 de Saussure noted that the front of the Mer de Glace had withdrawn since 1777 and was 300 m as the crow flies behind the moraine of 1600, about 1000 m forward of its 1958 position (Ladurie 1971: 343). Retreat in the last two decades of the century was swiftly followed, around 1820, by one of the main Little Ice Age maxima. Already in 1816 the Glacier des Bois had been described as 'every day increasing a foot, closing up the valley' (Shelley). J. D. Forbes wrote some twenty years later that 'the hameau des Bois . . . is almost in contact with the glacier, and, indeed, in 1820 it attained a distance of only sixty yards from the house of John Marie Tournier, the nearest in the village, where its further progress was providentially stayed' (Forbes 1843: 61) (Plate 4.1). But Forbes's interest in dating

Plate 4.1 The Mer de Glace reached out on to the floor of the Arve valley in 1823 when it was painted by Samuel Birmann (*Source: Au village des Prats*, Öffentliche Kunstsammlung Basel, Kupferstichkabinett, Inv. Bi. 30. 125)

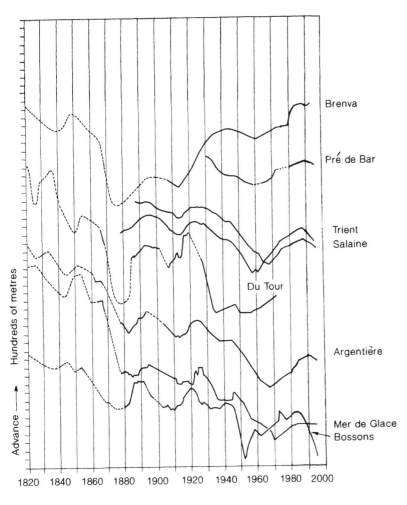

Figure 4.7 Advances and retreats of the major glaciers of Mont Blanc since 1820 (*Sources:* based on Mougin 1908/9, 1910, *Les Variations des glaciers Suisses,* nos. 101–2, Guex 1929, Capello 1941, Bouverot 1957, Vanni 1942–71, Reynaud 1984 and the Annual Reports of the Swiss Glacier Commission)

glacier variations was only secondary to his main interest in the physics of ice, and despite the distinction of his work in other respects, he incorrectly took 1820 to be the date when all the glaciers in the area reached their maximum extent. According to Tournier, a local guide, the Mer de Glace reached its maximum in 1825 (Forel 1889: 462–4), and Venance Payot, who began systematic observations of the fluctuations of its frontal position, dated its furthest advance to 1826. The Tour glacier, after reaching the village fields and closely threatening the village itself in 1818 (Mougin 1910: 6), was stationary between 1823 and 1826 (Figure 4.7).

In 1842 Forbes surveyed the tongue of the Mer de Glace and described how

when we approach the foot of the glacier . . . we are at no loss to perceive that the ice has retreated. The blocks of the moraine of 1820 . . . lie scattered almost at the doors of the houses and have raised a formidable bulwark at less than a pistol shot of distance where all cultivation and all verdure suddenly cease.

(Forbes 1843: 62)

His map showing the terminus of the glacier, 370 m less extended than in 1826 but 1300 m outside its 1960 position, was so well executed and so detailed as to provide a useful basis for future work. It also gave valuable impetus to French topographical mapping as it was thought disgraceful that Forbes, a foreigner, should have found it necessary to do the work himself (Adams-Reilly 1864).

By 1850 the Glacier des Bois had expanded again. No detailed measurements were made, but Vallot (1908) summarized the condition described to him by Couttet of Chamonix, who was 'capable de donner des renseignements précis', a sufficient testimony from Vallot, the quality of whose observations was far ahead of his time. The tongue was again only 50 m from Les Bois: in its extension it had destroyed a larch wood on the valley bottom and the frontal moraines were so swollen that boulders fell down their outer side. Above the snout the ice rose up as high as the steps of the Mauvais Pas, and because of the convexity of its surface it was impossible to see across the glacier. Favre (1867) gives 1855 as the date of maximum extension, but by then the upper part of the snout was already thinning and it was necessary to go down about 15 m over newly exposed moraine to get onto the ice near the Cabane de Burret.

The neighbouring Bossons glacier waxed and waned in much the same way. By 1812 the tongue was advancing again. In that year a series of cold summers began (Mougin 1910: 10) and by 1817 or 1818 the ice covered 4 or 5 hectares of cultivated land and a wood, and menaced the village of Monquart. The population, as had their forebears in 1643, processed to the snout and placed a cross on the frontal moraine: it proved a most useful marker for later observers. There

was a second advance in 1845 but not so marked as that of the Mer de Glace, and it only interrupted briefly the recession that had begun in 1820. Mougin was able to construct a detailed table of the fluctuations of the tongue of the Bossons from 1818 to 1908, based mainly on the observations which Payot made so faithfully until his death (Mougin 1908–9).

The recent history of the Brenva (Figures 4.2 and 4.8) can be traced more accurately than that of the other Italian glaciers of Mont Blanc (Orombelli and Porter 1982). At the beginning of the nineteenth century it began to swell rapidly and there is more information to be gleaned about this advance than about those contemporaneously affecting the French glaciers. D'Aubuisson (1811) reported that in 1811 the ice front stood 1440 m above sea level and was a thousand metres from Entrèves, which it had menaced much more immediately some years earlier, presumably around 1780. Canon Carrel of Aosta was reported by Favre (1867: 74) to have said that the glacier was 2000 metres from the village in 1810–12 and did not reach to within 1000 metres till 1818, but it seems safest to accept the contemporary account of d'Aubuisson from which it appears that the glacier had already reached almost as close to the village in 1811 as it did in 1818. In the course of its advance the Brenva felled two trees, one 200, the other 220 years old (Venetz 1833: 18),

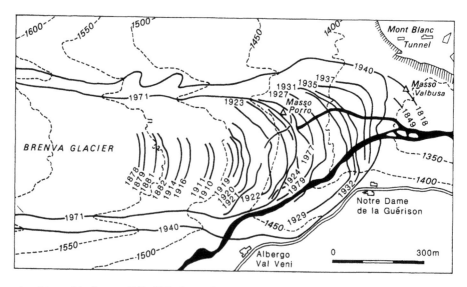

Figure 4.8 Frontal positions of the Brenva, 1818–1979 (*Source:* from Orombelli and Porter 1982)

so this expansion was more extensive than any of those in the preceding two centuries. The Brenva tongue enters Val Veni at right angles and by 1818 it had expanded to reach the opposite southern slopes and rise some way up them, so that it was above the level of Notre Dame de la Guérison (Figure 4.8). This chapel, still a notable place of pilgrimage today, had been reconstructed as lately as 1781, so we may conclude that the 1780 advance had not affected the Brenva to the same extent as the French glaciers. Forbes (1843: 206) was told that in 1818 'the hermitage connected with the chapel was supplied with water from a conduit which descended from the ice of the glacier which was then at a higher level'. The pressure of the ice on the chapel's foundations caused such serious cracks that it had to be demolished in 1820 and rebuilt a short distance away. By the time that Forbes visited it in 1842 (Figure 4.2), the glacier had thinned considerably but continued to cover the whole floor of Val Veni, with the Doire flowing beneath the ice. Forbes concluded that it was still much more extensive than it had been in 1767 when de Saussure (1779–96, vol. 2, plate III) sketched it. After 1842 the glacier advanced rapidly, the enlarged tongue approaching the 1818 moraines again. This advance culminated in 1849 (King 1858: 42). The hermit of Notre Dame de la Guérison described how, in 1850, the surface of the ice was only a few metres below the floor of the new chapel and the top of the ice higher than the crest of the right-hand lateral moraine (Sacco 1918: 55). After this the ice receded year by year, although only 10 km away the Allée Blanche continued to expand for another decade.

The pattern of advances with culminations around 1780, 1818 and 1850 was common to all the Mont Blanc glaciers. But while the mid nineteenth-century advances of some of the Italian glaciers brought them close to their frontal positions of 1818, some of the French glaciers had reached lower altitudes in 1780 than in 1820. Later nineteenth-century advances scarcely interrupted the general retreat which had now set in.

The differences in the behaviour of the glaciers on the two sides of Mont Blanc may perhaps be attributable to differences in aspect, but the greater covering of moraine and rockfall material on the Italian glaciers, especially on the Brenva, must have played a part in making them less responsive to climatic variations.

4.5 THE GREAT RECESSION OF THE GLACIERS – FROM THE MID NINETEENTH CENTURY TO THE PRESENT

All over the Alps mountaineering huts stand high above the ice, often entailing an ascent of a hundred meters or more vertically, up steep moraines, fixed ladders or even ropes. They were not built deliberately to ensure an awkward scramble at the end of the day; their isolation is due to the wasting of the ice during the last century or so. The glaciers, large and small, sprawling around the mountains of Mont Blanc, have withdrawn high into their valleys, leaving their moraines and debris spread behind them. The gleaming tongues of Argentière and Mer de Glace (Plate 4.2) have disappeared from the view of travellers almost as abruptly as they appeared in the sixteenth century. The swelling tongue of the Brenva has melted away from the chapel of Notre Dame de la Guérison.

Regular measurement of glacier fluctuations began during the deglaciation. Payot led the way with his observations of the Tour, Argentière, Mer de Glace and Bossons, published in the *Revue Savoisienne* and *Revue Alpine*. His efforts were augmented by the photographic record made by Tairraz of Chamonix. In 1892 widespread damage, including the destruction of part of the town of St Gervais, was caused by the bursting of a pocket of water in the little Tête-Rousse glacier. The incident was investigated by the Administration des Eaux et Forêts (Mougin and Bernard 1922) and the outcome was that this organization undertook the regular observation of the larger Chamonix glaciers, excepting at first the Mer de Glace where Vallot was already working (Vallot 1900, 1908). Charles Rabot succeeded in putting the programme on a more permanent basis in 1907. While Mougin was in charge he published the very useful *Etudes glaciologiques en Savoie*. After 1930 the results appeared in a series of reports on the variations of European glaciers collected by Mercanton and published in the snow and ice volumes of the *Hydrological Section of the International Union of Geology and Geophysics*. Unfortunately around 1960 observation on many of the glaciers monitored by the Administration des Eaux et Forêts ceased, but in 1977 it was announced that regular aerial surveys of twenty French glaciers were to be made at three-year intervals. They were to include the glaciers of Tour, Argentière, Mer de Glace,

Plate 4.2 The shrunken Mer de Glace seen from Montenvers, August 1958
Photo: A. T. Grove

Bossons, Taconnaz, Bionassay and Tre-la-Tête (Valla 1977).

In Switzerland Forel started the excellent series *Les Variations des glaciers Suisses* in 1880. These annual reports were published in the *Annuaire de Club Alpine Suisse* till 1924 and afterwards in *Les Alpes*, and have included figures for the frontal position of Trient, Orny and Saleina, the largest glaciers in the Swiss sector of the Mont Blanc massif. The Guexes, father and son, interested themselves in the Trient (Figure 4.7), and Jules Guex was able to build up a complete account of the fluctuations of its tongue for the period 1878 to 1928 (Guex 1929) (Figure 4.9).

The Italian glaciers also attracted greatly increased attention towards the end of the last century, but observations were less systematic. The Comitato Glaciologico Italiano was set up in Turin in 1890; Porro's work (1898, 1902, 1914) on the Miage was some of the most important that followed this impetus. Revelli (1911, 1912) and Valbusa (1921, 1931)

both made notable contributions during the next few decades, but regular observations of the fluctuations of the Italian glaciers did not begin till 1927 (Vanni 1958). Work in the Mont Blanc area was influenced after that especially by Capello (1936, 1941, 1957–8, 1966, 1971). Publication of reports on the variation of the Italian glaciers was undertaken by Vanni for several decades (e.g. Vanni 1942, 1970, 1971).

When the data for the larger Mont Blanc glaciers are plotted from these various sources, a surprisingly consistent pattern appears (Figure 4.7). Retreat was general between 1850 and 1880. The fronts advanced again from about 1880 to 1895, though not so much as they had previously retreated. From 1895 to 1915 recession continued but was not so rapid as it had been from 1850 to 1880. Advances from 1915 to 1925 were generally insufficient to bring the snouts back to their 1895 positions. After that, recession continued more rapidly in many cases, with a stationary period and even slight advances about 1945.

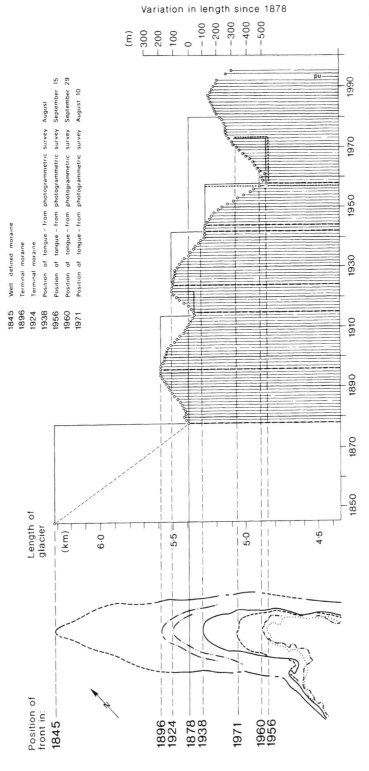

Figure 4.9 Glacier de Trient, 1845–1990. Yearly observations made by the Swiss Glacier Commission make it possible to trace the last three oscillations of the tongue of the Trient Glacier (*Source: after Les Variations des glaciers Suisses*, nos. 98–102)

The most obvious anomaly that emerges is the very well-authenticated advance of the Brenva from 1925 to about 1940, when its neighbours were all in retreat (Capello 1941). The Italian side of the Mont Blanc chain is especially prone to rock and ice falls (Capello 1957–8, Porter and Orombelli 1981). No one who has stood at Entrèves and looked up at the majestic rock wall beneath the Tour Ronde, Aiguille du Géant or Grandes Jorasses can fail to recognize this. On 14 November 1920 a mass of rock and ice, including the whole of a small hanging glacier, slid from the side of Mont Blanc de Courmayeur, fell 2800 m onto the upper Brenva, and shot another 5000 m down to the tongue (Valbusa 1921, 1931). Five days later another fall swept practically the whole glacier even more dramatically. This slide, of which a much higher proportion was rock, was about fifteen times greater in volume than its predecessor. Valbusa (1921) estimated that something of the order of 5 million cubic metres of rock were deposited on the lower tongue alone. Much of the old forest of Pertud was buried beneath masses of ice and rock, which had overshot the high lateral moraine. The Doire was dammed for a time, forming a temporary lake that practically lapped the walls of the chalets of Pertud. The Brenva was already advancing in 1920, in common with the other glaciers of the area, but was alone in continuing to swell after 1925. By 1931 its ice was in contact with the cliff below Notre Dame de la Guérison, and the tongue went on enlarging till 1940. By 1943 waning had begun, and this continued until the 1960s. As the Brenva was in phase with the other Mont Blanc glaciers until 1925, there is no doubt that it was the protective effect of debris from the avalanches and rockfalls of 1920 which caused its anomalous advance from 1925 to 1940.

The Miage too has a protective debris cover. Névé from the great ice mass of the Dôme de Goûter is canalized into the Miage's deep trough valley which cuts three-quarters of the way through the Mont Blanc massif. It is not perhaps surprising that, even now, the resultant icestream shoots out across Val Veni and then, swinging round through 90 degrees, pushes far down the main valley. The enormous scale of the Miage moraines (Plate 4.3), even in comparison with those of Brenva, is particularly evident from the *téléferique* station at the top of Mont Frêty, below the Col du Géant. The tongue is debris-covered for more than 5 km above the forked snout. Sacco thought that the Miage was, because of this, particularly insensitive to climatic fluctuations. Revelli found that by 1911 it had retreated only a few tens of metres from the 1879 position mapped by Marengo (Revelli 1911, 1912). There was a slight advance in the early twentieth century, for Sacco in 1917 found the glacier swelling and overriding some of its moraines. Subsequently the Miage slowly retreated until the late 1960s. The debris cover may well have damped down frontal oscillations during the last century or so; certainly its presence has made observation of such changes as have occurred more difficult.

The Tour glacier has been characterized by extraordinarily large seasonal fluctuations in its length, both during advance and retreat, because for much of the period since 1818 its tongue has hung over the great south-facing rock step which dominates Le Tour village. Sudden advances, like that of 267 m in July 1916, have been counterbalanced by equally drastic retreats, for example of 228 m in three months of 1915 (Mougin 1925: 2–9). During the 1920 advance the glacier was prolonged from time to time by the incorporation of a reconstituted ice mass at the foot of the step. Instability continued as the ice withdrew up the step during the subsequent retreat phase, which was therefore unusually swift. On 14 August 1949 a crack formed across the hanging tongue, and half a million tons of ice, equivalent to about forty years' accumulation, moved forward, broke into blocks, and poured down the precipitous face to the moraine below, killing a party of picnickers (Glaister 1951). The snout had probably retreated out of a gully, where it had been firmly entrenched, and the crack formed as it hung unsupported on the slope above (Messines du Sourbier 1950). The instability, so characteristic of a glacier of this type during the periods when the front is hanging, clearly limits the value of individual observations of advance and retreat, but it is striking how well the general pattern of fluctuations of the Tour fits in with those of others in the massif.

The Glacier des Bossons also advances and retreats very swiftly. A withdrawal of nearly 600 m between 1943 and 1951 was succeeded by an expansion of 350 to 400 m between 1952 and 1958, which was measured photogrammetrically. Finsterwalder (1959) suggested that this advance was probably due to increased accumulation in the firnfield during the period

Plate 4.3 Miage moraines with the bed of the Lac de Combal, August 1958
Photo: A. T. Grove

1953–6 causing a kinematic wave to reach the tongue in only three years, travelling three times as fast as the glacier itself. Lliboutry (1958) was able to use Vallot's observations (1900, 1908) to demonstrate the development of such a pressure wave on the Mer de Glace between 1891 and 1895 (Figure 4.10). Movement of such pressure waves down the Bossons is especially rapid as the bed is so steep. It is clear from Figure 4.7 that each general advance of the Mont Blanc glaciers has been preceded about ten years earlier by a forward surge of the Bossons (Grove 1966, Vivian 1971).

The Bossons glacier began to advance, together with its consort the Tacconaz, in 1953 following a period of rapid retreat. The advance halted briefly around 1960 but by 1976 the terminus had moved forward again by over 700 m (Veyret 1974). The other glaciers of the massif duly followed its example, each in its own time.

The swelling of the upper parts of the glaciers (Vivian 1971, 1975) which caused these advances has been particularly closely observed in the cases of the Mer de Glace and Argentière, which were amongst the last to advance in 1970 and 1971 (see Figure 4.7 and Reynaud 1977). Vallot surveyed not only the front but also five cross-profiles of the Mer de Glace. A sixth profile was added by Mougin in 1923, and a seventh by the Administration des Eaux et Forêts in 1970 (Figures 4.3 and 4.10), but thinning and withdrawal of the front in the mid twentieth century was so great that the two lowest profiles, at Le Chapeau and Mauvais Pas, have been lost. Figure 4.10 shows three periods during which the surface of the ice rose and a wave passed down the glacier, eventually causing an advance at the front. The first occurred between 1890 and 1900, the second between 1910 and 1930, and the third started in the 1960s and still continued in the 1980s (Lliboutry and Reynaud 1981, Reynaud 1977, 1984).

Seismic observations and surveys made on several of the Italian glaciers have allowed us to follow the passage of waves and the thickening preceding advances (e.g. Lesca and Armando 1972, Lesca 1972). The first

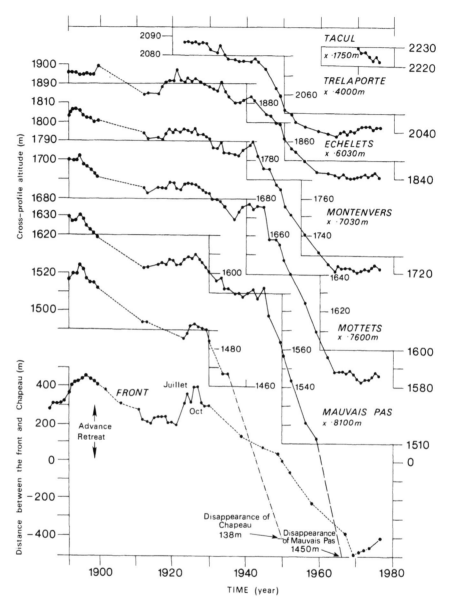

Figure 4.10 Changing height of cross-profiles at specified positions indicates the passage of waves down the Mer de Glace since 1900, which accounts for the episodes of advance shown on Figure 4.7. Vallot's observations were far more precise than those of contemporaries; no other such long series of observations are available elsewhere to demonstrate the relationship between variations in mass balance, glacier flow and frontal advance (*Sources:* after Reynaud 1977 and Reynaud et al. 1984)

glaciers to react on the Italian side were the cirque glaciers with steep slopes, such as those of the Grandes Jorasses and Planpincieux, which in 1963 had already covered markers installed 20 m in front of them in 1961 (Vivian 1975). Particularly spectacular evidence of renewed activity was provided by the Brenva, which by 1968 had destroyed a road built across the frontal moraines in 1960. The front of the Glacier de l'Allée Blanche was already 600 m in advance of its 1954 position by 1973 (Figure 4.11). However, the Pré de

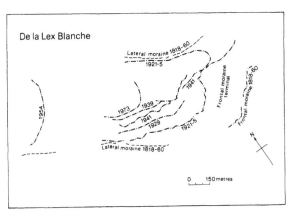

Figure 4.11 Frontal positions of L'Allée Blanche (Lex Blanche). Note the great retreat of the mid twentieth century and the advance between 1954 and 1973 (*Source: after Cerruti 1977*)

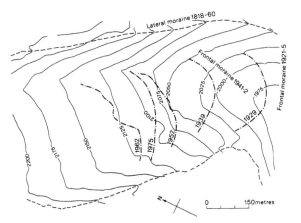

Figure 4.12 Frontal positions of the Pré de Bar in the twentieth century (*Source: after Cerruti 1977*)

Bar had still not managed to regain its 1952 position in 1975 (Figure 4.12). The difference in the speed of re-action of these two glaciers is no doubt associated with the steepness of the slope on which the tongue of the Allée Blanche lies compared with that of the Pré de Bar. The Trient glacier also advanced rapidly and by 1976 its terminus had regained over a third of the ground lost since the maximum of 1896 (Figure 4.9).

The switch from retreat to advance caused the gla-ciers to modify their forefields during their reoccupation of ground abandoned several tens of years earlier and

also to change their appearance quite markedly (Veyret 1971, 1974, 1981). Retreating tongues are character-ized by gentle profiles, caused by the excess of melting over delivery of ice to the front. Advancing tongues bulge or may even be cliffed. By the autumn of 1970 the general state of advance was immediately obvious to the informed observer.

It is generally accepted that the advances of the 1960s were associated with lower temperatures, especially in the summer ablation season (Vanni 1970), and also heavy snowfall. Corbel (1963) pointed out that since most precipitation falls in summer in the Chamonix–Mont Blanc area, a lowering of temperature in the May to September period leads to an important increase of snowfall at high altitudes. Between 1940 and 1950 the rate of glacial retreat was already lessening and some fronts were stationary. Glacial expansion was promoted by the conditions of the 1960s. In 1962, for instance, May temperatures were the lowest for a century. The average temperature for June was 2.1 °C below the normal for the previous thirty years and minimum temperatures the lowest on record. July remained cold and snowy, and then, after a brief fine, dry period in August and September, temperatures fell again in October. There were exceptionally heavy snowfalls in December. These low temperatures and heavy snow-falls in summer, typical of a period of glacier expan-sion, were reinforced by voluminous winter snowfall in 1962–3. The repetition of such sequences led to a situ-ation in 1971 when, despite heavy ablation in summer, none of the glaciers on the Italian side of Mont Blanc were retreating (Vanni 1971) and the general advances which followed were maintained into the 1980s (Sessiano 1982). The Pré de Bar was still advancing in 1990, as was the Argentière in the Arve valley, but by the mid 1980s the Bossons was retreating. It fore-shadowed, as usual, a general retreat in which, by 1990, the Brenva, Trient and Salaine were taking part.

Earlier intervals of expansion of the Mont Blanc gla-ciers also took place when temperatures were below average. Cerutti (1971) looked into the relationship between the glacier fluctuations and climatic conditions using meteorological data from the Great St Bernard Pass assembled by Janin (1970). She found that glacial advances had followed colder periods with more pre-cipitation, and retreat followed warmer periods. The dominant influence of temperature is well exemplified

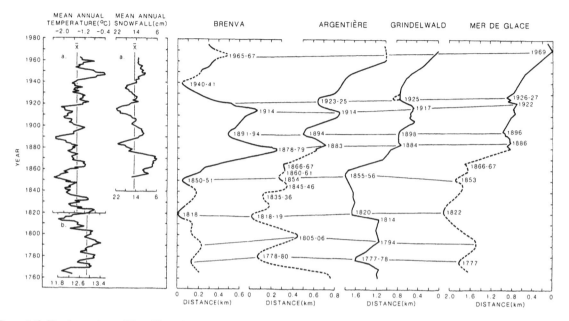

Figure 4.13 The fluctuations of Mont Blanc glaciers are closely paralleled by those of the Grindelwald Glacier (see Chapter 6). Note the relationship between periods of advance and retreat and the incidence of temperatures above and below the mean (measured at Great St Bernard between 1820 and 1980 and at Milan between 1763 and 1820) and decrease and increase of snowfall (measured at Great St Bernard between 1850 and 1960) (*Source:* from Orombelli and Porter 1982)

by the advance period of the 1880s and the retreat period of the 1920s and 1930s. Annual average precipitation on the pass was 1903 mm in 1876–90, while it was 2183 mm in 1920–40, but the greater precipitation was insufficient to counteract the rise in mean annual temperature of 0.4 °C. The general relationship between glacial behaviour and mean temperature is shown in Figure 4.13.

Orombelli and Porter (1982) noted that the temperature record, kept by the Osservatorio Astronomico di Brera in Milan, closely parallels over the last 160 years the record from the Gt St Bernard. If glacial variations are compared with temperature variations as recorded in Milan for the period 1763 to 1820, it emerges that the advances culminating in the 1770s and around 1820 coincided with periods of low mean annual temperature, while in the intervening retreat period temperatures in Milan were as much as 1.5 °C higher.

The history of climate during the Little Ice Age has been traced back further and in more detail for Switzerland, but before discussing this (in Chapter 6) the behaviour of the Mont Blanc glaciers is compared with that of certain glaciers in the Alps further east.

4.6 SUMMARY OF THE LITTLE ICE AGE RECORD FROM THE MONT BLANC GLACIERS

There is indirect evidence that the first Little Ice Age advances were in late medieval times. Major advances certainly occurred in the decades either side of 1600 and again in the 1640s. Later advances between 1740 and 1780, in the 1820s and about 1850 did not reach as far forward as those of the seventeenth century. A general retreat beginning in the 1850s was interrupted from 1880 to 1895, 1915 to 1925 and 1960 to 1985.

5

THE ÖTZTAL, EASTERN ALPS

The largest group of glaciers in the eastern Alps lies at the head of Ötztal, in the Austrian Tyrol, where mountains on the frontier with Italy rise to over 3600 m (Figure 5.1). The main névéfields extend from the Weisskugel, northeast towards the Mittelbergferner, and east towards the Hochwild. Many of the glaciers with which we are concerned flow southeast into the Rofen section of Ventertal, and most of the rest northwards into the Rofen and Spiegler sections of Ventertal, or into the head of Gurglertal (Figure 5.2). Both Ventertal and Gurglertal are steeply sloping with small, cultivated basins separated by deep gorges.

When the Vernagtferner advances across the floor of Rofental it obstructs the flow of streams draining the glaciers further up. The Gurgler expanding down its valley blocks the mouth of Langtal. In both cases lakes form and given sufficient time are liable to spill violently. The consequent floods have caused much damage in Ötztal and subsequent inquiries made by government authorities, the earliest at the end of the sixteenth century, are the main sources of information about the early part of the Little Ice Age in the eastern Alps.[1] These documents give a very clear impression of the problems and dangers faced by people living in the area and their reactions. The literature will therefore be discussed in some detail, so that the human implications of the more extreme events involved in Little Ice Age glacier expansion may be revealed more explicitly than was possible in earlier chapters dealing with northern Europe or the western Alps.

Several glaciers in Ötztal were selected for detailed observation towards the end of the nineteenth century by Finsterwalder, Blümcke and Hess (Finsterwalder 1897a, b, Blümcke and Hess 1895, 1897, 1899). In the second half of the twentieth century, mass balance studies were made of some of the key glaciers; investigations of the relationship between glacier fluctuations and meteorological controls by Hoinkes (1970) and his associates are particularly noteworthy. The voluminous literature was listed by Rudolph (1963); the Ötztal clearly provides a concentration of material for the eastern Alps fully comparable in richness with that available for the Mont Blanc massif in the west.

5.1 THE VERNAGTFERNER 1599–1601

For this period there is detailed information for only one glacier, the Vernagtferner. It provides only a very slender basis for a glacial chronology, as up to the mid nineteenth century it surged a number of times, the

---

1 The history of the Rofen and Gurgl lakes was first placed in the general context of Alpine glacier fluctuations by Richter (1891). He later provided full information on the documentary sources in the Gubernium at Innsbruck and in the Ferdinandeum (1892), and in the same paper corrected many details mentioned in his book of 1888. The sources consist mainly of reports from the Wardens of Petersberg and Castellbell, letters from local clergy, reports from special emissaries, and submissions from Innsbruck to the Imperial Court. Richter underestimated his own contributions, writing 'As these documents had been seen by other authors, I did not expect any new information from my investigations. Nor did I find any.' In fact Stotter, writing in 1846, had not seen all the documents, whilst Sonklar (1860) made some serious errors in his account (Richter 1891: 3).

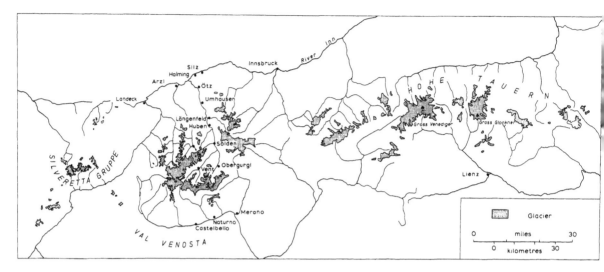

Figure 5.1 The principal glacier concentrations in the Austrian Alps

tongue thickening to an abnormal extent and advancing with exceptional speed for a European glacier. The probable reasons for these characteristics will be discussed later in this chapter. Meanwhile, it will be of interest not only to follow local reactions to the consequent hazard but also to examine the timing of its surges in relation to the major advance phases of the glaciers around Mont Blanc discussed in Chapter 4.

The Vernagtferner flows southeast to unite with the adjacent Guslarferner. Entering Ventertal at right angles, it has on occasions dammed back the rivers coming from the Hintereisferner, Kesselwandferner and Hochjochferner, to form a lake in Rofental, the Rofensee. An account of the first outburst of the Rofensee is given in the chronicle of Benedikt Kuen,[2] according to which

> in 1600, so our ancestors tell us, the big glacier behind Rofen after it had come into the valley according to its habit, broke out on the feast of St James [25 July], did great damage to the fields in the Elzthal [Ötztal], spoilt the roads and streets and carried away all the bridges. In the parish of Langenfeld [Figure 5.1] the water flooded the ground from Rethlstain to Lener Kohlstatt.

Kuen's wording suggests the possibility that this advance of the Vernagt was not unprecedented, but as he was writing after several later advances it may well have been these that he had in mind. He made no mention of the situation in 1601 about which Richter (1892) found a great deal of evidence surviving.

The ice was still in an advanced position in 1601, and the Rofensee threatened to bring even greater destruction not only to Ötztal but also to the Inn valley. A report was made by the government of Tirol and Vorarlberg at Innsbruck, dated 30 July 1601, and sent to Emperor Rudolph II.[3]

> In the morning of the 11th inst. the said Clerk of the Works [Jäger] reported to us verbally how he and each of the others . . . had found the lake and the glacier, and had found the situation so alarming that the said lake might break out on the 13th of the said month, and that no human means, help or advice could avert the danger, for all would be in vain, and the destruction that would follow would necessitate a great sum of money being required for the damage. Bearing in mind that the glacier which had advanced and broken out before this one, also in Ötztal, had caused damage of over 20,000 gulden, it is easy to imagine what this glacier, which is at least six times

---

2   The 'Chronicle of Benedikt Kuen', written by Johann Kuen in 1683 and his son Benedikt in 1715, was the only known account of the 1601 advance until Richter unearthed a mass of contemporary documents. There is an account of an earlier outburst of the Rofensee in the 'Fugger Korrespondenzen' (Klein 1968). The Fugger family owned land in Inntal.

3   The report, together with various supplements, was discovered by Richter in the Innsbruck State Archives, AVIII 19.

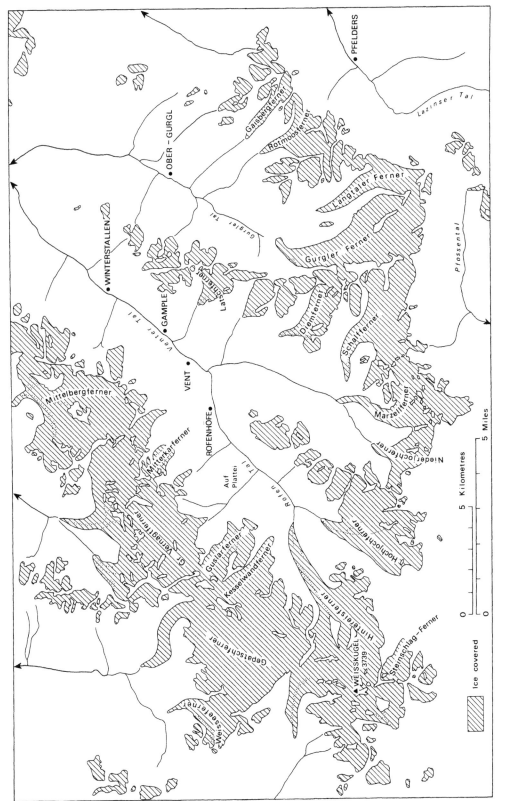

Figure 5.2 The glaciers of the Ötztaler Alpen. Note the very short tongue of the Vernagtferner in the mid twentieth century (*Source:* from 1:25,000 maps 'Gurgl' and 'Weisskugel-Wildspitze', published by Österreichischer Alpenverein 1951, based on photogrammetrical survey of 1942/3, revised 1950)

as big as the above-mentioned one, would result in, as it would doubtless run down through Ötztal to the river Inn and tear away some 200 houses, much fine land as well as bridges, levees, roads and paths.

(Richter 1892: 359)

It is clear that the glacier that had broken out before this one was not the Vernagt. 'Also in Ötztal' strongly suggests that two different glaciers and lakes were involved. It is reasonable to assume that the earlier disaster was caused by an outbreak from the Gurglersee a few years earlier; the wording suggests a recent date but not the previous year. As the Gurgler does not surge it can be taken that valley glaciers in the Ötztal area were already enlarged by 1600.

The report of 30 July 1601 on the Vernagt included supplements by Jäger, who noted that

the glacier fills the valley like a big round pate . . . it is not like others, smooth ice, but is full of cracks and crevasses with strange colours, so that one cannot marvel enough. Furthermore, this glacier has swelled and dammed a lake where there used to be nothing but a beautiful alp and grazing land. It is 625 fathoms [1312 m] long; its greatest width from one mountain to the other is 175 fathoms [332 m] and its depth according to Hans Rothen is 60 fathoms [114 m] and it is rising a man's height in 24 hours.

(Richter 1892: 361)

The Rofensee, he adds, 'has come in the course of two years, 1599 and 1600, and still grows higher, longer and broader with each day'. A number of streams were flowing into the lake but no water was escaping from the ice dam and therefore 'a great danger is to be feared'. Orders were given on 12 July to 'clear the banks of the river and remove dangerous heaps of wood and put out guards'. The authorities of the upper and lower Inn valley were directed 'in so far as they were exposed to any danger, to take measures to have processions with the Cross arranged everywhere and further to do all that is humanly possible to prevent too much damage caused by a possible eruption of the lake' (Richter 1892: 362–3). A procession was arranged to the abbey at Wilten, near Innsbruck, where a special sermon was given by a preacher who, it may be noted incidentally, had been provided beforehand with a copy of Jäger's report.

The Burgomaster and town council of Innsbruck wanted to have more information about the situation and they selected two experienced men, Peter Puppel,

a butcher of Innsbruck, and Martin Griessetter, chief miner of Hetting. They visited the Vernagt on 14 July 1601, finding it 'easy to get up to the ice' even though the valley downstream of the glacier was 'no wider than one fathom . . . quite a deep valley, running between hard mountains and rocks, so that one cannot get up anywhere but has to go through the defile'. They recorded that the previous week, 'according to the Rofen farmer and others living near the glacier, a piece of the glacier 200 fathoms [380 m] wide had fallen into the lake. Therefore . . . the water was pressed out.' On 12 July the lake had

run over a bit, then between morning and midnight the lake found an outlet. A stream . . . ran out between the mountain and the glacier, visible for about 30 paces . . . then . . . down through the glacier so that it can no longer be seen but only heard. Then much further on it comes out again under an avalanche and runs on, quite muddy, through the Öz valley. . . . Almost in the middle of the lake, on the left bank, looking upward there is a post, so . . . people can see with their own eyes that in one hour, while they have gone round it, the lake has fallen a handsbreadth. . . . It seems that the outlet will become deeper so that the lake will fall little by little. As the glacier is so crevassed and cracked, it is expected that it will melt away soon.

(Richter 1892: 363–5)

They estimated that almost twice as much water was running out of the lake as the tributaries were bringing in.

Most of the Ötztal above Sölden came under the jurisdiction of the Court of Petersberg at Haiming in the seventeenth century. The Warden of Petersberg was ordered by the government to make investigations of the dangerous glacier and lake but excused himself and commented that 'all human help is impossible'. He asked the local Rofen farmer to report if anything should happen and petitioned the Margrave of Burgau, asking to be relieved of the commission as he did not feel experienced or skilled enough to avert the danger. Furthermore he claimed that there was no one in Petersberg whose advice could be asked in this matter (Richter 1892: 365–6).

The Margrave wrote to the government of Upper Austria explaining the Warden's unwillingness to approach the glacier and the government ordered Count Daniel Felix of Spaur and Valor to bring pressure to bear on the Margrave. Count Daniel Felix encountered he Margrave on the bridge at Aichwaldel on 23 July 1601, as he was returning from a fishing expedition, and showed him Jäger's sketch and account of 11 July. The Margrave laughed at Jäger's assertion that it was dangerous to approach the glacier; he had visited it himself. He pointed out that the glacier was in the jurisdiction of Castellbell, not Petersberg. His need to do this is significant, for if the previous outbreak in Ötztal had been from Vernagtferner, the Petersberg people would have known it was not within their jurisdiction. The Margrave recommended that people with long sticks and sharp blades should be sent to widen the outlet of the lake and to crush the bits of ice blocking the course of the water. Advice should be taken, he emphasized, from suitably qualified people, those conversant with mining, building of sluices and dealing with water generally, for 'they could much sooner and much more easily find a suitable remedy'. 'We do not mind our Warden helping you', he concluded, 'but he must not be troubled by large costs.'

The Warden of Castellbell, Maximillian Hendl, went into action as soon as he received orders to do so, taking local people, but he 'could not in his hurry find any good miners, either for working on the ice or the rocks or sluices'. With a party of six, including an ecclesiastical judge, Christian Mayr, and his steward, Adam Rainer, he arrived at Rofen on 11 August 'in a terrible storm, with snow and rain'. The judge and the steward had been to Rofen only three weeks earlier to inspect their cattle, oxen and sheep, which had been grazing above and below the lake. They found since then, that the water had fallen by '13 good meat fathoms' leaving behind 'blocks of ice as big as houses and also smaller ones, which had been floating on the water and are now melting little by little'. He estimated that a third more water was flowing out of the lake than was entering it. The outflow had excavated a hole in the ice forming an arch one and a half times the height of a foot soldier's pike (a *Landsknecht*'s pike at this time was about 4 m long). 'The course and outlet of the glacier seem to be getting easier and more comfortable all the time.' The construction of sluices would not be any great help and, as he pointed out, the cost would have been considerable. Nevertheless the peril might not have passed entirely and 'we must pray to God Almighty and try to placate him with pious processions' (Richter 1892: 366–7).

Despite more rain and snow Abraham Jäger returned to Vernagt on 9 September 1601. He found the glacier no longer the same shape as it had been. The lake was 'half emptied and since the 5th of this month it has become 12 mining feet lower. . . . From this it is to be supposed that this lake will disappear completely in a short time.' Georg Wurmbser, carpenter of the salt works, accompanied him and also Michael Grasel and Melchior Wanner, 'who know about sluices'. Jäger concluded that it was not possible to place sluices in a useful position and 'it would be much better to leave everything to God'.

The order to put out guards had been revoked as early as 30 July, in view of the reported emptying of the lake. For the present the threat was over. The next severe flood did not come until 1678, and by then the witnesses of the 1600 disaster were dead.

5.2 THE ÖTZTAL GLACIERS 1678–1725

Disquiet in Ötztal may have existed for a year or two before the 1678 flood. Johann Kuen of Langenfeld, who was prominent locally and was often consulted by visiting emissaries and officials, wrote a short account in 1683 according to which he had heard in 1676 that the Vernagt was advancing; by the autumn of 1677 'it had reached the mountain opposite . . . so that it dammed the water'. However, a Capuchin friar, dispatched to Vent by the Bishop of Brixen in 1678 to calm the population and hold services, had written to this same Kuen of Langenfeld on 1 July 1678, immediately after he had visited the Rofensee with his congregation, and he dated the formation of the ice dam to March or April 1678 and the beginning of the lake to May (Plate 5.1). The contemporary account is likely to be more reliable than Kuen's, written five years later; the congregation had recently repaired the path which was in common use by travellers and herdsmen and had watched the formation of the lake 'with sad hearts'. The friar had taken a knotted measuring rope with him and found the lake to be 744 ellen (588 m) long and 250 paces broad alongside the ice.

> At a point 120 ellen [95 m] from the end of the lake which abuts on the glacier, we measured the depth of the lake and found it to be 10 ellen [18 m]. From this it can be deduced how deep it must be in the middle or at the end near the glacier. We have also observed that during

Plate 5.1a The advancing Vernagtferner, drawn by a Capuchin friar, Georg Respichler, portraying the glacier as it was on 16 May 1678, from the manuscript 'Über den Ausbruch des Vernagtferner in Vent im Jahre 1678', dated 28 July 1678 (*Source:* Landsmuseum, Innsbruck, Fb 3631)

the two hours we spent there, the water rose more than one span. So it is easy to see how dangerous the situation is especially as we fear that the glacier has not yet stopped advancing.

(Richter 1892: 375–6)

Within a week, on 6 July 1678, the lake was visited again by a Father Sabinus with a small party including an official from Petersberg. They found the lake much enlarged. According to a letter written by the Capuchin friar to his superior on 18 July (Richter 1892: 376–9) the ice dam had broadened to 840 ellen (654 m), the lake was 1380 ellen (1074 m) long, and men who knew the place said it was as much as 100 fathoms (190 m)

deep. The water level rose half an ellen and three breadths of a finger (44 cm) in the three hours the party was at Rofen and 'the glacier grows on all sides except the side where the water touches it'; here the water had undercut the ice. Nevertheless the dam was believed to be '78 ellen [60 m] lower than it had been years ago'. Presumably the traces of the level reached by the 1600–1 lake were still recognizable.

A regular series of visits was arranged to give warning of imminent danger, Kuen recording that 'every week a suitable person is sent to the glacier to bring back a report to the authorities'. On 14 July the friar walked up beside the glacier for three hours and was

Plate 5.1b The original key to Respichler's drawing

much impressed by its size: 'No messenger would be able to run round its circumference in one day, for it goes incredibly far and only the Good God knows how far it extends up the mountains.' So, as the weekly reports came in, 'the hay was mown somewhat earlier and brought in, the cattle sent into the alps and other preparations made'. Then, after several days of rain, the ice dam gave way.

'On 17th July the glacier split and the water broke through, preceded by a stinking fog and a terrible roar'.[4] Before dawn the flood crest had reached Huben (Fig-

ure 5.1) and it passed Langenfeld early in the morning. In a letter of 28 July 1678 Georg Respichler wrote that all the bridges except for the high one at Rofen 'have carried away so that nothing looks as it did before'. Some villages such as Hopfgarten, Östen and Tumpen were almost wiped out; others suffered only slightly less damage. Winkl and Platten lost eight or nine houses, together with barns and byres as well as goats and pigs. The damage to the land itself was so great that Respichler doubted whether the place would ever again be habitable. 'The roads and paths from Hochlrain to

4   In Richter (1892: 381–91), from J. P. Hurn (1770).

Vendt . . . have been spoilt completely in many places so that one can neither drive, nor ride, nor walk on them, and in some places it will be very difficult to mend them.'

Despite this trail of devastation there were few casualties, though in one group of villages, where the people were surprised by the flood before dawn and isolated from help for four days, two children were lost. Providence had been kind. There had been processions with the cross in various places and, as Kuen observed in his account of the flood,

> We can observe the beneficial results of these devotions in that, despite so much danger to life, what with the over-flow of water during the night, and with the many dan-gerous jobs at the bridges, paths and levees, we have all preserved our lives, and although some people have fallen into the water, they have got out again, which is certainly a miracle and a wonder.

The cost of the 1678 floods was high; 45,000 florins for Umhausen, 115,000 florins for Langenfeld, and 22,000 for Sölden. Benedikt Kuen, son of Johann, writing forty-four years later, thought the cost had been unnecessarily high; he considered that more should have been done in earlier times to control the streams and rivers. But there had seemed to be little need for strenu-ous measures for very many years. The advanced posi-tion of the glaciers from 1678 and for nearly half a century afterwards presented a new challenge for him and his generation which they faced with energy and determination.

While work was beginning on repairing the damage of 1678 there was another flood in 1679 which did not seriously interrupt the work. Next year, however, the water broke out from behind the ice and did 'terrible damage to houses, barns, lands and bridges all over the Ez valley and beyond the mountains . . . if all the water had broken through at once, not only the lower Ez valley but also the lower Yn valley might have been completely ruined' (Richter 1892: 383, see also Nicolussi 1993, plates 1–3).

According to the Kuen chronicle, in 1681 'the gla-cier came very low, it was hard and blue', and another outburst was expected.

> Therefore on 8th July twelve men were sent out from Langenfeld to cut a ditch through the ice, which, thank God, went well, so that the water ran out little by little

through this ditch. So this time we could live without danger or care, repair the damaged things and lands in comfort and pleasure though with hard work.

Kuen's account was written two years after the cutting of the ditch, which was evidently a complete success. Such measures have not always been so; a similar attempt to deal with a nineteenth-century hazard at Mauvoisin in Valais ended in disaster (Mariétan 1959 and Chapter 6).

The 1681 Rofen flood, as on previous occasions, was followed by an official inquiry. A judge, Jeremias Rumblmayr, the court architect and master of the Board of Works, Martin Gump, and several others made a survey of the glaciers in June. They reported that the lake had been dammed back in the course of the win-ter, between October 1680 and July 1681, so that the water rose '50 to 60 fathoms and spread a good hour's walk up the valley, here 1000 paces wide' (Richter 1892: 401). The spillway through the ice had been cut by thirteen men in three days under the direction of Johannes Kain, Steward of Langenfeld. The water ran away between the ice and the mountain in two or three days; Rumblmayr was inclined to think that the same thing would have happened if no action had been taken. The remaining ice was calving rapidly into what was left of the lake and a medium-sized stream was dis-charging underneath it. The situation could still be-come dangerous but at any rate there did not seem to be any threat to the Inn valley, which was Rumblmayr's main concern. He reckoned that a flood reaching the Inn would take ten hours to traverse the Ötztal and therefore lose its power.

Rumblmayr noted that the damage of the preced-ing years had been accentuated by tributary streams flooding and causing landslides. Whether their flooding was due to rain or snowmelt is not clear. At Langenfeld and Ästan, where fields had been cut away in 1678 and 1679, old flood protection works had been revealed and the river 'had resumed a course it had taken a hun-dred years or more earlier'. The countryside had suf-fered severely. 'All the bridges have been pulled down for fear of the overflowing of the lake and the ordinary roads are impassable and the people have partly left their houses on the flat land and retired with their cat-tle to some miserable cabins at the foot of both moun-tain slopes'.

The Commission was anxious to find a permanent solution to the flood hazard and Martin Gump, in one section of the report (Richter 1892: 402–7), considered the possibility of cutting a long gallery or tunnel through the mountain side. He saw the difficulties there would be in providing light and air for the excavators and the high cost of upkeep. The worst problem, he thought, would be prevention of blocking by ice from water trickling through the rock and freezing, or the stream itself freezing in the tunnel and not melting in summer 'because it is always colder in such places'. He concluded that a gallery would not be advisable.

Another member of the Commission of Inquiry, a joiner called Paul Heuber, suggested that a dam might prevent flooding. It could be built in one of three places; at Lenersegg below Zwieselstein where the Ötztal narrows, at Gämpel (Figure 5.2) where the stream flows through a narrow defile, or behind the Rofen fields. These proposals were put to local farmers, who found objections to all of them but eventually came down in favour of the dam at Gämpel. The cost of this was too high and some members of the Commission were not convinced that it would hold back the floodwater. Cutting through the ice was judged to be impractical. In fact no remedy commended itself.

By this time 'the glacier seemed to be retreating all the time, and the ice dam had shrunk to half its maximum height'. With some relief the Commission reminded the communities of Ötztal of their promises and those of their forefathers to make devotions for the remission of floods: 'For just as God Almighty has had such a monster come into existence during one winter so His Divine Goodness could be moved by earnest prayers so that he let it retreat in the same or a shorter time.' It recommended that the money that might have been spent on one or other of the suggested remedies should be devoted to relief of those who had suffered from the floods; 2000 florins already granted for projects were immediately available. It was also proposed that the province of Tirol should remit taxes for a certain period: 'the Abbess of Kiemsee, and especially the convent of Stams which gets most tithes, could convert the tithes into corresponding aid for a period of some years; and the government might give the sufferers new land instead of the spoilt land and houses' (Richter 1892: 398).

The Vernagt tongue remained extended into the Rofental for some thirty years, until 1712 when the remains of the ice dam finally melted away. Very much earlier, by 1683, we learn, 'the people can now live without danger or worry' (Richter 1892: 394).

Scarcely had the last remnants of the Rofen ice dam melted away than in 1717 the Gurglerferner, advancing down its valley, blocked the Langen tributary (Figure 5.2) and a lake formed in Langtal. This seems to have been the culmination of an advance that had been under way for many years. J. Cyriak Lachemyr, who visited the glacier on 2 July 1717, mentioned that an old hunter 'who had worked the area for more than forty years affirmed that in the course of that time the Ferner had driven further down into the valley' (Richter 1892: 414). Again there was concern about the danger of flooding in Ötztal. The Warden of Petersberg reported to Innsbruck that the 'Gurglerferner had formed a lake quite unexpectedly and without anyone knowing anything about it until three days ago'. However, Sonklar (1860), who based his account on Walcher (1773),[5] states that the Gurglerferner had extended past the junction with the Langtal for the first time in 1716. At any rate, by 1717 the ice was holding back a lake, 1600 paces long, 500 paces wide and about 70 fathoms deep. Avalanches of snow made it difficult to tell whether or not the water was escaping between the ice and rock, so guards were posted and prayers, processions and masses were arranged.

The Petersberg Warden was told by his messenger, a native of Sölden, that there was a story that the Gurglerferner had erupted three centuries earlier, causing damage. Kuen had also heard the story but was unable to find any support for it. Richter was not inclined to believe it. Nevertheless, it is quite possible that there is some substance in the story. Walcher (1773: 4)

---

5 Josef Walcher, a Jesuit, Professor of Mechanics at the University of Vienna, visited Ötztal in 1772 and afterwards published an interesting little book, *Nachrichten von den Eisbergen in Tirol*. Many of the documents unearthed by Richter were unknown to him and he based his account of the events of the earlier part of the eighteenth century on Kuen.

heard it said that glaciers appeared in the Tirol for the first time in the thirteenth century, when several cold winters followed each other and so much snow and ice collected in the high mountains that the sun was unable to melt it. Stoltz (1928: 19) cites a text of 1315 according to which farms in the Ötztal were granted respite from their taxes in that year as their lands had been 'ravaged ex alluvionibus et inundationibus'. There is no direct evidence that such damage was connected with glaciers advancing, but in the light of evidence of ice advances associated with tax remissions in Norway about this time, as well as direct evidence of glacier advances in other parts of the Alps, especially Switzerland, it is likely that the same was happening in the Tirol (Chapters 3 and 6, pp. 61 and 153).

As a consequence of the governor's report, a Commission from Innsbruck arrived on 2 July 1717, only to find that the lake had emptied the previous day. The damage was not very great, though a number of bridges had been wrecked. Lachemayr, a member of the Commission, pointed out that the Gurgler, not the Vernagtferner, was to blame: 'your Excellency might graciously note that this one is not the glacier of which there is a model at Innsprugg'.[6] The Rofental, 'according to reliable information . . . is now entirely free of water'. Lachemayr visited the lake held back by the Gurglerferner and reported that it was '1600 paces long and 500 paces wide'. He went past the lake and up to the foot of the glacier in Langtal. He hoped that the two glaciers might soon be confluent, thereby obliterating the lake.

The Commission recommended that a relief channel should be cut to drain the lake, that guards should be posted, and the river channels of the Ötztal and Inn valleys cleared. In the event it proved too costly to dig the relief channel, but observers were certainly posted and orders were given to expedite protective works, the managers of the salt works at Hall being told to give free wood to the people between Zwieselstein and Gurgl so they could build fortifications and bridges. The lake had emptied by 3 August, but on 16 September the Warden of Petersberg wrote that he was expecting another outburst within a year.

Guards visited the lake trapped by the Gurgler through the winter of 1717–18, receiving 30 florins on each occasion. On 1 April the Warden of Silz reported that, according to Erhard Prugger, 'the glacier has grown both longer and broader, not so much or so suddenly as last autumn, but it might advance if the weather grows warmer or there is a long period of rain'. Prugger's experience had evidently led him to a good understanding of glacier movement.

By May 1718 the lake was 1100 paces long and only 540 paces from the Langtal tongue. On 2 July, Langenfeld and Umhausen sent representatives to join a procession from Sölden to the lake. Jacob Kopp, vicar of Sölden, said mass and the representatives met to discuss possible remedies. On 6 July the vicar wrote to Lachemayr,

> I have already said mass there on the ice for the third time and done everything spiritual for the prevention of this evil. But I have not been able to find any means, nor have other competent people of whom many have been up there, with which to improve the situation or remove the imminent danger by human power.

However, he went on to suggest that it might be possible to widen the narrow gorge above the village of Gurgl using gunpowder.

By 14 July 1718 the lake was rising 2 inches every two hours and was expected to overflow in six days' time. But two days later water began to escape under the ice on the right-hand side, and on 1 August Kopp was able to tell Franz Lachemayr, a government secretary, that 'at the lake where I say mass every week 17 fathoms, or half the water, has run off without causing any trouble. And we have good hopes that the rest will run away between the ice and the rock without causing any damage at all.'

The sequence of the lake filling and emptying seems to have become a recurrent event, at least until 1724. After that there is no information about the Gurgler and its lake for fifty years, except that in 1740 when 'Christian Gstrein measured the Ferner at Gurgl . . . the breadth of the lake was 145 fathoms, the depth 66 fathoms and the length 500 fathoms' (Richter 1892: 388).

---

6   As far as is known this early model has not survived, nor is there any record of its maker.

5.3 THE ADVANCE EPISODE 1769–74

The Ötztal glaciers were said by Stotter (1846: 26) to have retreated for a while around 1750, but whether or not this was in fact the case, by the 1770s both the Vernagt and Gurgler glaciers, which had been out of phase earlier in the century, were in advanced positions. There had been several floods in the preceding decade in the Ötztal and other parts of the Tirol. In 1763 the Härlachbach fed by the Grasstallersee, the Stralkogler and the Grieskoglerferner, extensively damaged Umhausen, carrying away or covering with debris sixty-two houses, 'so that the next morning their owners could no longer recognize the place where the houses had been' (Walcher 1773: 62). In 1769 the Farstrinnerbach, which was usually very small, covered the streets of Oesten, near Umhausen, with debris, so that one had to descend to the door of the church instead of climbing six steps up to it.

Walcher (1773) gives a detailed description of the condition of Ötztal in the third quarter of the eighteenth century. Maize, wheat and other crops were grown around Ötz and Umhausen; Langenfeld specialized in flax. Further up the valley, barley and oats were grown but ripened completely only in warmer years. Grass and herbs could be cut only every second year towards the head of the valley, though the pastures near the ice were particularly rich 'so that it is difficult to find richer milk or better butter than at Fender [Vent] or Rofen'. Rofen was the last inhabited place.

By 1769 the Gurglerferner's advance was causing disquiet. It held back a lake that threatened to overflow and, although it was only 500 fathoms long in early September 1770, when Johann Peter Hurn, a district inspector of highways, visited it, he noted that it had been three times that length not long before (Richter 1892: 422). Walcher wrote that the glaciers in the area were increasing all the time. When he visited Ventertal he found that even the lesser glaciers such as Latschferner were swollen, much crevassed and with steep tongues reaching far down into the main Ventertal (Plate 5.2). The tongue of the Vernagt advanced more than 100 fathoms between the spring and autumn of 1770 and a further 25 fathoms in a few weeks in 1771. Not only the Vernagt but also the neigh-

Plate 5.2 Latschferner in August 1772. This illustration shows the enlarged tongue reaching far down towards Venterdal. A and B are mountain peaks, between them seracs or 'ice pyramids' C and E are shown. Walcher noted that the space between D, the source of the Latschbach and the ice pyramids was partially covered by old snow and firn. The tongue was thus protected by snow cover very late into the ablation season (*Source*: from Walcher 1773)

bouring Guslarferner were advancing and the local people were well aware of the serious implications of this conjunction of events (Walcher 1773: 35–7). An inspection made on 12 June 1771 showed that the gap through which water could drain down the main valley was only a hundred fathoms wide and a crevasse 800 fathoms long separated the 'Fernaggerferner' from 'the mother ice' (Richter 1892: 424–5).

In Innsbruck on 21 July 1771, the Governor of Upper Austria conferred with various experts including a professor of mathematics, Weinhart, the judge of Petersberg, Kirchmayr, and a clerk of the works called Nenuer. It was decided to prevent at all costs the threatened devastation. A new commission with extensive powers was set up and provided with a copy of a new

Plate 5.3 The Gurglersee in 1772 (A). The Langthalerferner on the left of the picture is marked B. The mountain in the centre is the Schwarzerberg (C). A number of travellers are making their way over the Gurglerferner (D) on one of the recognized long-distance routes (F and G) (*Source*: from Walcher 1773)

map of the Tirol.[7] Included on the commission were administrators, technicians and people with local knowledge: the manager of the mint and salt works, a Jesuit father, the clerk to the board of works, the judge of Petersberg, district highway inspector Hürn, a carpenter and a wood merchant. They reckoned the glaciers were advancing as a result of wedges of ice breaking away from the crevassed masses on the upper slopes and it was not considered that much could be done to prevent this.

On 29 July 1771 the commissioners went to inspect the lake held back by the Gurglerferner and found it to be 1600 paces long, 500 wide and 50 fathoms deep (Plate 5.3). The water was running away more slowly than usual and there were fears the outlet might become blocked altogether and freeze up. Cairns were erected to mark the position of the ice front and a post was set in place for observing the changing level of the water. If it were to rise a few fathoms more, than it was agreed that twenty or thirty workmen should dig a trench through the ice 280 fathoms long and 6 fathoms deep to let the water out. Reports made in July, August and September have been lost. It is known that Hürn was given the task of hiring men to prevent blocks of ice obstructing the stream flowing out of the lake. They carried on with this work until the end of October, though by then Hürn had died.

The main threat was from the Vernagtferner, which had continued to grow through the winter of 1771–2

---

7 Walcher noted that not only the larger villages, but 'also most of the houses, streams, bridges and roads' were marked on 'a new map of Tirol' then shortly to appear, which had been made with great diligence and exactitude by Ignos Weinhard, SJ, official lecturer in Mathematics and Mechanics at the University of Innsbruck, and Peter Annich and George Heuber, two Tirolese peasants (Walcher 1773: 2).

Plate 5.4 The Rofnersee in August 1772 as seen from a position on the Zwerchwand. The seracs of the Vernagtferner in the right foreground block the Rofental and hold up the lake (A). Note that the Vernagt tongue is not only very steep but fills its valley to a high level. The Plateikogl is marked H and I. The Kesselwandferner is marked N and the Hochjochferner D. F is the higher part of the Vernagtferner, M is the Guslarberg (*Source:* from Walcher 1773)

and by April was 60 to 70 fathoms high and 400 to 460 fathoms broad. It had obstructed the Rofen stream since November 1771 and by April the lake behind was 900 fathoms long and 30 to 40 fathoms deep. Prantl, Steward of Sölden, mentioned that the glacier was making a great cracking and rushing sound that could be heard at times in the houses of Rofen (Richter 1892: 429). By 8 June the lake was rising fast and a crack had formed 10 to 15 fathoms from the side of the valley. It was found impossible to dig an outlet channel because the ice was constantly moving. In a report to the government, Menz proposed using artillery to batter a way through the ice. But guns could not be brought to such a remote place, so bridges were torn down and a path was built down the valley above the level of any possible flood. By 3 July 1772, a flood was expected within a few days and Menz regretted that people had not been persuaded to work on dig-

ging a canal. On 20 July the ice barrier was as high as it had been in 1678 and the lake was 130 fathoms wide. Kirchmayr reported on 20 July that 'in all three valleys the ice is more crevassed right up to the passes than it was last year' (Richter 1892: 430–1). By mid August, when Walcher visited Rofen, he was astonished by the size of the mass of ice damming back the lake (Plates 5.4 and 5.5).

> The whole of the Vernagt valley is filled with especially big blocks right up to the path, and more still are coming. Standing on the Plattei summit . . . an incredible amount of ice stretches as far as the eye can see and it will certainly come down as soon as the other ice clears away.
> (Walcher 1773: 26)

Walcher attributed the many crevasses to the very hot summer. The ice continued to advance into the Rofental and by 19 September it reached 17 fathoms beyond the marker posts that had been erected. Then to

Plate 5.5 The Vernagt ice (E) in 1772 seen from a distance. The Rofental gorge below the dam (S) is in the centre foreground. The gentler slopes of the Platei (H) were used to collect hay which was stored locally (T). Stone cairns (P) mark the routeways. I is the Plateykogl (*Source:* from Walcher 1773)

everyone's surprise and relief the water found a way out between the ice dam and the side of the valley and over a period of eight or nine days drained quietly away.

In the summer of 1773 the lake rose 5 or 6 fathoms higher than in the previous year, according to Prantl, and then emptied suddenly, falling 13 fathoms between 11 and 12 July, and 30 fathoms in five hours on 23 July. But the other streams happened to be low and, surprisingly, 'the flood did no damage worth mentioning' (Richter 1892: 422).

In June 1774, the Vernagt was well forward, 'reaching the old marks in many places' and even exceeding them. By 22 June it was only 10 paces below its highest recorded level of 1681. Two days later, heavy rain

caused all the rivers and streams to rise and on 26 June the water found an outlet. Until 4 July it drained away slowly and then at six in the evening Prantl at Sölden saw the river rise suddenly and continue to flow strongly until next morning. The lake fell 31 fathoms in 12 hours but no damage was done. The danger was over for that year, but 'what will happen next year God only knows'.

An anticlimax; in 1779 the bookkeepers of the Imperial Government recorded that 1920 florins had been spent on the dismantling of twenty-two bridges in the early 1770s in Ötztal and for rebuilding them in 1779. The Vernagtferner had evidently melted back and the danger of flooding had disappeared.

5.4 GLACIAL ADVANCES IN ÖTZTAL 1845–50

For over thirty years little information is forthcoming about the Ötztal glaciers. Sonklar (1860, cited Richter 1891: 31) tells us that the years from 1812 to 1816 were cold and damp, and 1816 and 1817 were years of famine (Richter 1891: 31). The Niederjoch and Gurglerferner advanced (Rohrhofer 1953–4: 58). The Hintereisferner reached a maximum in 1818 (Richter 1891: 28). The Vernagt advanced with the others but not nearly as far as in 1770–4. Ensign Hauslab, later to become a prominent cartographer, was sent to survey the Vernagt area in 1817 and produced a fine map (Plate 5.6) which shows the Vernagt tongue enlarged but still 1.4 km away from the Rofental. Its advance continued until 1820–2 but the terminus did not reach into the main valley (Stotter 1846).

The advances of the Ötztal and other Tirolese glaciers in the early nineteenth century were less important than those that took place in the Mont Blanc massif about the same time. However, the mid century advances were as important as those of the Brenva and others on the Italian side of Mont Blanc and the expanded condition of the glaciers in 1850 is shown on a sketch map published by the Schlagintweit brothers (Plate 5.7 and Figure 5.3). The Marzellferner, Mutmalferner and Schalfferner came together at the head of Niedertal; the Hintereis joined the Kesselwand and the Vernagt came down across the Rofental.[8] The

---

8   Some confusion can arise because the names used for glaciers in Ötztal vary. Schlagintweit called the Schalff–Marzell the 'Stock–Marcell', the Mutmal the 'in den-Schwarse', and the Gurgler the 'Grosser Ötzthaler', whilst Sonklar called the Marzell the 'Marzoll'.

Plate 5.6 Hauslab's map of Vernagtferner in 1817 (*Source:* from Finsterwalder 1897a)

Plate 5.7 The end of the Vernagtferner in 1847 (*Source:* from Schlagintweit and Schlagintweit 1850, figure 55)

Niederjoch reached down beyond the 1770 limit in 1845, retreated a little and then, in 1850, returned to its 1845 position, remaining there for six years (Schlagintweit and Schlagintweit 1850, Rohrhofer 1953–4: 59). The Gurgler pushed well past Langtal and the emptying of the Langtalersee for a time became an annual event (Schlagintweit and Schlagintweit 1850: 145). Both the Schalfferner and the Hintereis resembled the Niederjoch in reaching a first maximum position about 1845–8 and a second about 1855–6 (Schlagintweit and Schlagintweit 1850, Richter 1891: 34).

The behaviour of the Vernagtferner as usual demands special consideration. Between 1822 and 1840 it had retreated a good deal 'more than one hour's walk' writes Stotter (1846: 28). Then it began to advance and for the first time measurements were made of the rate at

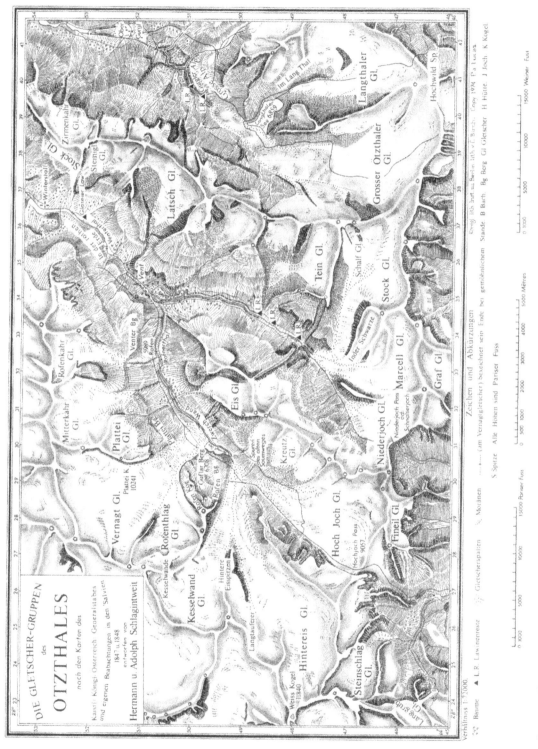

Figure 5.3 The Schlagintweits' map of the mid nineteenth century. Note the extended state of the Vernagtferner (*Source:* Schlagintweit and Schlagintweit 1850)

which this was happening. A farmer of Rofen called Nikodemus Klotz, who was also a chamois hunter, had noticed in 1840 that the névé of the Gurglerferner had thickened and that there were enormous quantities of snow on the upper Vernagt. Encouraged by the parish priest of Vent, he kept records which show that the terminus of the Vernagt advanced 3 m between 2 and 9 April 1843, and 200 m between 18 June and 21 August, its tongue filling the old lateral moraines (Frignet 1846: 88, and Figure 5.4 here). It continued to advance through the following winter, pushing a wad of snow in front of it and gradually accelerating (Table 5.1).

On 1 June 1845 the schoolmaster at Vent reported that the Vernagt had blocked the Rofental. A technical commission was set up 'to examine the situation at the Vernagtferner scientifically and practically and to make proposals for preventing or at least controlling the threatened danger'. The party of officials and experts assembled near Vent on 13 June 1845 and was joined by representatives from Sölden and other villages. They set off for Rofen but 'did not dare venture into the crevasses and seracs at the Zwerchwand', so it seemed that they would only be able to make a cursory survey. However, a small party of local mountaineers came to their assistance, and, crossing the Rofen ravine on a snow bridge, 'skilfully and surely climbed the often almost vertical rock. We shuddered to watch them as they went. Hr Bergratt Zottl had instructed them in the use of instruments and had indicated the lines they were to measure' (Stotter 1846: 49).

The party had completed its measurements of the lake and was crossing the Rofen meadows to reach the bridge when the stream suddenly changed colour and became dark brown, bringing with it lumps of ice. The cry 'the lake is breaking out' went from mouth to mouth. They hurried to the bridge and found the water rising gradually at first but soon much more rapidly. 'The lake with all its power had suddenly broken through the ice dam' (Stotter 1846: 51). Next day Klotz went up to the ice dam and returned to Vent by the time of Sunday mass to tell them that the water had not cut through the ice barrier but had found a way underneath it; the opening was already blocked and the lake beginning to refill.

As they returned down the valley they saw the damage that had been done already. 'From Vent to Umhausen scarcely one tenth of the track along which we had ascended was passable.' Out of twenty-one bridges between Rofen and Umhausen only three remained undamaged. In the rocky gorge between Zwieselstein and Sölden debris hung in the trees many fathoms above the stream bed. The floodwater had destroyed the dykes protecting the meadows and fields in the Sölden basin, carrying away everything in its path. Though the buildings were all on the edge of the valley or on slopes rising from its floor, they were quite badly damaged. Farmland had been flooded as far downstream as Langenfeld. The flood wave had reached the Rofen bridge shortly after five in the evening, Sölden at seven; it was between one and two the following morning that it reached Innsbruck; 'thus', wrote Stotter, 'the water traversed the 22 hour way from Vent to Innsbruck in about eight hours'.

The lake formed again in 1846 and drained away slowly. There was another sudden emptying on 25 May 1847 and again on 13 June 1848. Hoinkes (1969) has estimated that the volume of the 1847 lake was comparable with that of 1678, about 10 million m^3; that of 1848 contained about 3 million m^3 of water (Hess 1918).

The velocity of the Vernagt had greatly diminished. When the Schlagintweits measured the surface movement between 28 August and 20 September 1847, at two cross-profiles 483 m and 840 m up the glacier, they found it to be 0.13 m and 0.09 m per day respectively, a great deal less than the 2.3 m per day in the period July 1844 to June 1845. The surface of the glacier had bulged up and had been deeply crevassed (Sonklar 1860: 37). After 1847 the surface got lower and when he saw it in 1852 Sonklar was able to appreciate the change that had taken place:

> The bulge in the middle had disappeared and by looking at the sides of the valley it was possible to see that the surface of the ice had collapsed no less than 250 to 350 feet or 80 to 100 metres since the time of its maximum. . . . The strongest impression was produced by the enormous extent of the two lateral moraines . . . the glacier appeared like the bottom of a valley between these two moraines.

Since 1848 there have been no more floods. The catastrophic emptyings of the Vernagt lake took place between about 1600 and 1848 and in addition there were six occasions when there was reasonable cause

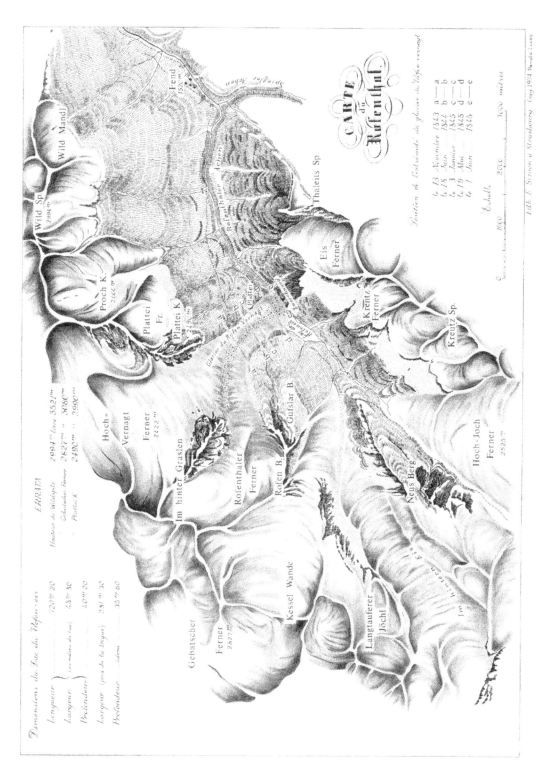

Figure 5.4 The 1843–5 advance of the Vernagtferner (*Sources*: this map was published by both Frignet 1846 and Stotter 1846)

Table 5.1 Mean daily rates of advance of Vernagtferner from 18 June 1844 to 1 June 1845

Period	Mean daily rate of advance (m)
18 June–18 October 1844	1.0
18 October 1844–3 January 1845	2.1
3 January–19 May 1845	3.3
19 May–1 June 1845	12.5

Table 5.2 Dates of filling and emptying of the Vernagt lake between 1599 and 1846

Lake first formed	Slowly emptied	Rapidly emptied
1599	1601	15 July 1600
1678	1679	16 July 1678
	1681	14 June 1680
1771	1772	23 July 1773
	1774	
1845	1846	14 June 1845
		28 May 1847
		13 June 1848

Source: Hoinkes 1969

for alarm in Ötztal but the water drained out slowly (Table 5.2).

5.5 THE RETREAT OF THE GLACIERS AFTER 1850

The glaciers of the Ötztaler Alpen continued to diminish in volume from 1850 until about 1964, though there were periods between 1890 and 1900 and again around 1920 when many advanced.

Although there is little precise information for the years between 1850 and 1890, all available accounts agree that the glaciers were retreating (Kerschensteiner and Hess 1892). Towards the end of the 1880s it became known that the Swiss glaciers were beginning to advance again and this resulted in a widespread effort to record the positions of the Tirolean ice tongues.

In the summer of 1883, Richter started a programme of mapping in the Ötztal. He intended to base his work on the 1:25,000 military map of 1870 but found that it was not of much use except for the Vernagtferner.

Table 5.3 Nineteenth-century retreat of the Vernagt, Mittelberg, Taschach and Gepatsch glaciers, Austria

Glacier and source	Interval	Retreat (m)	m/y
Vernagt (Richter 1885)	1847–83	2092.5	58.1
Mittelberg (Richter 1885)	1856–70	162.5	11.6
	1870–83	717.5	55.2
Taschach (Richter 1885)	1856–78	353.0	16.0
	1878–83	137.0	27.4
Gepatsch	1856–88	460.0	14.3
(Finsterwalder 1928)	1886–99	153.0	51.0
	1891–6	127.0	25.4

He therefore concentrated on making simple tape measurements of the positions of the fronts of nine glaciers in addition to the Vernagt. They showed substantial differences. The Hintereisferner, which terminated in a narrow gorge, had retreated only 150 m since the mid century maximum, whereas the retreat of the Mittelberg was quite obvious from a distance, for the terminus had receded 880 m since Sonklar saw it in 1856 (Sonklar 1860: 134) (Table 5.3). The Taschach had gone back 490 m between 1856 and 1883 whereas the Marzell–Schallferner tongue, ending in a ravine, had retreated only 72 m though it had thinned by about 100 m. The surface of the Gurgler was 20 to 25 m lower than it had been, though the terminus had not retreated very much. Richter was intrigued by these variations and realized that more knowledge was required about the relationships between the frontal positions of the ice, the height of the névéline and the form of the subglacial surface. Many of the glaciers seemed to have got thinner until about 1870 and then their fronts had begun to retreat rapidly. The Vernagt (Plate 5.8) was a law unto itself. In the 1880s its tongue was remarkably small in relation to its overall area. It occupied over 16 km² above the 2800 m contour and only 103 ha below that height.

Finsterwalder, Schunck and Blümcke from Munich selected the Gepatschferner for particular attention in 1886 and 1887 (Finsterwalder and Schunck 1888). Collaborating with Kerschensteiner of Nuremberg, they mapped the tongue of the Vernagt (Finsterwalder 1897b), and they also attempted, rather unsuccessfully, to make velocity measurements. The Hochjochferner was mapped in 1890 and the Hintereisferner in 1893–4 (Richter 1893–4).

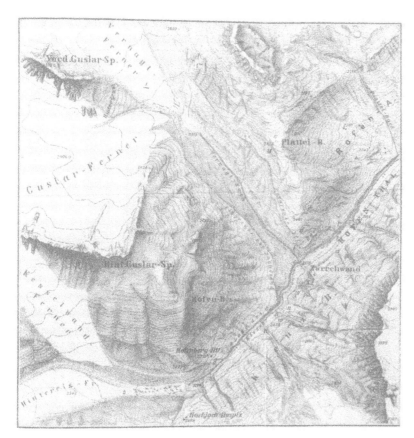

Plate 5.8 The Vernagtferner and Guslarferner in 1888 (*Source*: from Richter 1892)

As early as 1874 Richter advocated regular observations of the frontal positions of a large number of glaciers in the eastern Alps. His scheme was eventually adopted by the Deutscher und Österreichischer Alpenverein in the 1890s. By this time the recession that had begun in the 1850s was slowing down or had halted. The research committee of the Alpenverein requested its members and sections to collaborate in making regular observations and to continue with those already being made (Finsterwalder 1891). Observations on the Niederjoch tongue started in 1891 (Rohrhofer 1953–4), and the Breslau section of the Alpenverein began measurements of the Gurglerferner (Plate 5.9) in 1895, though this particular set of observations was discontinued within a few years because of the difficulties of the site (Srbik 1942a). The number of tongues measured varied from year to year and was not to reach eighty until 1924 (Patzelt 1970).

In 1891 about 40 per cent of the glaciers being measured in the eastern Alps were advancing, and by 1900 about 50 per cent; this compares with only 30 per cent of the glaciers in the Swiss Alps at the beginning of the century (Patzelt 1970). The majority of the glaciers in the eastern Alps then receded for a decade before advancing once more; by 1919 about 75 per cent were waxing. Then came a major recession, interrupted only briefly between 1925 and 1927. In the Ortler and Stubaier Alpen accumulation might have been enough to cause enlargement, had it not been for the dry, hot summers of 1928 and 1930, with high rates of ablation (Steinhauser 1957).

From a map study of the changes in the area and height of the surface of the ice in eight representative glaciers, R. Finsterwalder (1953) was able to show that the ice had continued to diminish in volume as well as in extent since the mid nineteenth century. An analysis

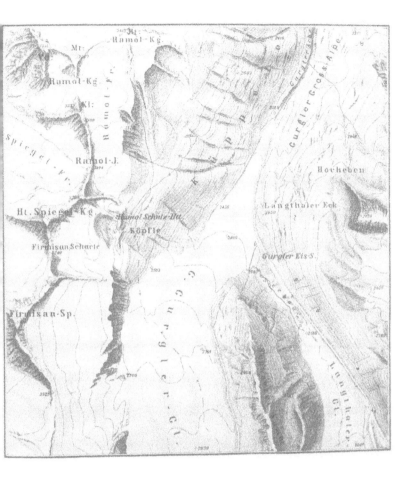

Plate 5.9 The Gurglerferner in 1888
(*Source:* from Richter 1892)

of the results from glaciers, which included the Gepatschferner and the Hintereisferner, indicated that rates of lowering were similar for the periods 1850–90 and 1920–50, about 61 cm per year; between 1890 and 1920, wasting was only half as fast, about 30 cm per year. The variability of rates of lowering in time and space is indicated by Table 5.4.

The retreat of the glaciers at the head of the Spiegler Ache (Figure 5.2) in Ötztal has been traced by Srbik (1935, 1936, 1937, 1941, 1942a, b) and by Rohrhofer (1953–4). The withdrawal of the Niederjoch, Marzell, Mutmal and Schallferner between 1850 and 1950 is portrayed in Figure 5.5. The waning of the glaciers of the eastern Alps, and of the Ötztal in particular, was fastest between 1938 and 1951 (Figure 5.6). It continued into the 1960s, with the wastage in 1963 and 1964 being particularly severe. Then heavy snowfall in

the spring of 1965 was followed by a cool, damp summer. That same year, for the first time since 1921, less than 50 per cent of the glaciers being measured in the eastern Alps were receding (see Chapter 6 and Figure 5.6). From 1966, more than 25 per cent were recorded as advancing (Patzelt 1993). The Rofentalferner and the adjacent Kesselwandferner in the Ötztal, which had advanced in the early 1890s and again about 1926, again responded quickly (Patzelt 1970). On average from 1974 to 1985, as many glaciers in the eastern Alps were advancing as were retreating (Patzelt 1993, figure 2). Then nearly all of them began to retreat.

5.6 MASS BALANCE STUDIES IN THE ÖTZTAL

The cartographic studies made of the Ötztal glaciers in the closing years of the nineteenth century and the early

Table 5.4 Annual changes in the surface height at various altitudes of the Hochjoch, Guslar and Vernagt glaciers between *c.*1893 and 1940

Altitudinal zone (m)	Changes in height (m)					
	Hochjochferner		Guslarferner		Vernagtferner	
	1893–1907	*1907–40*	*1889–1912*	*1912–40*	*1889–1912*	*1912–40*
2500–600	−2.32	−1.50				
2600–700	−1.62	−1.06	−1.55		+0.05	
2700–800	−1.28	−1.04	−0.51	−0.57	−0.13	−1.07
2800–900	−0.84	−0.49	−0.42	−0.59	−0.53	−0.51
2900–3000	−0.92	−0.40	−0.20	−0.39	−0.26	−0.07
3000–100	−0.75	−0.10	−0.17	−0.29	−0.57	−0.03
3100–200	−0.47	+0.05	−0.24	−0.35	−0.41	−0.07
3200–300	−0.69	+0.14	−0.28	−0.35	−0.20	−0.10
3300–400	+0.23	+0.09	−0.36	−0.36	−1.12	−0.33
3400–500					+0.04	−0.40

Source: R. Finsterwalder 1953

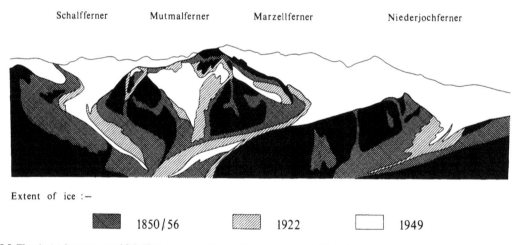

Schalfferner   Mutmalferner   Marzellferner   Niederjochferner

Extent of ice :—

■ 1850/56   ▨ 1922   ☐ 1949

Figure 5.5 The diminishing extent of Schalfferner, Mutmalferner, Marzellferner and Niederjochferner between the mid nineteenth and mid twentieth centuries (*Source*: after Rohrhofer 1953–4)

decades of the twentieth by Finsterwalder (1888, 1889, 1897a, b and 1928) and Hess (1918, 1924 and 1930) provided a useful basis for the studies of mass balance that have been made since the Second World War.

Mass balance studies of a systematic kind began in Rofental in 1952, when Schimpp (1958, 1960) inserted a network of stakes in the Hintereisferner. This was extended by Rudolph in 1954 and kept under continuing observation, in a judiciously reduced form,

from 1954 onwards (Hoinkes and Rudolph 1962a, b, Hoinkes 1970). Mass balance observations were extended to the Langtalerferner in 1962 and to the Kesselwandferner and Vernagtferner in 1965. The earlier maps were used in conjunction with new surveys by the Bavarian Academy of Sciences to produce a series of five maps of the Vernagtferner, at a scale of 1:10,000, for the years 1889, 1912, 1938 and 1969. These accompany *Fluctuations of Glaciers 1965–70* and

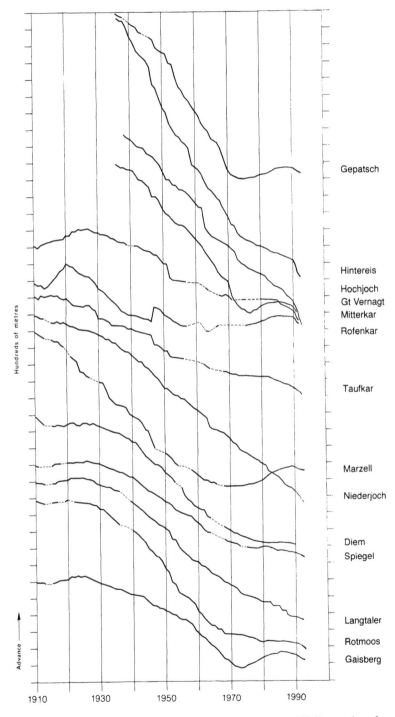

Gepatsch

Hintereis

Hochjoch
Gt Vernagt
Mitterkar

Rofenkar

Taufkar

Marzell

Niederjoch

Diem
Spiegel

Langtaler

Rotmoos

Gaisberg

Hundreds of metres

Advance

1910   1930   1950   1970   1990

Figure 5.6 The fluctuations of the Ötztal glaciers since regular measurements began until 1992 (*Sources*: from figures published in *Zeitschrift für Gletscherkunde und Glazialgeologie* by Patzelt and others)

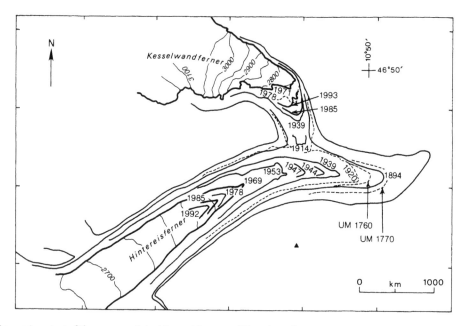

Figure 5.7 Changes in extent of the tongues of the Hintereisferner and Kesselwandferner since 1847 (*Source:* from Kuhn *et al.* 1985)

Fluctuations of Glaciers 1970–1975 (Müller 1977) in the series published for the International Commission on Snow and Ice by IAHS and Unesco. They allow the variations in the positions of the ice fronts and the surface contours to be compared in detail in the successive periods between the maps.

Of the first twelve years of observations of the Hintereisferner, all except two revealed negative budgets, the exceptions being 1954/5 and 1959/60. The mean specific balance for the 12-year period was −48 g/cm², that is an overall lowering of about half a metre annually. During this period of predominant loss between 1952 and 1964, more than twice as much was removed by ablation as was added by accumulation, and the average area of net nourishment exceeded the average area of net wastage by 47 per cent. The average loss for the period was very greatly influenced by two years of high budget deficits, 1957/8 and 1958/9. Temperature and radiation values were high in the summer of 1957, removing not only the snow of the previous winter but also exposing to ablation the firn layers of at least five years. There were few falls of snow or cloudy days to afford protection. The summer of 1959 had much lower radiation and lower average temperature, but mass loss was again high as there had not

been much snow the previous winter, and a large area of low albedo (low reflectivity) was consequently revealed. Over the two years 1957/8 and 1958/9 net ablation removed seven times as much mass as was gained by accumulation. The area of net accumulation was only a half that of the area of net ablation.

A sequence of years with predominantly positive mass balances came after 1964/5, when winter snowfall was above average and the following summer was cool and cloudy with not more than five days in a row having sunny weather. At the end of the ablation season in 1965 the old snowline was much lower than usual and the area of winter snow surviving was three times the area of the glacier surface from which it had been removed. Mass balances on the Hintereisferner were positive in each of the four years until 1967/8; then came two years with a negative balance. The six years with a positive balance, 1954/5, 1959/60 and 1964/8, more than counterbalanced the effects of twice as many years with a negative balance, for when Finsterwalder and Rentsch (1976) compared photogrammetric surveys of eight glaciers in the eastern Alps for 1950 and 1969, they found a mean rise in surface level of 0.1 m per year over the whole period. This compares with a mean annual lowering for the

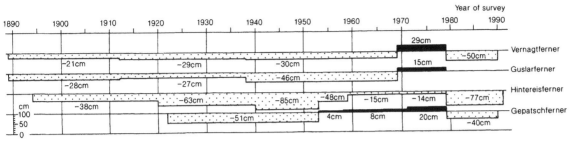

Figure 5.8 Annual changes in elevation of four glaciers in the Ötztaler Alpen, 1889–1990 (*Source*: from Haeberli 1985)

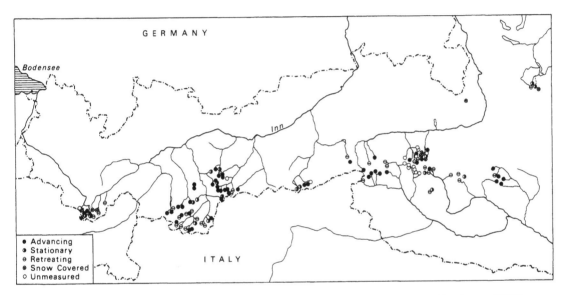

Figure 5.9 Distribution of advancing, stationary and retreating glaciers in Austria amongst those observed for the year 1983/4 (*Source*: from Patzelt 1984)

period 1920 to 1950 of 0.6 m. Between 1969 and 1979 (Finsterwalder and Rentsch 1981), there was again a mean rise in the surface level of all the glaciers in the sample, except for the Hintereisferner, but the rise was less than in the previous decade. The higher snowfields did not show much change, however, suggesting that the period of expansion was tailing off. But the surface of the Gepatschferner rose considerably between 1971 and 1973, although the tongue was still receding in 1977 (R. Finsterwalder 1978).

The first three years of mass balance studies of the Vernagt were in the period of budget surplus of 1965–8. In those three years 12.6 million cubic metres of water equivalent were added to the glacier, thereby increasing its bulk by 2.5 per cent. In the succeeding six years the general tendency for ablation to exceed accumulation was resumed, but during the 1970s positive mass balances became the rule once more and the upper part of the Vernagtferner increased in volume.

5.7 TREE RINGS MARK GLACIER ADVANCES AND RETREATS

Variations in tree-ring thicknesses can help to date glacier advances rather precisely. Around 1816, 'the year without a summer' which followed the eruption of

Tambora in 1815, when the rings of *Pinus cembra* in the Ötztal Alps thinned abruptly and then after five or six years returned to their normal width (Nicolussi 1995, figure 2), the rings of trees growing close to the 1805 moraine of the Gepatschferner thinned in the same manner as those of the Ötztal generally. A few years later, ring thicknesses in these same trees diminished much more markedly and over a longer period than those from other parts of the Ötztal Alps when the approach of the glacier's advancing tongue lowered temperatures in their immediate vicinity (Nicolussi and Patzelt 1996, figure 3). The tree-ring record confirms that by 1855 the Gepatschferner had advanced somewhat further than the 1805 moraine as it is portrayed in a painting between 1801 and 1805 by F. F. Rank. Thin rings also denote the prolonged extension of the ice between 1580 and 1640. The date of the outermost moraine of the Gepatschferner forefield, marking that glacier's furthest Little Ice Age advance, is provided by a tree partly covered by the moraine in which the ring closest to the bark was formed in the early summer of 1679.

Reconstructions were made of the history of the Hintereisferner over the last six hundred years by using ring widths of *Pinus cembra* from stands about 15 km away. The sums of the widths from the early wood of the preceding year to the late wood of the succeeding year were related to annual mass balance values back to the 1930s as calculated from degree-day records. The resultant accumulated 'dendro-mass balance' values show peaks at times known to have seen culminations of Alpine glacier advances: around 1600/1640, 1680, 1720, 1820 and 1850 (Nicolussi and Patzelt 1996, figure 6, and plates in Nicolussi 1995). Between AD 1400 and 1600 the tree-ring record suggests that the glacier behaved much as it did between 1870 and 1940 with retreat dominant. As yet the record is too short to provide any sign of a fourteenth-century Little Ice Age advance.

5.8 THE LITTLE ICE AGE IN THE EASTERN AND WESTERN ALPS COMPARED

A comparison of events in the Ötztal and the Mont Blanc massif shows that the various phases of ice advance and retreat corresponded very closely in the two

regions, if it is accepted that the evidence from the Vernagtferner is indicative of climatic conditions. The first ice-damming of the Rofental at the end of the sixteenth century coincided with the extension of the Mer de Glace and its neighbours into the Arve valley. The formation of the second Rofensee in the late 1670s and early 1680s coincided with the flooding of the Lac du Combal, an extended Miage glacier and enlarged ice masses on both sides of the Mont Blanc massif.

The advance of the Gurgler in the early eighteenth century to a forward position in 1712 is not exactly matched in the western Alps but there were important advances elsewhere in Europe in the early 1700s, notably, as we shall see in Chapter 6, of the Grindelwald Glacier. The Gurgler remained enlarged until 1740 and during this first part of the eighteenth century the Mont Blanc glaciers were all quite extensive. The simultaneous advances of the Gurgler and Vernagtferner and many of the other Ötztal glaciers around 1770 coincided exactly with the vigorous advances of glaciers in the Arve valley to positions comparable with those they reached in the early seventeenth century.

Advances were general in both regions in the first half of the nineteenth century, around 1820 and again about 1850. The advances of the Ötztal glaciers about 1850 brought tongues to within a few metres of their moraines of 1600 and 1770. They remained at or near these maxima until 1855 by which time many of the French glaciers were already waning.

By the end of the nineteenth century the broad parallelism of the behaviour of the glaciers of the eastern and western Alps was well recognized and minor discrepancies were also beginning to be appreciated. It was noticed that the advance of the Ötztal glaciers in the 1890s continued until about 1900, by which time the Mont Blanc glaciers and their neighbours in Switzerland had been in retreat for some years. From the measurements of Klebelsberg (1920, 1926, 1943. 1949) and from diagrams showing the proportions of retreating and advancing glaciers in Switzerland (Kasser 1967, 1973), Austria (Patzelt 1970) and the French Alps (Vivian 1975), it is possible to date the glacial retreats and resurgences from one end of the Alps to the other (Figure 6.11).

It is clear from the comparison that in the late nineteenth and late twentieth centuries glaciers in the western Alps advanced a few years before those in the east. Whereas the front of the Glacier des Bossons began to move forward in 1952, the Brenva and the Allée Blanche in 1955, and the Tour in 1960, the most sensitive glaciers in the Ötztal, such as the Rofenkar, did not begin to advance until 1965 (see Figure 5.6). However, in the two preceding centuries, it seems possible that the Vernagtferner in the east advanced a few years earlier than the Mont Blanc glaciers. But the Vernagt is liable to surge.

5.9 THE SURGING OF THE VERNAGTFERNER

All the evidence available indicates that the Gurglerferner responds to fluctuations in climate in phase with the other glaciers in the Ötztal and it appears that the Ötztal glaciers have in general advanced and retreated at times comparable with those of Mont Blanc. However, the Vernagtferner, which as it happens is the only glacier in the Ötztal for which we have any detailed information before the eighteenth century, is an exception in being remarkable for its unconventional behaviour. Richter (1892: 354), though he mentions there was nothing very unusual about the timing of the Vernagt's advances, was well aware of its abnormality; indeed it was the singularity of the Vernagt's behaviour that decided him to choose to work on it. He knew the Vernagt had been out of phase with the Gurgler and other glaciers in not advancing in the years about 1712, whereas both had advanced in the 1770s and about 1820. Then there had been its remarkably rapid advances in 1599, 1678, 1770 and 1845 followed by long periods of retreat and slow surface speeds. Why should its fluctuations be on so much larger a scale than those of its neighbours?

Hoinkes (1969) argued that the Vernagt was a surging glacier. Such glaciers, sometimes known as galloping glaciers, are typified by catastrophic advances occurring with some degree of regularity, during which speeds of flow are abnormally high. These advances are not always related to variations in climate. Ice accumulates for some years and then for some reason spills out of the reservoir area and pushes its way forward, shearing at the margins and crevassing chaotically. Judging by the intervals between its catastrophic advances, the

Vernagt might have been expected to push forward again round about 1928. This did not happen, though, as Hess (1930) pointed out, the surface velocity did increase slightly between 1924 and 1929. But surges do not necessarily occur with great regularity and it would seem likely that both topographic and climatic elements contribute to the behaviour of surging glaciers.

Hoinkes (1969) emphasized the importance of the part played by topography. Seismic investigations of the rock floor underlying the Vernagt- and Gurglerferner show that a sill exists beneath both of them at an elevation of between 2900 and 3000 m, with basins behind about 50 m deep in the case of the Gurglerferner and about 20 m deep in the case of the Vernagtferner (Miller 1972). The main mass of both glaciers lies above the level of this sill, which in the quiet periods between surges is close to the height of the névéline. Hoinkes suggested that the glacier ice might become frozen to the bedrock at the sill at times when the ice at that altitude was thin, allowing the winter cold wave to penetrate to the bed. If the following summer were to be cool, no great quantity of meltwater would be released and the ice would remain frozen to the sill. Ice might accumulate behind the rock/ice barrier thus formed, until pressure melting point was reached at the bed, when large masses of ice would suddenly be released and move rapidly down the slope towards the Zwerchwand of Rofental.

Clearly surges resulting from any such mechanism would have to depend in large part on mass balance. This point was further developed by Kruss and Smith (1982). They pointed out that the Vernagtferner has passed through two radically different phases during the historic period, characterized respectively by cyclic surging and by shrinkage back into higher regions favourable for accumulation. Between the surge of 1845 and 1966 the glacier contracted in area from 13.8 km² to 9.6 km² and shrank to about half its former volume (R. Finsterwalder 1972). Kruss and Smith calculate that the surging mode could only operate so long as the glacier remains large enough to create substantial basal melting, through a combination of high basal stresses and rapid flow-rates. They considered that the great retreat since 1848 resulted from a climatically led negative mass balance over the entire surface, of the order of 0.2 m per year, the glacier having been particularly vulnerable to ablation at that time because of its very great extension into lower altitudes.

Should positive mass balances reoccur for a sufficiently long period the combined Vernagt–Gurslarferner would surge again. Negative mass balances in the first half of the twentieth century lowered the surface of the glacier; between 1938 and 1954 its surface fell by an average of 70 cm annually. In the late 1960s and again in the late 1970s, the balance was positive and from 1964 until 1984 recession of glaciers in the eastern Alps generally was slight. Between 1979 and 1982 the surface of the accumulation area of the Vernagtferner sank somewhat while parts of the tongue were elevated by 3.5 m per year, rising in some places by as much as 8 m. An interpretation of the volume shift as being the outcome of a surge is supported by a short-lived doubling of the glacier's flow velocity (Reinwarth and Rentsch 1994).

Mass balances of the Vernagtferner and other glaciers in the Ötztal and elsewhere in Austria were persistently negative from 1980 into the 1990s, mass losses being especially heavy in 1995/6 when an extremely dry winter was followed by a warm early summer.

5.10 SUMMARY OF THE VERNAGTFERNER RECORD

Most information about the Little Ice Age in the Ötztal comes from the Vernagtferner, a glacier subject to surging but with a record of advances and retreats which corresponds quite closely with those of other glaciers in the Alps. An advance of the Vernagtferner in the late sixteenth century culminated around 1600. Retreat was followed by another advance in the 1670s and a retreat in the early 1680s but with the ice remaining in a forward position until 1712. After a renewed advance in 1717 which lasted about seven years, there may have been a retreat. Another advance in the 1760s lasted until about 1775 and was as great as that a century earlier. A much less important advance around 1815–22 was followed by retreat, but a renewed advance in the 1840s brought the terminus forward to the position it had reached in 1680 and 1770. General retreat after 1850 was interrupted, but only slightly, between 1890 and 1900 and again around 1920. Then came a major recession, fastest between 1938 and 1951 and continuing until 1964. For the next twenty years recession was only slight but has since accelerated.

Snow-capped Alpine ranges appear on the horizon almost everywhere in Switzerland; routes over the major Alpine passes skirt glaciers; many villages occupy sites threatened even today by avalanches and flooding. No wonder documents relating to glacier advances are abundant. Glaciers have commonly extended below the treeline, overriding trees and leaving their stumps in place. With wood and other organic material suitable for radiocarbon analysis, moraines have been more effectively dated here than in many other parts of the world. Weather diaries and phenological records date back several centuries. The Swiss were the first to make systematic annual measurements of the changing positions of glacier fronts.[1] The climatic implications of glacier fluctuations, first recognized in the early nineteenth century, have attracted renewed scientific attention in recent years (Pfister 1979a, 1980b, 1992a, b, 1994a, b, Pfister et al. 1994a, b, 1996, 1998a, b). As a result of all this activity, the timing and sequence of glacier fluctuations during the Little Ice Age have been traced in more detail in the Swiss Alps than anywhere else.

6.1 EVIDENCE OF GLACIAL EXPANSION IN MEDIEVAL TIMES

The strongest documentary evidence of glacial advances prior to the sixteenth century comes from the eastern part of Valais and from the Bernese Oberland (Walliser Alpen and Berner Alpen, Figure 6.1). The Allalin-gletscher, on the eastern flank of the Mischabel group, is known to have advanced across Saastal, one of the Rhône's left-bank tributary valleys, on several occasions between the sixteenth and nineteenth centuries, each time holding back a lake that eventually spilled more or less violently down the lower valley. The Allalin also seems to have blocked Saastal in the thirteenth century, for a document dated 13 April 1300 (Grémaud 1878) concerns the tenancy of pastures at Distelalp, near the head of the valley and upstream of the site of the Mattmarksee, the name given to the lake that is held back by the Allalin when it advances across Saastal. Two of the local farmers requested the Mayor of Visp 'to grant us tenure of the pasture in question from the glacier upwards, according to the custom of the Saaser Visp valley, so that the said men do not prevent us grazing as far down as the glacier'. This text was interpreted by Lütschg (1926), Ladurie (1971) and Delibrias et al. (1975) as showing that the Allalin was sufficiently enlarged in 1300 to block Saastal and separate the upper mountain pastures from the lower valley of the Visp, as it did again in the nineteenth century. Nowadays the glacier overlooks the valley from the west, not far downstream of the dam that holds back an artificially enlarged version of the Mattmarksee that submerges the old chalets of Distelalp. If the Allalingletscher blocked the valley in 1300, it was much larger than today but probably smaller than in 1822, when the Schwarzberg, a much smaller glacier than the Allalin, extended so far down into Saastal as to bisect the Mattmarksee

---

1 They are now published annually by the Glaciological Commission of the Swiss Academy of Sciences and by the Laboratory of Hydraulics, Hydrology and Glaciology at the Federal Institute of Technology, Zurich.

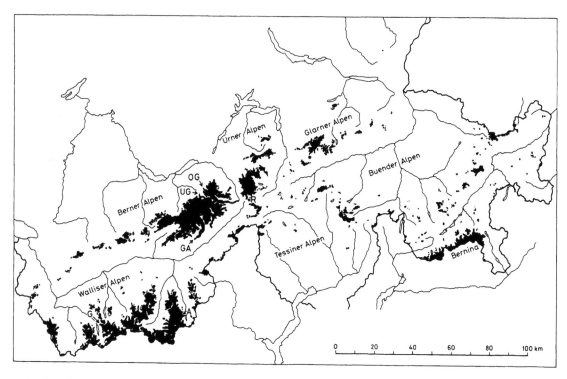

Figure 6.1 Distribution of glaciers in Switzerland. A = Allalin, R = Rhône, G = Gíetroz, GA = Grosser Aletsch, OG = Oberer Grindelwald, UG = Unterer Grindelwald

Plate 6.1 The dam formed by the Allalingletscher in the background, with the extended Schwarzberggletscher entering the Mattmarksee in the middle distance. An etching by Thalés Fielding, from an aquarelle by Maximilian de Meuron, dated 1822 (*Source*: from Lütschg 1926)

(Plate 6.1); in the document there is no mention of two ice dams.

If the Allalin was much enlarged in 1300, we can assume with some confidence, despite the dearth of documentary evidence, that the same was true of other Swiss glaciers, for a great deal of support has been obtained by dating trees killed by the glaciers as they advanced in late medieval times.

The retreat of the Grosser Aletschgletscher in the twentieth century uncovered the remains of a number of larch trees still in the places where they had grown. Tree-ring counts showed that the trees, when they were overwhelmed by the advancing ice, were about 150 years old. Radiocarbon dates from two of them (B-71: 800 ± 100, and B-32: 720 ± 100) which calibrate to AD 1160–1300 and AD 1220–1400 respectively, show that they were growing in the thirteenth century (Oeschger and Röthlisberger 1961, Rothlisberger et al. 1980).[2]

Kinzl (1932) had earlier provided evidence that the Aletsch glacier had begun to advance well before 1385. The Valais is the driest part of Switzerland; both pastures and croplands are watered by networks of irrigation channels (bisses) bringing water long distances down the mountain sides from glaciers and snowfields. Many of them are several centuries old (Brett 1995). The remains of the Oberriederin bisse, which is recorded as having been abandoned in 1385, were exposed in 1961 by the retreating ice of the Aletschgletscher. Lamb (1965) concluded that it had been constructed during the Medieval Warm Period[3] and had to be abandoned in the fourteenth century when glacier ice advanced

over it.[4] Here it might be noticed that the Theodulepass in 1330 was not recorded as having been ice-covered (Wills 1856) whereas in 1528 the historian Aegidius Tschudi di Glarus (1538) stated that 'on its crest there extends for the space of 4 Italian miles a great field of ice that never melts or disappears'.

The Unterer Grindelwaldgletscher has long been known to have advanced into woodland. Gruner (1760) wrote that 'in the central part of' the glacier, 'all the way up on the flanks of the Vischhorn and Eiger, a good number of larch trees (Pinus larix) protrude from the ice. Since the wood of this species is known to harden in wet conditions those trees may have been living there centuries ago.' Gruner seems to have been correct, for samples of larch and other trees collected from moraines in front of the glacier have given ages ranging from (BM-95: 1280 ± 150) AD 640–900 to (Gif-2977: 640 ± 80) AD 1280–1400 (Delibrias et al. 1975). The results provide clear support for advances of the Unterer Grindelwaldgletscher well before the sixteenth century, but as the dated material may not have come from trees in situ, they do not otherwise assist reconstruction of the fluctuation history.

Enlargement of the Allalin, Aletsch and Grindelwald glaciers in the thirteenth and fourteenth centuries could hardly have occurred without expansion of other Alpine glaciers. The chronology of over twenty moraine sequences in the Canton of Valais investigated by research students from the Geographical Institute of the University of Zurich[5] has confirmed that the oscillations of the Allalin, Aletsch and Grindelwald were not

---

2 Calibration of these dates, and all others mentioned here, is according to Stuiver and Reimer 1993.

3 The existence of the Medieval Warm Period has been a matter of controversy but Pfister et al. (1998a) have shown, using verified documentary evidence, that winter temperatures in western Europe from AD 1150 to 1299 were similar to those of the twentieth century. 'The warm and stable climate of the thirteenth century supported sub-tropical plants such as olive trees in the Po valley (northern Italy) and fig trees around Cologne' (1998: 535). This study was based on the many volumes of the *Monumenta Germaniae Historica Scriptores*, the first edition of medieval sources which was critical in method. In all, 8850 meteorological texts were abstracted for the period AD 750–1300.

4 For a more detailed discussion see Holzhauser's chapter on 'Le grand glacier d'Aletsch': 142–65 in Zümbuhl and Holzhauser's (1988) 'Glaciers des Alpes du petit âge glaciaire'. This special number of *Les Alpes* includes colour copies of numerous old maps and paintings of the Aletsch, Rhône, Rosenlaui and Unteraar glaciers.

5 Research in Austria, particularly by Patzelt and Bortenschlager (e.g. 1973) helped to stimulate important programmes of work in Switzerland concerned with glacial and climatic history in the Late Glacial, Holocene and Little Ice Age. Research students including F. Röthlisberger, Schneebeli and Zumbühl, working under Professor Furrer at the University of Zurich, produced a series of diploma projects and theses (listed in Röthlisberger et al. 1980) which, together with those from other Swiss universities, especially Bern, constitute the most detailed history of the Holocene for any part of the world (see Chapter 15).

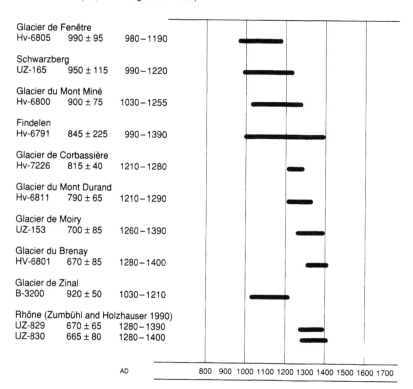

Dates from soil samples beneath moraines except those from Glacier de Zinal, which came from
larch wood.

The Valaisian Alps (Röthlisberger *et al.* 1980)

Glacier de Fenêtre Hv-6805	990 ± 95	980–1190
Schwarzberg UZ-165	950 ± 115	990–1220
Glacier du Mont Miné Hv-6800	900 ± 75	1030–1255
Findelen Hv-6791	845 ± 225	990–1390
Glacier de Corbassière Hv-7226	815 ± 40	1210–1280
Glacier du Mont Durand Hv-6811	790 ± 65	1210–1290
Glacier de Moiry UZ-153	700 ± 85	1260–1390
Glacier du Brenay HV-6801	670 ± 85	1280–1400
Glacier de Zinal B-3200	920 ± 50	1030–1210
Rhône (Zumbühl and Holzhauser 1990) UZ-829	670 ± 65	1280–1390
UZ-830	665 ± 80	1280–1400

AD   800 900 1000 1100 1200 1300 1400 1500 1600 1700

Figure 6.2 Calendar ages (AD) indicating the initiation of the Little Ice Age in Switzerland

Notes: The numbers in the margin give the laboratory identification number, the ^{14}C age and the calibrated age for 68% probability. Horizontal lines represent the calibrated ages

exceptional. The maximum ages obtained by radiocarbon dating of organic samples from beneath many moraines were listed by Röthlisberger *et al.* (1980). The calendar ages of some of these, obtained by calibrating them according to Stuiver and Reimer (1993), are shown on Figure 6.2. All the data on this table were obtained from soil samples, except those from the Glacier de Zinal which came from larch wood. Radiocarbon dates obtained from soils are less reliable than those from wood but the results show conclusively that the glaciers advanced well before the sixteenth century. The positions of the moraines formed by thirteenth- and fourteenth-century advances turn out to be close to those reached by the glaciers in the middle of the nineteenth century.

Much more reliable information can be obtained by radiocarbon dating wood found within moraines; the most accurate of all comes from the stumps of trees killed by advancing ice found in their growth positions (Holzhauser 1984b, 1997). Such *in situ* stumps can provide precise calendar dates of death by comparison of their annual ring sequences with dendrochronological series. Figure 6.2 shows some examples of dates relating to the beginning of the Little Ice Age.

Intensive field investigations made since 1980 have yielded extraordinarily detailed chronologies for a number of Swiss glaciers (e.g. Holzhauser 1983, 1984b, 1995, 1997, Zumbühl and Holzhauser 1988, Pfister *et al.* 1994a, b). The most detailed history of all is that of the Grosser Aletsch Glacier (Figure 6.3) over the last 3200 years (Holzhauser 1980, 1984a, 1997). Reconstruction of the Grosser Aletsch fluctuations was based on a range of evidence including the positions of absolutely dated *in situ* trees, as well as radiocarbon dated samples, many of them also *in situ*. The great majority was found in lateral positions above the present glacier; the heights above the ice surface at which they were found forming an essential element of the evidence

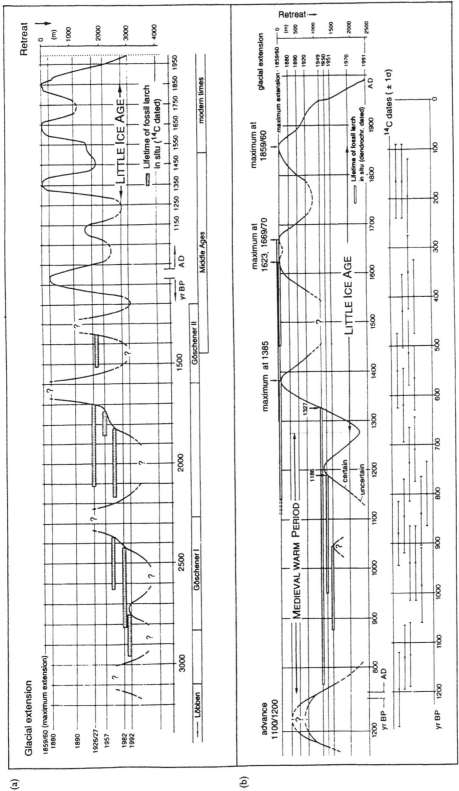

Figure 6.3 The fluctuations of (a) the Grosser Aletsch Glacier over the last 3200 years, and (b) the Gorner Glacier over the last 1200 years (*Source:* (a) after Holzhauser 1984a, revised, (b) Holzhauser 1997)

used in reconstruction. Most of the dates were taken from the outer rings of trees overrun by the ice. The position of the Oberrieden Bisse was taken into account. Additional information was derived from historical documents, paintings, drawings, photographs and direct measurements of the glacier front position made since 1880.

An advance of the Grosser Aletsch preceding the Medieval Warm Period culminated about AD 850–900. The glacier then retracted and was smaller during the Medieval Warm Period than it has been at any time before the present century, apart from a minor advance culminating between around AD 1100–50. This was not on a similar scale to the advances of more recent centuries, and therefore has to be considered as a short-lived interruption of medieval warmth, rather than part of the Little Ice Age. Holzhauser concluded that the beginning of the Little Ice Age 'has to be put in the second half of the 13th century'. The advance culminating in the fourteenth century brought the front forward as far as did those of the seventeenth and nineteenth centuries. The three main culminations were separated by withdrawals which did not suffice to return the front to positions near those of the preceding warm period.

Since the publication of Holzhauser's initial study of the Grosser Aletsch, further investigations have served to reveal portions of the histories of other glaciers in the Swiss Alps including the Reid and Zinal (Holzhauser 1985), and more particularly the Gorner and Grindelwald glaciers (Pfister *et al.* 1994a, b, Holzhauser 1995, 1997, Holzhauser and Zumbühl 1996). Since 1990, when Zumbühl and Holzhauser published a substantial catalogue of their findings, the assemblage of radiocarbon dates and absolute ages from fossil trees has been much extended as have the data on historical and pictorial sources.

The fourteenth-century advance of the Gorner Glacier can be traced from the positions of *in situ* larch stumps killed by the ice, together with absolutely dated tree rings (Holzhauser 1997: 45–50). The front was already far in advance of its present position by 1327; it then advanced by about 10 m per year until 1341 and at an increased average rate of about 21 m per year to reach a maximum in 1385 (Figure 6.4). The position of the tongue in 1385 was practically identical

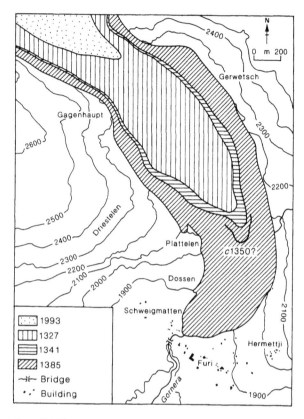

Figure 6.4 The advance of the Gorner Glacier in the fourteenth century (Source: after Holzhauser 1997)

with that reached in 1859–65 during its nineteenth-century culmination.

The three Little Ice Age advance phases of the Gorner and Grosser Aletsch were very similar in both timing and extent (Figure 6.5). However, the initial expansion of the Grosser Aletsch at the end of the Medieval Warm Period was faster than that of the Gorner. It advanced at a speed of 40 m per year, to reach its maximum during the 1370s. The Rhône glacier also reached a maximum between 1350 and 1400, in this case slightly more extensive than advances which have occurred since.

The intensive studies of glacier forefields that have been made in Switzerland, especially by Holzhauser, have thus demonstrated that the Little Ice Age advances were under way in the thirteenth century and culminated for the first time in the fourteenth century, and

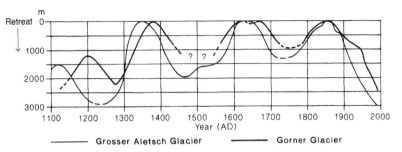

Figure 6.5 A comparison of the fluctuations of the Grosser Aletsch and Gorner glaciers during the Little Ice Age (*Source:* from Holzhauser 1997)

that the fluctuations of the Grosser Aletsch can be taken as typical of the behaviour of large valley glaciers in the Alps.

6.2 THE CLIMATE OF MEDIEVAL SWITZERLAND

The climatic circumstances responsible for the advances of the first part of the Little Ice Age are not fully known. This would require information about precipitation and temperature throughout the year. Advances can only occur after a run of years in which snow accumulation is sufficient to increase volume more than ablation decreases it. Though the Little Ice Age was certainly not continuously cold, the balance between temperature and precipitation was evidently such as to sustain glacier volumes at relatively high levels. The quality of the winter seasons must have been one of the important factors in this balance.

The quantity and quality of documentary data available from the early centuries of the second millennium is less than for the period after 1500. It has nonetheless been possible to trace the general characteristics of winters in fourteenth-century Europe during the first Little Ice Age glacial advances (Pfister *et al.* 1996). Dated and verified documentary sources relating to winter conditions, such the freezing of lakes and rivers, or references to 'hard' winters recorded in explanation of unusually large amounts of grain fed to manorial animals, or mentions of standing water in connection with the profits from winter pastures, were assembled and used to form temperature indices.[6] Indices deviating from 'average' were attributed only to those winters documented in at least two contemporary sources. The system used is shown in Table 6.1.

Although the data cannot sustain extremely detailed analysis, or distinguish the Swiss Alps from the rest of

Table 6.1 Rating of temperature indices from documentary evidence

Index	Winter type	Descriptive data monthly	Proxy indicators (lowlands)
−3	Severe	3 'cold' months and	extreme duration of snow cover and water bodies ice covered for several weeks
−2	Cold	2 'cold' months or	ground snow covered for several weeks
−1	Cool	1 'cold' or 2 'cold' and 1 'warm' month	without
0	Average	offset of 'cold' and 'warm' months	without
1	Mild	1 'warm' or 2 'cold' and 1 'warm' month	without
2	Warm	2 'warm' months or	little or no snow and activity of vegetation
3	Very warm	3 'warm' months and	little or no snow and activity of vegetation

Source: Pfister *et al.* 1996, table 1

6 Most of them came from the many volumes of the *Monumenta Germaniae Historica.*

the western and central European region,[7] they paint a general picture of winter conditions across Alpine Europe in the fourteenth century, a picture which helps to provide at least a partial explanation of the circumstances which led to rapid and substantial glacial expansion. Average winter temperatures were somewhat below twentieth century values throughout the period. A run of cold years occurred from 1303 to 1328, during which only the winter of 1303/4 was 'warm' by present-day standards, but 11 were cold, 4 of them severe. The winters of 1305/6 and 1322/3 were comparable in intensity and cold with the hardest winters of the last 300 years (Pfister *et al.* 1996), and the period as a whole was comparable with several occurring later in the Little Ice Age, such as 1561–75 when winter temperatures in Switzerland have been estimated to have been 1.6 °C below the 1901–60 average (Pfister 1985). Average winters evidently predominated from the end of the 1320s to the mid 1350s, though there was still an excess of cold seasons. Winter temperatures were notably variable from 1355 to 1375, ranging from the mildness of 1359/60, when it remained warm and dry from the New Year until the end of March, to the bitter winter of 1363/64, when frost persisted from early December to the end of March. In the Bernese Oberland, the Lake of Brienz froze, which it has never done again even in the most severe winters of recent centuries.

A fuller picture should eventually be forthcoming from detailed explorations of tax rolls and other local documents such as those used by Titow (1969, 1972) who obtained from them valuable information which can be interpreted in terms of the weather conditions in summer as well as winter in fourteenth-century England.[8] Proxy indicators such as tree rings hold information about the summer growing seasons in various forms including their isotope values which are likely to provide in future quantitative data on summer temperature.

6.3 GLACIER OSCILLATIONS IN THE LITTLE ICE AGE OF EARLY MODERN TIMES

The Rhône Glacier, in full view of travellers crossing the Grimsel and Furka passes, must be one of the most frequently observed and well-documented glaciers in the world. On 4 August 1546, when he was riding towards the Furka, Sebastian Munster relates that he came to 'an immense mass of ice . . . about 2 or 3 pikes' length thick and as broad as the range of a strongbow'.[9] Evidently the ice reached down to the valley floor and ended in a steep front 10–15 m high and some 200 m broad. Three centuries later in 1836 when it was sketched by a Norwegian artist Thomas Fearnley it still, or again, reached down to the valley floor (Plate 6.2b). On 28 July 1868 Gerard Manley Hopkins wrote in his journal (1953) that he had seen the Rhône Glacier, ending in

a broad limb opening out and reaching the plain, shaped like the fanfin of a dolphin or a great bivalve shell turned on its face, the flutings in either case being suggested by the crevasses and the ribs by the risings between them, these being swerved and inscaped strictly to the motion of the mass. . . . We went into the grotto and also the vault from which the Rhône flows. It looked like a blue tent and as you went further in it changed to lilac.

The following year, 1869, the Swiss Alpine Club and the Swiss Academy of Sciences jointly founded the

---

7 In order to assess whether data from different parts of Europe could properly be combined in the same series, a careful investigation was made. Regions with similar year-to-year fluctuations were identified using cluster analysis of some twenty-seven long instrumental temperature records. Fluctuations were first assessed for the period 1901–60, and then the stability of the pattern found was tested by applying the same method of analysis to the records for three earlier 60-year periods. It emerged that the spatial pattern for winter changed little despite the considerable extent of climatic changes over the past 300 years. It was consequently argued unlikely that the pattern in the fourteenth century would have been fundamentally different. It was concluded that data originating south of a line from Birmingham to Malmö, that is in a region including the Swiss Alps, might be amalgamated into one single series. There remains, of course, the probability of there having been marked sub-regional differences in individual years. But the general outline is quite clear.

8 Titow himself was not disposed to believe in the value of his own results, but investigation of other sources relating to southeast England has both corroborated and extended his data, providing information, for instance, about the extreme wetness from August 1313 to February 1317 (Grove 1996).

9 The sixteenth-century Swiss pike measured between 4.6 and 5 m.

Gletscherkollegium (later to be called the Glacier Com-mission). The Rhône Glacier was chosen in 1874 to be the special subject for study, the aim being to under-stand the historical development of the glacier and its flow. A few years later, measurements were being made of the rate of ablation, precipitation and firnline eleva-tion. Data were collected annually on the volume as well as the area of the glacier, and the position of the tongue was surveyed every month, winter and summer from 1887 to 1910, to produce maps on a scale of 1:5000. This research programme, so thorough for its time, culminated in the publication of 'Mensurations au Glacier du Rhône 1874–1915' (Mercanton 1916). One of its diagrams (Figure 6.6), showing the changes in the position of the terminus from 1602 until 1914, based on a collection of fourteen descriptive and pictor-ial records, was still being used in later syntheses such as that of Furrer *et al.* (1980).

The oscillations of the Allalin Glacier over the past four centuries have attracted particular attention be-cause of the repeated flooding caused by the spilling of the Mattmarksee, the lake held back by rocks and ice when the glacier advances across Saastal (Mariétan 1965). In early modern times, according to Lütschg (1926) who collected together many of the relevant documents,[10] the valley was first blocked in 1589. In 1633 the lake broke through the ice dam and in Saas 'half the fields were buried in debris and half the inhab-itants were forced to emigrate and find their miserable bread in some other place' (Ruppen *et al.* 1979).

In the Val de Bagnes, another left-bank tributary valley of the Rhône, similar catastrophes were associ-ated with the Glacier de Giétroz (Figure 6.7). A flood in 1595 penetrated the galleries of newly opened silver mines, overwhelmed the thermal baths at Bagne, de-stroyed over 500 buildings and drowned 140 people. A similar outburst occurred in 1640. Particularly well

documented is the flood of 1818 (Sion archives); the two preceding summers had been very cool and the Giétroz, like most of the other glaciers in the Alps at the time, began to advance. Blocks of ice from the tongue fell over a rockstep (which the glacier overlooks at the present day from above the Mauvoisin hydro-electric dam) and amalgamated to form an ice cone stretching across the valley floor.[11] When it was reported to Sion in April 1818 that a lake had formed behind the ice, the authorities sent the canton engineer, Venetz,[12] to deal with the situation and artists to record the scene (Plates 6.3 and 6.4). By the middle of May the ice barrier was a kilometre long and 130 m high, holding back a lake 2 km long and 55 m deep. Starting on 11 May, Venetz, with local volunteers, worked day and night to dig a trench 200 m long across the ice. Eventually, in spite of late snowfalls and further great falls of ice, water began to escape along the trench on 13 June and about 6 million cubic metres, about a third of the lake's volume, had drained away before the ice dam collapsed at 4.30 in the afternoon of 16 June. Then the rest of the lake rushed down the valley as a wall of water 30 m high, reaching up as much as 100 m in narrow gorges, tearing away the soil here, deposit-ing vast spreads of gravel there. It almost caught an inquisitive English visitor (his mule was washed away). It carried away 42 chalets at Bonatchesse, 30 at Brecholey and 57 at Fionnay. At Lourtier, the highest permanent village in the valley of the Drance, 60 build-ings were destroyed. At Chamsec, where 58 buildings were ruined, a man aged 92 saved himself by climbing an oak tree growing on a debris mound that had been left by the flood of 1595. Sawmills, fulling mills, flour mills and an iron works were engulfed as the water sped down to Martigny. At 6 p.m. it reached Martigny Bourg, flooding the streets to a depth of over 3 m, and went on to inundate the floor of the Rhône valley,

---

10 Lütschg was unaware of some of the safeguards of more modern methodology. He included a number of extracts from Ruppen's 'Die Chronik des Thales Saas' but very great caution must be adopted in using chronicles of this kind (Bell and Ogilvie 1978).

11 Bridel 1818a, b, Sion archives Fasc. 1. R.21.28.

12 Ignace Venetz (1788–1859) of Visperterminen came from an old-established Valaisanne family. He turned early to the study of natural science and mathematics. As an engineer, first working for his own canton of Valais and later for Vaud, much of his professional life as well as his private interest was in glaciers and glacier-fed rivers. He played a leading role in the foundation of glaciology and glacial geology, finding time for original research despite the demands of his profes-sional career and the need to support a large family (Mariétan 1959).

(a)

(b)

Plate 6.2a (p. 162, top) The Rhône Glacier in 1777 by Alexandre-Charles Besson, a colleague of H. B. de Saussure (1725–1809) (*Source*: from a private collection, Bern)
Photo: H. J. Zumbühl
Plate 6.2b (p. 162, bottom) The Rhône Glacier in 1835 sketched by Thomas Fearnley from a site above the glacier (*Source*: Nasjonalgalleriet, Oslo, Inv. no. A.3412)
Photo: Jacques Lathion
Plate 6.2c The Rhône Glacier in 1848 in an aquarelle by Hogard showing the series of moraines left during retreat. The outermost set has been dated to 1602, the second set to 1818, and the third to 1826. The picture indicates clearly that a fourth set was forming in 1848 (*Source*: Zürich Graphic Collection, E.T.H. Case 648)
Photo: H. J. Zumbühl
Plate 6.2d The Rhône Glacier was still across the valley floor in 1862 when painted by Eugen Adam (*Source*: Zürich Graphic Collection, E.T.H. Case 648)
Photo: H. J. Zumbühl

(c)

(d)

Plate 6.2e By 1912 the Rhône Glacier tongue was hanging and the valley bottom was clear of ice
Photo: from Mercanton 1916

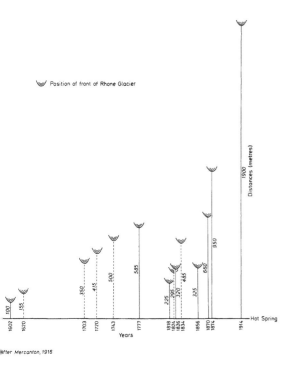

Figure 6.6 Variation in position of the front of the Rhône Glacier between 1602 and 1914 (*Source*: from Mercanton 1916)

until the years 1894 to 1899. Then the retreat of the ice at the head of the Val de Bagnes allowed a lake to accumulate between the Otemma and Crête Sèche glaciers which violently spilled over each summer between 1894 and 1899, causing floods that destroyed the bridges downstream. Eventually engineers excavated a trench in the ice and rock barrier to allow the water to drain away harmlessly.[13]

A particularly significant study in relation to Little Ice Age climate is that of the Grindelwald Glacier (Messerli *et al.* 1975, 1978, Pfister *et al.* 1994a). In tracing the fluctuations of the Oberer and Unterer Grindelwald over the period 1590 to 1970, Zumbühl (1980) used as evidence not only written documents and maps but also relief drawings, sketches and paintings collected from libraries in half the countries of Europe. In all, he examined 300 views of the Unterer Grindelwald and 186 of the Oberer, dating from 1640 to 1900. He had to date the illustrations and establish the artists's viewpoints precisely. Some masterpieces turned out to be quite unsuitable for the purposes of the glacier historian. Particularly accurate representations were the oil paintings of Caspar Wolf (1735–83) and a watercolour of 1820 by Samuel Birmann (1793–1847).

Both the Oberer and Unterer Grindelwaldgletscher were more extensive between 1600 and 1870 than they have been since (with the possible exception of a few years around 1685). Although the upper glacier was found to respond up to a decade earlier than the lower one, the two generally behaved in a similar way and the fluctuations of the Unterer Grindelwald (Figure 6.8), which can be traced more accurately, hold for the Oberer (though during much of the period 1960–80, when the Oberer Grindelwald was advancing, the Unterer was retreating, a difference in the response time of the two glaciers greater than had previously been recognized).

From about 1590 to 1640, the Unterer Grindelwald advanced to reach its maximum extent, some 600 m

spreading a layer of silt over the marshes of Guersay (Guercet). Before midnight, debris was being washed into Lake Geneva. Had Venetz and his men not released much of the water before the ice dam broke the damage would no doubt have been even greater, possibly exceeding 1595 (Tufnell 1984).

There were ice falls in the Val d'Hérens in that same year, 1818, which took six years to clear; ice from the Weisshorn destroyed the village of Randa in December 1819 (Sion archives T.P. 135/1). In the Val de Bagnes, the Glacier de Corbassière continued to advance down its valley until 1822 and the threat of renewed blockage of the Mauvoisin valley persisted until 1824. But for the rest of the century there was no serious threat

---

13 The practical advantages of obtaining detailed histories of sensitive and dangerous glaciers such as the Allalin and Giétroz, and of keeping a strict watch on them, were unhappily underlined by the disastrous collapse of ice from the Allalin onto contractors' buildings erected below it while the hydroelectric dam was being built in 1965. There was much loss of life (Vivian 1966). The tongue of the Giétroz now hangs far above the waters, held back by another dam in the Val de Bagnes, the Mauvoisin, and is the subject of especially strict surveillance by the Swiss Glacier Commission.

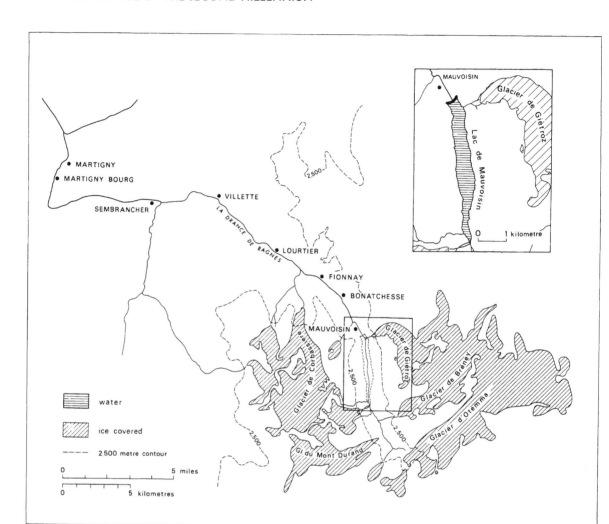

Figure 6.7 The Glacier de Giétroz and its setting (*Source*: Zumbühl, in Messerli *et al.* 1978)

forward of the rock band called the Unterer Schopf. This natural marker remained under the ice until 1870, though it was very nearly exposed by a retreat towards the end of the seventeenth century. In 1718 and again in 1743 the ice advanced to reach 100–300 m below the Unterer Schopf, retreating between and after these dates almost as far as the marker. Another advance between 1771 and 1780 brought it forward about 400 m; there it formed a moraine, since removed by erosion, which appears in several oil paintings of the time. By 1794 the ice had melted back to the Unterer

Schopf. Twenty years later it was expanding and it continued to do so until 1820–2 (Plate 6.5c). Another advance brought it further forward in 1855–6 than it had been since the seventeenth century. Then came rapid thinning and retreat which has continued to 1980, except for a halt from 1882 to 1898, a minor readvance between 1915 and 1925 and a pause from 1925 to 1935. The total withdrawal from 1856 to 1970 amounted to 1800 m.

Systematic measurement of the frontal positions of the Grindelwald Glacier began in 1880 as part of the

Plate 6.3a The Glacier de Giétroz (2) spilling over (3) into the upper Val de Bagnes. The dam is still incomplete; the Vallon de Torembec (5) is not yet flooded (*Source*: from Bridel 1819)

Plate 6.3b The Giétroz dam in May 1818. Venetz and his men are at work on the trench, while members of the public observe from a safe distance (*Source*: from Bridel 1818)

Plate 6.4 Ignace Venetz about 1815, painted in oil by Laurent Ritz (*Source:* Musée de Marjorie, Sion)

Swiss Glacier Commission's national programme.[14] Figure 6.9 shows the variations in position of the ice fronts of the Grindelwald and other glaciers in various parts of Switzerland which have continuous records since 1890. Differences in detail, such as the rapid re- treat of the Grosser Aletsch in the 1960s and 1970s when the Oberer Grindelwald and the Allalin were advancing, can be attributed to differences in altitude, aspect and size, and to differences in amount of moraine cover and topographic setting of the snouts.

---

14   A general account of the work of the Swiss Glacier Commission is given in Portmann (1975, 1976, 1978, 1980a, b). The first fifteen annual reports were produced by F. A. Forel (1841–1912), who took a leading part in the organization of the Commission and the direction of its researches. The programme was ambitious in its aims from the first, as well as outstanding in the degree of its continuity, and has latterly become much more sophisticated. The first two reports were published in *Echo des Alpes* 17 and 18 (1881, 1882), reports 3 to 44 in *Annuaires du Club Alpine Suisse* 18 to 58 (Bern 1883–1924) and report 44 onwards in *Les Alpes*, starting with no. 1 in 1925. Since 1954 the collection of data and preparation of annual reports have been undertaken by staff of the Laboratory of Hydrology and Glaciology of the Federal Institute of Technology of Zurich.

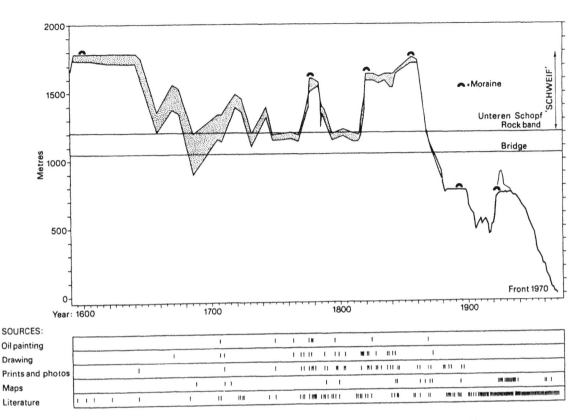

Figure 6.8 The frontal positions of the Unterer Grindelwaldgletscher, 1590–1970 (*Source:* after Messerli *et al.* 1978)

Figure 6.10 shows the fluctuations of the Unterer Grindelwald since 1600 in comparison with those of the Rhône, Fiescher and twenty-three other Swiss glaciers. In spite of differences in cumulative mass balance values, the behaviour of the termini has been rather consistent over the long term (Reynaud 1980, 1983).

The Swiss example of regular recording of glacier fronts was followed in Austria, France and Italy, so that a comparison can now be made of glacier-front behav-iour over most of the Alps for over a century. The phases of advance and retreat correspond very closely through-out the region, as Richter (1891) recognized long ago and as Ladurie (1971) and Bray (1982) later confirmed (Figure 6.11). In the course of the twentieth century, most glacier fronts advanced only in the intervals be-tween about 1915 and 1920 and again, sixty years later, between 1974 and 1981. Between the mid nineteenth century advance and 1973 (the baseline year of the

The work of the Swiss Glacier Commission gave rise not only to similar programmes in other parts of Europe, but also to worldwide activity. Internationally co-ordinated observations had started with the foundation of the International Glacier Commision in Zurich in 1894, and was later followed by the Permanent Service on the Fluctuations of Glaciers, which in turn became the World Glacier Monitoring Service. The volumes of *Fluctuations of Glaciers*, published at 5-year intervals, have widened their regional coverage from the Alps, Scandinavia and Iceland to include glaciers in North and South America, and several parts of Asia, and even a few in Antarctica. The sophistication and the range of information reported has also increased. The *Fluctuations of Glaciers* are published under the wing of several international organiza-tions, but significantly the World Glacier Monitoring Service is still located in Zurich. Since 1990 it has also been responsible for publishing the *Glacier Mass Balance Bulletin*, designed to speed up access to data by reporting values from selected glaciers at 2-year intervals.

Plate 6.5a The Unterer Grindelwald, painted by Joseph Plepp (1595–1642). The exact date is unknown. The descending glacier (A) is pushing everything before it with a great noise. The river (B) streams out from under the ice. The houses (C) have had to be abandoned because of the glacier (*Source*: Zumbühl 1980, K1.11, p. 198)
Photo: H. J. Zumbühl

Plate 6.5b The front of the Lower Grindelwald in 1762, painted by Johann Ludwig Aberli (1723–86) (*Source*: Zumbühl 1980, K13.1, p. 205)
Photo: H. J. Zumbühl

Plate 6.5c Samuel Birmann (1793–1847) visited the Unterer
Grindelwald in 1826 and portrayed the glacier with well-developed
seracs (*Sources*: Original in the Kunstmuseum, Basel. Zumbühl 1980,
K60, p. 193)
Photo: H. J. Zumbühl

Plate 6.5d Grindelwald and the Unterer Grindelwaldgletscher
photographed in 1885 by Jules Beck (1825–1904) (*Source*:
Zumbühl 1980, K139, p. 247)

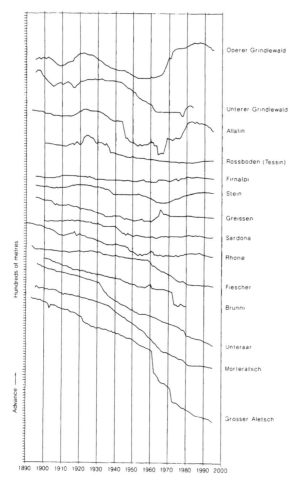

Oberer Grindlewald

Unterer Grindlewald

Allalin

Rossboden (Tessin)

Firnalpi

Stein

Greissen

Sardona

Rhone

Fiescher

Brunni

Unteraar

Morteratsch

Grosser Aletsch

Figure 6.9 Advances and retreats of Swiss glaciers in various parts of the country since *c.* 1890 (*Source*: based on figures published in the Annual Reports of the Swiss Glacier Commission)

official Swiss inventory) glaciers in Switzerland lost approximately 185 km² or 40 per cent of their total area and equilibrium line altitudes rose by about 80 m (Maisch 1988).

The record of glacier fluctuations extending back over the last four centuries and more would seem to afford opportunities for extending the climatic record back beyond the beginning of the instrumental period. The position of a glacier terminus fluctuates according to the cumulative mass balance and the dynamics of the glacier, which in turn depend on its shape and the form of its bed. The mass balance depends primarily on accumulation of snow and losses by melting. In the

case of a single glacier, then, the climatic controls are the amount of snow falling on the glacier, particularly in the winter season, in relation to the ablation losses caused by insolation and warm, moist winds particularly in summer.

A comparison of cumulative mass balance curves from all the observations that were made in the Alps in the twentieth century shows that trends of increase and decrease were in concert (Figure 6.12). Surfaces of glaciers were lowered by about 60 cm per year in the period of negative mass balances between 1920 and 1950, causing the great retreat of glacier fronts in mid century. Positive mass balances gave mean annual increases in elevation of glacier surfaces of several centimetres from 1910 to 1920, 21 cm from 1959 to 1969 cm, and 14 cm from 1969 to 1979 (Patzelt 1985). The consequent advances of glacier fronts lagged behind by a few years or even decades, the interval depending on the size and steepness of the individual glacier.

Studies of mass balance on the Rhône and other glaciers have shown that a very large part of the variance can be explained in terms of a few climatic variables (Reynaud 1980, 1983). In the case of the Rhône glacier, the correlation coefficient in a regression formula relating specific net balance to annual precipitation (October–September) and mean summer temperature (May–September) at Reckingen (9 km to the south) is remarkably high: 0.90 (Chen and Funk 1990). Between 1881/2 and 1986/7 mass balances for the Rhône glacier were positive only between 1905 and 1920. The contraction of the glacier between 1882 and 1985, which amounted to a kilometre, according to this model can be understood in terms of the 1921–85 mean annual precipitation having been greater by 47 mm (i.e. about 4 per cent) and summer temperature higher by 0.9 °C than in 1882–1920. Chen and Funk conclude that small temperature fluctuations greatly influence mass changes, while the influence of precipitation changes is smaller. That may have been the case in the twentieth century; however, set in a longer historical context, the difference in temperature was large and that of precipitation quite small.

In general, the cumulative mass balances of Alpine glaciers have been found to depend primarily on summer temperature and secondly on the amount of precipitation in June. Most important is the mean daily

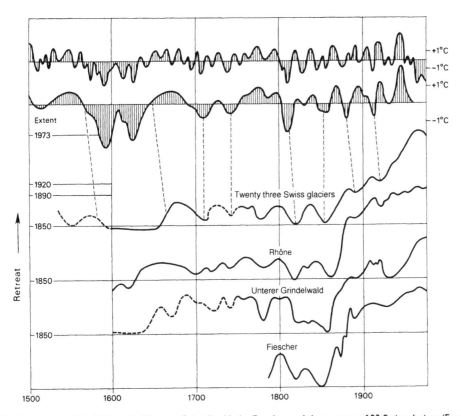

Figure 6.10 The fluctuations of the Rhône, the Unterer Grindelwald, the Fiescher and the average of 23 Swiss glaciers (Furrer *et al.* 1980), compared with tree-ring density curves (*Picea* at Laurens, Berner Alpen). It can be seen that if the density curves are shifted a few years forward (to the right), there is a clear correspondence between high tree-ring density (low summer temperature) and ice advance; the time lag depends on the size and conformation of the glacier (*Source:* from Röthlisberger *et al.* 1980)

temperature for July and August, which accounts for 58 per cent of the variance. June precipitation accounts for 16 per cent, and total precipitation between October and May for 5 per cent. Regression formulae allow mass balances to be reconstructed over the period of weather records. This underlines the importance of consistent and continuous weather recording, a consideration which is not always appreciated by cost-cutting bureaucracies. Such modelling also allows estimates to be made, though less securely, of the climatic fluctuations likely to have been the cause of glacier advances and retreats before the period of instrumental records, thereby indicating the value of studying glacier history (Oerlemans 1994).

Hoinkes (1968), who had demonstrated a good correlation between the fluctuations of Swiss glaciers and the 5-year running means of deviations of summer temperatures and precipitation from 1851–1950 means, went on to investigate the relationship between glacier fluctuations and meso-scale weather situations. He showed that for the period 1953–65 there was a negative correlation between the mean specific mass budgets of the Grosser Aletsch and the Hintereisferner on the one hand and the deviations from the 1951–60 average height of the 500 mb surface over Munich and Payerne from May to September on the other (the latter being a useful indicator of the prevalence of cyclonic or anticyclonic conditions over the Alpine region). Hoinkes considered that well-defined deviation patterns of the 500 mb surface would be related to *Grosswetterlagen*, that is 'fairly persistent synoptic situations which determine the form and sequence of

a) Austrian glaciers

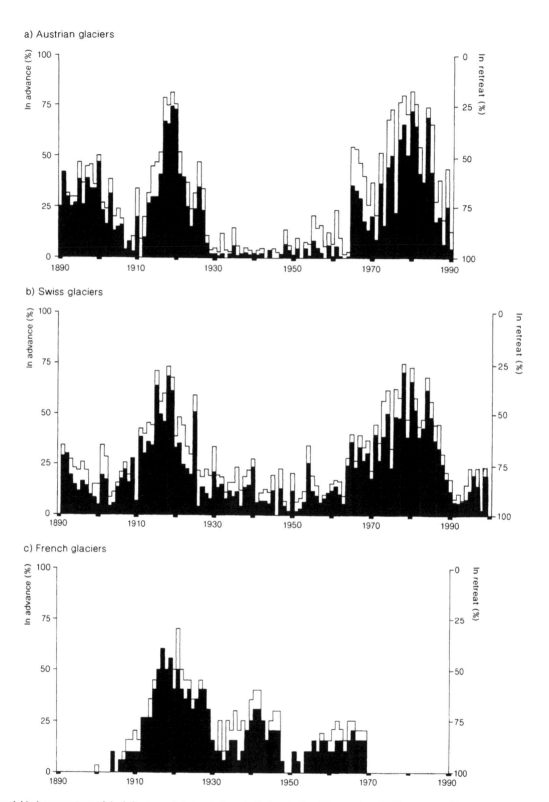

b) Swiss glaciers

c) French glaciers

Figure 6.11 A comparison of the behaviour of glaciers in Austria, Switzerland and France since 1890, in terms of the percentage of those observed each year which were found to be advancing, retreating and stationary

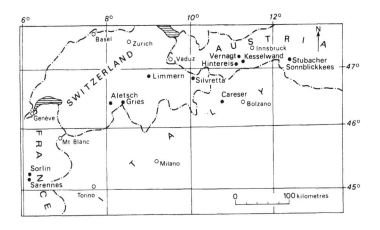

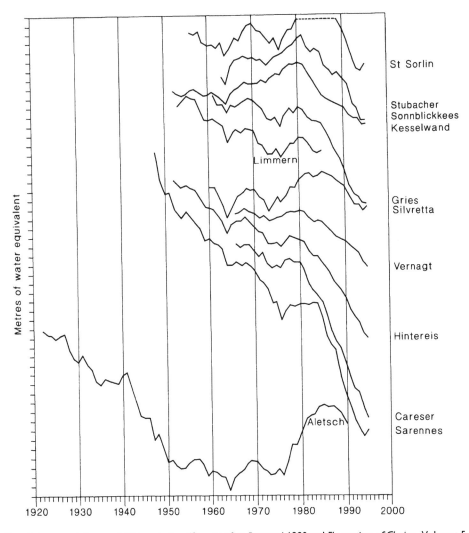

St Sorlin

Stubacher
Sonnblickkees
Kesselwand

Limmern

Gries
Silvretta

Vernagt

Hintereis

Careser
Sarennes

Aletsch

Metres of water equivalent

1920   1930   1940   1950   1960   1970   1980   1990   2000

Figure 6.12 Cumulative mass balances of Alpine glaciers (*Sources*: after Reynaud 1980 and *Fluctuations of Glaciers*, Volumes 5, 6 and 7 (Haeberli and Müller 1988, Haeberli and Hoelzle 1993, Haeberli et al. 1998))

Table 6.2 Grosswetterlagen with a strong influence on glacier nourishment

---

Favourable influence

W1   Ocean-type in winter – winter precipitation
F4   Cold spells in spring – solid precipitation and/or
     late starting ablation season
S6   Monsoon-type in summer – reduced radiation,
     increased albedo

Unfavourable influence

S7   Fine weather in September – long ablation season
S5   Anticyclonic weather in summer – high radiation
     and temperature, low albedo

Source: Hoinkes 1968

weather events for a period of several days or even weeks',[15] and identified five types of *Grosswetterlagen* likely to have a strong influence on the nourishment or shrinkage of glaciers (Table 6.2).

Hoinkes made an initial assumption that a single day with favourable *Wetterlagen* would compensate for one day with unfavourable *Wetterlagen*. The algebraic sum of daily frequencies (W1 + F4 + S6) – (S5 + S7) over a year then gives an annual index and the indices can be summed over periods of years and compared with observed glacier fluctuations. The period of stationary or advancing glacier fronts in the 1890s corresponds to an average annual index of +100 for the period 1890–4, whereas for 1947, when all the Swiss glaciers were receding, the value of the index was –27. Hoinkes concluded that for a general advance of the Swiss glaciers the index would have to exceed 90 for several years.

6.4 SWISS CLIMATE DURING THE LITTLE ICE AGE

Ignace Venetz, the engineer from Visperterminen who had experience of coping with the hazards presented by the Giétroz and Allalin glaciers, addressed the Société Helvétique des Sciences in 1821 (Venetz 1833). Probably for the first time, he used glacier enlargement and diminution as evidence to show that temperature 'rises and falls periodically but in an irregular manner'. That the climate of Switzerland was becoming more rigorous had been conjectured at a meeting of the society in 1817, a year after its members had turned their attention to the study of glaciers (Mariétan 1959) and two years after the eruption of Tambora (Chapter 18, p. 637). They already had a number of weather records at their disposal. The Economic Society of Bern had set up a network of weather stations with uniform instrumentation in 1759, and regular observations of temperature and pressure had begun at Basel in 1755 and were started at Geneva in 1768.

The climatic record can be extended back beyond the middle of the eighteenth century, by utilizing a whole range of other kinds of evidence (Pfister 1975, 1977, 1978a, b, c, 1979a, b, 1980a, b, 1981, 1992a, b, 1994a, Pfister *et al.* 1994b) As a result of a systematic search of Swiss libraries and archives, Pfister collected together over 27,000 documentary records relating to past climate, including 70,000 daily weather observations. The frequency of such records increases towards the present; between 1525 and 1550 there were about thirty records per year; from 1550 to 1657 information is available for nearly every month, and after 1659 for every month. After the middle of the eighteenth century the quantity of data becomes so large that Pfister limited his attention to the more readily quantifiable material, not only instrumental measurements but also the monthly number of rainy and snowy days noted in weather diaries, phenological records,[16] and comments on snow cover, snowfall on Alpine pastures, floods, and dates of lakes freezing. These data have been brought together to provide a coherent story of changing weather patterns in Switzerland over the last four centuries.

---

15   Hess and Brezowsky (1952) published a daily catalogue of *Grosswetterlagen* for mid Europe from 1881 to 1951. Since then, details have been published monthly by Deutscher Wetterdienst, Zentralamt Ossenbach.

16   Phenological calendars were used thousands of years ago in both China and the Roman Empire (Hopp 1974). The term 'phenology' was introduced by a Belgian botanist, Charles Morren, in 1853, but it was the Swedish botanist Carolus Linnæus (Carl von Linné) who initiated modern phenology and phenological networks. In his *Philosophia Botanica* he outlined methods for compiling annual plant calendars of leaf opening, flowering, fruiting and leaf fall, together with climatic observations 'so as to show how areas differ'. Phenological networks involve the co-operation of regular observers working in selected site areas and recording the dates of occurrence of specific growth stages or phenophases of

Diaries of the weather are especially valuable sources. An instance is the daily record kept by Wolfgang Haller; his daily entries allow estimates to be made of the temperature and of the frequency of sunny and rainy days near Zurich in 1545–6 and 1550–76. In the same area, regular observations were made by a parson, Heinrich Fries, from July 1683 to October 1718. From 1721 to 1738 a Winterthur baker, Rudolph Reiter, was so obsessed that he noted the weather hour by hour not only through the day but also through much of the night.

The first systematic observations of precipitation in Switzerland were made by Johan Jacob Scheuchzer; his records, starting in 1708 and continuing to 1731, are the earliest in central Europe. Hans Jacob Gessner, who like Scheuchzer was a Zurich clergyman, observed the rainfall from 1740 to 1746 and also from 1750 to 1754 (Pfister 1978b). In Basel a professor of law, Johann Jacob d'Annone, kept a diary of instrumental readings for nearly fifty years from 1755 to 1804.

Pfister (1980a) has made particularly effective use of phenological data in reconstructing past climate and weather. The growth patterns of plants in Switzerland, as in other temperate latitudes, respond mainly to temperature. Most of the records relate to fruit trees, grapes and field crops. The dates of the vine harvest are especially well documented and significant; in Switzerland they are to be found in local police records.[17] Grape yields depend mainly on summer weather, being high when July and August have been hot and dry, low after cool and rainy summers (Primault 1969). For the derivation of factors for converting phenological information into climatic values, Pfister considers that corresponding data over a period of fifteen to twenty years are needed, covering a wide range of weather conditions.[18] The statistical treatment has to take into account the influence of altitude and exposure; an increase in altitude of 100 m corresponds to a delay of 3.6 days in the opening of vine flowers (Becker 1969). Departures from mean values are the significant features of phenological records and it is important to know whether the mean for a particular species has varied through time for other than climatic reasons, such as the introduction of new varieties.

Studies of this kind require not only a knowledge of plant ecology but also involve the application of modern methods of historical analysis to a wide range of data (Bell and Ogilvie 1978). Some well-known European weather compilations[19] have been shown not to

---

chosen indicator plants. These networks can provide important supplements to climatological networks. The time of occurrence of phenophases of many plants is to a large extent controlled by local climate. Phenological observations of selected indicator plants are recognized as being of assistance in forecasting or estimating the occurrence of succeeding phenological events in the same or other species. Maps of long-term averages of occurrence of phenophases can be used to identify zones of similar climate in the same way as meteorological maps. However, individual phenological observations must be interpreted with caution, because plants respond to the full range of environmental factors and are affected by such matters as soil type, exposure and topography in addition to micro-meteorology. Observations should therefore be considered in relation to data from neighbouring sites.

If maps of long-term phenological averages can be related to meteorological conditions, it follows that if meteorological conditions change this will be reflected in the timing of phenophases. If phenological series of adequate length are available for a period covered by good meteorological observations there is a good possibility that recorded phenophase data from the past can be decoded in terms of their dominant meteorological control.

17 Possibilities offered by vine harvest dates were recognized by a Swiss scientist, Dufour, in 1870 and have been put to good use since especially by Ladurie (1971) and Ladurie and Baulant (1980).

18 Most of the data used by Pfister resulted from vineyard share-cropping agreements which were held by public institutions and authorities. The division between landlord and tenant being known, the landlord's portions are listed and so provide useful series. Complications result because the acreage is not always known. Pfister compiled four regional series from different parts of the Swiss Plateau, each compiled from a variety of local series of unknown acreage. These four show correlations between 0.56 and 0.76. The main series, computed from the regional series, is highly correlated (r = 0.87) with a fifth series compiled from the several local series for which the acreage is known. It is recognized that yields can be affected by changes in techniques of cultivation, manuring, pruning and so on, or by a change in the variety grown. Climatic factors will be important in explaining short-term fluctuations.

19 The first of these sources was a compilation by Hennig (1904) published by the Royal Prussian Meteorological Institute. It exhibited many of the common weaknesses of historical works produced by people with little understanding of the need for identification of the sources of all individual items of information and of the value of source analysis. Bell and

be verifiable (Ingram *et al.* 1981) and partly as a consequence of this there has been a tendency to discount prematurely the possibility of climatic fluctuations having been influential in human affairs (Parry 1978). Data verification involves not only the selection of reliable evidence and the rejection of dubious material but also the identification of the time and place of an observation and, when descriptive terms are used, an appreciation of the linguistic conventions current at the time of writing (Ingram *et al.* 1981). The characteristics and shortcomings of instruments and the peculiarities of the places where they were housed have to be taken into account. Pfister (1981) had special dating problems in Switzerland because, although the Gregorian calendar was adopted by the Catholic cantons in 1583, the majority of the Protestant cantons did not follow suit until 1701, and some continued to use the Julian calendar for varying times into the eighteenth century.

In order to handle his data effectively Pfister coded it for computer processing, distinguishing five main categories: daily non-instrumental weather observations, miscellaneous observations from chronicles and annuals, precipitation values, monthly frequency counts of various kinds, and temperature measurements. He distinguished between precise reports and interpreted data. A ten-day interval was taken as the smallest reporting unit, except in the case of phenological material, when allowance was made for individual days. Some records could not be coded and were stored verbatim, but for the most part the documentary information was quantified in such a way that temperature and rainfall could be compared with 1901–60 mean values to produce thermal and wetness indices.

Computer listing allows data from various sources referring to the same time period to be rapidly assembled and compared. For the winter of 1572–3, for instance, the most severe winter of the last 500 years, Wolfgang Heller's record of twelve days of snow in December is supplemented by Savion's report of Lake Geneva having frozen. There follow observations from independent sources of milder weather about the middle of January in Bern and Zurich and breakup of the ice on the lake about the same time (Pfister 1981).

The phenophases, the various stages in the development of plants, have been related to monthly temperatures using stepwise multiple regression. It has been found, for example, that if the sweet cherry flowers more than two weeks earlier than the mean date of flowering, either February and March have been warmer than average or else January was extraordinarily warm and February and March somewhat warmer than average. If flowering is three weeks late, April has been about 5 °C cooler than the mean for 1901–60. The date of the appearance of the first vine flowers usually proves to depend on May's temperature but a very warm April can also give early flowering. Distinguishing conclusively the timing of a weather abnormality is not easy but additional information, even if it is only a remark in a diary, say about unusual warmth at a particular time, can provide a vital clue and allow the timing to be determined more closely.

Vine harvest dates have generally been taken to reflect mean summer temperatures (Ladurie 1971, Ladurie and Baulant 1980) but Pfister's (1980a) analysis shows that temperatures in spring and early summer are much more important than those in August and

---

Ogilvie (1978), in their discussion of the nature and use of weather compilations, noted that while Hennig distinguished between information he had gathered from newspaper articles, town guides and travelogues and references taken from historical sources, he was wholly uncritical and drew his information almost entirely from secondary sources and unreliable editions. Illustrative of the results of his approach is his statement that there were many hot dry summers between AD 988 and AD 1000. He cited eighteen references in support of the conclusion, but the two earliest and most nearly contemporary of these record only one single hot summer. All the other sources quoted are much later, many of them seventeenth- and eighteenth-century compilations. The idea of a succession of hot summers arose from Hennig's own summary of unreliable information. Some more recent workers, such as Britton (1937), claimed to distinguish authentic from dubious material but still used a high proportion of unreliable sources, including that part of the 'Historia Croglandensis' which has been shown to be a forgery made in the interests of claiming rights to disputed land. Britton did exhibit a far more critical attitude to some other matters, demonstrating for instance the way in which events recorded by contemporary medieval writers may be magnified in the course of time. He traced the way in which the record of a severe storm flood which inundated some villages in 1099 was transmuted over time into an extraordinary storm in which the Goodwin Sands were formed.

Table 6.3 Indicators used for the construction of thermal indices for individual months

Month	Cold	Warm
December–February	Uninterrupted snow cover Freezing of lakes	Scarcity of snow cover Signs of vegetation
March	Long snow cover High snow frequency	
April	Frequent snow Beech tree leaf emergence Tithe auction date Sweet cherry first flowers (± 1.3 °C)	
May	Tithe auction dates (± 0.6 °C) Vine first flower (± 1.2 °C) Date barley harvest begins	
June	Tithe auction dates (± 0.6 °C) Vine full flower (± 1.2 °C) Vine last flower	
July	Vine yields (± 0.6 °C) Colour/maturity of first grapes Vine harvest dates (± 0.6 °C)	
August	Wine yields (± 0.6 °C) Tree-ring density (± 0.8 °C)	
September	Wine quality Tree-ring density (± 0.8 °C)	
October	Snow cover Snow frequency	Reappearance of spring vegetation (cherry blossom, etc.)
November	Long snow cover Snow frequency Freezing of lakes	No snowfall Cattle in pastures

Source: Pfister 1981

September: the correlation coefficients he gives are May −0.54, June −0.59, both with 0.001 significance, and July −0.34 with 0.03 significance. He found no significant relationship between harvest date and temperature in April, August and September. These findings are supported by the discovery that vine harvest dates are more strongly correlated (r = +0.74) with indicators of weather conditions in the early summer, notably the dates of tithe auctions[20] which are dependent on the time when the grain ripens, than with tree-ring densities, which reflect temperatures in late summer. Table 6.3 gives an impression of the range of the different kinds of indicators that can be used in estimating thermal indices over the last 450 years and the standard errors of temperatures derived from the phenological data.

20 Until the nineteenth century the right to collect tithes in certain parts of Switzerland was sold by auction to tithe farmers. They undertook to collect the tithes, being allowed to retain a small proportion of the proceeds. The dates of the auctions and the inspections which preceded them depended upon the ripening of the grain. Mean dates were shown to be a function of altitude, a difference of 100 m causing 4.6 days delay. Pfister found forty-two series of auction dates covering the period 1611–1825 which he aggregated into a unified series. The advance or delay of the auction by seven days was found to correspond to a deviation of 1 °C from the 1901–60 average.

While the dates of individual phenopauses can give an indication of conditions in an individual month, the pattern of phenopauses over a longer time period can give additional information. Thus, a lengthening of the interval between phenopauses indicates below-average temperatures in the course of the year, as compared with the 1901–60 means. Delay of phenopauses indicates that at least four months between March and July had temperatures below the 1901–60 averages. Advance of all phenopauses indicates that two or more of the spring months as well as June temperatures were above the modern average and that July temperatures were average or higher.

Pfister's archival studies (1980a, b) enabled him to derive precipitation as well as thermal decadal indices for Switzerland over the period 1550 to 1820 (Figure 6.13). Each month was given negative values if it was cold and dry, positive values if warm and wet, and zero values if it approximated to the 1901–60 means. Monthly values are weighted by factors of 1 to 3 according to the magnitude of the departures from the mean, a factor of 3 indicating extreme conditions by twentieth-century standards. Then both precipitation and thermal indices for each season are obtained by summing three monthly indices to give values that range from −9 to +9.

The rings of trees near the upper limit of forests provide another source of information about past temperatures (e.g. Schweingruber *et al.* 1979). The density of the summer wood, as measured by X-ray methods, varies with July to September temperatures. Earlier in the Little Ice Age, summer temperatures were about 2 °C lower than those of the late nineteenth and early twentieth centuries. In the intervening period conditions varied in much the same way that Pfister's results present in more detail.

The period between 1570 and the end of the sixteenth century was on average cooler and wetter in all seasons of the year than the preceding half century (Table 6.4). The percentage of cold winters doubled and of wet winters trebled. Between 1525 and 1564 there was no winter in which snow persisted in the Swiss lowlands for more than eighty days, yet there were four such winters in the decade 1564–74. In Zürich, the proportion of total precipitation in the form of snow rose from 44 per cent in 1550–63 to 63 per cent between 1564 and 1576. Autumn and springtime were

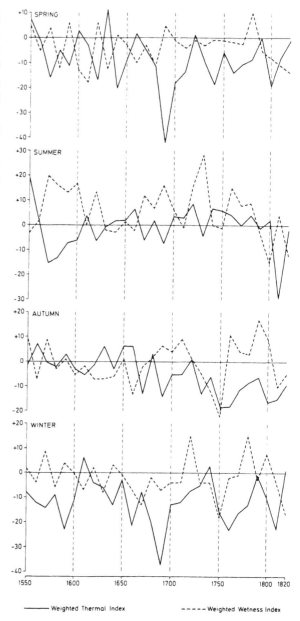

Figure 6.13 Seasonal weighted thermal and wetness decadal indices for Switzerland, 1550–1820 (*Source:* constructed from data in Pfister 1981)

cooler as well, but the chilling was less marked than in winter and summer. Between 1525 and 1569 there were twice as many warm summer months as cold, but between 1570 and 1600 half the summer months are

Table 6.4 Warm, cold, wet and dry months, 1525–69 and 1570–1600

	Percentage of months which were undoubtedly:			
	Warm	Cold	Dry	Wet
Summer				
1570–1600	28	47	12	38
1525–69	36	18	18	24
Difference	–8	+29	–6	+14
Autumn				
1570–1600	24	19	13	18
1525–69	24	7	20	11
Difference	0	+12	–7	+7
Winter				
1570–1600	12	44	16	18
1525–69	15	24	9	6
Difference	–3	+20	+7	+12
Spring				
1570–1600	18	33	16	17
1525–69	20	24	16	16
Difference	–2	+9	0	+1
Year				
1570–1600	20	36	14	23
1525–69	24	18	16	15
Difference	–4	+18	–2	+8

Source: Pfister 1981

rated as having been cold. The summer of 1573 was a particularly bad one and, in general, extreme conditions were more pronounced than they are today. For instance, the summers of 1588 and 1596 were extremely wet, with rain falling on 77 out of the 92 days of June, July and August in the Lucerne area.

Pfister's studies allow the changing character of individual months in Switzerland to be traced from 1550 to 1820. As Figure 6.13 shows, the spring months were colder and drier throughout most of the period than in the first half of this century, the coldest spring decades being the 1640s, 1690s and 1740s; only the 1550s and 1730s were warmer than the 1901–60 mean. Whereas May's temperature and precipitation were near the early twentieth-century mean and April, though

drier, was no cooler except slightly between 1525 and 1720, March was persistently colder throughout the whole period, except for about twenty years at the beginning of the seventeenth century. In the 1640s, 1690s and 1760s, March temperatures were particularly low, 2 °C below the 1901–60 mean; together with persistent dryness these low temperatures are believed to indicate the prevalence in early spring of northerly winds and blocking anticyclones over north–central Europe.

The high frequency of cool, wet summers in the last three decades of the sixteenth century as compared with 1525 to 1569 mainly involved June and July; June remained generally cool and wet right through the seventeenth century in spite of the fact that summers in the middle decades were not much different from those of the early twentieth century. July was frequently wet but not markedly cool in the first half of the eighteenth century and then, after 1760, it was cool as well as wet especially between 1810 and 1820, when July temperatures, 1.2 °C below the 1901–60 mean, were largely responsible for the overall coolness of the summers.

In the three decades before and after 1600 the cool, dry springs and wet summers were followed by autumns with mean temperatures similar to those of the first sixty years of the twentieth century. Then autumn seems to have become cooler, especially in the 1750s and 1760s and the early years of the nineteenth century. The autumns of the 1730s, 1740s and 1750s were wet, as were those of the early nineteenth century.

While cool, wet summers were features of certain decades in the Little Ice Age, winters throughout the period were on the whole cold and dry. The decade with the coldest winters in Switzerland, as in much of the rest of Europe including Iceland and Scandinavia, was from 1690 to 1700, though in the Swiss Alps this cold phase did not lead to substantial glacial advances because winter precipitation was insuffecent[21] to allow positive mass balances (Pfister 1992a). But in London more days with snowfall were recorded in that decade than in any other before or since (Manley 1969). The coldest winter of the decade in Europe generally was that of 1694/5. In Switzerland the Bodensee (Lake Constance) froze over completely with a layer of ice thick enough to bear fully laden wagons (Lindgren and

21   The very low incidence of floods in the Swiss Alps in the period 1651–1700 is noteworthy (Pfister 1992a, figure 7).

Neumann 1981). For an ice cover over the entire lake surface a temperature anomaly of at least −4 °C is needed (von Rudloff 1967). Never since that year has Lac de Neuchatel frozen over so abruptly. The winter of 1784/5 was also extraordinarily long and severe, with snow lying in Bern for 154 days, 60 of them between the beginning of March and end of May. The corresponding figures for the hard winter of 1962/3, it might be noted, are 62 and 12 (Pfister 1978a), though the city was warming itself to a much greater degree by this time.

In years of heavy snowfall, when the snow cover persisted well into the late spring on upper Alpine pastures, winter crops were heavily damaged by a snow mould (*Fusarium nivale*). This was notably the case in 1785 and also in 1757, 1770 and 1789 (Pfister 1978a). No such outbreaks have occurred in the last 120 years; *Fusarium nivale*, it seems, was a peculiar feature of the Little Ice Age in Switzerland.

The end of the Little Ice Age in Switzerland was marked by a run of warm, dry summers in the 1860s and 1870s. This was the golden age of Alpine mountaineering when so many Swiss peaks were climbed for the first time. Over the century before 1860 the summer months had 2.5 more rainy days than in the century following, a statistically significant difference. Temperatures in summer in the century after 1860 rose by only 0.1 °C, but the warming in the other seasons was much greater, as Table 6.5 shows, with a rise in mean winter temperature of 0.6 °C.

The later part of the Little Ice Age in Switzerland can be seen in perspective as having been the outcome of a climatic fluctuation involving cold winters extending into March, with wet summers and short growing seasons, as compared with the period before 1560 and since 1860. When it is examined in much more detail it

Table 6.5 Differences in mean seasonal temperature between 1755–1859 and 1860–1965

	(1755–1859)–(1860–1965)
Spring	+0.3
Summer	+0.1
Autumn	+0.4
Winter	+0.6

Source: Pfister 1978c

is seen to have been made up of a number of minor fluctuations each lasting some decades. There was, for example, the period between 1560 and 1630, with cool, wet summers and cold winters and springs; this came to an end when both summers and winters became warmer and drier. Even the minor fluctuations were not usually homogeneous. The 1690s as a decade stand out as having been cold at all seasons, with heavy precipitation in spring, summer and autumn. The 1720s included hot summers in 1723, 1727 and 1729, when there were booms in wine production in spite of the frequent thunderstorms. Following warm summers in the years 1759 to 1763 there came a sequence of fourteen cool summers and then seven warm ones. The last of these was followed by the long, hard winter of 1784/5. Switches in the characteristics of individual months could be abrupt; June was rather cool and wet from 1760 to 1800 and again in the decade 1810 to 1819, but in 1820 June was remarkably warm, 4 °C above the 1901–60 mean. In fact, it is not easy to make general statements about the climate of the Little Ice Age; the weather was variable as it is today, but there were more frequent cold winters and wet summers than there have been since 1860 and these were often bunched together.

What can we say now about the climatic conditions associated with the advances and retreats of the Swiss glaciers as they are exemplified and indeed represented by the Unterer Grindelwald Glacier? The Unterer Grindelwald reached its forward position at the end of the sixteenth century following three decades when temperatures in all the seasons were low and precipitation in the form of snow was more abundant than usual. Within this period, clusters of cold, wet summers occurred in 1560–4, 1569–79 and 1585–97 (Figure 6.14).

Glacier retreat in the 1640s followed the arrival of warmer summers after 1630. A readvance starting about 1686 preceded the frigid decade of the 1690s, though this may well have provided the momentum that carried the glaciers forward until 1718. A retreat of the Grindelwald Glacier back to the Unterer Schopf coincided with the hot summers of the 1720s (which were experienced in England as well as Switzerland). A slight readvance of the Grindelwald about 1740 seems to have been associated with increased summer wetness, and then from 1756 to 1768 the glacier front was back

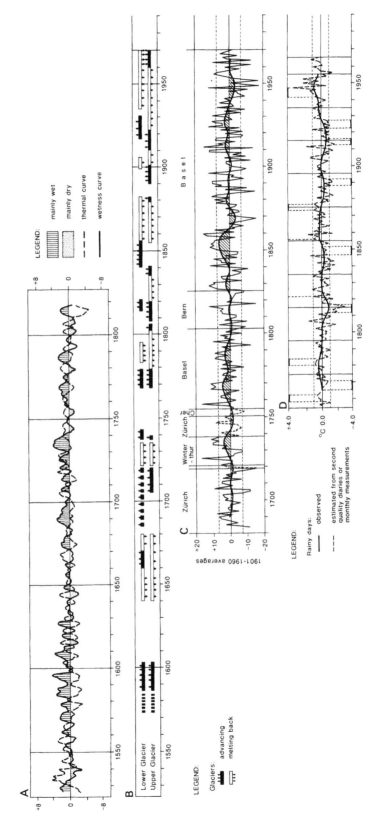

Figure 6.14 Movements of the terminal snouts of the Grindelwald Glacier (b), in comparison with 5-year moving averages of weighted thermal and wetness indices for the summer months (June, July and August) (a), number of rainy days in the summer months as departures from 1901–60 averages (c), and summer temperatures in Basel as departures from 1901–60 averages (d) (Source: from Pfister 1980a)

near the Unterer Schopf. Quite suddenly it then advanced rapidly for ten years, stimulated by abundant snowfall and low spring and summer temperatures in 1767–71. The retreat that followed, from 1784 to 1794, immediately followed a sequence of four warm summers; then for twenty years the glacial terminus was back again at the Unterer Schopf.

An advance starting in 1814 carried the Unterer Grindelwald further forward in 1820 than it had been for 170 years, while the Oberer Grindelwald almost reached the 1600 moraine. This expansion was generated by a long sequence of cold, wet summers and cold autumns, starting with that of 1812 and continuing to 1817. The expansion came to a sharp halt with the hot summer of 1820. The ice remained in an advanced position and its tongue crept forward a hundred metres between 1840 and 1855 assisted by the low temperatures in all seasons between 1847 and 1851 and high precipitation as indicated by numbers of raindays at Basel (Figure 6.14). At length, in 1861, the Grindelwald Glacier began to retreat; within six years the Unterer Schopf was visible and by 1880 the tongue

had retreated a kilometre from the 1856 moraine. Temperatures were above average in all seasons for the period from about 1860 to 1875 but not by a remarkable amount. However, according to Pfister (1980b), the deficit in precipitation was greater than had been known since 1680 and probably since 1560.

The glacier maintained its position for the last twenty years of the nineteenth century, with the aid of marked cooling in all seasons from 1887 to 1890 but not, it seems, accompanied by much increase in precipitation. The retreat was resumed for a few years until the cold summers preceding and during the First World War took effect and the ice pushed forward almost to the 1890s moraine. There the Unterer Grindelwald remained more or less stationary for the rest of the interwar period, rather anomalously it would seem, for the Oberer Grindelwald retreated over 200 m and the Stein and other glaciers much more rapidly in the same period, presumably in response to above-average temperatures in all seasons from 1927 to 1935. Eventually, in 1940, the Unterer Grindelwald followed the example of the majority of the other Swiss glaciers, and

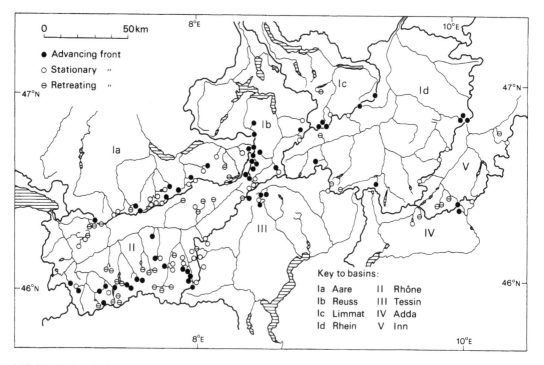

Figure 6.15 Distribution of advancing, retreating and stationary Swiss glaciers amongst those with fronts observed for the year 1974/5 (*Source:* from the **96th Annual Report of the Swiss Glacier Commission**)

Alpine glaciers generally, by retreating. It shrank steadily by an average of about 40 m annually until 1964 in response to a succession of warm dry summers culminating in the decade 1943 to 1952.

While the 1960s, 1970s and 1980s saw the Oberer Grindelwald advancing in the company of about 40 per cent of the other Swiss glaciers (Figure 6.15), the Unterer Grindelwald lagged behind again until it also began to advance in the early 1980s. In general glacier fronts, as in the years around 1920, were less decisive in their movements than they had been between 1930 and 1960. Some were advancing while others were retreating. By 1980 most glacier tongues were far higher up their valleys than they had been sixty years earlier. Some glaciers, notably the Corbassière and Giétroz, began to flow twice as fast as usual in response to the positive mass balances of the 1970s, but soon decelerated to their normal pace. In the 1980s, especially in 1982/3 with its hot 'El Niño summer', negative mass balances comparable with those of the 1940s returned. Annual losses from the Silverette and Gries for example were of the order of a half-metre to one metre annually between the mid 1980s and mid 1990s. Although ice losses diminished in the early 1990s, the fronts of 90 per cent of Swiss glaciers were in retreat in the years following the hot summers of 1994 and 1995. By the end of the century the fronts of some of the more rapidly responding glaciers, including the Unteraar, had reverted to their mid century positions (Hoelzle *et al.* 1999). Even the fronts of the Fee, Trient, Oberer Grindelwald and Tschierva, which had been advancing for some decades, retreated. The tongue of the Aletsch, with its long response time, had retired 2 kilometres from its 1950 position. Its front at 1554 m is still well below the timberline, hidden in a gorge (Herren *et al.* 1999).

The Little Ice Age seems to have receded into history; the climbing huts of the Victorians built a few tens of metres above the ice now look down on the glaciers from a height of a hundred and fifty metres or more. The glory of the Alpine glaciers is departing.

6.5 A SUMMARY OF SWISS GLACIER AND CLIMATE FLUCTUATIONS IN THE LITTLE ICE AGE

Evidence from historical documents and from trees felled by the advancing ice and dated by their rings and by radiocarbon shows that Swiss glaciers advanced in the decades around 1300 as far as they did in the nineteenth century. In the intervening period they advanced in the 1580s, about 1640, 1720, 1740, in the 1770s, about 1818, 1850 and the 1890s. Twentieth-century retreat was interrupted by advances between 1915 and 1920 and again between 1974 and 1981. It is not easy to make general statements about the climate of the Little Ice Age; the weather was as variable as it is today, but there were more frequent cold winters and wet summers than there have been since 1860 and these were often bunched together.

7

SOUTHERN EUROPE

The Pyrenees, Maritime Alps and Apennines

The history of the many small glaciers that still survive in the remote recesses of the Pyrenees and the southern Maritime Alps of Italy has received comparatively little attention. Only in the 1990s was the presence of ice on the Macizo Central of the Picos de Europa recognized, when it was revealed by melting of the formerly perennial snow cover (Suárez and Alonso 1994). Some ice may remain on the Sierra Nevada in southern Spain but information about it is sparse. The Calderone, the only glacier in the Apennines and the southernmost in Europe, clings to the Corno Grande, 2912 m, the highest peak of the Gran Sasso in the Abruzzi east of Rome.

The fluctuations of the southern European glaciers failed to impinge directly on the lives of local people who consequently kept no documents relating to them. Not until scientists began climbing mountains towards the end of the eighteenth century were records kept of the ice extent. This also happened to be the time when many glaciers reached their Little Ice Age maxima. Since then the story has been one of ice retreat interrupted by temporary advances in some decades.

7.1 THE PYRENEES

The Pyrenees stretch for 250 km from the Bay of Biscay to the Mediterranean, forming a barrier between France and Spain about 42 to 43 °N. Small glaciers at altitudes above 2500 m are scattered across the central part of the range on major massifs from Balaitous in the west to Maladeta in the east (Figure 7.1). The regional snowline, lower on the northern than the southern side, rises from about 2900 m in the west to over 3100 m in the east. The positions occupied by the glaciers and the forms they take are strongly influenced by local topography and microclimates. Some, such as those overlooking Gavarnie, stretch along rock ledges. Others occupy niches or shallow basins. The largest are in cirques. Valley glaciers are rare. At the end of the twentieth century only the Glaciar de Aneto, on the southern side of the range, had an area of more than a square kilometre. Many glaciers are remote; none has threatened villages or farms and so fluctuations of ice fronts have not achieved archival record. Here an attempt is made to put together the main features of Pyrenean glacial history since the late eighteenth century by assembling the written, cartographic and photographic evidence from the various ice-bearing massifs.

7.1.1 Historical sources

The main sources of information about the Pyrenean glaciers are the published observations of travellers, scientists and mountaineers, together with their sketches and photographs.[1] No evidence has been found predating the records made in the late eighteenth century by Ramond de Carbonnières (1789, 1792, 1801). Of outstanding importance are the deliberate observations, detailed records and precise maps made more recently by the Abbé Ludovic Gaurier (1912, 1921, 1933a, b,

---

1 Many early photographs and sketches are preserved in museums and private collections, notably the Muséum d'Histoire Naturelle, Toulouse and the Musée Pyrénéen, Château Fort, Lourdes.

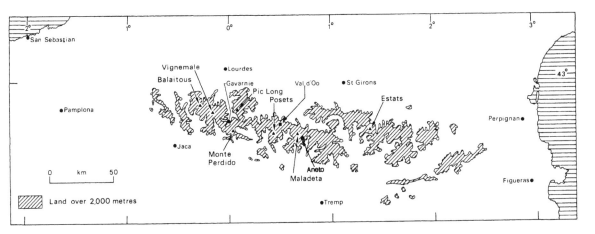

Figure 7.1 The principal massifs of the Pyrenees

1934), who had a life-long interest in the Pyrenean glaciers (Lassere-Vergne 1989).

A Cassini map from the second half of the eighteenth century is too inaccurate to be useful.[2] The first triangulation was made by a French military survey team in 1825–7, but the first map of value for glaciological purposes is the '*Carte du Mont Perdu et de la Region Calcaire des Pyrénées Centrales*' on a scale of 1:40,000 made by the geographer Franz Schrader in 1874. His six 1:100,000 contoured maps of the central Pyrenees, completed in 1882, covered the Spanish as well as the French side of the mountains. In 1914 they were followed by a 1:20,000 map '*Le Massif de Gavarnie et Mt Perdu*'. Detailed maps of the Pic Long and Seuil de la Baque[3] areas, based on surveys by Eydoux and Maury, were published by Leon Maury early in the twentieth century. An excellent map of the Vignemale area was produced by Alphonse Meillon in 1929; unfortunately, revisions of glacier extents were not included in later editions. The first deliberate photography of Pyrenean glaciers from the air was undertaken on the Abbé Gaurier's initiative (Gaurier 1934b). Institut Géographique National maps on scales of 1:25,000, based on air photographs, provide good modern coverage of the French side of the Pyrenees, the more recent ones covering parts of the southern versant. Modern Spanish maps are insufficiently accurate to be useful for glacier

studies except in the most general way. Even the *Mapa Topografico Excursinista*, 1:25,000 sheet of Maladeta–Aneto is not accurate enough to serve as a base for quantifying ice extent.

7.1.2 Pyrenean glacier monitoring

Forel, who had taken the lead in establishing regular monitoring of Swiss glaciers (see Chapter 6), wrote to the French Alpine Club as early as 1886 stressing the 'grand intêret de posseder des observations précises, et comparable avec celles des Alpes, dans d'autres chaînes de montagnes, et especialement dans les Pyrénées' (Forel 1886). Prince Roland Bonaparte (1890, 1891) attempted to respond to Forel's call to monitor the Pyrenean glaciers, employing local guides. He selected seventeen glaciers for regular surveys but was unable to arouse sufficient enthusiasm to maintain the effort and it soon faded out (Plandé 1939: 168). A few of the markers he had placed on glacier forefields were later rediscovered and taken into account.

An entire century was to elapse before data began to be published relating to a few of the glaciers on the Spanish side of the mountains although some measurements began earlier (Haeberli *et al.* 1993). Glaciers on the southern side of the range were monitored by a group that catalogued the ice of the Iberian Peninsula

---

2   The two maps included in Ramond's *Observations fait dans les Pyrénées* (1789) were probably based on Cassini.
3   Also Seil and Sehl de la Baque.

for the World Glacier Survey (Serrat 1980). An inventory of the characteristics and dimensions of the Spanish glaciers was made in the 1980s by Martinez de Pisón and Parra (1988). Field surveys on the northern and southern sides of the Pyrenees were made between 1988 and 1990 by parties led by Anne Gellatly as part of a European Union Programme 'Reconstruction of Mediterranean Climate' (Grove and Gellatly 1992, 1995).[4] Repeat photographs were taken, observations were made on the state of many glaciers on both the north and south sides of the Pyrenees, and glaciers mapped by Gaurier were resurveyed. Moraines were plotted from air photographs and dated by radiocarbon or by lichenometry.

Since 1998 the University of Zaragoza's Geography Department has been organizing field surveys of Spanish glaciers. They are intended to continue on a regular basis with the results appearing in the *Boletín Glaciológico Aragonés*.

7.1.3 Pyrenean glacier fluctuations in recent centuries

7.1.3.1 Late eighteenth-century expansion

The earliest useful accounts of Pyrenean glaciers were by Louis-François Ramond de Carbonnières (1755–1827). He had acquired an active interest in glaciers when he visited the Alps in 1777.[5] A decade later he explored the Pyrenees, deliberately visiting several of the major massifs and noting the appearance of their glaciers at a time when ice fronts were evidently as much as 200 m lower than at the end of the twentieth century.

In 1787 Ramond climbed to the Brèche de Roland in the Cirque de Gavarnie to find out whether or not there was ice under the snow lying on the great limestone ledges. Digging until he found true ice at the top of the little Glacier de la Brèche (Figure 7.2, Plate 7.8b), he confirmed the existence of glacier ice in the Cirque. He was told by his guide 'm'apprit qu'avant les premièrs jours de Septembre, ces glaciers ne seroient point

découverts & que même ils ne se découviroient point de tout, si le fin d'Aout n'étoit pas chaude' (Ramond 1792: 71). The extent of late lying snow at that time is confirmed by contemporaries such as Pasumot (1797: 177).

In August of that same year 1787 Ramond explored the granite massif of Maladeta, south of the main crest of the Pyrenees, entirely in Spain (Figure 7.3). He found the snout of the Glacier de la Maladeta a little way above the Lago de la Paderna,[6] 'une épaisse & vaste calote de glace, traversée de grandes crevasses toute dirigées de haut en bas' (Ramond 1789: 23). Entering one of the crevasses he estimated its depth as 40 pieds (13 m).

Also in August 1787 he climbed the granite Luchonnais massif, at the head of the Val d'Oo (Figure 7.4). The view far surpassed what can now be seen:

> ce fut donc avec un mouvement de surprise que je vis au-dessus d'une lac totalement glacé, tout environné de neiges que percoient trois bandes de glace, voisines de la superficie, qui paroissoient appartenir à un seul glacier, dont la surface ne se découvre peut-être jamais en entier & qui paroit être lui-même le prolongement d'une très grande bande de glace, que l'on voit même-temps, dans les neiges de la pente opposée à celle ou nous trouvions. Celle-ci s'étend au loin vers les montagnes du port de Clarabide, dont on voit les vallons couvert de neiges éternelles, à un aspect où le soleil du midi devoit s'apposer plus efficacement à leur accumulation & ces neiges tapissent presque toutes les hauteurs qui se présentent à la vue. C'étoit, le plus beau desert de ce genre que j'eusse trouvé dans les Pyrénées. . . . On donne le nom Sehl de la Baque au lieu où nous étions. Le lac glacé est designe par le même nom.
>
> (Ramond 1789: 149)

As Ramond went up from the Lac Glacé to the Port d'Oo, he followed a line of cairns, evidently an established route used by smugglers. This suggests that, unless the cairns were rebuilt regularly, the Sehl de la Baque had been in a stable condition for some time before 1787 (Ramond 1789: 149).

Ramond investigated the Monte Perdido (or Mont Perdu) massif twice in 1797 (Figure 7.5, Plate 7.1). Glaciers, which sit in a row on the great ledges on the

---

4 Support was provided by EC grant EV4C-0044-UK(H) made to J. M. Grove.
5 The year when 'Sketches of Switzerland in letters from Wm. Coxe' was published. Ramond was sufficiently interested to translate it into French with his own extensive comments (Ramond 1789 or 1792).
6 Also Lac de Rencluse.

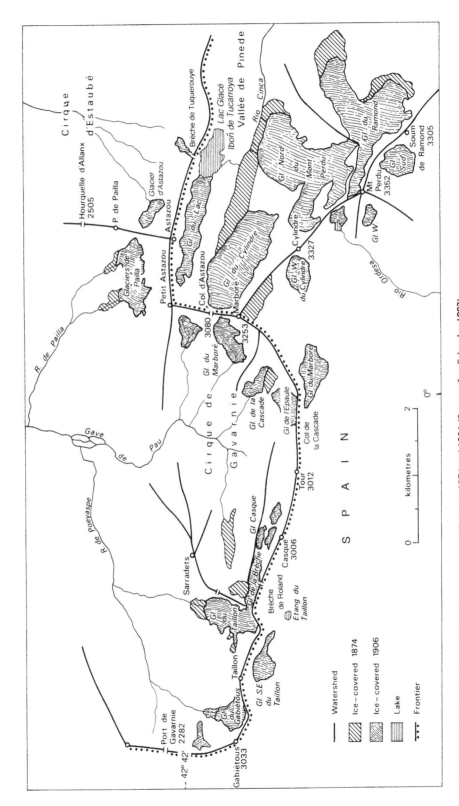

Figure 7.2 The Cirque de Gavarnie showing areas covered by ice in 1874 and 1906 (*Source: after Schrader 1882*)

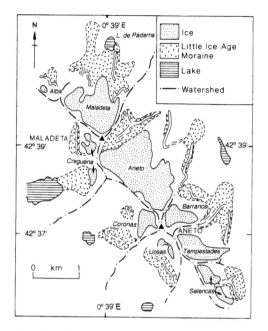

Figure 7.3a Sketch map of the glaciers and Little Ice Age moraine on the Maladeta massif

north side, join up when the ice thickens and after a period of negative mass balances separate into discrete units. On 12 August Ramond approached them from the Cirque d'Estaubé by way of the Brèche de Tuquerouye.[7] The ascent over snow was easy, taking only two hours. From the Brèche he had a good view of the northern side of the ledges and saw ice completely covering them.

> Taillé du même ciseau qui a façonné les étages du Marboré, il présente une suite de gradins, tantôt drapés de neiges, tantôt herissés de glaces qui débordent et se versent les uns sur les autres en larges et immobiles cascades, jusques aux bords d'un lac dont la surface encore glacée mais déjà dégagée de neiges, brillait d'un éclat sombre.
>
> (Ramond 1801: 59–66)

He walked round the lake on the ice and obtained a close view of the lower glacier which he saw reaching down towards the edge of the lake, the Lac Glacé or Ibon de Tucarroya.

He returned on 8 September, taking the same route as before. Lacking a snow cover, it was now a much

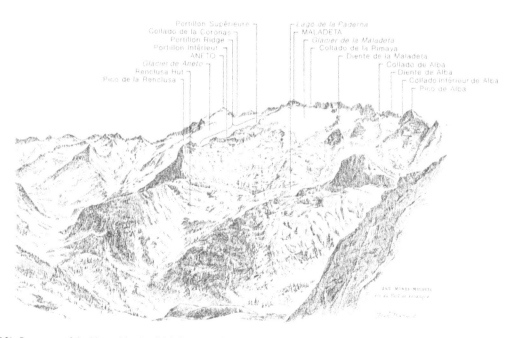

Figure 7.3b Panorama of the Monts-Maudits (Maladeta massif) as seen from the Port de Vénasque to the northeast, by Jean Dehais

7   Also Tucarroya.

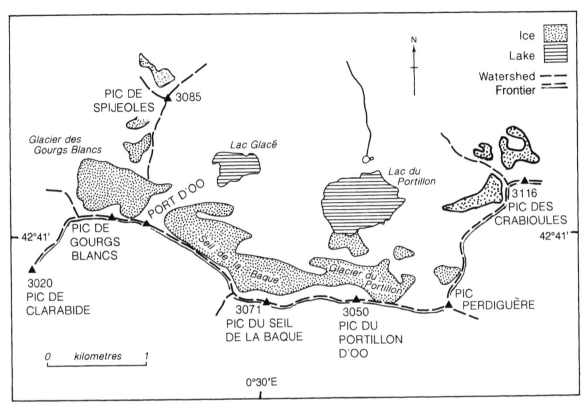

Figure 7.4 Sketch map of the Luchonnais glaciers

more difficult ascent and it took five hours instead of two. The view from the Brèche de Tuquerouye had greatly changed; now the Glaciar de Monte Perdido (Glacier Nord du Mont Perdu) reached the edge of the lake.

> D'immense rochers de glace accablent de leur poids les gradin démembrés de la montaigne, et trois de ses étages sont chargés de ces amas de pointes bizarres qu'on ne peut comparer qu'a des vagues solides. Leur base est d'une épaisseur enorme; elle plonge verticalement dans le Lac, et les cavernes dont elle percée y vomissent l'eau par torrents. Une de ces vôutes se fendit devant nous: en é clatant, elle rendit le seul son qui ait frappé ici notre oreille; et son était un coup de tonnerre.
>
> (Ramond 1801: 168)[8]

Ramond's account of the ice reaching the shore of the lake has to be accepted. He had been able to walk on

the ice during his August visit and must have been able to see exactly what the situation was. His account of the 'immense rochers de glace' seen in September gives a vivid impression of a major icefall. This was probably triggered by rapid ablation since his first visit, clearing the snow below the Brèche de Tuquerouye.

Though Ramond did not explore the Vignemale massif himself, he recorded that in 1798 his friend La Beaumelle found an 'immense glacier' between Vignemale and Montferrat dividing into two branches separated by rising ground. The tongue of the Glacier d'Ossoue Ramond saw for himself from the bottom of the Val d'Ossoue and made a detailed sketch of it (Figure 7.6). His writings demonstrate that Ramond was an accurate observer, interested in detail. They show clearly that the Pyrenean glaciers were more enlarged

---

8   Schrader (1894: 406) did not accept Ramond's account of the glacier reaching the lake, but appears not to have read the account of his second visit, and perhaps underestimated the scale of glacier fluctuations in recent centuries.

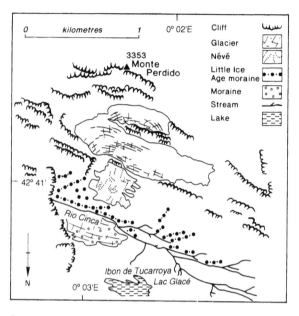

Figure 7.5 The Glaciar NE on Monte Perdido (south towards the top) (*Source*: after Martinez de Pisón and Parra 1988)

Plate 7.1 Monte Perdido, September 1981
Photo: Instituto Geografico Nacional

in the late eighteenth century than they have ever been since.

7.1.3.2 A mid nineteenth-century advance

Little direct evidence is available about the behaviour of the Pyrenean ice fronts in the early nineteenth century, though there are clear indications that retreat soon followed the period of enlargement witnessed by Ramond and his contemporaries. In the middle of the nineteenth century the glaciers enlarged again, but not to the same extent as when Ramond saw them. Particularly clear evidence comes from the Glacier du Pays Baché on the eastern side of the Néouvielle[9] massif (Figures 7.7a, b and c, Plates 7.2, 7.3a, b). No accounts of conditions on or around Néouvielle are to be found before that of Pierre-Toussaint de la Boulinière, a mountaineer who saw the main summit 'avec ses vaste champs de neiges' on 26 July 1812 (Beraldi 1898, vol. 1). When Vincent Chausenque made the first successful ascent of Néouvielle on 10 July 1847 he crossed on his way 'un grand plan neigé', and ascended the peak by way of steeply inclined snow reaching almost to the summit. Clearly, extensive snow cover lasted through the middle of summer in the early and mid nineteenth century.

Michelier, an engineer in charge of works for Lac d'Orédon (east of the Lac de Capdelong) which was to be used as a reservoir (Figure 7.7a), made observations on the Pays Baché, between 1869 and 1885. His interest was such that he wrote a long paper on the connections between the glacier fluctuations and the variations of climate which caused them (Michelier 1887). The Glacier du Pays Baché, to the east of the Pic de Long, is fronted by a complex moraine consisting of at least

---

9   Also Neoubielhe.

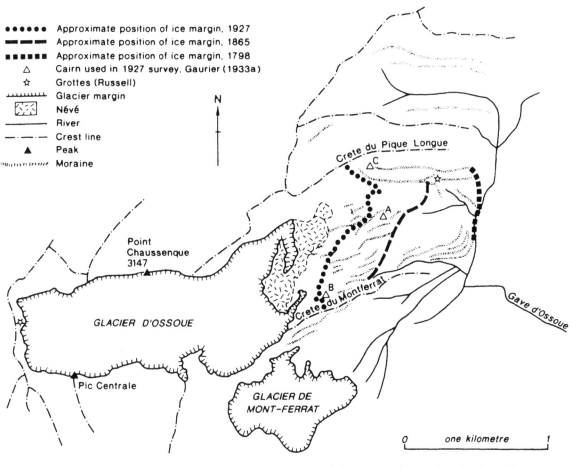

Figure 7.6 Sketch map of Glacier d'Ossoue to show former ice margins, compiled from aerial photographs and field mapping (*Source:* Grove and Gellatly 1995)

two ridges, carrying lichens but neither vegetation nor soil (Figure 7.7c). Evidently advance periods had brought the ice front to roughly the same position at least twice in the latter part of the Little Ice Age. Outside this main frontal ridge lies an older, well-vegetated moraine, probably deposited at some time in the Holocene.[10] Michelier was told by local *isard*[11] hunters that in 1856 the ice had still been in contact with the moraine fronting the Glacier du Pays Baché, and he speculated that it might have been deposited during

10 Lichen measurements, made by A. F. Gellatly in 1989, from the main frontal moraine yielded average sizes of 4.8 ± 1.65 mm from a sample of 32 Rhizocarpon *sensu lato,* a surprisingly low value considering that the moraine is known to have been deposited in the middle of the nineteenth century. This slow growth rate may be the result of the high altitude and exposed position. There was a substantial amount of late lying snow in both late July 1988 and September 1989. On the outer crest of the main moraine a sample of 40 thalli had an average size of 22.7 ± 7.6 mm, while the five largest had an average size of 35.2 ± 7.6 mm. Sixty metres outside this compound frontal moraine, a prominent, well-vegetated moraine carrying a thick peaty organic topsoil about 55 cm thick, with a well-developed B/C horizon just under 20 cm thick, is unquestionably older than the other unvegetated moraine.

11 Wild goats then common in the Pyrenees.

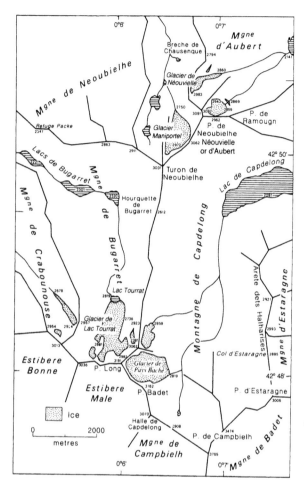

Figure 7.7a Sketch map of the glaciers of the Néouvielle massif

Figure 7.7b Pays Baché Glacier, 1906 (*Source*: redrawn from Eydoux and Maury 1907)

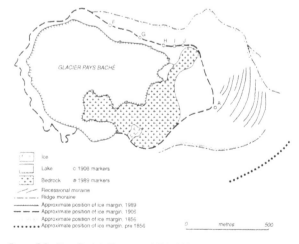

Figure 7.7c Pays Baché Glacier in 1856, 1906 and 1989 (*Source*: from Latham and Gellatly 1989)

an advance period starting around 1812 when, according to Boulinière's account, snowfields were more extensive.

Vallot (1887) compared Ramond's sketch of the tongue of the Glacier d'Ossoue with his own made from the same viewpoint (see Figure 7.6). The glacier had evidently been longer in 1798. Vallot concluded that it had then made a rapid retreat interrupted by a stationary period or one of slow advance when it had deposited a moraine. This pause in its retreat had ended at least twenty years earlier according to Henri Passet, the guide with whom Vallot had explored the glacier. As Passet remembered it, in 1865 the ice reached to the summit of the left lateral moraine, and Vallot cal-

description accords with the distribution of Little Ice Age moraines around the Coronas lake as recently mapped by Martinez de Pisón and Parra (1988). It seems that the Glaciar de Coronas must have been near its Little Ice Age maximum when Franqueville saw it in the 1840s.

7.1.3.3 Retreat in the 1860s and 1870s

After 1864 great sheets of névé and snow disappeared in the Néouvielle area, the glaciers contracting especially rapidly between 1864 and 1870 (Michelier 1887). Snow cover was also diminishing rapidly on Vignemale. Henri Russell[12] (1908: 80) wrote that there was so little snow in the winter of 1870 the Glacier d'Ossoue was 'entiérement dénudé et dechiré dans tous les sens par d'effrayantes crevasses, comme au mois de Septembre'. Michelier was probably the first to base estimates of loss of snow and ice thickness on the extent of light-coloured zones free of lichens surrounding the diminished névés. 'La base des crêtes sur les flancs desquelle s'appuyaient les névés est blanches sur une zone de 10 m a 15 m de hauteur-dessus de cette zone, le granit parait noirâtre.' He reckoned the névés had lost 10–15 m in thickness, wastage being greatest on the eastern and northern slopes. While in 1869 the Pays Baché ice was 25 m from the frontal moraine, in 1883 the distance was 120 m. Michelier estimated the loss of volume at about 8,400,000 m³, attributing it to reduced precipitation and a series of warmer winters after the middle of the century which continued until the 1880s.

Trutat, Conservateur of the Musée d'Histoire Naturelle de Toulouse, was the first scientist to attempt to set up regular measurements of a Pyrenean glacier. He had direct experience of ice thinning, for having installed markers on the Glacier de la Maladeta in September 1873, he returned on 1 September 1875 to find his markers had disappeared; 'depuis 1873, le glacier a subi un énorme retrait' (Trutat 1876: 444). Estimating that the front had retreated about 50 m since his previous visit, he noted that retreat was general:

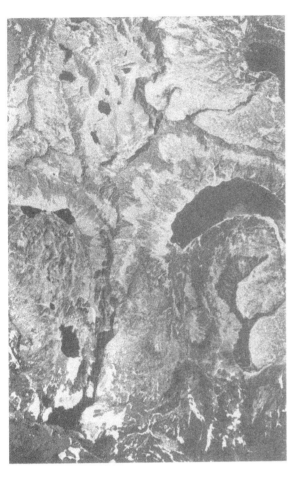

Plate 7.2 Aerial photograph of the Néouvielle Massif in 1983
Photo: Institut Géographique National

culated that the glacier must have been 100 m further forward and 10–20 m thicker. A sketch made by Anne Lister shows that the glacier was still extended in 1838 but provides no information about the time when the moraine was deposited.

The little Glaciar de Coronas, on the south side of the Maladeta massif (see Figure 7.3), was visited by Franqueville in 1842 (Franqueville 1977 edition). It then reached the northern edge of the Lago de Coronas 'and extended in a semi-circle round it'. This

---

12 Henri Russell was an Irish-French eccentric who spent much of his life in the Pyrenees. He became so enamoured of Vignemale that he had four caves excavated in the mountain between 1882 and 1894 so that he could live in them for long periods without intruding on the natural landscape. He was given a ninety-nine year lease of a large area of the Vignemale in 1889 by the Syndicate of the Vallée de Barèges (Meillon 1987: 133).

Plate 7.3a Glacier du Pays Baché, 13 August 1963
Photo: P. Hollerman 1968

Plate 7.3b Glacier du Pays Baché, 10 September 1989
Photo: A. F. Gellatly

Plate 7.4 'Le Glacier de la Maladeta vu de la Pena Blanc' (*Source:* Trutat 1876)

'depuis j'explore les Pyrénées, je vois . . . les glaciers fondre sous mes yeux' (Trutat 1876: 483). He did not consider it possible to establish whether retreat had been continuous or whether periods of advance had complicated the pattern, but one of his photographs provides evidence (which he did not notice himself) that withdrawal had not been unbroken. His photograph of the Maladeta published in 1876 (Plate 7.4) shows the lateral moraine of the Maladeta, below the Dent de la Maladeta, with three subsidiary half arcs, lying one below another and departing from the main crest line in sequence. These represent a series of frontal positions caused by minor advances or halts in recession. The date of formation of the main lateral is unknown.

The glaciers in the Cirque de Gavarnie had also retreated by the 1870s (Vallot 1887). In the early 1880s 'Les gradins du cirque de Gavarnie, superposés dans un ordre si parfait, ont perdu cette brillante parure de neige qui les ornait si bien, et montrent la roche à nu' (Degrange-Tousin 1882: 575). Prince Roland Bonaparte began his attempt to measure the variations of

the Pyrenean glaciers in 1890. In the early 1890s, the guides he sent out to report on the state of the glaciers found all of the Gavarnie glaciers either stationary or retreating (Bonaparte 1891: 515–16).

In 1894 Schrader reckoned that the total ice cover in the Gavarnie area was 3.48 km^2. It is tempting to use his figures as a basis for comparison with values obtained more recently, as has been done for the glaciers on the Spanish side of the Pyrenees (Arenillas *et al.* 1992). But Schrader himself was conscious that his surveys had been made over the course of several years and that precision had been handicapped by snow and névé obscuring the margins (Schrader 1894: 405). It is only possible to use his figures as a basis for comparison in a general sense.

7.1.3.4 A halt in recession followed by expansion in the 1880s and 1890s

Bonaparte (1891) found numerous instances of the disappearance of névés[13] and glaciers and of reduction

---

13 He hoped his project would come to include studies of the altitude of snowlines, as well as tongue positions, because he wished to identify links between glacier fluctuations and meteorological conditions and so deduce a law (Bonaparte 1890: 446).

Plate 7.5 'Villa Russell' in the 1890s (*Source:* Trutat 1876)

of those that remained, but he was aware that a marked increase in accumulation had begun by 1885 and expected to see the glaciers lengthening (Bonaparte 1890: 447). He compared photographs of the Glacier Nord du Mont Perdu taken in 1880 by Schrader, in 1886 by Vallot and in 1890 by Regnault.[14] The series revealed that by 1886 the thickness of ice on the uppermost terrace had increased and by 1890 had further increased at lower levels.

Unusually detailed information about the period of increased ice volume in the late nineteenth century comes from a surprising source. Russell had caves excavated on Vignemale above the accumulation basin of the Glacier d'Ossoue and observed the variations in level of the névé year by year (Russell 1908: 582–600). The Villa Russell cave at 3201 m was excavated in 1881; the Grotte des Guides at 3202 m in 1885, and the Grotte des Dames at 3205 m in 1886 (Plate 7.5). The three together were known as the Grottes de Cerbillona. Eventually two more were cut near the front because

of the rapid increase in height of the névé on the upper glacier which was associated with notable changes in volume and appearance of the whole ice stream; Russell wrote in the summer of 1890:

> depuis un grand nombre d'années, jamais personne n'avait vu le glacier si boulversé, si déchiré, si disloqué . . . de sa base à son sommet . . . complètement transformé. N'était plus, vers le fin de l'été, qu'un enfer de séracs et des gouffres (comme ceux du nord du Mont Perdu), un chaos de crevasses inconnues. . . . jusque sur la plaine de neige, toujours unie, toujours immaculée, qui s'etend comme un lac sous les portes de mes grottes supérieures.
>
> (Russell 1908: 154–6)

Russell was aware that glaciers elsewhere in Europe were in a healthier state than they had been for some time. In 1886 he wrote:

> Il est heureux que la troisième [of his excavations], celle des 'Dames' depassé d'au moins quatre mètres le niveau des deux autres, car dans les Pyrénées, comme dans les Alpes, certains glaciers commencent à remonte, après avoir

---

14   It has not proved possible to find these photographs.

baissé et reculé pendant plus de trente ans. S'ils n'allongent pas, ils se boursouflent. . . . Le glacier d'Ossoue, que je mesure depuis plusieurs étés aux mêmes epoches, c'est élévé de cinq mètres en deux ans sous le seuil de mes portes.

(Russell 1908: 595–6)

In 1888:

Le Glacier monte toujours, et bientôt mes pauvres grottes s'ensevelissent sous les neiges d'une manière permanente . . . . La seul qui soit sortie de son tombeau de glace pendant cinq ans (1889–1894) est celle des dames. . . . L'énorme crue du glacier m'ayant forcé de me fixer ailleurs, et de beaucoup descendre, je fait creuser l'année suivant, 1888, deux autres cavernes . . . a l'altitude modeste de 2400 métres, au bas du grand glacier.

(Russell 1908: 595–6)

Russell recorded the level of the surface of the névé in relation to the Grottes de Cerbillona from 1882 to 1895. The period of increased accumulation seems to have begun in 1884–5 when snow quite covered the entrance to the Villa Russell and perhaps ended in 1895, though Russell recorded 'la neige tomba comme en hiver' in July 1896, 4 metres of snow in front of the Grottes de Belle-Vue in July 1897, and 'tant de neige sur le glacier' at the end of July 1897. In the absence of more precise information about ablation it is reasonable to take at least the whole period 1885–95 as one of generally increased accumulation.[15]

7.1.3.5 Fluctuating retreat in the early twentieth century

Although recession dominated during the twentieth century, it was not continuous. Retreat halted or was interrupted by slight advances in 1906–11, 1926–7, 1944–5, 1963–4 and 1978–9. Our knowledge of the first two of these interruptions depends on Gaurier's observations, which were of noteworthy accuracy. Without his writings and more air photographs, it would be difficult to compare twentieth-century events in the Pyrenees with those elsewhere in Europe.

Gaurier started systematic observations in 1904, just in time to register the end of the retreat phase which ended late nineteenth-century expansion. Typical of the observations he made during this period were those he made on the Balaitous (Figure 7.8). The Glacier de Plaa (now known as the Pabat) had lost most of its snow by 26 July 1905 and all the lower part of Las Néous was bare by 9 August while the Névé de las Clottes had melted away except in a few places at the foot of cliffs. After 1905 there came a period of ice expansion (Plate 7.6). All Gaurier's observations between 1906 and 1911 were of increasing snow cover and glacier thickness. On 15 September 1909 Las Néous was 'totalement chargé de la neige de hiver et touche les moraines qui sont précédées grand névés . . . le mouvement de progression en épaisseur s'est donc très accentué' (Gaurier 1912: 64).

Gaurier took up Russell's observations of the level of the névé in relation to the Grottes de Cerbillona, visiting the caves in either August or September of each year from 1902 to 1909, except in 1906 and 1908 when he arranged for others to take his place (Gaurier 1912: 105). He concluded that these measurements showed a second period of increase between 1906 and 1909 (Table 7.1).

In front of the Glacier des Oulettes, about 2 km northwest of Russell's caves in the névé of the Glacier d'Ossoue, Gaurier in 1905 discovered markers Bonaparte had installed in 1892 and measured a retreat of the ice front over the intervening thirteen years of between 125 m and 145 m (Table 7.2, Figure 7.9). He placed three reference markers in front of the Glacier d'Ossoue and the Glacier des Oulettes, two in front of the Petit-Vignemale, and one in front of the Glacier de Clot de la Hount. Returning in August 1905, after a summer with very little snow, he found the Vignemale glaciers suffering from unusually marked ablation in their accumulation basins, and general regression of their tongues. The Glacier d'Ossoue was no longer acquiring névé from the col beneath Pique Longue. The upper surface of the Ossoue was 4 m below the entrance of Villa Russell (Table 7.1). The left side of the tongue had retreated and thinned; the right side was stationary. On the bedrock slabs of the forefield

15 Gaurier (1921: 105) took this period of increase as lasting from 1889 to 1894. He considered that the 'petit crue' began in the winter of 1888/9 because at the end of August 1888 the snow surface was level with the sill of Villa Russell and below the entrances of the higher caves. There was a minor interruption in the incidence of increased accumulation, but Gaurier appears to have ignored Russell's account of the situation at the end of the summer of 1896.

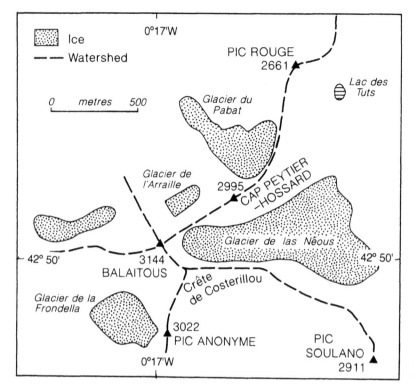

Figure 7.8 The glaciers and snowfields of the Balaitous Massif

Plate 7.6 The SW face of Balaitous showing extensive snow cover on the Glacier Frondella (left centre) which may have impeded mapping during the late nineteenth century (*Source:* Trutat 1876)

Table 7.1 Observations of the level of the névé of the Glacier d'Ossoue in relation to the Grottes de Cerbillona, 1902–9 (m)

Date	Villa Russell 3201 m	Grotte des Guides 3202 m	Grotte des Dames 3205 m
18 Aug. 1902	−3.5	−4.5	−7.5
1 Sept. 1902	−4.5	−5.5	−8.5
12 Aug. 1905	−4	−5	−8
28 Sept. 1906	−3.75	−4.75	−7.75
2 Aug. 1907	+4	+3	0
18 Oct. 1908	+4	+3	0
20 Sept. 1909	+6	+5	+2

Source: from Gaurier 1921: 105
Note: The figures for the Grottes des Dames and the Villa Russell for 1907, 1908 and 1909 were transposed, probably during printing of Gaurier (1921)

Table 7.2 Distances of markers from fronts of des Oulettes, d'Ossoue and Le Petit-Vignemale glaciers (m)

	Glacier des Oulettes					Glacier d'Ossoue			Le Petit-Vignemale	
	Bonaparte's markers		Gaurier's markers							
	B1	B2	R1	R2	R3	R1	R2	R3	R1	R2
1892	60	82				35	46	0	0	0
1904		222	23	30	115	58	42	n	+3	0
1905	180	227	24	60#	120	58	42	n	n	n
1906			23	60	120	58	42	n	n	n
1907			n	n	120	58	n	n	n	n
1908			23	60	120	n	n	n	n	n
1909			n	n	n	n	n	n	n	n
1910			n	n	n	n	n	n	n	n
1911			n	n	120	58	42	0	n	n

Source: from Gaurier 1912
Notes: n = no measurements, marker hidden by snow; # = apparent retreat due to removal of névé which had masked the front, the ice had not withdrawn

Gaurier found a line of black paint which he concluded had been drawn across the rock by Bonaparte's people in 1892, 'et si, comme je le suppose, la ligne de peinture noire qu'on suit sur le rocher a été tracée pour marquer le Front du glacier, la rive droite est absolutment stationnaire depuis 1892' (Gaurier 1912: 119).

Between 1904 and 1911 it appears from Gaurier's measurements that although the glacier fronts retreated it was by no more than about 30 m and the ice was beginning to thicken. The winter of 1905/6 had seen heavy snowfall (Table 7.1). At the end of September 1906 the level of the snow surface was 0.25 m higher than it had been the previous year. From that time

onwards it was above the sill of Villa Russell by a matter of metres, and at or above that of the Grottes des Dames. In late August 1910 all the Grottes de Cerbillona were buried; even those down at Bellevue lay beneath 3 m of snow (Gaurier 1912: 132). The state of the Glacier d'Ossoue was typical of the others on Vignemale. Gaurier's annual descriptions tell of extending névé and thickening glaciers. Fronts were protected from ablation by snow or névé; some like that of Petit-Vignemale were bulging. Even the Clot de la Hount, believed in 1905 to be a dead glacier, was 'donc nettément en progression' (Gaurier 1912: 31). Gaurier judged that heavy snowfall with late springs

●●●●● Approximate position of ice margin, 1911-27
━━ ━━ Approximate position of ice margin, mid 19th century
■■■■■ Approximate position of ice margin, late 18th century
☆ Reference point EF1
⊥⊥⊥⊥⊥ Glacier margin
  Rockfall
  Névé
  River
━ ━ ━ Crest line
  Moraine

CRÊTE DU PETIT VIGNEMALE

CLOT DE LA HOUNT

GLACIER DU PETIT VIGNEMALE

GLACIER DES OULETTES

N

0   one kilometre   1

Figure 7.9 Sketch map of the Glacier des Oulettes, compiled from air photographs and field observations by A. F. Gellatly in 1988, to show the positions of former ice margins

and cold summers were responsible for the positive mass balances, the principal effect of which was to thicken rather than lengthen the glaciers. In 1911, after a winter with less accumulation came a warm, dry summer. The winter snow melted away altogether and also some from the previous year, but the glacier fronts were still protected by névé (Plate 7.7). All the markers except one in front of the Glacier des Oulettes remained hidden by snow, and as yet there was no sign of retreat. No observations were made during the next decade which included the First World War and so there is no direct evidence of the time when expansion ceased and retreat began.

Between 1906 and 1911, Gaurier also observed the Gavarnie glaciers (Plate 7.8a). In 1906 he found no signs of advance. The Glacier du Casque and the Brèche de Roland had thinned and withdrawn from their moraines, but the Taillon had a swollen front, cut by deep crevasses. While establishing one of his own markers, Gaurier found two of the markers placed during Bonaparte's 1893 programme. These were now 160 m

from the ice margin, giving the first proper measurement of the Taillon's retreat over a definite period of time (Gaurier 1921). The roches moutonnés and moraines filling the valley, he noted, pointed to a considerable recent regression. The upper part of the Glacier Sud-Est du Taillon was very concave, but the front bulging. In 1907 all the glaciers showed signs of increasing thickness, and the markers in front of the Taillon were covered by snow. Enlargement continued through 1908. In 1909 and for two more years the terraces of the Cirque de Gavarnie were full of snow, hiding the glaciers except for the front of the Taillon. On 16 August 1911 the Cirque de Gavarnie was still full of snow with ice visible only where the bulging fronts of the Taillon and Gabiètous showed through. In Spain the little glacier Sud-Est du Taillon was advancing strongly, its left side encroaching onto its moraine (Gaurier 1912: 149–52). Gaurier's records make it clear the glaciers experienced a sequence of years with positive mass balances between 1907 and 1911, leading to discernible thickening.

Plate 7.7a Glacier d'Ossoue, Vignemale, 19 September 1911 (*Source*: from Gaurier 1912)

Plate 7.7b Glacier d'Ossoue, Vignemale, 12 August 1988
Photo: A. F. Gellatly

Plate 7.8a The glaciers of the Cirque de Gavarnie in 1906 (*Source:* from Gaurier 1921)

Plate 7.8b The glaciers of the Cirque de Gavarnie, 10 August 1989
Photo: A. F. Gellatly

In 1906 Eydoux and Maury made a theodolite survey of the Glacier du Pays Baché (Figure 7.7b).[16] It had a concavo-convex profile and was crevassed, with a front 210 m from the terminal moraine, having retreated 90 m since Michelier's observations of 1883. Gaurier (1921) published a general account of all the Pic Long and Néouvielle glaciers and what he saw during his visits, concluding that all the glaciers had probably expanded slightly between 1906 and 1909.

Between 1901 and 1911 Gaurier also made observations in the Monte Perdido area. The three sections of the Glaciar de Perdido were still connected and in 1901 and in 1907, when the Perdido was bare of snow in August, he could see that all of it was much crevassed and that one section had swollen. By September 1908 the increase in nourishment and volume was even more evident. Ablation was restricted in 1909 and 1910, and in 1911 ablation was slight during the summer and the glaciers continued to thicken. Gaurier's records demonstrate that between 1907 and 1911 climatic conditions caused general increases in volume and discernible thickening of the Pyrenean glaciers.

7.1.3.6 Halts and advances in the 1920s and 1930s

When Gaurier returned to Vignemale in 1921 he found all the Grottes de Cerbillona were now above the level of the snow surface and the accumulation area of the Glacier d'Ossoue had thinned. The Glacier des Oulettes was practically unchanged, its front at around 2150 m (Gaurier 1934a). He mapped the fronts of the Oulettes and Ossoue in 1921, remapped them in 1923 and 1927 when both had advanced, and calculated that the area of the tongue of the Ossoue had increased by '1 hectare 80 ares 20 centianes' between 1921 and 1927 (Figure 7.6). The area of the Oulettes had also increased but only by expansion at the sides (Figure 7.9 and Plate 7.9).

In 1924 the Commission de Glaciologie et d'Hydrologie des Pyrénées arranged for the first vertical air photograph to be taken of the Glacier du Taillon and Brèche de Roland. The short period of enlargement watched by Gaurier during his previous set of observations had ended. In September 1925 he mapped

the Glacier de Brèche de Roland and discovered that the distance between the front and his 1906 markers was unchanged. In 1926 the glaciers were hidden by snow in the late summer, and when he remapped the front on 10 August 1927 he found that it had advanced (Figure 7.10). The Casque was also enlarged and contemporary photographs of the Taillon show both thickening and advance. The end of expansion probably came in 1928, though Gaurier saw no signs of retreat in this his last year of observation. Abnormal dryness which had begun in April continued into September and the glaciers were much denuded of snow (Gaurier 1934a).

In 1928 Gaurier returned to the Luchonnais and the Sehl de la Baque (Figure 7.4). A single sheet of ice still extended from the Port d'Oo to the Port du Portillon, but thinning was on the point of separating it into two. Impressed by the importance of the thinning, Gaurier took the trouble to measure small changes in area. The Glacier des Gourgs Blancs is shown on Gaurier's 1928 map separated from the Sehl de la Baque by a morainic ridge, considerably wider than it had been in 1912. He returned for the last time in 1930. The Lac du Portillon had been artificially lowered by about 19 m in connection with a hydroelectric power scheme, revealing a large delta. He observed seven loops of moraine, 2–8 m high, above water, and others under water. These he thought represented successive retreat positions (Gaurier 1934c).

7.1.3.7 Retreat interrupted in the 1960s and late 1970s

The most precise information about fluctuations in the 1960s and 1970s comes from the Spanish side of the Pyrenees, and from mass balance measurements of the Glacier d'Ossoue made in 1977/8 and 1978/9. In 1979 glaciers on the southern side of the range were monitored by a group cataloguing the ice of the Iberian Peninsula for the World Glacier Survey (Serrat 1980, Serrat and Ventura 1993). Another survey was initiated in the 1980s by the Ministerio de Obras Publicas y Urbanismo, MOPU (Martinez de Pisón and Parra 1988).

---

16 Eydoux and Maury surveyed in the Pyrenees between 1899 and 1906, using the methods recommended by Vallot in his *Manuel de Topographie Alpine.*

Plate 7.9a Glacier des Oulettes, 3 August 1924 (*Source:* from Gaurier 1934b)

Plate 7.9b Glacier des Oulettes, 15 August 1988
Photo: A. F. Gellatly

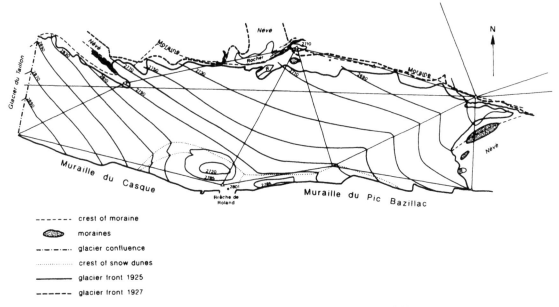

------- crest of moraine
moraines
-·-·-·-·- glacier confluence
.... crest of snow dunes
—————— glacier front 1925
—————— glacier front 1927

Figure 7.10 The glaciers near the Brèche de Roland in 1925 and 1927 (*Source:* from Gaurier 1934a)

In 1979 both the Frondella and the Brècha Latour were classified as snowfields and no other permanent snowfields were found (Serrat 1980, Serrat and Ventura 1993). The front of the Frondella had now separated from its moraine. In 1988, although it was still regarded as on the way to extinction, the Frondella was once more classified as a glacier (Table 7.3).

The Monte Perdido massif in 1979 carried three glaciers and two important névéfields with a combined area of 87 ha (Serrat 1980, table 1, Serrat and Ventura 1993). The Marboré and Celindro Glaciers were confined to the lowest of the three structural terraces. The Monte Perdido ice occupied both the upper terraces, being connected by *siracs*. It was concluded from air photographs and field examination in 1979 that the period between 1957 and 1979 had been one of general stability, but the build up of névé below the terminal siracs suggested that rejuvenation of the glacier on the lowest terrace occurred sometime after 1955, probably during the late 1970s. In 1988 the Glaciar de Marboré–Celindro covered 0.35 ha and extended from a maximum altitude of 2957 m to a minimum of 2625 m. It was 100 m wide, with a maximum length of 300 m (Martinez de Pisón and Parra 1988).

The Maladeta glaciers were virtually unchanged between 1957 and 1979 (Serrat 1980, Serrat and Ventura 1993). Observers were handicapped by a lack of good maps and in 1979 by an abundance of fresh snow. Exceptionally, the Tempêstades had increased in area by 14 ha since 1957, its advancing snout forming a push moraine in front. Accuracy ratings were provided for the dimensions given in 1979 because of difficulties of delimitation in the field. The 1988 team was favoured by better conditions with less lying snow; this has to be remembered when comparing the two sets of figures. It appears that the Maladeta, Aneto[17] and Tempêstades all increased slightly between 1979 and 1988, but the Alba, Barrancs, Salencas, Coronas and Llosas all decreased, the smallest glaciers dwindling, the larger ones tending to gain. But there are discrepancies in the Tables, probably attributable to differences in field conditions in 1979 and 1988, making it difficult to draw firm conclusions about minor changes in the intervening decade. Between 1979 and 1988

---

17   For photos of Aneto in 1910 by J. Soler and in 1992 by D. Parker see p. 117 of *Weather* 1993.

Table 7.3 Recent changes in the areas of Pyrenean glaciers

| | Balaitous and Monte Perdido massifs | | | | | |
| | Frondella | | Brécha Latour | | Monte Perdido | |
	1979	1988	1979	1988	1979	1988
Area ha	7	8	4.3	7	48	49.5
Max. alt. m	2860	2900	2903	2950	3180	3175
Min. alt. m	2720	2765	2720	2800	2690	2700
Length m	300	300	200–300	250	700	600
Width m	300	350	100	350	1200	1250
Slope	25–30°		30°			

| | Maladeta massif | | | | | | | |
| | Alba | | Maladeta | | Aneto | | Barrancs | |
	1979	1988	1979	1988	1979	1988	1979	1988
Area ha	3	15	60	75	132	136	28	18
Max. alt. m	3005	3035	3240	3180	3330	3280	3290	3240
Min. alt. m	2900	2950	2780	2720	2780	2800	2900	2870
Length m	100	190	900	1100	1200	1200	900	750
Width m	200	260	900	1200	1600	1800	400	470
Accuracy	3	2	1		2			

the front of the Aneto receded by 12 m and the Maladeta by 16 m, according to markers painted on the bedrock. The period of minor expansion had evidently ended.

In 1977 the Glacier d'Ossoue was one of two Pyrenean glaciers chosen for mass balance measurement by the Parque National des Pyrénées and the Division Nivologie de CTGREF of Grenoble (Pont and Valla 1980). A simple volumetric method was chosen, involving measurement of winter accumulation and summer ablation at stations at altitudes of 2715 m, 2830 m, 2985 m, 3100 m and 3180 m at the start and end of each summer melt season. Though the method is approximate, giving values estimated as accurate only to about a third of a metre, the results were sufficient to reveal positive mass balances in both of the years 1977/8 and 1988/9 (Table 7.4).

7.1.3.8 Overview of twentieth-century retreat

Few deliberate observations were made after Gaurier's death, but examination of photographs and field

Table 7.4 Mass balance of the Glacier d'Ossoue

| | Metre thickness of water equivalent per year | | |
	Accumulation	Ablation	Balance
1977/8	2.67	1.86	+0.80
1978/9	3.30#	2.75	+0.55

Source: from Pont and Valla 1980
Note: # estimated value

mapping have helped to trace the recession during the twentieth century and to demonstrate the extent of glacier loss since the Little Ice Age maximum. In many cases percentage loss of volume was greater than loss of area, thinning being more important than frontal retreat.

When Dr Gellatly's field parties began their observations in the Pyrenees in 1988 the minor advances of the 1970s were over and retreat was well under way once more. The Las Néous Glacier (Figure 7.8) was

Plate 7.10 Glacier Las Néous, 1905, showing bare ice surface and approximate ice extent in 1989 (*Source:* from Gaurier 1912 and A. F. Gellatly)

Table 7.5 Heights of Las Néous Glacier (m)

	1926		1989	
	North	South	North	South
Max. alt.	3000	3000	3000	3000
Min. alt.	2510	2330	2507	2340

one of those chosen for detailed examination in 1989 because of the possibility of comparing it with Gaurier's map of 1926 and his 1905 photograph (Plate 7.10). The results are in Table 7.5.

The altitudes of the top and bottom of the glacier altered very little over the sixty-three-year interval, but the upper section of the southern lobe separated from the rest and in 1989 the upper ice round the headwall

was thin and disintegrating. In the central basin the ice was no more than 30 m thick, a half to a third as thick as when Gaurier saw it. Debris derived from the headwall and over-steepened lateral moraines provide some protection only around the lower margins.

One of Bonaparte's markers by the Oulettes and several of Gaurier's survived, despite the advances of the 1920s. After Gaurier mapped the front in 1927 a general and rapid retreat began. Survey markers, dated 1964, presumably placed by Les Eaux et Fôrets, were found in 1989 on bedrock on the east side of the glacier indicating a retreat of 84 m. In all, the ice front had retreated about 236–50 m since 1892, and 96–110 m since 1927. The apparently minor change in the appearance of the Glacier des Oulettes since it was photographed in 1904 is an illusion; its area is half what it was a century ago (Figure 7.9).

The estimated positions of the front of the Ossoue since 1798, shown on Figure 7.6, are based on written accounts and photographs. The 1988 front was over 200 m behind its 1927 position, and since 1789 its western edge has melted back about 750 m. The loss of area is estimated at 31 ha.

The front of the Glacier du Pays Baché in the Néouvielle massif, examined and mapped by Gellatly in 1989, had retreated over 370 m from the position recorded by Michelier in 1856, and the surface had downwasted 20–30 m. Since Eydoux and Maury (1907) mapped it in 1906 it had retreated 150 m, and when it was photographed by Hollerman in 1963 the profile had become concave. In 1989 the glacier was 480 m long, and its lowest altitude 2890 m. Eydoux and Maury had put the lowest altitude at 2817 m, so the terminus had risen at least 67 m in eighty-three years. Twentieth-century withdrawal, though very substantial, was not continuous, for retreat was interrupted after 1900, leaving a series of recessional moraines. Eydoux and Maury stated specifically that they had found no small frontal moraines within the great outer moraine to indicate any interruptions of retreat, and Gaurier made no mention of the well-preserved low-lying transverse ridges found by Gellatly; they must all have been deposited between 1906 and 1988. If the glacier readvanced between 1906 and 1910, some of the ridges may relate to that period.

Only the Glacier de Gourgs Blancs and the Sehl de la Baque, in the Luchonnais group (Figure 7.4), were

Table 7.6 Loss of area of the Portillon glaciers

Year	Area (ha)	% loss since 1912
1912	103	
1928	80	22
1945	64	37
1951	50	51

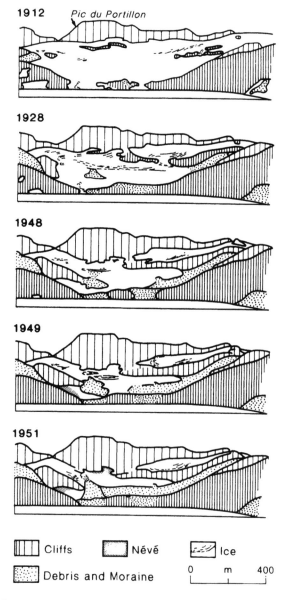

Cliffs	Névé	Ice
Debris and Moraine	0 m 400	

Figure 7.11a Recent history of advance and retreat of the ice extending from Port d'Oo to the Portillon d'Oo (*Source*: after Brunet 1956)

observed in any detail during the mid twentieth century. Some observations were organized by Les Eaux et Fôrets and carried out with the assistance of geographers from Toulouse University. Brunet (1956) used old photographs to trace the course of the separation into five separate glaciers of the 'seule nappe de glace s'etendant de Port d'Oo au Portillon d'Oo' which Gaurier (1934c: 16) had insisted should carry the single name 'Sehl de la Baque' (Figure 7.11a).[18]

In 1912 the eastern wing of the Sehl de la Baque (the present Glacier du Portillon) occupied the whole of the floor of the cirque, reaching the large moraine bordering the Lac du Portillon. By 1928 the front had almost completely withdrawn from the moraine and thinning of the upper part of the glacier was far advanced. By 1950 thinning had resulted in separation into four separate bodies each on its own rock terrace. One of these no longer flowed. The western wing of the Sehl de la Baque had withdrawn from the Lac Glacé, leaving a substantial frontal moraine, but no recessional moraines. In 1954 complete separation of the Portillon glaciers was imminent. Profiles surveyed by Sennec in 1951 and the two following years revealed that the surface of the central glacier had been lowered by between 1.5 and 4 m in the middle. Thinning of the centre of Glacier Oriental du Portillon was only 0.5 to 1 m. This probably resulted from its situation below the slopes of the highest peak in the region, where accumulation was greater than further west. Comparison of Gaurier's measurements of 1912 and 1928 with those of Les Eaux et Fôrets made in 1952/3 and 1954 showed increasingly rapid shrinkage towards the middle of the twentieth century.

18   Photographs taken by Gaurier in 1912 and 1928, by Gaussen in 1945, and in 1945 and 1950 by Maison Alix de Luchon were collected by Les Eaux et Fôrets and used by Brunet to trace changes.

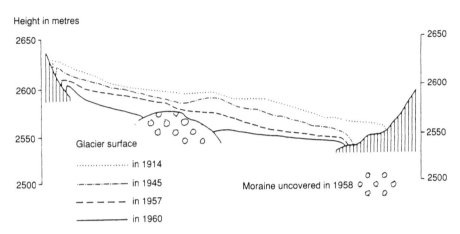

Height in metres

Glacier surface

.... in 1914

–·–·–·– in 1945

– – – – – in 1957

————— in 1960

Moraine uncovered in 1958

Figure 7.11b Cross section of the Glacier du Taillon to show decrease in thickness during the twentieth century (*Source*: from Mounier 1962)

While the area covered by ice had been reduced to less than half by 1952, the proportional loss of volume was much greater. Brunet estimated that less than a quarter – and perhaps much less – of the volume of 1912 remained in 1956. Loss has continued since; the lower tongue of the Portillon Glacier has disappeared and the remaining ice is confined to rock terraces and small plateau ledges.

The thinning and recession of the Glacier du Taillon in the Cirque de Gavarnie has been traced in unusual detail (Figure 7.11b) and can be compared with observations of temperature and precipitation recorded from 1878 to 1984 at the Pic du Midi de Bigorre (2862 m), 10 km north of the main divide of the Pyrenees (McGregor *et al.* 1995). Periods of glacial advance during the first half of the twentieth century followed periods of below average seasonal temperature and higher than average winter precipitation. Dominant retreat since the 1950s is attributable to warming and less snow.

7.2 THE ITALIAN ALPS

In the Italian Alps, Holocene glacial advances seem to have reached a maximum in the Little Ice Age. Most glaciers extended furthest between 1800 and 1820, obscuring evidence of earlier, lesser advances. By 1981, after a moderate positive fluctuation in the 1970s,

glaciers in the Valtournanche and Valpelline valleys leading down to the Val d'Aosta (Figure 7.12) had diminished to between about 60 and 70 per cent of their early nineteenth-century areas and mean equilibrium altitudes had risen by about 140 m, equivalent to a rise in temperature of about 0.8 °C over the last 200 years (Vanuzzo and Pelfini 1999, Vanuzzo 2001). Further east, in the Valtellina, glacier termini retreated by as much as a kilometre between 1940 and 1965 in response to very warm years between 1925 and 1952. Here, again, there were modest advances in the 1970s and early 1980s before retreat was resumed (Pelfini and Smiraglia 1992).

Pelfini (1999) concluded from the scarred wood and thin rings in trees on a lateral moraine close to the edge of the Madaccio glacier, 60 km west of Bolzano, that the ice had advanced between 1600 and 1620 and subsequently retreated. A larch still growing on the moraine showed diminished growth rate indicative of cooling about 1586 and again between 1597 and 1612 with two particularly thin rings in 1605 and 1608. Narrow larch rings indicate other intervals with slow growth rates, from 1670 to 1720, 1750 to 1785 (especially 1770 to 1782) and 1792 to 1821. The glacier reached its most advanced position about 1770. A spruce growing at the foot of the outer slope of the 1770 moraine germinated at the latest by 1790 after which the cold phase attenuated before renewed

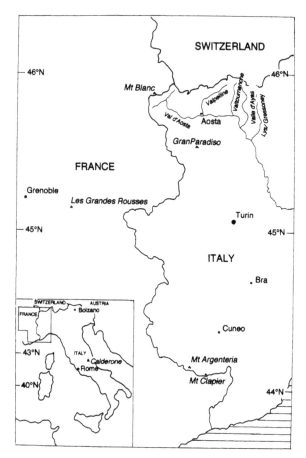

Figure 7.12 Location of Italian and French glaciers mentioned in the text

cooling brought the glacier forward to reach another advanced position about 1820.

In the Valle d'Ayas, about 240 km west of the Madaccio, a Little Ice Age advance of the Grande de Verra glacier dammed back a stream creating the Lago Blu. Larches close to the lateral moraine record not only cooling from the 1740s until 1790, especially marked in 1785–6 and 1795, but also the glacier's advance from 1790 to 1819, its retreat over the next 20 years and a short-lived advance to an inner moraine, dated 1855–6.

A deliberate attempt was made in the Gressoney valley, immediately to the east of the Valle d'Ayas, to calculate the response times of tree rings and frontal variations of the Lys glacier to weather by comparing a local dendrochronological data series from 216 larch

cores, glacier snout variations and the 1935–86 climate record from a meteorological station at Lake Gabiet (2350 m) 3 or 4 km away. The tree-ring thicknesses were found to respond almost immediately to summer temperature, trees outside the 1821 moraine giving the best fit. Ice front variations responded to summer temperature rather than precipitation (as seems to be the case generally in the southern Alpine glaciers), the delay being five years.

7.3 MARITIME ALPS

7.3.1 Glaciers in Les Grandes Rousses range

Glaciers in Les Grandes Rousses, 30 km east of Grenoble, are believed to have reached their maximum extent around 1850. In 1905–6 the area was mapped by Grenoble University and since 1948 glaciologists there have made observations of the mass balance of the Glacier de Sarennes (Valla and Piedalla 1997). In 1891, when its area was about one square kilometre, the glacier was visited by Prince Bonaparte who noted that its tongue was stationary. The mass balance back to 1880 has been reconstituted with the aid of the meteorological records for Grenoble plus a digitized version of the 1905 map, aerial photos taken every three years since 1970, and both seismic and radar soundings. It appears that since 1850 the glacier's area has shrunk from 150 ha to about 40 ha; its average thickness is now only 30 m, a third of the 1850 maximum.

Shrinkage of the St Sorlin glacier, also in the Grandes Rousses, ceased from 1964 until 1989 and then recommenced (Vincent *et al.* 2000).

7.3.2 Glaciers in the Clapier and Maledia–Gelas massifs

The Clapier and Maledia–Gelas massifs, at the extreme end of the Southern Maritime Alps, carry small glaciers. Both are within the Mercantour National Park, between 44°8′–44°5′N and 7°23′–7°27′E. Currently six glaciers just about survive there (Figure 7.13). Others have disappeared in recent years or have been reduced to semi-permanent snow patches (Hansse 1970, Vivian 1975).

Map coverage is incomplete. The most accurate is the Institut Géographique National (IGN) series at

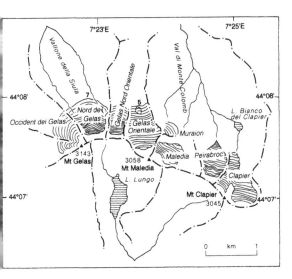

Figure 7.13 Mt Maledia and Mt Clapier and the glaciers in the Mercantour National Park

1:25,000 and 1:50,000. Portrayal of the glaciers is either not attempted or inaccurate, which is understandable as the ice is not easily recognizable on air photographs on account of the small size of the glaciers and the extent of surface debris (Plate 7.11).

The alpine passes have been used by travellers, pilgrims, armies and smugglers for many centuries. Some limited indications of conditions for such travellers can be gleaned from various sources (Bicknell 1913). For instance, military campaigns at the end of the sixteenth century are known to have experienced unusually late snow and 'salt routes' through the alpine passes were abandoned, notably, between 1583 and 1589, the Pagari Pass (2819 m), which was probably blocked by the Maledia Glacier (Pappalardo 1994). Harsh conditions recurred in the early nineteenth century, heavy snowfalls in 1838 destroying the chapel at La Madone de Fenestre at 1903 m (Paschetta 1980). Meteorological measurements were made by the Servicio Idografico Italiano in valley sites near Cuneo, including Cuneo itself (556 m) from 1857 onwards and at Bra (290 m) from 1859 onwards. Since the late nineteenth century

temperatures have risen intermittently by over 0.6 °C while the trend of annual precipitation has been downwards, which together would readily explain the predominance of negative mass balances and glacier shrinkage.

The earliest systematic descriptions of the glaciers were made by Mader between 1896 and 1908 (Mader 1896, 1909) and by Viglino (1896). The earliest photographs date from their visits. The glaciers on the northern slopes are depicted in the enormous collection of photos assembled by Chevalier Victor S. de Cassole between 1896 and 1930. These were taken after Italian regulations concerning access and photography in the region had been relaxed in 1897 (Bicknell 1913).[19] Many other photographs of the glaciers are held by the Italian Glaciological Society in Turin or have been published in the *Bollettino Comitato Glaciologico Italiano* by Rocatti (1912, 1914, 1917, 1927), Camoletto (1928, 1932) and Rachetto (1962).

Studies were made in connection with an EEC project of 1990 and 1992, when the glaciers and moraines were mapped (Grove and Gellatly 1992, Gellatly *et al.* 1994a). Lichenometry based on *Rhizocarpon geographicum s.l.* using three historical calibration points was used to date moraines and trace the timing of fluctuations before the late nineteenth century (Gellatly *et al.* 1994a).[20] More recently Federici and Pappalardo (1996) and Federici and Stefanini (2001) have extended the lichenometric studies measuring the shorter axes of the five largest *Rhizocarpon* section *Rhizocarpon* lichens on the control points for the growth curves and on the moraines. Their results show that the moraines fall into five groups, the first involving moraines probably dating to between about 1260 and 1400, the second including moraines dating mainly to the latter half of the seventeenth century, the third belonging to the last period of the Little Ice Age between about 1760 and 1825, and the remaining two being younger still, the earlier one possibly formed between 1850 and 1855 following a pronounced cold interval.

At one time the Clapier and Pierbroc glaciers beneath Mt Clapier (3045 m) and its subsidiary peak Cime

19   This catalogued collection is held in the Musée Massina, Nice.
20   The calibration points were, first, a memorial to the second Alpini Regiment, erected in 1927 at Terme di Valdieri (altitude 1365 m), second, granite curbstones used in 1853 to construct the mountain road above Terme di Valdieri (altitude 1465 m) and, third, the Col de la Ruine (2724 m).

Plate 7.11a Glacier Maledia, Italy, from terminal moraine above Rifugio Pagari, August 1949
Photo: P. Rachetto

Plate 7.11b Glacier Maledia, Italy, from terminal moraine above Rifugio Pagari, 30 August 1974
Photo: P. Rachetto

Pierbroc (2947 m) were united and flowed into a basin below Lago Bianco del Agnel (*c*.2160–200 m). By the 1990s the two glaciers were completely separate, a series of moraines marking their retreat. In 1992 the Clapier Glacier, extending from 2790 m down to 2660 m at its snout, had an area of 6.2 ha. Snow drifting from the southern, French side of the range accumulates below the cliff face of Mt Clapier and ice from this upper basin falls over a bedrock step into a shallow, elongated cirque. Early in the twentieth century the glacier occupied a whole series of small basins. Now, steep-sided lateral moraines on the eastern side of the cirque, standing over 40 m above its floor, and poorly preserved moraines on the west side mark the glacier's former extent.

7.4 THE APENNINES

7.4.1 The Calderone Glacier

The Calderone (42°28′15″N, 12°27′08″E) lies within a cirque facing NNE, carved into the Corno Grande (2912 m), the highest peak of the Gran Sasso massif (Figure 7.14). When, at the instigation of Arturo Desio, a detailed topographic survey was made in 1990, the ice extended from a maximum altitude of about 2839 m against the headwall to a minimum of 2670 m (Gellatly *et al.* 1994a, b). Below 2760 m the surface is buried under a layer of debris about a metre thick, making identification of the ice edge, and thus the length of the glacier, very difficult.

The compound terminal moraine at 2700 m, lies on a prominent bedrock ridge at the lip of the cirque. Meltwater seeps through the fissured bedrock under the moraine. Earlier phases of ice expansion are recorded by meltwater channels breaching the limestone bedrock of the rim. Ice-cored moraines and ice-deformed talus lie on the slopes above the tongue. Two well-defined trimlines highlighted by contrasts in weathering cross the head and sidewalls of the cirque. The uppermost trimline rises from about 2750 m, above the terminus, to over 2880 m on the headwall. The lower and fresher trimline rises from 2720 m at the lip of the cirque and eventually merges with the upper trimline below Corno Grande (Gellatly *et al.* 1994b). They mark two relatively long periods of expansion when the glacier was thicker and more extensive. The

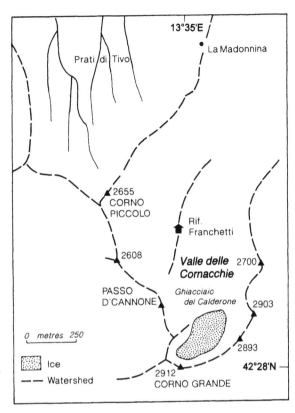

Figure 7.14 The Calderone Glacier

compound nature of the frontal moraine and the breaches in the lip of the cirque may relate to briefer periods of expansion. No evidence allowing dating of any of these features has been found. A weakly developed ridge of clastic material runs obliquely across the Valle delle Cornacchie above the Rifugio Franchetti (Figure 7.14) and descends to about 2500 m (Gellatly *et al.* 1994b). This is best explained as indicating that the glacier expanded outside the cirque, presumably during the period of its maximum extension during the Little Ice Age.

Only two brief historical accounts have been found. The first, by De Marchi (1573, quoted by D'Orefice *et al.* 2000), comes from an archival description of the first recorded ascent of the Gran Sasso in August 1573. The view of the glacier from the top of the mountain was of 'a great valley about 1500 m in length where snow and ice lie perpetually'. In July 1794 Delfico made the first ascent of the Corno Grande and, looking from

Table 7.7 Maps of Ghiacciaio del Calderone

Year	Scale	Source	Contour interval (m)
1884–5	1:50,000	Istituto Geographico Militare	10
1916	1:5000	Marinelli and Ricci	5
1934	1:1000	Sforzini and Tonini	5
1954–5#	1:25,000	Istituto Geographico Militare	25
1958	1:1000	Caloi, Zuccarini and Vaccanti	5
1960	1:1000	Balducci, Pesavento, Di Fazio and Tiberio	5
1966	1:1000	Tonini	5
1981–2#	1:10,000	Abruzzi Region	10
1990	1:750	Gellatly, Tomkin, Parkinson and Latham	10

Source: after D'Orefice *et al.* 2000, table 1
Note: # based on aerial photographs

the northern slope, saw a 'plain almost completely surrounded by high peaks, forming a majestic circular depression, and always covered by very solid snow' (Delfico 1794, quoted by D'Orefice *et al.* 2000).[21] Although these accounts are very brief, their contents are informative, for the cirque basin is no longer filled with ice and snow as it was when De Marchi and Delfico saw it; snow cover in summer is now much reduced or absent. Neither of the early climbers saw any signs of the debris cover, which was subsequently to increase in extent and importance, becoming a major feature during the twentieth century.

Delfico's wording suggests that the ice tongue in the late eighteenth century reached the frontal moraine on the lip of the cirque at about 2700 m. D'Orefice *et al.* (2000) regard the IGM map of 1884 as showing the upper glacier much as it was when Delfico saw it in 1794. The history of the Calderone's fluctuations since the late nineteenth century has to be pieced together from a miscellany of maps, photographs and surveys (Bonarelli 1983, Smiraglia and Veggetti 1991, 1992). The maps used by D'Orefice *et al.* (2000) in their study of the history of the Calderone are listed in Table 7.7.

A comparison of maps made by Marinelli and Ricci and by Sforzini and Tonini suggests that the area of the glacier diminished by 14–15 per cent between 1916 and 1934. However this gives only a rough guide to the extent of change in the intervening period on account of lack of precision, especially of the earlier map, and the difficulty of identifying the edge of the ice beneath its debris cover. That the Calderone is shown as having been 40–50 m longer in 1934 than in 1916 may be due to the lack of accuracy of the cartography, but could alternatively indicate that advances occurred after 1916 as they did in the Pyrenees and the Alps.

The course of events later in the twentieth century can be followed more closely, in large part because of a series of systematic observations by Tonini (1934, 1936, 1955, 1961) who calculated that between 1934, when the ice still reach the lower trimline, and 1960 the glacier thinned on average by 7 m, and in the lower ablation zone by nearly 20 m. The loss of volume over the same period he put at 420,000 m³. Tonini recorded frontal advances of 5.5 m in 1936 and 7.2 m in 1938 so retreat was interrupted from time to time. But wastage between 1966 and 1990 reduced the surface area by about 68 per cent and the ice volume possibly by as much as 90 per cent (Gellatly *et al.* 1994b) (Plate 7.12a and b). Losses were most pronounced between 1960 and 1990, with the area of clean ice decreasing by approximately 20 per cent (Smiraglia and Veggetti 1992).

Geolectric surveys by Smiraglia and Veggetti (1992) have found the thickness of ice on the cirque floor to be 10 m and up to 16 m on exposed ice faces, but in crevasses measurements show that in some places it is no more than 3–8 m thick. Observations from the

---

21 It is possible that further information might be gleaned from descriptions of later ascents. It is not clear whether thorough searches have been made.

Plate 7.12a Calderone Glacier in 1949 when conditions were unfavourable to glacier growth and old snow had virtually disappeared
Photo: D. Tonini

Plate 7.12b Calderone Glacier in 1990 with a substantial debris cover extending high up the glacier and no old snow remaining by mid August
Photo: A. F. Gellatly

Corno Grande ridge suggest that the ice on the upper slopes was little more than 5 m thick in 1990 (Gellatly *et al.* 1994b). Large quantities of ice, possibly 40 per cent of the total volume of the glacier, lie concealed beneath the frontal moraine and under rockfalls in the lower basin. On this assumption, only about 120,000 m³ of ice remained in the cirque, so the Calderone's chances of surviving for many more years are slim (Smiraglia 1994).

7.5 GLACIER EXTINCTION

The glaciers of southern Europe were evidently very much larger than they are now towards the end of the eighteenth century and, probably, in the decades around 1600. Their retreat, in response to temperatures rising by about 1 °C over the last 150 years, has been interrupted at intervals of a decade or two by cool years or sequences of years with greater snow accumulation. Fragmentary data from the Pyrenees as from the Alps point to the years around 1850, 1890, 1910 and 1925 as having seen minor advances, but the situation has varied from one part of the region to another. Many areas saw retreat halting for some years between 1960 and 1980. Then, in the 1990s, recession resumed at a pace that, if it is maintained, will soon result in the disappearance of most of southern Europe's glaciers.

8
ASIA

In many parts of the world deliberate measurements of the fluctuations of glacier fronts and mass balances have started only recently. Historical records are sparse, so the data available are very uneven in amount and quality. Of the first three IAHS/Unesco volumes published on the fluctuations of glaciers (Kasser 1967, 1973, Müller 1977), the first contained only European material and even the third had data for only one Himalayan glacier, yet the Himalaya contain by far the most substantial area of ice outside the arctic and antarctic regions, and variations in the volume of glacier runoff there are directly important to the economy of densely settled, irrigated areas adjacent to the mountain chain. Regional coverage was further improved in the fourth volume (Haeberli 1985) but it still contained data for only seventeen glaciers in the whole of China, compared with 120 in Austria. By the time the fifth volume appeared (Haeberli *et al.* 1993) the Chinese coverage had been reduced to ten glaciers, but reports on ten Indian and seven Nepalese glaciers were included, in addition to eighty-two from the USSR. The sixth volume contained data on frontal variations of ten glaciers in China, six in Pakistan, four in India and seven in Nepal. With the possible exception of China, European historical records are more plentiful and more informative than those of other continents; in consequence, elsewhere greater reliance inevitably has to be placed on moraine dating.

A number of sample areas are examined in the following review, which is not exhaustive but is believed to represent the state of knowledge as it existed in the 1990s. In certain areas where studies so far made have been in the nature of reconnaissances, new findings will no doubt modify the picture we have at present.

8.1 THE GLACIERS OF ASIA NORTH OF THE HIMALAYA APART FROM CHINA

The principal regions with glaciers in this territory were, until 1989, in the USSR. Fifteen institutions there were engaged in making observations on 200 mountain glaciers, with mass balances being measured on nearly forty of them. Now, seven independent states are responsible for monitoring the glaciers though several of them lack the necessary expertise and equipment. However, of the 78,000 km^2 covered by ice, some three-quarters are on the Russian arctic islands discussed in Chapter 10, and much of the rest, in the Urals and on the Pacific coast, is in Russia. Of the ten glaciers in the former USSR which in 1996 had been observed for twenty years or more, work on eight continues, though in some cases on a reduced scale (Kotlyakov *et al.* 1996). Attention is directed here to the glaciers of the Urals, the Caucasus, central Asia and the Pacific coast.

8.1.1 The Urals

The glaciers scattered along the 2100 km length of the Urals are all quite small; together their total area is only about 28 km^2. Lying on the eastern, leeward side of the crestline, many are in cirques 700 m to as much as 1000 m below the regional snowline (Grosswal'd and Kotlyakov 1969). They depend for their nourishment on westerly winds driving the snow and concentrating it in leeside hollows. Close to equilibrium in the second half of the nineteenth century, their tongues were almost stationary. During the last 150 years they have shrunk only slightly, their fronts a few metres higher

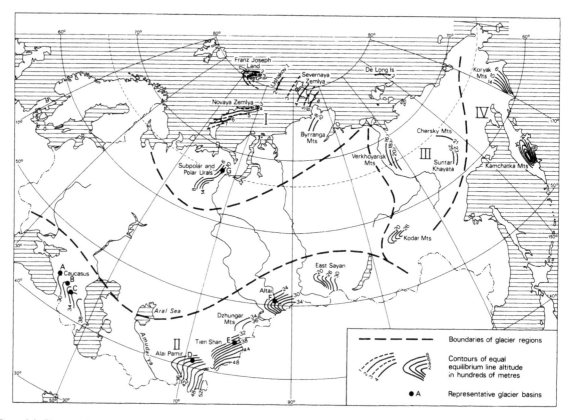

Figure 8.1 Glacierized areas and snowline elevations in Asia outside China and the Himalaya. Glaciers mentioned: E = Tsentral'nyy Tuyuksu, F = Aktru, G = IGAN (*Source:* after Grosswal'd and Kotlyakov 1969)

and back about 200 m (Mikhalenko and Solomina 1996).[1]

Changes in the net balance of the glacier called Lednik Instituta Geografii exemplify a general trend. (The values given in Table 8.1 are estimates for which the basis was not indicated by Grosswal'd and Kotlyakov 1969.) From Table 8.1 it seems that on the whole glacier wasting increased from 1900 onwards, though termini were stationary in 1905–10, 1921–30 and 1946–50. The IGAN and Obruchev glaciers in the Polar Urals were also wasting until the mid 1960s, but

precipitation increased in the early 1950s and mass balances became positive in the late 1960s and early 1970s. The IGAN, for instance, was expanding for several years before 1980. The wasting of the Obrucheva merely slowed down (Figure 8.2).

The results of investigations in the Polar Urals were reviewed by Serebryanny and Solomina (1996) who concluded that lichenometric dating of moraine systems provides evidence of many advances during the last millennium, four main generations being identified, with ages of 740–700, 370–340, 200–195 and

---

1 A reconstruction of the mass balance of the IGAN over the period 1817–1959, based on precipitation and air temperature at the nearest meteorological station, calibrated with measured mass balances from 1957–81, indicated that twentieth-century volume loss in the Polar Urals was substantially less than in other areas, especially the Caucasus and Kamchatka. Accumulation here depends heavily on drifting and avalanching. Consequently, dependence on weather stations for precipitation figures could have involved a lack of precision in the reconstruction. This result is substantiated by the present position of the glacier fronts near their Little Ice Age moraines.

Table 8.1 Estimated annual net balance, Lednik Instituta Geografii, 1850–1963

	Acc./g/cm	Abl./g/cm	Net balance
1850–75	178	174	+4
1876–1900	168	166	+2
1901–25	170	192	−15
1926–50	170	203	−33
1951–63	220	288	−68

Source: Grosswal'd and Kotlyakov 1969

100–70 lichen years BP. Treelines, which had occupied their highest positions in the second millennium AD at altitudes of 390–420 m from the early eleventh to the late thirteenth centuries, fell by 80–100 m between the late thirteenth and fifteenth centuries. Warmer conditions in the sixteenth century are represented by two new generations of larch; then came strong cooling in the seventeenth century leading to treelines falling 20–30 m. Warming in the mid eighteenth century caused a rise of 40–60 m. The maximum temperature decrease during the cold intervals of the Little Ice Age, in the early seventeenth century, 1810–35, 1850, and the 1880s has been put at 1.0 °C to 1.2 °C (Shiyatov 1981, cited Solomina 1996). Treelines all over the Urals fell during cold conditions in the late nineteenth and early twentieth centuries. Glacier fronts remain quite close to the inner Little Ice Age moraines, the ice having receded by hundreds of metres rather than by kilometres in the course of the twentieth century. Many if not most of the glaciers, of which the IGAN is the best known, lie below the regional snowline and are nourished by snow drifting and avalanches (Mikhalenko and Solomina 1996). Glacial advance episodes indicated by moraine ages correlate well with cold periods in the thirteenth, late seventeenth and late nineteenth centuries indicated by tree rings. Solomina (1996) has pointed out that the reality of temperature depression before AD 1400 remains to be well established because of the limited amount of data for this early period.

A reconstruction of mean summer temperature, based on the density of rings in *Larix siberica* from a region on the eastern side of the Urals indicated that

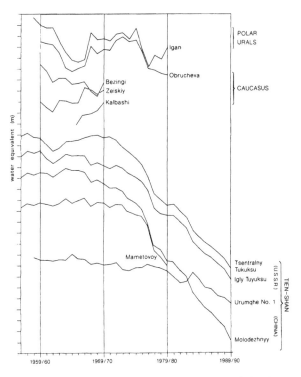

Figure 8.2 Changes in mass balance of representative glaciers in the Polar Urals, Caucasus and western Tien Shan from 1958/9 to 1989/90 (*Source*: from figures published by Kasser 1973, Müller 1977 and Haeberli 1985)

the eleventh and twelfth centuries were cool, though not so cold as the sixteenth and seventeenth centuries (Briffa *et al.* 1995). This reconstruction suggested that the mean temperature of the twentieth century was higher than in any period of similar length since AD 914 though, as the authors noted, the density of the data covering the medieval warm period is very much less than for the cooler centuries which followed[2] and the relative warmth through much of the thirteenth and fourteenth centuries means that the statistical significance of the twentieth-century Urals warmth is equivocal. Moreover, summer temperatures alone cannot provide a complete key to variations in glacier mass balance, especially in a region where topographical effects promote wind drifting and eddying.

2 The proposition that it did not occur in the northern Urals therefore cannot be taken as proven. It might be noted that there is evidence of important regional variations elsewhere in Asia during this period (see Chapter 15, pp. 454–7).

8.1.2 The Caucasus

The Caucasus are 20° south of the Polar Urals and very much higher. They lie between mid latitude steppe and the subtropics. According to Kotlyakov and Krenke (1979), there were 2047 glaciers in the Caucasus, covering 723 km².

In medieval times the snowline was higher and the glaciers smaller than they are now, and a number of mountain passes then in use are now covered with ice. As in the Valaisian Alps (Röthlisberger 1974), tracks leading to them can still be distinguished (Kotlyakov and Krenke 1979).

Moraines, horseshoe-shaped and well spaced out, mark ice advances at various times during the last millennium. The most recent are close to the present ice fronts, those from the early Little Ice Age are further downvalley (Kotlyakov et al. 1991). Two main glacial advance phases within the Little Ice Age have been recognized, the first in the second half of the thirteenth to the early fourteenth centuries, and the second in the early seventeenth to mid nineteenth centuries (Serebryanny and Solomina 1996). During the twentieth century, Caucasian glaciers withdrew by between 700 m and 3 km. For the earlier cold phase an equilibium line depression of 145–60 m is suggested.

Lichenometric studies in the Baksan basin indicate the cessation of mudflow activity in the fourteenth and fifteenth centuries. This is attributed to prevailing low temperatures (Skinova 1991). In the second cold phase, a depression of the equilibrium line by 50–60 m was associated with mean annual temperatures estimated to have been 0.3 °C below present and mean annual precipitation 120 mm above current values.[3] Tree-ring studies point to large-scale avalanching about this time (Kotlyakov and Krenke 1979) when glacial advances were frequently reported in contemporary Russian travel literature.

Details of the advance phases or stationary periods of the glaciers on the northern slopes of the Caucasus during the Little Ice Age have been identified from lichen dating of moraines to 1270–1310, 1475–85,

1675–83, 1790–1802, 1846–60, 1911–13, 1927–30 and 1946–49 (Kotlyakov et al. 1991). Advances on the southern slopes were for the most part practically synchronous.[4] It was judged that variations in the beginning and end of advanced phases were within a range of 4–10 years, though no estimates were provided of the accuracy of the lichenometric dating or the extent to which this may have varied over the period between the thirteenth and twentieth centuries.

One of the foreign travellers who ventured into the Caucasus in the nineteenth century, at a time when they were scarcely known outside Russia, was Douglas Freshfield (1869). He wrote in the account of a visit in 1868 that the Caucasus were less known than the Andes and Himalaya. Merzbacher (1901, cited Horvath 1975) found evidence of glacier advances in recent centuries in a valley of the central Caucasus. Ruined buildings lay close to glacier tongues and local legends and songs told of a glacier near Ushkul, probably Lednik Khalde, then six miles away from the village, having advanced and destroyed all of it but for the church. The people still held an annual festival in thanksgiving!

The Caucasian glaciers continued to advance, according to Horvath (1975), until about 1849 and then in 1860 began to retreat. Freshfield and Sella (1896: 53) commented that 'the movements of the Caucasian and Alpine glaciers have of late years shown a general correspondence. In 1863 the Caucasian ice was in retreat. About 1875 the tide seemed to turn, and in 1887–9 many glaciers were slightly advancing.' Since then retreat has been predominant though there have been stationary periods or slight advances in 1910–14 and 1927–33, when summer temperatures were lower and there was some increase in winter snowfall. By the 1970s the ice-covered area had diminished by about a quarter and glacier volume by a similar amount (Kotlyakov and Krenke 1979). The shrinkage of the ice is said to have been more marked on the northern than on the southern slopes, though according to Serebryanny and Solomina (1996) the timing of glacier oscillations, indicated by lichenometry (based on *Rhizocarpon*

---

3 The maximum Little Ice Age summer temperature decrease in the Caucasus, for comparison, has been put at 1.5 °C to 3.0 °C (Turmanina 1971, cited Solomina 1996).

4 On the southern slopes moraine formation has been dated at 1258–1300, 1379–1400, 1637–78, 1777–80, 1804–8, 1852–6, 1885–7, 1898–1902, 1927–32, 1953–5 and 1966–7 (Kotlyakov et al. 1991, citing Golodkovskaya 1982).

geographicum growth curves), historical evidence and ^{14}C dating, was synchronous in most cases. Glaciers heavily covered in debris, such as Lednik Shkheldy, started to retreat some 50–70 years after the rest.

Towards the middle of the twentieth century, retreat of the glaciers came to a halt. Firnlines fell by 200 to 300 m between 1959 and 1963 as the result of a 30 per cent increase in winter precipitation and a lowering of summer temperature by 0.3 to 0.6 °C as compared with the previous decade. There was a further increase in precipitation of 20–30 per cent in the 1960s. Accumulation zones thickened by 15 m to 18 m and many glaciers advanced (Kotlyakov and Krenke 1979). Lednik Alibek, which had been receding at a rate of 2–7 m per year between 1904 and 1959, advanced on average 2.2 m per year between 1960 and 1968. Irregular increases in mass balance of some glaciers continued until 1970 (Figure 8.2). Large-scale photogrammetric surveys of the El'brus glaciers show them continuing to expand into the 1970s, the Bol'shoy Azau advancing 27 m and growing thicker and wider than it had been in the 1960s. However, in 1979, of twenty-six glacier tongues measured in the central Caucasus only six were still advancing and the rest retreating (Kotlyakov and Krenke 1979, 1981).

The Djankuat[5] suffered a marked loss of volume between the late 1970s and mid 1980s, but this was more than counterbalanced by gains in the succeeding period, especially in 1986/7 and again in 1992/3. In both years snowfall was much above average and huge avalanches distorted the usual snow distribution patterns (Haeberli *et al.* 1994, Popovnin 1996). In 1993, despite high summer air temperatures, total ablation was 16 per cent below normal. The resulting gain in mass for the year was 1.13 m water equivalent (mwe), the second highest since direct measurements began.[6]

In 1996 the front of the Djankuat advanced and formed a moraine (Mikhalenko and Solomina 1996). The following year saw heavy snowfalls and a mass gain of 0.28 mwe (Haeberli *et al.* 1999). Palaeoclimatic studies in the Caucasus continue to be intensified with Khar'kov University investigating past climatic conditions by using X-rays to measure the varying density of tree rings.

8.1.3 The Alai Pamir

The glaciers of the Pamir, 'the roof of the world' (the region between the Oxus River and the Alai Range), were reviewed by Horvath (1975) who made use of a translation of a comprehensive account by Zabirov (1955) as well as referring to more recent sources. The ice-covered area in the mid twentieth century occupied 8041 km², about 11 per cent of the mountains, and the snowline was above 4800 m over 60 per cent of the region. Glacial meltwater, most of which goes into the Amudar'ya, was seen as vital for irrigation and for the level of the Sea of Aral. In 1988 the ice was described as covering more than 6000 km² (Haeberli *et al.* 1989), but the imprecision of this value means that the difference between the two figures cannot be taken as an accurate measure of ice recession.[7]

The Little Ice Age history of the Pamir glaciers has not been seriously investigated. Tree-ring studies agree that growth rates were low from the late seventeenth to the early eighteenth centuries, and also during the first half of the nineteenth century. Travellers in the second half of the nineteenth century reported several advancing glaciers (Serebryanny and Solomina 1996). Because of their economic importance the glaciers have received much attention in recent decades, with programmes including monitoring of frontal positions,

---

5   Or Dzhankuat is a northwest-facing valley glacier with an area of 3.1 km² on the northern slope of the main section of the Caucasus.

6   Mikhalenko and Solomina (1996), concluded from the mass balances for 15 glaciers, including the Dzhankuat, that 'Up to 1964 the glaciers of all mountain systems (in the former USSR) retreated intensively. Since 1965 a positive mass balance is observed.' Their conclusion differs from the account of Kotlyakov and Krenke (1979). Some of the difficulties involved in mass balance reconstruction in the Caucasus are illustrated by the positive balance of the Dzhankuat in 1992/3, a year in which above average accumulation was promoted by anomalously high spring snow densities, high winds and active avalanching, as well as low temperatures, and ablation below normal, despite warm summer temperatures (Haeberli *et al.* 1994).

7   Konovalov (1996) gives a total area of 7500 km² for the 11,500 glaciers in the Pamiro-Alai.

flow measurements, and mass balance studies. It is generally concluded that shrinkage of the Aral Sea, which has attracted much attention since Horvath wrote, has almost certainly been caused primarily by over-abstraction of water from the rivers, rather than by reduced supplies of glacial meltwater.

The northwestern part of the Pamir is the most heavily glaciated. The snowline rises southeastward to reach a maximum of 5200–40 m on the peaks of the Shakhdarinskiy Khrebet, which include Pik Lenina, 7134 m, Pik Karla Marksa, 8726 m, and Pik Engel'sa, 6510 m, and then falls again. The best-known glacier, the Lednik Fedchenko, 71 km long, is one of the largest valley glaciers outside the subpolar regions and the largest in the former USSR. A major source of nourishment of this complex system is the névéfield of the Khrebet Akademii Nauk, which intercepts moisture-laden westerly airstreams. Regular observations of the Fedchenko, the best-known system, began in 1959 and were later extended to monitor the differing behaviour of its various components. These included tributaries such as Lednik Medvezhiy, which surged in 1973 and 1989, and glaciers fed by large-scale avalanches which are possibly associated with the earthquakes to which these ranges are subject. Though there were minor advances of some glaciers in about 1914, between 1927 and 1935, and again between 1946 and 1958, recession has predominated in the twentieth century.

Recession in the interior of the Pamir has been less than in the outer parts (Mikhalenko and Solomina 1996). Between 1959 and 1964, the budgets of a number of the glaciers there were positive and many of them advanced, though some were retreating again by the late 1960s. On the northern slope of the Alai Range, seven glaciers slowly advanced in the years 1963–73 and then retreated more quickly between 1978 and 1991 (Kotlyakov et al. 1996, figure 2). The mass balance of the north-facing Abramov,[8] in the Amudar'ya basin, Kirghizstan, with a surface area of 26 km², has been measured since 1967/8 (Haeberli et al. 1994) (Figure 8.2). In two successive years, 1991/2 and 1992/3, mass balance was positive for the first time

since observations began. The causes were temperatures lower than normal together with above average snowfall in June 1991/2 (the Pinatubo effect) and in the winter of 1992/3. Hot summers in 1996 and 1997 resulted in strong negative balances (Haeberli et al. 1999).

In spite of the prevalence of earthquakes in the Pamir and the great altitude of the glaciers, their general behaviour in the twentieth century was similar to that of glaciers in Alpine Europe, involving retreat interrupted by short-lived advances particularly around 1930, and again in the early 1960s and early 1990s.

8.1.4 The Tien Shan

The marginal ranges of the Tien Shan intercept moist air masses and the snowline lies around 3000 m, whereas on the inner ranges it rises to 4000–200 m, and exceptionally to 4600 m. Mean annual precipitation in the area with glaciers has been estimated at 300–600 mm, reaching 1000 mm and more in the Central Tien Shan (4200–400 m altitude). Precipitation declines from east to west. Near Rybachye the annual precipitation is only 150 mm, while further east it is above 800 mm. About 11 per cent of the area is ice-covered (Serebryanny and Solomina 1989).

Investigations in the central Tien Shan have been concentrated in the Kirghizskiy, Zailiyskiy, Kunghei and Terskei Alatau and the Ak-Shiyrak mountains, where nearly a hundred forefields have been examined (Kotlyakov et al. 1991). As many as eight to ten closely spaced moraines are found near the glacier snouts, the innermost in the majority of cases dating from the early twentieth century. The best-known glaciers are those in the Zailiyskiy Alatau, the most northerly of its ranges, which stretches along the 43rd parallel near Alma Ata and reaches up to 5000 m. The ice, covering 500 km², is concentrated in the upper basins of the Bol'shaya and Malaya Almatinka and other large rivers that flow into Lake Balkhash.

Observations of the Tsentral'nyy Tuyuksu,[9] a small glacier in the Almatinka basin began in 1902. A cold,

---

8   The Abramov has a temperate accumulation zone but cold ice near the surface of the ablation area. Annual mean temperature at the equilibrium line, around 4200 m, is −6.5 °C to −8 °C. Average annual precipitation, measured at 3840 m above sea level, is about 750 mm (Haeberli et al. 1994).

9   Also known as the Tsentralniy Tuyukuyskiy.

Table 8.2 Retreat of the Tsentral'nyy Tuyuksu, 1923–61

	Total retreat (m)	Mean annual retreat (m/y)
1923–37	113	7.8
1937–47	170	16–17
1947–53	30	5
1953–6	18	6
1956–9	25	8
1959–60	11–13	11–13
1960–1	16–17	16–17

Source: Makarevich 1962

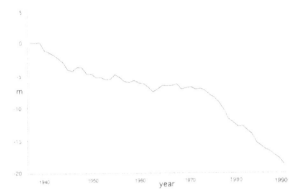

Figure 8.3 Accumulated mass balance of the Tuyuksu Glacier as measured from 1957 to 1990, and reconstructed from 1937 to 1956. Accumulation almost balanced ablation losses from 1950 to 1970 and then shrinkage of the glacier recommenced (Source: from data in Dyurgerov et al. 1996)

north-facing glacier at a height of about 4000 m and with an area of 2.86 km² in 1995, it is surrounded by continuous permafrost (Haeberli et al. 1994, Dyurgerov et al. 1996). A late nineteenth-century advance of the glacier continued until 1923. Makarevich (1962) suggested that its history was attributable both to earthquakes[10] and also to a period of high precipitation known to have caused a rise in the level of Lake Balkhash in the late nineteenth century. It was a time when the Zhangyryk and possibly the Bogatyr glacier on the northern slopes of the neighbouring Kunghei Alatau also advanced (Table 8.2). Between 1905 and 1958 the Tsentral'nyy Tuyuksu lost 3 per cent of its total area and its surface was lowered 53 m along a profile near the snout. In the late 1950s it was stationary or advanced slightly.

Mass balance observations began on four glaciers in the late 1950s and were extended to five more in 1965. The results show short periods of positive balance both in the late 1950s and late 1960s; Lednik Partizan gained over 5 m of water equivalent between 1964 and 1976. But the general trend of the majority of the nine glaciers between 1970 and 1990 was strongly negative, with some having water equivalent losses of as much as 5 or 6 m thickness (Reynaud et al. 1984) (Figure 8.2). The ice-covered area on the northern slopes of the Zailiyskiy Alatau decreased from 305 km² in 1937 to 208 km² in 1989 (Dyurgerov et al. 1996). In 1992/3, as a result of abundant precipitation and a cool sum-

mer, the mass balance of Tsentral'nyy Tuyuksu was strongly positive for the first time since 1963/4 and the equilibrium line was 158 m lower than in the previous year (Haeberli et al. 1994). But after winter snowfall close to the mean in 1996/7, the summer ablation season was so warm and protracted there was no net accumulation anywhere on the glacier and the net mass balance of −1.47 m was the largest negative value recorded since 1978 (Haeberli et al. 1999). Between 1960 and 1995 the front of the glacier steadily retreated a total distance of half a kilometre.

The front of the Shumsky glacier on the northern slopes of the Djungarski (Dzhungar) Alatau Range, lay about a kilometre behind the latest Little Ice Age moraine in 1965. Over the next 35 years it retreated 238 m, the ice thickness diminishing at a rate of about 0.1 m per year (Cherkasov et al. 1997).

Twentieth-century recession affected other glaciers on the Zailiyskiy Alatau Range, including those in the Bol'shaya Almatinka basin 10 km to the west of Tuyuksu, and in the Talgar basin 10 km to the east. Short periods of stabilization or slight advances interrupted retreat in the mid 1930s, late 1940s and early 1950s (Horvath 1975). A significant increase in precipitation and lowering of summer temperatures

---

10   The attribution of glacial advances to earthquakes has not proved to be well founded elsewhere (see p. 277), and can probably be discounted.

began in the 1950s, leading to periods of positive mass balance in the following decade, notably in 1954. Comparison of air photographs taken in 1977 with those taken in the International Geophysical Year of 1957 revealed that not all frontal changes have been in the same direction (Kotlyakov 1980a, b).

Systematic investigations of the moraines of the Tien Shan were made by the Institute of Geography of the Russian Academy of Sciences during the 1980s (Serebryanny and Solomina 1989, 1996). A great many lichenometric age determinations have been made in this region but their accuracy is relatively low.[11] Current knowledge suggests that moraine sets were deposited during advance periods in 1210–15, 1340–90, around 1440, 1540–50, around 1590, 1650–60, 1680–1710 and 1730–1910. The majority of the moraines were formed between the second half of the eighteenth and the first decade of the twentieth century, but the lowest temperatures, according to studies of tree-ring width, occurred between 1370 and 1420. Several moraines were formed at this time by glaciers that reached their maximum extent early in the Little Ice Age, including some in the Ak-Shiyrak mountains, as well as the Terskei Alatau. However only a very few remnants from the earlier periods survive, the greater part having been washed away or overlapped by later advances.

According to Kotlyakov *et al.* (1991), maximum advances of the Tien Shan glaciers can be identified more precisely by dendrochronology than by lichenometry. Correlations between annual increment indices of *Picea schrenkiana* and *Juniperus turkestanica* and available instrumental records show considerable variation in both summer temperatures and precipitation in the Issyk-Kul region and on the northern slopes of the Kirghiz range during the last few centuries. The pattern of summer temperatures along the northern slopes of the Kirghiz range and southern slopes of the Kunghei Alatau indicates a succession of warm and cool intervals, the warm intervals often similar to present-day conditions. The results resemble those from the Alps discussed earlier, glacial advances alternating with significant withdrawals and each new advance in more recent times beginning from a position considerably behind that of the preceding advance. Summer mean temperatures over the last 200 years in the Terskei Alatau are believed to have varied within a range of 2.5 °C, and on the northern slopes of the Kirghiz range within about 2.0 °C.

In the Issyk-Kul region, precipitation was relatively high from the eighteenth to the beginning of the nineteenth century, while the nineteenth and twentieth centuries experienced much lower precipitation. This conclusion from dendroclimatalogical studies, is reflected in historical and archaeological records. Thus the Issyk Kul lake overflowed to the Chu river from the mid seventeenth to the early nineteenth centuries, whereas in the middle of the nineteenth century, when according to lichenometric data glaciers were close to their maxima positions, there was insufficient runoff to allow Issyk Kul to maintain its outflow. This has been taken to imply that the lake level in the seventeenth and early eighteenth centuries together with its subsequent lowering was controlled by changes in the precipitation regime, not by fluctuations in meltwater runoff. Kotlyakov *et al.* (1991) consider that glaciers reached their maxima in Tien Shan primarily on account of increased summer precipitation, often in the form of snow, though they noted that their data indicate that precipitation increases decline with altitude.

8.1.5 Altai and Sayan

The Altai and Sayan Mountains prolong the great ranges of the Pamirs and Tien Shan (Figure 8.1). The snowline rises from 2300 m in the north and west to 3400 m in the southwest and 3500 m in the east. The range of conditions is wide because of the location of the mountains at the junction of three climatic regions, Mongolian, central Asian and west Siberian.

11 *Rhizocarpon geographicum*, the most widespread lichen genus, only develops in this region 50–70 years after substrates have been exposed, and is rarely dominant in the plant communities. *Caloplaca* has to be used for dating in the last 100–50 years and *Aspicilia* and *Placolecanora* for older substrates. Moreover not one but several species within the genera *Caloplaca* and *Aspicilia* have had to be used as indicators, and these have different growth rates. Lack of lichens on moraine surfaces further hampers dating (Solomina 1996). As there are only a few reliably dated surfaces in the Tien Shan, reference sites are widely dispersed over a range of climatic regions (Kotlyakov *et al.* 1991).

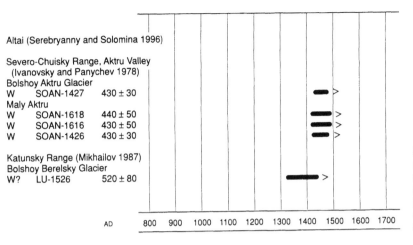

Altai (Serebryanny and Solomina 1996)

Severo-Chuisky Range, Aktru Valley
  (Ivanovsky and Panychev 1978)
Bolshoy Aktru Glacier
W   SOAN-1427 430 ± 30
Maly Aktru
W   SOAN-1618 440 ± 50
W   SOAN-1616 430 ± 50
W   SOAN-1426 430 ± 30

Katunsky Range (Mikhailov 1987)
Bolshoy Berelsky Glacier
W?   LU-1526 520 ± 80

AD   800 900 1000 1100 1200 1300 1400 1500 1600 1700

Figure 8.4 Key radiocarbon dates relating to Little Ice Age advances in the Altai (*Source:* from Grove 2001b)
Notes: The numbers in the margin give the laboratory identification number and the ^{14}C age. Horizontal lines represent the calibrated ages. > = Advance after date; W = Wood

Glaciers in the Altai, like those in the Tien Shan, have recent moraines of different ages within a few kilometres of their snouts that have been attributed to the Little Ice Age (Revyakin and Revyakina 1976). The second of three moraines in front of the Berelsky glacier, on the southern slope of the Katunskiy Range, has been dated by buried wood to 520 ± 80 BP (AD 1320–1450) and a moraine in front of the Maly Aktru has similarly been dated to 430 ± 30 BP (AD 1440–80) (Figure 8.4). Dendrodata suggest that cool periods occurred in the intervals 1560–1600, 1690–1715, 1785–1805 and 1810–50 (Ademenko 1985, cited Kotlyakov *et al.* 1991). Judging from maximum lichen diameters on the moraines of the Maly Aktru Glacier, the last three of these, at least, were associated with glacial advances. Mean summer temperatures, reconstructed from tree-ring data from the Bish-Iirdu massif of the Chuysky Range (50°05′N, 87°45′E), were about 2.0–2.5 °C lower than the present during the coldest period, and equilibrium lines 65–70 m lower.

Lichenometric dating in the Aktru valley suggests that fifteenth-century advances of the Bolshoy and Maly Aktru were synchronous but differed in scale (Serebryanny and Solomina 1996). The Bolshoy advances culminated in the seventeenth century; those of the Maly Aktru at the end of the eighteenth or beginning of the nineteenth centuries. (No estimate of the reliability of the lichenometry is given.)

In 1842 Chikhachev observed that there were extensive snowfields in the eastern Altai and the glaciers were in advanced positions. Since then retreat has predominated but systematic data collected at the end of the nineteenth and beginning of the twentieth centuries indicate that it has been on quite a modest scale (Serebryanny and Solomina 1996). A reconstruction of the mass budget of the Maly Aktru from 1838 to 1980, based on the assumption of a close relationship between temperature and precipitation indices at the Barnaul and high mountain Aktru weather stations on the one hand and multiyear budget observations by observers from Tomsk University on the other, indicates a progressive reduction in glacier mass balances from the 1880s to the second decade of the twentieth century, with minor but fluctuating increases.

The frontal position of Rodzevic, the largest glacier in the Akkem basin (about 10 km²), first observed in 1850 (Smiraglia 1985) retreated between 1850 and 1974, especially rapidly in the 1930s and less so in the 1950s (Table 8.3 and Figure 8.5).

Intermittent observations of the termini of the Katunskiy, Akkemskiy and Sapozhnikov since the 1880s show that wasting was particularly swift in the mid twentieth century and continued until 1970. Some small glaciers disappeared entirely and the majority lost 30–60 per cent of their areas. Valley glaciers survived better, diminishing in area by only 5–13 per cent (Narozhnyi 1986, cited Kotlyakov *et al.* 1991).

Mass balances have been measured since 1962 on Maly Aktru and more recently on the other five Aktru

Table 8.3 Retreat of the snout of the Rodzevic Glacier, 1850–1974

Years	Retreat (m)	m/y
1850–97	156	3
1897–1927	350	12
1927–32	49	10
1932–52	986	50
1952–66	156	11
1966–7	14	14
1967–8	17	17
1968–73	98	20

Source: after Smiraglia 1985, based on the Soviet Glacier Inventory

glaciers (Kotlyakov 1980a, b).[12] Since the late 1960s cumulative mass balances have fluctuated sharply. A general increase in volume after 1983 led to accelerated flow in the central and lower glaciers. Then, in

1987/8, accumulation on Maly Aktru was 30 per cent above normal, a value without precedent for the whole observation period, and frequent snowfalls in the following summer reduced ablation (Haeberli and Herron 1991: 49). By 1993 the fronts of the Aktru glaciers were advancing (Haeberli *et al.* 1994). In 1995/6 the mass balance of the Maly Aktru was slightly negative, −0.13 mwe. Then came the anomalous year 1996/7 when accumulation was 21 per cent above the average and melting, on account of the length and warmth of the summer, 19 per cent more than usual, resulting in a slightly negative balance (Haeberli *et al.* 1999).

Until recently investigations in the Altai failed to disclose evidence of early Little Ice Age advances (Kotlyakov *et al.* 1991). However new equations of growth rate for *Aspicilia tianshanica* and *Rhizocarpon geographicum*, using mudflows dated by radiocarbon as controls, indicate that advances took place in the fourteenth century. Equilibrium line depression is

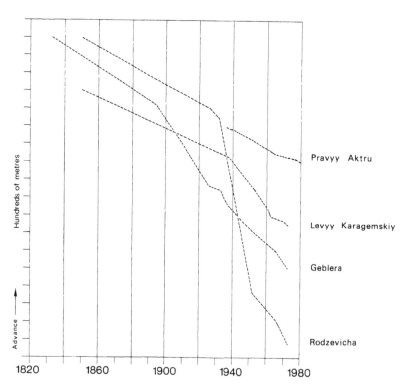

Figure 8.5 Retreat of Altai glaciers in the nineteenth and twentieth centuries (*Source*: from figures published in Revyakin and Revyakina 1976 and Haeberli 1985)

12   Average annual precipitation is about 520 mm, measured at 2130 m above sea level, and annual mean temperature at the equilibrium line (about 3130 m) is −9 °C to −10 °C (Haeberli *et al.* 1994).

stimated to have been 150–70 m as compared with .00–20 m in the nineteenth century (Solomina *et al.* .994).

A cold interval in the Sayan from the mid seventeenth century to the first half of the twentieth century, identified on the basis of stump samples in the Tunkinskie Goltsy (50′N, 100–101°E) and Munk-Sardyk (51°43′N, 100–101°E) ranges, has been confirmed by historical evidence of increased snowfall and glacial advances at that time. Ice cover on the Sayan greatly diminished during the twentieth century (Serebryanny and Solomina 1996).

8.1.6 Kamchatka

The Kamchatka peninsula is about 1250 km long and up to 500 km wide, with the Sea of Okhotsk to the west and the Pacific and Bering Sea to the east. Two longitudinal mountain ranges are separated by a tectonic trough. Over a hundred volcanoes, many of them active and rising to heights of 3000 to 5000 m, are mostly situated on the coastal side of the eastern range. Winter brings bitterly cold winds from Siberia which pick up moisture from the Sea of Okhotsk and occasionally dump heavy falls of snow. More precipitation is brought by cool summer winds from the Pacific. On the lowlands, alder, poplar and willow are widespread; tundra vegetation covers the higher regions.

Lichenometric dating of eight moraines in the Kamchatka peninsula, using *Rhizocarpon geographicum* curves has indicated glacial advances in 1850–70 and 1910–20, with one moraine date pointing to an advance in the 1690s (Solomina *et al.* 1995).[13] Results using an improved growth curve for a Rhizocarpon subgenus have indicated that in eastern Kamchatka the main advances of four glaciers occurred in the mid sixteenth century, the second half of the eighteenth century, around 1830, 1850–60 and 1890. The maximum stage of the Little Ice Age was reached in the mid to late nineteenth century, when glacier fronts generally

reached 100–200 m lower than at present.[14] Mass balance measurements of the Kozelskiy, a south-facing valley glacier on the eastern slope of the volcano Avachinskiy, began in 1973.[15] After a short period of positive balances in the early 1970s, negative balances predominated until the early 1990s when slight advances were recorded (Figure 8.2) (Haeberli *et al.* 1994).

8.2 THE HIMALAYA

The Himalaya extend for 2000 km from the Hindu Kush and Karakoram and the sources of the Indus in the west to the headwaters of the Dikang–Brahmaputra in the east (Figure 8.6). The main east–west arc is composed of several parallel ranges, the Lesser Himalaya in the south rising to 4500 m and the Greater Himalaya further north to over 5500 m. Of the thirteen mountains on the globe rising to over 8000 m ten, including Everest, are amongst the giant peaks. To the north lies the high plateau of Tibet and central Asia. The ice-covered area in the 1950s was 33 times that of Europe, according to Wissman (1959). This may be an overestimate, but it remains the largest such area outside the polar regions (Vohra 1980).

The Himalaya separate the monsoon lands of the Indian subcontinent from the deserts and steppes of central Asia. Most of the precipitation is brought by the southwest monsoon. Glaciers south of the main watershed are much larger than those to the north, on account of the greater snowfall, and terminate at much lower levels, reaching down in some cases into forest (Plate 8.1). In Sikkim in the east, the firnline is at 5500 m and the treeline reaches an elevation of 3900 m; on the Nanga Parbat massif in the northwest, the firnline comes down to 4700 m and the treeline is at 2000 m. Glacier meltwater irrigates the fields of mountain villages and contributes to the flow of the great rivers of India and Pakistan, generating power and allowing the cultivation of immense areas.

---

13 The growth curve used was based on seven points on lava flows and moraines, dated by tephrochronology or historical records between 15 and 300 years old (Solomina *et al.* 1995).
14 But according to Mikhalenko and Solomina (1996) glacier fronts in Kamchatka are currently 50–60 m higher than their Little Ice Age moraines.
15 The Kozelskiy is believed to be temperate. Average annual precipitation at 1850 m asl is about 2100 mm, and mean annual air temperature −3.4 °C at the equilibrium line (Haeberli *et al.* 1994).

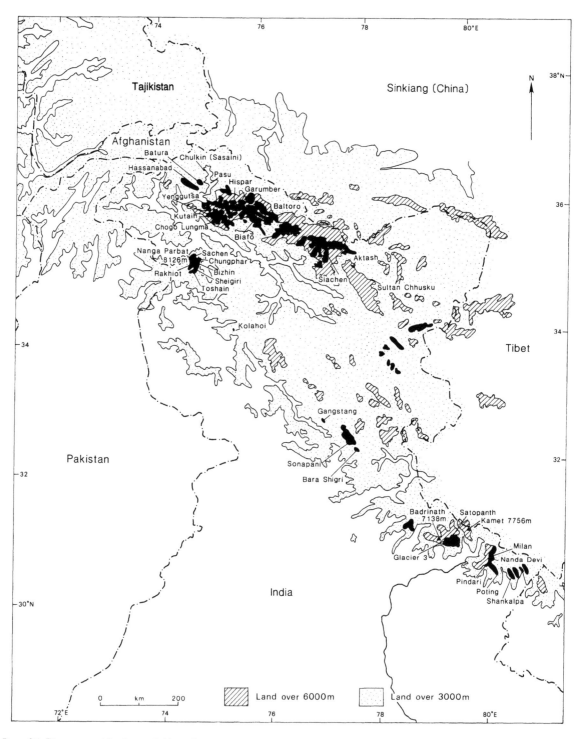

Figure 8.6 The western Himalaya with Nanga Parbat and Karakoram

Plate 8.1 The Himalaya between Mt Everest and Sikkim (Paiku Hu, 29.0°N, 86.5°E), 6 October 1984. Kathmandu is in the bottom left and La-Sa in the top right of this handheld photograph taken from the Shuttle and covering an area of about 75,000 km². Note the contrast between the well-watered zone south of the range and the aridity of the Tibetan Plateau
Photo: NASA Astronaut Linhof

Practically no documentary records are known relating to the extent of the glaciers before the nineteenth century. However, a Hindu scribe was quoted by Smythe (1932) to the effect that:

He who thinks of Himachal [the Himalayan snows] though he should not behold him, is greater than he who performs all worship in Kashi [Benares]. And he who thinks on Himachal shall have pardon for all sins, and all things that die on Himachal, and all things that on dying think on his snows, are freed from sin. In hundred ages of the gods, I could not tell thee of all the glories of Himachal, where Siva lived and where the Ganga falls from the foot of Vishnu like the slender thread of a lotus flower.

This quotation is not merely a consolation to glaciologists; the glacier source of the Ganges plays a very important part in Hindu mysticism. Thousands of pilgrims visit the temples near the sources of the Ganges every year. The temple of Gangotri is nowadays 30 km downstream from the true source at Gomukh. It seems likely that when the temple was built the glacier at the source of the Ganges was much nearer to Gangotri; documents or traditions may yet tell us more. Certainly many glaciers have tracks or trade routes running along their lateral moraines and even crossing the ice, but no written records are known to refer to them. Access by long approach roads is time-consuming and commonly dangerous on account of landslips and avalanches; political boundaries present problems. Where maps exist, and they are few, they are often inaccurate. Monitoring of glaciers is consequently expensive and has to be confined to a short snow-free season between June and September.

Vivid descriptions of some of the Himalayan glaciers are to be found in published and unpublished papers of explorers and climbers who gradually unravelled the topographical relationships of the ranges. Much of this work was accomplished by officers of the Survey of India such as Godwin-Austin, but the remarkable contributions of travellers such as Sir Martin Conway and the redoubtable wife-and-husband team of Fanny Bullock Workman and William Hunter Workman, with their many expeditions around 1900, deserve special mention. The single-handed achievements of Aurel Stein and Kingdon Ward were on a heroic scale.

A strong thread of Alpine experience ran through the early approaches to the glaciers. The Workmans were accompanied on their 1908 Hispar expedition by Calciati and Roneza, surveyors recruited through the good offices of Charles Rabot (Chapter 4). The three Schlagintweit brothers of Munich, Rudolf, Herman and Adolf, who despite their youth were already well known for their glacial work in the Ötztal (Chapter 5), were engaged by the East India Company in the 1850s, on the recommendation of Alexander von Humboldt, to carry out a geomagnetic survey of India. Adolf, the one most concerned with glacial matters, was murdered at Kashgar in August 1857. His brothers produced four volumes out of the twenty that had been projected, recording the results of the expedition's work. They recovered Adolf's journals and 'a considerable number of his drawings and collections' (Schlagintweit and Schlagintweit 1860–6, vol. 4: 565), but failed to publish them. Eventually they found their way into the Bavarian State Library where they rested in oblivion until R. Finsterwalder (1937) recognized their potential value, especially that of the paintings with the detail of the glaciers recorded by the hand of an experienced glaciologist. Kick (1960, 1967, 1969) later made use of the written material but it seems likely that the journals have not yet been fully exploited (see Plate 8.2).

At the beginning of the twentieth century the British Alpine Club, recognizing the importance of glacial oscillations as indicators of climatic variation, initiated the collection of records relating to them in all parts of the world (Freshfield 1902). After the founding of the Commission International des Glaciers, Freshfield, as the British member of the Commission, drew the attention of the Trigonometrical Survey of India to the importance of collecting data on the secular movements of the principal Himalayan glaciers (Holland 1907). Difficulties were presented by the great distance of the glaciers from permanent stations as well as from 'the extreme complexity of the departmental system in Calcutta' (Freshfield 1902), and eventually the Geological Survey of India undertook responsibility for the work (Holland 1907). The first twelve glaciers to be selected were visited in 1906; sketchmaps were made, observation stations set up, and photographs taken (Hayden 1907, Walker and Pascoe 1907, Cotter and Coggin-Brown 1907). From 1907 onwards, detailed accounts of glaciers and groups of glaciers appeared from time to time in the Records of the Geological Survey, some officers spending their short leaves extending the observational network (e.g. Auden 1935). Well-known travellers such as Smythe (1932), Shipton (1935) and Ward (1934) brought back information about glaciers not previously described. After a period of reduced activity at the time of partition, new stimulus was given by the International Geophysical Year of 1957 (Tewari 1971). Lynam (1960) directed his 1958 expedition to the Kulu–Lahul–Spiti watershed partly because the Survey of India map sheet 524, depending heavily on nineteenth-century surveys, was known to be accurate as far as the main valleys were concerned but to have side valleys like that of the Bara Shigri sketched in with 'more imagination than accuracy'.

By the 1970s a few observations of mass balance were beginning to be made in India (e.g. Raina et al. 1973). Japanese glaciologists were working in Nepal and a joint Japanese–Nepal research project was set up. As a result, Higuchi et al. (1980) were able to compare the terminal positions of the glaciers in the Khumbu region of east Nepal in 1975 with those in 1960 reported by Müller (1970) and found there had been a general retreat. Japanese teams surveyed Rikha Samba and other glaciers in Hidden Valley on the arid north side of Dhaulagiri Himal in 1974 and again in 1994 (Fujii et al. 1977, 1996). They found that all the glaciers had thinned and retreated in the 20-year interval, the rate of thinning having increased between 1958 and 1963.

Mass balance values are now available for a few Indian glaciers, results from four of them being published in Volume 6 of *Fluctuations of Glaciers* together with data on the frontal positions of two (Haeberli et al. 1993). With increasing economic development in the Himalaya region, involving hydroelectric power stations using meltwater from ice and snow, glacier mass balances have become of practical as well as academic interest (Kulkarni 1992). In such rugged and difficult terrain, satellite images are the logical means of obtaining information, so long as some of the basic relationships are understood. Data for the Gara Glacier for 1977–8 to 1982–3 and for the Gor-Garong in Himachal Pradesh from 1976–7 to 1983–4 have been used, together with less complete field data from seven other glaciers in the western Himalaya, to assess the relationship between mass balance values and accumulation area ratios (AAR) (Kulkarni 1992).[16] The results indicated that in the western Himalaya zero mass balance corresponds to an AAR value of 0.44.[17] The relationship may differ for glaciers in Sikkim and Arunachal Pradesh in the eastern Himalaya. Kulkarni's results, obtained from Landsat images combined with topographic maps, suggest that mass balances in the western Himalayas were positive in 1986–7 and negative in 1987–8.

Himalayan cedars growing at 2730 m in the western Himalaya have provided a tree-ring record going back from 1987 to 1390 (Yadav et al. 1999). A correlation of −0.53 was found between ring width and spring temperature as measured at Simla 150 km to the west and Mukteshwar 200 km to the south over the period 1876–1987. Temperature reconstruction back to 1390 on this basis failed to show any prolonged periods of cooling during the Little Ice Age, though anomalously cool periods were indicated as having occurred in 1507–36, 1758–87 and 1813–42.

8.2.1 Kenchenjunga–Everest

When Freshfield visited the Kenchenjunga area in 1899, he disparagingly referred to the glaciers as they were delineated on the official maps as no more than 'a few worms crawling about the heads of valleys'. Today the maps of the Mount Everest area are markedly better than elsewhere in the Himalaya. Müller (1970, 1980), when he compiled an inventory of 1936 glaciers in the region, was able to make use of Royal Geographical Society 100,000 maps of 1961 and 1975 and also

---

16   These data were in part drawn from unpublished reports of the Geological Survey of India.
17   Detailed studies have shown that the accumulation area ratio of glaciers, that is the ratio of the area above the equilibrium line to the total area, is closely related to specific net budget or mass balance. It has been generally accepted that for a glacier with a linear increase in net budget with altitude and a symmetrical distribution of area to median altitude, an accumulation area ratio (AAR) of 0.5 would indicate that the glacier was in equilibrium and neither advancing nor retreating. It has emerged from fieldwork in North America that AARs vary regionally from between 0.5 and 0.6; see Chapter 9 (Meir 1962, Meir and Post 1962).

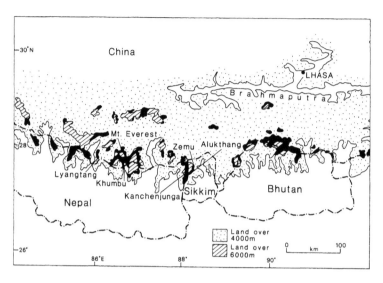

Figure 8.7 The Everest area

numerous photographs taken by climbing expeditions. The largest glaciers are on the south side of the main mountain ranges where precipitation is brought from June to mid October by the southwest monsoon. The floors of the main valleys become increasingly arid towards the north. Evidence is accumulating that on the peaks and ridges of the Nepalese, eastern Himalaya, precipitation reaches a maximum on south-facing slopes at altitudes of between 5000 and 7000 m. The accumulation area ratio is very low, on average about 0.41, probably because the extraordinarily high relief promotes very large-scale avalanching, adding to accumulation at the higher levels; also, abundant rock debris resulting from slope failures mantles many glacier tongues, protecting them from ablation (Müller 1980).

The Mount Everest region is one of the few areas in the Himalaya where moraine sequences have been examined in sufficient detail to give any useful information about glacier front fluctuations in recent centuries (Figure 8.7). In the Imja Khola basin of the upper Khumbu, Müller (1961, 1980) distinguished a sequence of fresh-looking moraines which he thought had been left quite recently, lying within older and much larger moraines, lichen-covered and partly vegetated, which form ramparts bulging out of side valleys to partition the main valleys. These older features, called Dughla by Müller, and Thukhla by Fushimi (1978), have been taken to belong to the Little Ice Age on a

basis of lichen size and a single radiocarbon date of 480 ± 80 (B-174) which calibrates to AD 1390–1500. Four of Röthlisberger's samples from the Khumbu region also give dates relating to the Little Ice Age (Röthlisberger and Geyh 1986) (see Figure 8.8). Several kilometres downvalley there are still larger and older moraines (see Chapter 15). The topographical relationship of the Dughla to these Pheride moraines is comparable with that of Little Ice Age moraines to those of the Egeson and Daun stages of the Late Glacial in the Alps.

Mayewski and Jeschke (1979) found specific data on the terminal positions of only five glaciers amongst the many hundreds in the region. Of these only the Zemu and Alukthang had as many as three measured points between 1860 and 1968. The Zemu, the largest glacier in the eastern Himalaya, originates on the precipitous eastern slopes of Kenchenjunga (8536 m) and flows north and then east. Moraines on the north side of the tongue, extending down the valley for at least 600 m below the terminus, are covered with trees which provide opportunities, as yet unexploited, to obtain a minimum age for the ice advance which created them.

The Zemu was first visited in 1891 by the political officer in Sikkim, Claude White, and a photographer from Calcutta. They found the snout of the glacier, ending in a cliff about 150 m high, was at an altitude of

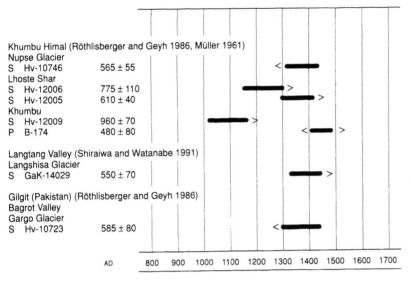

Khumbu Himal (Röthlisberger and Geyh 1986, Müller 1961)
Nupse Glacier
S Hv-10746   565 ± 55
Lhoste Shar
S Hv-12006   775 ± 110
S Hv-12005   610 ± 40
Khumbu
S Hv-12009   960 ± 70
P B-174   480 ± 80

Langtang Valley (Shiraiwa and Watanabe 1991)
Langshisa Glacier
S GaK-14029   550 ± 70

Gilgit (Pakistan) (Röthlisberger and Geyh 1986)
Bagrot Valley
Gargo Glacier
S Hv-10723   585 ± 80

AD   800 900 1000 1100 1200 1300 1400 1500 1600 1700

Figure 8.8 Key radiocarbon dates relating to Little Ice Age advances in the Everest area (*Source:* from Grove 2001b)

Notes: The numbers in the margin give the laboratory identification number and the ^{14}C age. Horizontal lines represent the calibrated ages. > = Advance after date; < = Advance before date; < > = Advance before and after date; P = Plants; S = Soil

4200 m. Eight years later, Freshfield (1903: 205) spent two days making a plane-table map (which does not seem to have survived) of the lower part of the glacier, 'a huge grey, billowy stream' flowing towards him, the moraines not arranged in lines but all over the place, giving some excuse to the official cartographer who refused to recognize the glacier below.

> A little plain hard by was covered with the traces of the recent passage of yaks. We had touched the 'high level' route connecting Sikkim through the Talung valley with Tibet. . . . This particular route serves chiefly for the transport of salt and timber; salt out of Tibet, timber into it. It crosses the Zemu glacier above its snout, much as the route of the Gries Pass in the Alps crosses the glacier of the same name. Ascending steeply we followed the faint tracks, soon lost, which led into the glen or hollow on the south side of the Zemu glacier and then, bearing to our right, climbed several grass-grown moraines until we found ourselves close to the edge of the retreating ice.

The Geological Survey of India, recognizing the need to record the changing positions of the glacier termini, and urged by Freshfield, sent La Touche (1910) to map the position of the Zemu snout. He did so in 1909, finding little evidence of change over the preceding decade. The Zemu was scarcely visited again until 1965 (Raina *et al.* 1973) when it was reported to reach down to 4260 m, having retreated 440 m in the course of the century. It ended in an ice cliff reduced

to 65 m in height, with moraine covering most of the tongue.

The Alukthang Glacier in the same region, according to a Major Sherwill, had by 1909 retreated half a mile since 1861. Mayewski and Jeschke (1979) present a diagram showing the glaciers of the Kenchenjunga–Everest area retreating in the latter half of the nineteenth century and then retreating more slowly or remaining stationary in the twentieth century, but the observational base is not very secure. Both the Zemu (Bose *et al.* 1971) and the Jungpu (Mayewski and Jeschke 1979) had thinned recently, as had the Khumbu (Müller 1970). Higuchi *et al.* (1980) reported that between 1960 and 1975, 90 per cent of the glaciers free of debris had retreated, but it should be noted that this estimate is based on the change in altitude of termini since Müller's (1970) report. The Zemu has been mentioned as one of the glaciers, and there are many others, where ablation is limited by debris covering the tongue.

For the region to the east of Sikkim we have very little information, though Kingdon Ward (1934) makes some mention of the Ka-Gur-Pa Glacier on the Mekong–Salween divide.

> Stretching down the glacier valley for half a mile beyond the snout is a lateral moraine, its summit and far side covered with trees, while the flank facing the glacier is almost

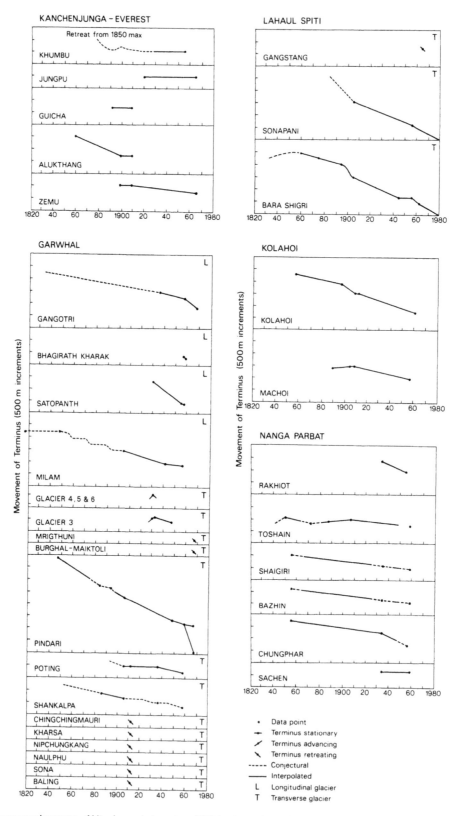

Figure 8.9 Advances and retreats of Himalayan glaciers since 1820 (omitting glaciers known to surge, or for which information is minimal, and with the curve for the Bara Shigri altered) (*Source*: after Mayewski and Jeschke 1979)

bare below or clothed with plants trying to establish themselves. The summit of the moraine is about 350 feet [100 m] above the glacier and shows a sort of step structure as though there had been periodic fluctuations in the retreat of the ice.

He concluded that the retreat had been due to a change in precipitation.

Glacial fluctuations from the 1970s to 1989 were investigated by the Japanese Glaciological Expedition to Nepal (Yamada *et al.* 1992). In the Dudh Kosi and Imjo Khola basins of the Khumbu region, surveys were made of the terminal regions of six tongues free of debris and one, the Khumbu Glacier, flowing from Mt Everest (or Sagarmatha as it is known in Nepal) with a heavy debris cover. All of them had retreated, and the rate of retreat of some of those visited in 1978 (Fushimi and Ohata 1980) increased in the 1980s. The Kongma Glacier, on the southern flank of Mehra Peak (5817 m), probably advanced sometime between 1981 and 1989, as one of the earlier reference points was found buried by till. The Chukhung advanced 3.5 to 6.8 m per year between 1976 and 1978 (Fushimi and Ohata 1980) but receded about 100 m during 1978–89. In the Shorong region, Glacier AX000 was found to have retreated about 100 m and to have lost one of its two tongues since 1978, though the other was little changed. The front of AX010, a small valley glacier, investigated more intensively than others in Nepal (Higuchi *et al.* 1980, Fushimi and Ohata 1980, Yamada *et al.* 1992) had retreated about 30 m during 1979–89, and thinned near the terminus by 12 m.

Reports on meteorological observations made during the last decade or so emphasize the great variability from year to year of winter temperature and snow cover in the Khumbu Himalaya (Bertolani and Bollasina 2000, Ueno *et al.* 2001). In particular it was observed that non-monsoon disturbances cause greater year-to-year change in annual precipitation than does monsoon precipitation.

In the Langtang region of Nepal, about 100 km west of the Khumbu area, Ono (1984, 1985) had demonstrated how, under Himalayan conditions, a chronology of advances and retreats can be derived from a study of the detailed morphology and stratigraphy of glacier forefields. In front of the Yala Glacier he distinguished six till sheets, differentiated by their surface fluting and texture. Each till sheet is bounded by a moraine similar in form but higher than other morainic ridges running across the till sheet which, it was argued, are annual push ridges formed by advances at the beginning of the post-monsoon seasons and left behind as the ice melts back again the following year. By analogy with push moraines formed in 1976 by the Gyajo glacier in the Khumbu Himal (Fushimi and Ohata 1980), and assuming that the time span required for readvance is proportional to the height of the moraines bounding the till sheets, it was estimated that the last important advanced position of the Yala was attained in 1815 and that since then minor readvances have taken place in 1843, 1867, 1887, 1903, 1921, 1953 and 1976, i.e. about every twenty-two years. An earlier undated Little Ice Age extension of the Yala occupied a position similar to that reached by the glacier in 1815.

The Yala front advanced between 1982 and 1987, and then retreated over the next two years, though by less than 10 m (Yamada *et al.* 1992). The debris-covered Lirung Glacier in 1989 had changed very little since its position was ascertained by triangulation in 1987. All the glaciers in Nepal for which measurements were obtained by the 1989 team had retreated, but the magnitude of recession varied from one to another. The rates of retreat found by Yamada's team in 1989 were generally greater than those that had been reported by Fushimi and Ohata (1980), but the evidence from both the Yala and Kongma indicates that retreat has not been continuous, but interrupted by advances. An assessment by terrestial photogrammetry of the overall changes in position between 1980 and 1991 of the debris-free tongues of five south-facing glaciers in the Langtang Khola basin found little change,[18] though two north-facing tongues had both advanced, the Gangtsa La by about 50 m and the Gangchenpo West by about 30 m (Kappenberger *et al.* 1993). Future monitoring of the ice in this region will be assisted not only by the results of the work of several Japanese field teams and by the Langtang glacial inventory completed

---

18   Yamada *et al.* (1992) had reported a mean advance of 2.6 m between 1982 and 1987, and mean retreat of 4.0 m between 1987 and 1989.

by Shiraiwa and Yamada (1991), but also by the publication in 1990 of 1:50,000 maps of the area by the Austrian Alpine Club. Recent studies confirm that shrinkage of glaciers, both those free of debris and those covered with debris, is accelerating (Ageta *et al.* 2001).

Observations in the more arid part of Nepal, less affected by the southwest monsoon than the Khumbu, Shorong or Langtang areas, were first made in 1974 (Fujji *et al.* 1977). All six glaciers in the northward draining Mukut Himal revisited in 1994 had retreated, and new bedrock was showing through in their accumulation basins (Fujji *et al.* 1996). The front of the Rikha Samba, the largest, had retired by about 200 m since 1974, and the ice had thinned. A core 23.25 m long extracted from its accumulation area indicated a decrease in accumulation from an annual mean of 25–30 cm between 1958 and 1963 to only 18 cm annually over the period 1963–94.

8.2.2 Garwhal or Kumaon

The Garwhal area, some 700 km west of Everest–Kenchenjunga, contains some of the largest glaciers in the Himalaya. Some of them reach down to 3600 m. The Ganges rises in the complex massif topped by Kedarnath (6940 m), Badrinath (7138 m) and Kamet (7756 m) which overlook the great Hindu shrines of Kedarnath, Badrinath and Gangotri associated with the main sources of the river. To the south lies the great glacier-girt group of Trisuli (7120 m), Nanda Devi (7816 m) and Nanda Kot (6861 m). The main peaks were surveyed in the 1870s but the country was not examined in detail, the surveyors having been ordered not to waste time on uninhabited areas. Exploration of the Badrinath range between Alaknanda and Gangotri was one of the objects of Smythe's 1932 expedition and Shipton (1935) entered the sanctuary of Nanda Devi in 1934.

Thousands of pilgrims visit the shrines each year. Relatively few see the ice from which the Ganges issues and, so far as is known, none of them have left records that would allow its fluctuations in position to be traced. Shipton (1935) recorded a tradition that 'many hundreds of years ago' when there was no high priest at Kedarnath temple, the high priest of Badrinath used to

hold services at both temples in the same day, a story reminiscent of the Alpine tradition of the priests who crossed Mont Blanc from Val Veni in six hours to say mass in Chamonix. Local people believed the direct route from Badrinath to Kedarnath to be 4 km; Shipton, a fast mover, took two days to cover the distance, and he estimated it to be 40 km.

Mayewski and Jeschke (1979) plotted the frontal fluctuations of seven glaciers in Garwhal and found that over the preceding century all except No. 3 in the Arwa valley had retreated (Figure 8.9). They were able to establish that six more glaciers had retreated between 1910 and 1920, and that two others had done so in the mid 1960s. Their assertion that all the glaciers had been in retreat since 1850 remains unsubstantiated, and has been seriously undermined by a critical analysis of the available literature by Kick (1989). He emphasized that many of the glaciers of 'High Asia' are thirty times larger than those of the Alps and several of them are of the surging type, so it is not surprising to find they reached their maxima at various times between the early nineteenth and early twentieth centuries.

Sharma and Owen (1996) used geomorphological techniques, lichenometry and optical dating methods in an attempt to throw light on the Quaternary history of the northwest Garwhal Himalaya. During the Bhujbas event, which corresponds with the latter part of the Little Ice Age, they estimated that equilibrium line altitudes were 20 to 80 m lower than in the latter part of the twentieth century, the amount depending on aspect. Glaciers in the Bhagirathi catchment, including the Gangotri, were 40 to 130 m thicker and extended as much as 2.2 km outside their present limits.

The Milam glacier, nourished in large part by tributaries from the eastern slopes of the Nanda Devi–Trisuli ridge, and flowing 16 km southeast towards Milam village, is second in size only to the Gangotri amongst the Kumaon glaciers. According to local tradition, recorded by Cotter and Coggin-Brown (1907), the ice was in a forward position and reached the site of the village 'a thousand years ago'. In 1906 the terminus was about a mile from the village. In 1849 'according to Kishen Singh Rawat, Bahadur of Milam, known to science as "AK", the explorer of Tibet, the ice cave 52 years ago was about 800 yards in advance of its present position'.[19] In 1846 Eduard Madden was told by the Milam villagers that the Milam Glacier had already retreated by 2 or 3 miles. (Madden 1847). The Milam front was therefore well behind its 1850 position before Sir Richard Strachey saw it. He spent some time in the Milam valley in 1848 and measured the speed of flow of the ice at several points before crossing the glacier into western Tibet. He followed the well-established trade route used for the exchange of the salt and borax of Tibet for the grain of India. Strachey (1900) left an excellent account of his observations, finding the people of Milam intelligent and 'decently educated', even having some knowledge of Hindu literature, and was therefore disposed to accept their statement that the ice formerly extended to a *rask* or fortified wall 'which is now several hundred yards below the terminus'. In 1855 the snout of the Pindari was painted by Adolf Schlagintweit. This painting, reproduced by Kick (1960), showed the shape, position and landscape surrounding the glacier so clearly that it could almost certainly be used to reconstruct the 1855 characteristics of the tongue with considerable accuracy, especially as he also left descriptions and some measurements.

The Milam, together with the Skunkalpa and the Poting, were amongst the twelve glaciers originally selected for monitoring by the Geological Survey of India. Accordingly Cotter and Coggin-Brown made plane-table sketches of the positions of their tongues and the points from which they had taken photographs. They noted that the existence of a whole complex of lateral moraines indicated that retreat could not have been continuous.

The nineteenth-century retreat of the Milam is recorded in more detail than that of other glaciers in Kumaon, though there is also some information to be gleaned from the retreat of the Pindari which is fed principally from the slopes of the Nandakhat–Nandakot ridge. Madden (1847) was told that the Pindari had

retreated before 1846, though less far than the Milam. The left lateral moraine was then already much higher than the glacier surface. Strachey visiting it in 1847 found 'the distance between the ice caves and the median moraine is about 2 miles'. A Colonel J. W. A. Mitchell wrote in 'a book kept in the Phurkia dâk bungalow for the purpose of recording observations of the movement of the glacier' that it 'appears to have retreated about 100 yards since I visited it in 1884'. It is not known whether this book still exists. Other material of this sort may well survive in private papers. Cotter and Coggin-Brown (1907) noted that 'the moraine terminates about a mile from the ice cave'. So the retreat here was about a mile (1.6 km) from 1847 to 1906.

During the twentieth century, observations in Garwhal were made more frequently. Between 1906 and 1957 the Milam receded 617 m (Jangpani and Vohra 1962) and a further 23 m between 1963 and 1964 (Kumar *et al.* 1975). Observations of the Skunkalpa give no indication of advance (Cotter and Coggin-Brown 1907, Jangpani and Vohra 1962, Kumar *et al.* 1975). Auden (1935), spending his leave from the Geological Survey of India making the first map of the tongue of the Gangotri glacier in 1935, was impressed by evidence of its wasting and retreat over the preceding decades, but when Ross saw it the following year he could find no sign of any change in its position. A retreat of 600 m took place between 1936 and 1967 by which time some of the tributaries had retreated more than the main glacier and had separated from it (Tewari 1971). No records of even slight advances in the nineteenth and twentieth centuries have been forthcoming from any of the Garwhal glaciers that have been measured, though there are recessional moraines, so far undated, indicating that there must have been at least short halts during this period. However, Gilbert and Auden (1932–3) recorded that

19 Kishen Singh Rawat was the greatest of the remarkable Indian explorers known as the Pundits, who worked for and were trained by the Survey of India. In the mid nineteenth century Europeans and other foreigners were not allowed to enter Tibet, but exceptions were made for certain groups of semi-Tibetans, amongst them the Bhotias living in the Johar valley at the head of which Milam lies. Accordingly, certain members of the Rowat clan were selected for intensive training at the Survey Headquarters at Dehra Dun. They were not only instructed how to find directions by the compass, latitude by the sextant, and height by taking the temperature of boiling water, but also how to measure distances by counting paces – 2000 paces to the mile, with 31.7 inches to the pace. AK was thus a man who might be expected to give a reliable estimate of distance, though it has to be admitted that he must have been relying on his memory in this particular case.

Himalayan glaciers are for the most part retreating, but some are advancing in northern Garwhal. One glacier debouching into the Arwa valley has advanced so far that the valley is in danger of being choked by it. Were it to advance two hundred or three hundred yards further the valley stream will be dammed and as the valley is flat for several miles, a large lake will be formed. The bursting of the dam will be disastrous for Badrinath and other villages in the Alaknandra valley and might even result in serious floods far down in the plains.

The attention of the Public Works Department was attracted by the possible danger to Badrinath and to bridges and buildings on the heavily used pilgrim route (Gunn 1930). In the event, the glacier advanced no further and retreated 199 m between 1932 and 1956 (Tewari 1971). This is the only one of the small glaciers in the Arwa valley known to have advanced in the 1930s, though some thought that the Satopanth was also advancing.

8.2.3 Lahaul–Spiti

The transition zone between monsoon and arid climates crosses the Lahaul–Spiti region (Lynam 1960). Egerton (1864) found 'a good deal of rain and heavy clouds, and the air was damp and heavy. In Spiti the weather continued cloudy and drizzly, but the clouds were less constant and higher, while the air was delightfully pure and comparatively dry. Accordingly, ice is much more extensive in the southern part of the Lahaul–Spiti region than in the northern part of the Sutlej basin.' However, we have information about the fluctuations of only three glaciers, the Sonapani (Kurien and Munshi 1962), the Gangstang and the Bara Shigri, literally 'the debris-covered glacier'.

The records of the Bara Shigri are unusually long and detailed. Egerton (1864), crossing the glacier in July, found the ice extending far across the valley of the Chandra.

> The path lay across the Chota Shigri, a vast moraine left by some former glacier which, from some cause or another (perhaps the increase in temperature asserted by some Himalayan travellers) has melted away.... Beyond the Chota Shigri is a comparatively open piece of ground from which we ascend slightly onto what appears another moraine, like the Chota Shigri, but of much greater extent, being from its source to the Chundra river about four miles long, and in breadth some two miles. This is, in fact, the Shigri or Great Glacier, as you soon find out from

walking on it.... Little streamlets are everywhere trickling from the surface, exciting your wonder how there should be water on top of this mass of apparently porous rubbish, till suddenly opens before you a rent in the mass, the walls of which are clear green ice, and you see that you are travelling over an enormous glacier, coated with dirt and gravel, and sprinkled with huge rocks and boulders.... If you follow the glacier down to the river (at no slight risk of breaking your neck) you see a large torrent issuing from beneath the glacier itself to join the Chundra river and eating away the glacier till it forms a promontory jutting out between the torrent and the river, with the perpendicular wall some 150 feet high.... It seems impossible to do much to improve the road over this treacherous element. You cannot turn its flank, for below is the river and above is a chaos of crevasses.

According to Mayewski and Jeschke (1979) the Bara Shigri advanced across the Chandra valley about 1860 and dammed back a lake. This lake was mentioned by Egerton as being 'a mile or two northwards from a bend in the Chandra river'. Calvert (1873, cited Hughes 1982) had to cross the Bara Shigri as did Colonel Tyacke (1893, cited Hughes 1982). From Puti Runi to the next camping ground, Korcha, is nearly 10 miles. During this march the Shigri glacier is crossed. Calvert made no mention of a lake though he went into some detail about crossing the glacier on his way to Spiti. Tyacke (1893) was more specific, writing that 'The Shigri is remarkable for the fact that some five and forty years since, it burst from the mountains above, and bringing down millions of tons of debris with it, completely dammed up the Chandra River, which remained so dammed for some months before it burst through the barrier'. The effects were quite clear when Egerton walked through in 1863

> From the Shigri, for four miles up the river, you see unmistakable signs of the river having been dammed up by the glacier. Throughout this distance the bed widens out to some 1500 to 2500 feet. The bottom is quite flat and is covered with a deep deposit of sand and gravel; and, strange to say, for these few miles there is no rock or boulder to be seen, though below the glacier the river is full of them. The natives say that the river was pent up for eleven months.

It seems possible that the tongue advanced to dam the river more than once. According to Egerton (1864),

> It is known that twenty-seven years ago this glacier first burst out of the mountains above and . . . formed a huge dam, extending right across the river Chandra for many months. . . . I believe no large river in the region has been

seriously invaded by a glacier in the last twenty-five years, and according to the Schlagintweit brothers, snow is decreasing and the occurrence of such an event becomes yearly less probable.

There seems then to have been a glacial maximum around 1848 and possibly also in the 1830s, but by 1864 the glacier had started to recede. Egerton's forecast of its future behaviour was correct.

The position of the Bara Shigri was recorded on several later occasions; in 1906 by Walker and Pascoe (1907), in 1945 by Krenck and Bhawan (1945), in 1956 by Dutt (1961), and in 1963 by Skrikantia and Padhu (1963). Mean annual retreat rates have been estimated to have been 60 m from 1890 to 1906, 20 m from 1906 to 1945, up to 28 m from 1955 to 1963, and 6 m from 1963 to 1980, giving a total retreat of 2.6 km over a period of ninety years (Mackley and McIntyre 1980). These values cannot be regarded as precise because the ice margin is very difficult to identify at all accurately on account of the huge amounts of debris spreading from the glacier surface to the moraines bordering it.

Adolf Schlagintweit sketched a glacier on the left side of the Chandra valley which he called the Bhoru Nag, in June 1856, and in the same month a glacier north of Shinko La (5097 m) in the Num Kum group. (He generally adapted place names used by the local people.) The sketches have never been compared with present-day photographs (private communication, W. Kick, 19 August 1984).

8.2.4 Kolohoi

Mount Kolohoi (5024 m), on the northeast side of the Vale of Kashmir, was climbed and mapped by Neve and Mason in 1912. Neve (1907) had first visited the area in 1887 and in 1912 he opined that the glacier tongue had retreated about a quarter of a mile over the preceding twenty-five years and more than a mile since the Topographical Survey first mapped the area in 1857. Odell (1963) reported that when he visited Kolohoi in 1961 the ice had receded another half mile. It was on this basis that Mayewski and Jeschke (1979) indicate a recession of 800 m from 1857 to 1912, and 800 m more from 1912 to 1961. The Machoi glacier, also on Kolohoi, after advancing a little around 1900 (Odell 1963), then retreated 457 m between 1906 and 1957.

8.2.5 Nanga Parbat

The Nanga Parbat massif (Figure 8.6) carries sixty-nine glaciers with a total area of 302 km² according to Kick (1980). The northern side is the most heavily glacierized, the reasons being the greater extent of the accumulation areas overlooking the Indus headwaters and exposure to winds bringing precipitation from the northwest in winter. Above 4500 m, annual snowfall is estimated to exceed 8000 mm as compared with less than 120 mm below 2500 m. Because of the extreme steepness of the southern slopes the tongues of some of the fifteen glaciers, all heavily encumbered with morainic debris, reach down to grazing areas at 2800 m close to villages such as Tashing and Rupal.

Adolf Schlagintweit visited the area in September 1856. Finsterwalder (1937), comparing one of his paintings with a map made in 1934, showed that the ice cover had diminished over the interval. Schlagintweit's notes and sketches show that the Tashing or Chungpar glacier had advanced to reach a maximum at some time before 1856, when the ice surface was some 20 m below a recent moraine. It dammed back a lake, the water of which had escaped before 1872 when Drew (1875) reported that the ice, though thinner, still blocked the valley. The Chungpar retreated by about half a kilometre between 1856 and the mid 1930s.

The Sachen Glacier was painted by Adolf Schlagintweit in 1856 (see Plate 8.2). He described the lateral moraine as standing 18–28 m high above the adjacent lake. In 1958, Kick found the same moraine towering 180 m above the lake and judged that the lateral moraine must have been built up greatly after 1856, probably at the time that the glacier reached its maximum extension which was, according to the Survey of India triangulation records, around 1900. Since then the tongue has scarcely changed its position though its surface had dropped about 10 m by 1958 (Plate 8.3).

Knowledge of more recent conditions on Nanga Parbat is based mainly on German photogrammetric studies using ground photographs. Rakhiot, in the north, surveyed for the first time by Finsterwalder in 1934, was resurveyed by Pillewizer in 1954, while six other glaciers in the east and south, also originally surveyed by Finsterwalder in 1934, were resurveyed in

Plate 8.2 The Sachen Glacier, Nanga Parbat with Sango Sarr Lake, painted by Adolf Schlagintweit in 1856
Photo: Wilhelm Kick

Plate 8.3 The Sachen Glacier and its moraines with Sango Sarr in 1958. Compare Plate 8.2
Photo: Wilhelm Kick

1958 by Loewe (1961) and Kick. The Sachen Glacier was almost stationary between 1934 and 1958, but most of the other tongues receded. The Rakhiot retreated 450 m but the surface was lowered along only 10 per cent of its length. The retreat of the Chungpar, Bizhim, Sheigiri, Toshain and three unnamed tongues was less than 20 m per year. Three other unnamed glaciers were stationary. No glaciers were found to have advanced between 1934 and 1958. Pillewizer (1956) deduced that increased supplies of névé from the north side of Nanga Parbat had already replaced the post-1934 loss in the upper and middle parts of the glaciers, indicating that the tongue was set to advance. A survey made in 1985 revealed that the Rakhiot had advanced about 200 m since 1954 (Gardner 1986).

8.2.6 Karakoram

Separated from the rest of the Himalaya by the Indus valley, the Karakoram extend east to west for 800 km from India into Pakistan. With an average height of 5500 m (Desio 1977), the range includes the massifs of Rakaposhi, Haramoshi and Batura Mustagh. Inappropriately meaning 'black stones', the Karakoram give rise to several of the largest glaciers outside the polar regions, some over 50 km long, including the Hispar, Biafo, Baltoro and, largest of all, the Siachen. They are fed by a multitude of transverse glaciers flowing in directions at right angles to the grain of the country down to the main longitudinal valleys. Estimates of the ice cover range from 28 to 37 per cent (Goudie *et al.* 1984) and of its area from 13,500 to 15,000 km². The snowline is at about 5100 m in the south, 5600 m in the main Karakoram ranges, and 4700 to 5300 m in the north (Shi Yafeng *et al.* 1980).

The diurnal, seasonal and altitudinal contrasts in climate are even more marked here than elsewhere in the Himalayan region. At the higher elevations precipitation may exceed 1000 mm while below 4000 m the glaciers extend down onto semi-arid valley floors with mean annual precipitation at Skardu (2288 m) only 160 mm and at Leh (3514 m) 83 mm.

Mayewski and Jeschke (1979) were able to find records of the fluctuations of seventy-four glaciers in the Karakoram, though for nineteen of them there was only a record of the position of the terminus at a single time and an indication as to whether it was advancing or retreating. Further information was added in the course of the Royal Geographical Society International Karakoram Project, 1980, when the snout positions of seven glaciers in the Hunza valley were surveyed and their positions related to those found previously (Goudie *et al.* 1984). Goudie *et al.* accepted Mayewski and Jeschke's conclusions, including their finding that 'the glaciers of Nanga Parbat have experienced a general but rather minor retreat (average 6 m yr⁻¹) since 1850. The only significant departures from this low retreat rate exist for Rakhiot and Chungpar glaciers which between AD 1930 and 1950 had retreat rates as high as 30 m yr⁻¹.' But, as Kick (1960, 1967) showed, the position of the terminus in 1850 is known only for one of the sixty-nine glaciers on Nanga Parbat, the Chungpar. Details of the Chungpar were recorded in a sketch by Schlagintweit and by Drew (1875). The Rakhiot was not visited by Schlagintweit. Pillewizer measured a recession rate of 22.5 m per year from 1934 to 1985, and Gardner (1986) an advance of 5 m per year from 1954 to 1985. A photogrammetric resurvey of the Chungpar by Kick (1960) gave a recession of 0–4 m per year depending on location along the ice margin.

The glaciers of the Karakoram are unusually susceptible to surging (Figure 8.10). A good deal of attention was attracted to this characteristic by the flood hazard created by the Chong Kumdum glacier damming the Shyok valley in 1928 (Ludlow 1929–30). It was known that great floods had been caused by the collapse of ice dams in the past and Mason (1930) brought together details of Indus floods since 1780 and also took up the question of which glaciers were prone to repeated rapid advance. He was inclined to assume that all the ice-dammed lakes were the outcome of such advances. Hewitt (1969) questioned whether, in fact, this always was the case. Some of the glaciers mentioned by Mason, such as the Aktash which advanced 2.5 km in five weeks in 1935–6 and the Hasanabad which advanced 9.7 km in ten weeks in 1890 (according to Hayden 1907 – an average speed of 6 m per hour) certainly do surge. So do the Garumbar, Yengutsa, Kutiah and Sultan Chhusku which are amongst those listed by Mayewski and Jeschke. None of these is of much value for building up a picture of climatically controlled glacier fluctuations. On the other hand, some of the flooding could have been caused by glaciers advancing essentially in response to climatic

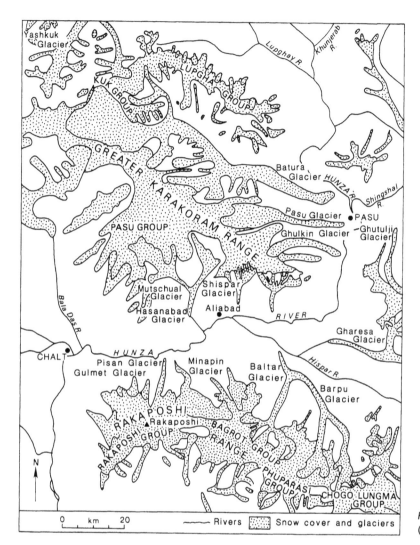

Figure 8.10 The Karakoram glaciers
(*Source:* after Goudie et al. 1984)

fluctuations. Furthermore, it is significant that over a period of over four decades in the middle of the twentieth century no major glacier dams are known to have formed (Hewitt 1982), until, in 1978, a lake with an area of 6 km² was held back by ice in the upper Yarkand.

The fluctuations of the Chogo Lungma, which ranks amongst the largest icestreams in central Asia, are known in more detail than is usually the case because of its nearness to the village of Arandu. Citing the Shogar chronicle which, as he mentioned, has not yet received the critical attention of historians, Kick (1962) argued that the village had been continuously occupied since

the twelfth or thirteenth century. A watchtower that survived into the nineteenth century is believed to have dated back to the sixteenth century. According to local inhabitants the glacier was 2 km from the village at the beginning of the nineteenth century. Vigne wrote (1842) that it was advancing in 1835, and in 1861 Godwin-Austen (1864) found the terminus only 400 m from Arendu with the tongue abutting 'against the mountainside the whole way down'. Over the next two years the ice retreated, but when the Workmans (1905) and Oestreich went up the valley in 1902 the terminus was still only 315 m from the village (Figure 8.9).

However, the ice was thinning and the glacier had melted back leaving a natural routeway for 30 km between the valley side and a massive lateral moraine. It should be noted that the diagram given by Mayewski and Jeschke (1979) does not accord with Kick's account. It seems likely that the glacier had attained its greatest volume between 1865 and 1900 but that the lower part of the tongue was protected by debris so that the retreat of the terminus was much delayed. Between 1900 and 1913 the front advanced again for 200 m to bring the ice only 110 m from the village to the alarm of the local people, but it advanced no further. By 1954 the terminus had retreated about 100 m, the ice was 30–60 m below the crest of the lateral moraine, and its velocity was half what it had been in 1903. Between 1954 and 1970 the front retreated 200 m and between 1970 and 1979 a further 70 m (Best *et al.* 1981).

A similar record of nineteenth-century advance comes from the 58-km long Biafo Glacier of the central Karakoram. It enters the Braldu valley at right angles and terminates opposite a prominent rock buttress. Godwin-Austen found the ice in 1861 wedged against the rock buttress, completely blocking the Braldu valley. In early 1892 it was a quarter of a mile back and Conway (1894) reported that it retreated a further quarter of a mile in August of that year. As the frontal moraine it left behind was vegetated, wasting must have started several years earlier. Seven years later the Workmans (1908) found the tongue hardly reaching into the Braldu valley. It advanced again by 1902 according to Pfannl, pushing the Braldu river across into a narrow bed, but by 1905 the snout was low and fissured and retreat had recommenced (Mason 1930). It may be worth mentioning that Neve (1907), Longstaff (1910) and also Conway and Godwin-Austen reported that glaciers in Kashmir were advancing in the first few years of the twentieth century. However, when the Workmans returned to the Biafo in 1908 they found it in much the same relatively shrunken state as that in which they had seen it in 1899. On the other hand, de Filippi (1912, 1932), on the Duke of Abruzzi's expedition of 1909, wrote of

the marvellous spectacle of the Biafo icestream, 300 feet high. . . . In its invasion of the Braldoh valley it has pushed the river up against the left wall of the valley. . . . It was only on our return journey when we ascended the left side

of the Braldoh valley on the Skoro La road when we clearly saw the river flowing under the open sky through a narrow gap between the valley wall and the steep front of the glacier. The latter showed no trace of frontal moraine.

In 1922, Featherstone (1926) found the glacier still right up against the river, forcing it into the valley side and causing great landslides. Desio (1930) reported the front was 180 m from the Braldu in May 1929. Auden (1935) supervised the making of a map of the Biafo snout on 1 and 2 June 1933, which shows the snout double-nosed and 134 m from the buttress at the closest. He was much impressed by the marked variations in the position of the ice front from one season to another and expressed some uncertainty as to whether there had been much change from 1861 until the time of his own visit. At the same time he pointed to recent thinning, shown by the exposure of a band of bare unconsolidated debris, 30 m wide and diminishing in width upglacier, that had also affected small hanging glaciers on Skoro La (5070 m) that were 150 m higher in July 1933 than they had been in July 1909. He noted that although the Biafo was no shorter than in 1861 it was narrower and thinner than it had been in 1909 and 'many of the lateral and hanging glaciers in Baltistan appear to be in retreat'. Hewitt (1967) observed the Biafo front continuously from September 1961 to May 1962, when it stood in the middle of the Braldu valley, and recorded fluctuations during that winter of up to 20 m, scarcely enough to invalidate observations of the kind mentioned above. There seems to be enough evidence to indicate that the Biafo and its neighbours were in advanced positions in the middle decades of the nineteenth century and then retreated. A sharp advance in the first few years of the twentieth century was followed by retreat, but over the last 120 years the changes have been quite modest.

The Batura, on the northern side of the Karakoram, flows NNW to SSE for 59 km towards the Hunza river. To the south lies the great wall of mountains topped by Batura Mustagh (7795 m). The ridge to the north has peaks ranging from 4000 to 5000 m and none over 6000 m. The total area of the Batura was 285 km^2 in 1975 and the accumulation area ratio 0.5, much of the ice and snow being supplied by avalanches. 'Nowhere else have I heard such uninterrupted avalanche thunder as in this part of the Karakoram. The incessant row provides the solution of the problem of how this long valley with its small névé fields gets its fodder', wrote Visser (1934). Furthermore, ablation is checked by moraine covering much of the tongue. The terminus was in full view of the ancient Silk Road, just as it is now from the Karakoram Highway linking Pakistan and China. Like the Biafo, its advances are constrained by the opposing valley wall. According to the Batura Glacier Investigation Group led by Shi Yafeng (1979), which worked there in 1974–5, about 200 years ago the Batura advanced into the Hunza valley, penetrating 2 km below the present terminus and overrunning five channels built across Holocene moraines by the people of Pasu village, so that it seems that the Batura reached a position comparable with an earlier Holocene advance and then retreated.

The Batura still reached down onto the floor of the Hunza valley from 1880 to 1930 (Mason 1930). On a map he made in 1885, Woodthorpe noted that 'the glacier reached the Hunza river which passes round the snout'. The ice had withdrawn somewhat by 1909 when Egerton saw it but reported that it still threatened to block the valley. It was described by the Vissers in 1925 as pushing its way across the river, 'blocking the whole width of the valley'. In the 1970s the inhabitants of Passu and Khailur told Chinese glaciologists (Zhang Xiangsong and Shi Yafeng 1980) that in the autumn of a year between 1910 and 1930 the glacier advanced to block the river, though the water could still pass under the ice, and that in the spring of the following year it withdrew. Between 1944 and the visit of a German–Austrian expedition in 1954 the terminus retreated 300 m, and by the time it was mapped in 1966 on a scale of 1:20,000, in connection with the construction of the Karakoram Highway, it had gone back another 527 m. In 1966 the ice front began to move forward again and by 1974, when the Batura Glacier Investigation Group (1979) made a photogrammetric survey of the terminus, the central ice cliff had advanced 90 m. It came forward another 10 m the following year and the Investigation Group forecast that the glacier would continue to advance into the 1990s.

The recent histories of the Pasu and Gulkin glaciers, flowing parallel and to the south of the Batura, were investigated by the International Karakoram Project (Goudie et al. 1984). Both were a few hundred metres further forward in 1913, when Mason saw them, than when they were visited by Woodthorpe in 1885. Price Wood found the Pasu had retreated 800 m between 1885 and 1907, so a marked advance must have taken place between 1907 and 1913. Both glaciers retreated in the course of the twentieth century; the Pasu was more or less stationary in the 1970s whereas the Gulkin is known to have advanced at least 200 m between 1974 and 1980.

The fluctuations of the positions of the tongues of the longitudinal glaciers of the Karakoram, apart from those that have surged, have been very modest in relation to their great length. The fragmentary nature of the historical record and the different years and seasons when observations have been made complicates the identification of regional trends and the detection of anomalous behaviour. It is unfortunate that it is the longitudinal glaciers and some of those which surge that have attracted the most attention, rather than the transverse glaciers that are generally more sensitive to climatic fluctuations. Despite these difficulties it is possible to discern important differences between the Himalayan glaciers and those of the Karakoram.

8.2.7 Himalayan summary

Throughout the Himalaya, glacial retreat has predominated since about 1880 though, as moraine patterns indicate, this retreat has not been continuous (Figure 8.11). Current small fluctuations can be identified by careful monitoring of sensitive glaciers such as that carried out on 14 small glaciers in eastern Nepal between 1970 and 1976 (Ikegami and Inoue 1978). Six of them were found to be retreating, three advancing and one fluctuated irregularly (Fushimi and Ohata 1980, Fushimi et al. 1981). The Gyajo glacier advanced in 1970, retreated, and then in 1976 advanced again, forming a push moraine that partly obliterated the 1970 moraine. Ageta et al. (1980) and Ageta and Kadota (1992) showed that temperature plays the decisive role in the mass balance of the small, debris-free glaciers of Nepal, summer temperature determining whether the monsoon precipitation falls as rain or as snow. Between 1978 and 1989 the terminus of Glacier AX010 in the Shrong region of east Nepal retreated about 30 m (Kadota and Ageta 1992) and in the succeeding decade 80 m (Fujita et al. 2001).

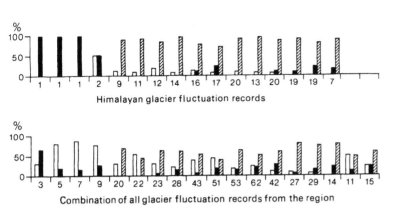

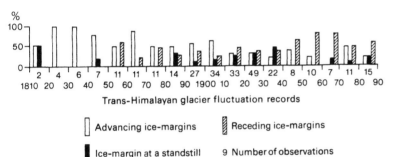

Figure 8.11 Comparison of the fluctuations of Himalayan glacier termini with those of the Trans-Himalaya since the beginning of the nineteenth century. Note that the majority of the observations were made between 1850 and 1960 (*Source*: after Mayewski *et al.* 1980)

The termini of Karakoram glaciers are known to have retreated from advanced positions reached in the mid or late nineteenth century. Between 1920 and 1940, when the density of the record was greater than before or since, more glaciers were known to be stationary or advancing than were known to be retreating. Retreat in the subsequent period was halted or reversed in the 1970s.

The advances of the Karakoram glaciers between 1890 and 1910 have been associated by Mayewski *et al.* (1980) with strengthened monsoonal airflow. Precipitation over lowland India was lower rather than higher than average in this period. In the Karakoram there were very few precipitation gauges but it is known that lake levels in nearby Tibet were high in about 1900 and snowfall may have been heavy (de Terra and Hutchinson 1934).

8.3 CHINA

The major glacierized areas of China are in the northwest of the country, the largest ice masses being in the Tien Shan, Kunlun Shan (Plate 8.4) and Himalaya (Figure 8.12). The mean annual precipitation of 300–1000 mm on the mountain ranges, is very much higher than in the intervening basins, where it is no more than a centimetre or two. The snowline increases in elevation towards the south from 2900–3400 m in the Altai, 3600–4400 m in the Tien Shan, and 4400–5100 m in the Qilian Shan, to 4600–6200 m in the Kunlun Shan and northern Himalaya. The glaciers of the eastern Hengtuan Shan and eastern Qilian Shan are fed by the southeasterly Pacific monsoon and are of a maritime temperate type. Those in the south, the Himalayan, Nyinchintangla and western Hengtuan Shan glaciers, face the southwest monsoon and are cold continental, as are those of the Karakoram, Kunlun Shan, Tien Shan, Altai and western Qilian Shan, all of which are fed by snow brought by westerly airstreams.

Research on glaciers and climatic history in China has been centred since 1958 at the Lanzhou Institute of Glaciology and Geocryology (Shi Yafeng *et al.* 1981, Zhang Xiangsong *et al.* 1980, 1981, Xie Zichu 1992) and at a glaciological research station at the head of the

Plate 8.4 Northwestern Tibet and the Kun Lun from the Shuttle in October 1984. In this photograph of the Chopanglik region, Hotien and Min-Feng at the southern edge of the Tarim basin are in the top left-hand corner of the Plateau of Tibet with lakes Aksu Chin in the bottom left, Ya-hsi-Chr-Hu at top right and Tsu-chia-Brh-Hu bottom right
Photo: NASA Astronaut

Urumqi valley in the Tien Shan (Hsieh Tze-chu and Fei Ching-shen 1980, Shi Yafeng *et al.* 1981, Shi Yafeng and Zhang Xiangsong 1984). By 1992, in spite of the extremely difficult and remote terrain in which they are situated, about 80 per cent of the Chinese glaciers had been inventoried. It was estimated that there were 45,375 glaciers with a total area of 52,735 km² (Wang Zongtai and Yang Huian 1992). In some of the most difficult regions information was extracted from maps based on air photographs and Landsat images (Table 8.4). Work remains to be done in the Himalaya and southeast Tibet (Xie Zichu 1992).

Glaciers are particularly concentrated in the Mt Qogir region of the Karakoram, the Mt Hantengri–

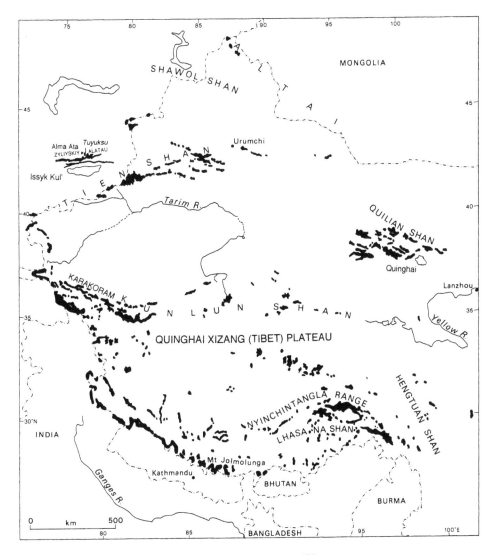

Figure 8.12 Glacierized areas in China (Source: after Shi Yafeng and Wang Jintai 1979)

Tomur in the Tien Shan, the Kunlun Peak area, the eastern peak area of the Nyinchintangla[20] and the Mt Everest (Qomolangma) area of the Himalaya (Wang Zongtai and Yang Huian 1992). This distribution is significant because the higher the ratio of ice cover to total area, the higher its potential as a meltwater source, and the greater the problems caused by glacier floods and debris flows.

Irrigation in the oases at the foot of the Tien Shan, Kunlun Shan and Qilian Shan depends on water from

20 Difficulties arise because Nyinchintangla has been transliterated from Chinese into English in several different ways in the literature. Thus Ageta et al. (1991) give it as 'Nyainquentanglha' but Wang Zongtai and Yang Huian (1992) give it as 'Nyainqêntanglha' while earlier authors gave as spelt on Figure 8.12.

Table 8.4 Distribution of glaciers in China

Region[a]	No. of glaciers	Ice cover km^2	Snowline m asl
Altai	403	280	2800–3350
Tarbagaty	21	17	3310–3380
Tien Shan	9128	9257	3620–4450
Pamir[b]	1823	2623	4400–5860
Karakoram[c]	1926	4769	4900–6010
Kunlun	7615	12,263	4500–6080
Altun	229	266	4800–5360
Qilian	2815	1930	4300–5160
Qiangtang Plateau[d]	2283	3355	5700–6200
Taggula Range[e]	1570	2206	5300–5800
Gandise Range[e]	2982	1615	5000–6200
Nyinchintangla[e]	3900	7536	4200–5700
Himalaya[f]	9000	11,000	5400–6250
Hengtuan	1680	1618	4600–5100

Source: after Wang Zongtai and Yang Huian 1992
Notes:
a Only those parts of each region within Chinese territory are included
b Does not include data from the territory disputed with the former USSR
c Does not include data from the source of the Shiyok River
d Does not include data from eastern tributary of Indus or source of Shiquan River
e Some data inventoried, some estimated from air photographic maps
f Estimated from air photographic maps and Landsat images

snowfields and glaciers. Meltwater from the Qilian Shan flowing through the Kansu Corridor is used for supporting fisheries as well as for irrigation. Monitoring of glacier behaviour in these regions is considered all the more important because groundwater levels have been falling in the Urumqi region and environmental degradation in Kansu has reduced surface and groundwater supplies (Wang Wenying 1983). It is little wonder that work at the Lanzhou Institute was first focused on glaciers as sources of water for the arid regions.

In recent years a greater emphasis has been put on glacier variations in relation to climatic and environmental change (Xie Zichu 1992). It has been realized that the mass balance of glaciers in monsoon regions, where the main periods of accumulation and ablation both occur in the summer, differs from regions where accumulation is concentrated in winter and ablation in summer. Co-operation with scientists from Japan, America, Switzerland, Russia and other countries has been increasing. Ice cores extracted from the Dunde and Guliya icecaps (e.g. Yao Tandong *et al.* 1991) and boreholes made in Glacier No. 1, in the Urumqi headwaters (Echelmeyer and Wang Zhongxiang 1987), have

thrown light on climatic history (see Chapter 15, pp. 554–6) and glacier mechanisms. Hydrometeorological investigations have continued to concentrate especially on the Urumqi basin (Higuchi and Xie 1989). Forecasting of snowmelt and runoff, making use of satellite images, has allowed continuous adjustment to the volume of water stored in large hydroelectric reservoirs in the upper reaches of the Yellow River (Xie Zichu 1992). The formation of ice-dammed lakes in the Himalaya, Karakoram and the mountains of southeastern Tibet has been monitored, and an inventory of dangerous moraine-dammed lakes has been completed (Xie Zichu 1992, Ding Yongjian and Liu Jingshi 1992), notably in the basin of the Yarkand River (Desio 1930). As many as 134 old shorelines there record the levels of lakes formed when the Keyajir entered and blocked the Kelequin valley (Zhang Xiansong 1992) causing disasters comparable with those in the Val de Bagnes in Switzerland (Chapter 6).

According to a 230-year record of droughts and floods in the Yellow River basin, the spell with the most severe drought occurred during the eleven years 1922–32, since when runoff has been greater than the

Plate 8.5 The growth of lichen on rocks is being measured in order to calibrate a lichen growth curve for the Urumqui region
Photo: A. T. Grove

long-term average. A period of anomalously high run-off was around 1820.

Ice cores from the Dunde Icecap provide incontrovertible evidence that the Little Ice Age affected China (Chapter 16). Modern investigations of the history of Chinese climate, based on documentary data, are filling in the detail. They reveal regional variations and show that from the late fourteenth to the late eighteenth century temperature has oscillated on the decadal scale. They have identified three long cold periods: in the mid fourteenth century, between the 1550s and 1690s, and in the nineteenth century. The coldest decade was around 1650, the second coldest in the nineteenth century. Mean annual temperature has varied from the coldest phase in the Little Ice Age to the warmest decades of the twentieth century through more than

1 °C. No part of China has been immune from fluctuations in precipitation or the incidence of lower temperatures, frost and snow, except perhaps the far western margin of Tibet (see Chapter 14, pp. 384–94).

Moraine dating has so far attracted less attention in China than in regions northwest of the Himalaya discussed in the previous section, though work on unravelling the historical record represented by the many thousands of moraine systems has begun. In south Tibet (Xizang), the moraines of the Kanbukoa glacier near Yadong have been dated by lichenometry as marking advanced glacier positions about 1818, 1871 and 1885.[21] A set of Little Ice Age moraines in front of Glacier No. 1 at the head of the Urumqi valley and close to the glaciological research station has yet to be dated (Plate 8.5).[22] The principal moraines were 600,

21 Calibration of the growth curve depends on the growth of *Rhizocarpon geographicum*, since boulders free of lichen were exposed in a mudflow in 1940.
22 Glacier No. 1, in the headwaters of the Urumqi Valley in Chinese Tien Shan, is a small valley glacier, consisting of two branches facing northeast, and extending from 4486 to 3740 m asl. It is a predominantly cold glacier surrounded by continuous permafrost, though it reaches melting point over wide areas of the bed. Accumulation and ablation both take place primarily during the warm season, the formation of superimposed ice being very important (Haeberli *et al.* 1994: 65).

Plate 8.6 Glacier No. 1, the most extensively studied glacier in China, August 1985
Photo: J. Stevenson

500 and 300 m from the tongue in 1984. Since the Little Ice Age maximum in the eighteenth century, the area of the glaciers in the Urumqi valley has decreased by 44 per cent and the snowline has risen by over 80 m. The negative balance was much greater in the 1980s than in the preceding two decades (Shi Yafeng and Zhang Xiangsong 1984, Shi Yafeng and Ren Jiawen 1990) (Plate 8.6).

The Gongga Mountains, extending north–south between the southeastern edge of the Tibetan Plateau and the Sichuan basin, carry some seventy-four glaciers. First described by Heim (1936), they are small enough to be sensitive to short period, low amplitude changes in climate (Su Zhen *et al.* 1992). Precipitation, brought by the southeast and southwest monsoons, is heavier on the east than the west side of the mountains. Snowlines range from 4800 to 5000 m on the east, where the lower slopes are subtropical, and from 5000 to 5200 m on the more arid western side.

Moraines, assumed from their fresh appearance to belong to the Little Ice Age, lie 2–3 km below the glacier termini on the eastern side of the mountains, but only 0.2–1.5 km downvalley on the western side. The 13-km long Hailuogou Glacier on the eastern side is probably temperate, the near surface ice in the ablation area having been at the melting point when it was observed in 1990. Decayed wood extracted from the outermost of three fresh Hailuogou moraines has been dated to 608 ± 70 BP (AD 1300–1400) (Zhou *et al.* 1991, Li and Feng 1984) and wood from the outermost moraines of the Da Gongba and Xiao Gongba glaciers, which were once confluent, to 620 ± 40 BP (AD 1300–1400) and 440 ± 50 BP (AD 1430–1490) (Figure 8.13).

Heim (1936) found the Hailuogou only 20 m from its nearest end moraine and, referring to a 1:200,000 map made by Imhof in 1930, concluded that many of the glaciers had either been stationary or had advanced

Plate 8.7 Holocene and Little Ice Age moraines in front of an unnamed glacier, near Glacier No. I, August 1985
Photo: A. T. Grove

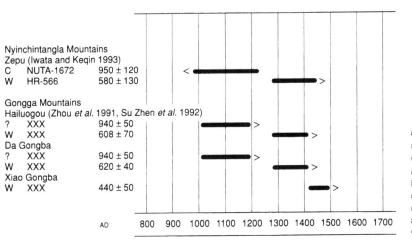

Figure 8.13 Key radiocarbon dates relating to Little Ice Age advances in China (*Source*: from Grove 2001b)
Notes: The numbers in the margin, where known, give the laboratory identification number and the ^{14}C age. Horizontal lines represent the calibrated ages. > = Advance after date; < = Advance before date; C = Charcoal; W = Wood

in the early twentieth century. From 1930 to 1966, retreat was dominant on both the eastern and western slopes, the Hailuogou retiring 1150 m from its position as described by Heim, and the Da Gomba 175 m. Between 1966 and 1981, the mean annual retreat of the Hailuogou diminished from 32 to 13 m, while some other glaciers, including the Da Gomba, were stationary or even advanced. After 1980, retreat was resumed and accelerated, in the case of the Hailuogou from 13 to 19 m per year.

The overall retreat since the mid twentieth century is believed to have been caused by a rise in temperature of between half and one degree (Shi Yafeng and Ren Jiawen 1990). The tongues of the longest glaciers have retreated several hundreds of metres and some several kilometres. The Musart in the Tien Shan retreated 750 m between 1909 and 1959, the Azda in southeast Xizang 700 m between 1933 and 1973, and the Skyany, northeast of Qogir Feng (better known as K2, 8611 m) retreated 5.25 km between 1937 and 1968. Retreats in the interior of Tibet were less than elsewhere, and in western Tibet, western Qilian Shan and eastern Tien Shan glacier fronts generally withdrew less than 10 m per year. Some debris-covered tongues were stable. The Rongbuk on the north side of Everest, surveyed in 1921 and again in 1959 and 1966, remained the same length but the ice had thinned. From 1956 to 1976 all 22 glaciers observed in the Qilian Shan were retreating; 8 termini in the eastern sector withdrew an average of 12.5 to 22.5 m a year, and 14 in the western section at only 1.1 to 2.7 m a year. Shrinkage slowed towards the end of the period (Shi Yafeng *et al.* 1980). The fragmentary data from the Tien Shan suggest a similar pattern there. The glaciers on Mount Mystagh and Mount Kungur to the east of the Pamirs withdrew at a rate of 1.7 to 3.7 m a year between 1907 and 1960. The Azar in the southeast retreated rapidly in the 1950s. In fact, wherever information is available, retreat seems to have been dominant from the 1930s until 1960.

Between 1950 and 1990, when annual mean temperatures in China as a whole increased by 0.2 to 0.4 °C, the north was generally warmer but there was some cooling in the southwest. The changes in temperature from year to year were most marked in the winter season and the annual range of variation was three or four times the range of variation of global mean annual values (Hulme *et al.* 1994).

After 1960 the retreat slowed or was halted for a few years. In the Qilian Shan average temperatures were 0.8 to 1.3 °C lower in the decade 1967 to 1976 than in the previous decade. Precipitation at three stations above 3000 m remained little changed, increasing from 345 to 355 mm. According to Shi Yafeng and Wang Jingtai (1979) the fall in temperature and a rise in precipitation extended all over western China. On the Tibetan Plateau, temperatures in the 1960s were everywhere 0.7 °C lower than in the 1950s and precipitation increased by about 5 per cent.

The retreat of the Musart Glacier slowed to 2 m a year between 1962 and 1978. Glaciers in the Nyinchintangla Range were still retreating but a kinematic wave passing down the tongue of the Aza Glacier indicated an advance was to be expected. Ten glaciers in the western part, east of the Yarkand River, the Chuanschuikon, were advancing at an average rate of 15.5 m a year from 1968 to 1976. During the 1960s montane lakes such as the Yangchayung rose and flooded roads built in the previous decade. Many of the glaciers of the Amne Machin Mountains, a 200-km long extension of the Kunlun Shan, advanced between 1966 and 1981. In 1981, of the forty glaciers in this range fifteen were known to be advancing and twenty-three were stationary or had variations within the limits of the photogrammetric techniques used for the glacier inventory (Wang Wenying 1983). In the Qilian Shan, mass balances on four glaciers were positive from 1974 to 1977 and though many tongues were retreating, rates of recession were reduced. Thus the Shui Guan Ho No. 4 glacier in the eastern Qilian Shan receded by 16 m a year between 1956 and 1975 but only 6 m a year between 1976 and 1978, and by 1981 two small glaciers were advancing. Comparison of air photographs of the 1950s and 1960s with 1980s' Landsat imagery reveals that in the 1970s glaciers were advancing not only in the Qilian Shan and other eastern areas but also in the Alai Pamir, Karakoram and Himalaya. In the Tien Shan over the period 1962–78, of thirty-eight glaciers observed, 13 per cent were advancing, 26 per cent were stationary and 61 per cent were retreating. Glacier No. 1 showed negative net balances predominating in the late 1960s and early 1970s followed by positive balances more recently. The Keqkar Glacier advanced in 1980 onto grass-covered moraine (Ersi 1985). The greatest advances of all measured by Chinese glaciologists in recent years were in the Karakoram, on the frontier with Pakistan.

The situation in the 1960s and 1970s was summarized by Wang Zongtai (1989) as in Table 8.5. Over the whole of China, 66 per cent of the glaciers for which measurements were obtained were retreating, 21 per cent were stationary and 13 per cent were known to have advanced. It is difficult to be sure that the proportion of glaciers in each range known to be retreating,

Table 8.5 Number of monitored glaciers in China, retreating, advancing and stationary, by region in the 1960s–1970s

Range	Total	Advancing	Retreating	Stationary
Altai	6	1	5	0
Qilian	28	0	24	4
Tien Shan	61	4	41	16
Pamir	16	4	10	2
Kunkun	38	10	17	11
Karakoram	11	1	7	3
Himalaya	6	1	4	1
Nyinchintangla	8	2	6	0
Gandise	2	0	2	0
Tanggula	1	0	1	0
Qiangtan Plateau	1	0	1	0

Source: from Wang Zongtai 1989, table 2

advancing or stationary was representative, or how the 178 glaciers specified in the table were selected from the 31,975 inventoried when Wang Zongtai was writing. He thought that retreat had slowed after the early 1960s, became minimal in the late 1970s, and then increased once more in the early 1980s. Recession was especially marked in the Hengtuashan, east Nyinchintangla, and A'nyamaqenshan, all of which are affected by the southwest monsoon and have relatively high temperatures and precipitation. On the Guxiang, for example, east of Nyinchintangla, the mean annual temperature is –4 °C, and annual precipitation 2600 mm at the snowline. Recession was less in the Tien Shan, Alai Pamir, Karakoram and western Kunlun where the climate is dominated by westerlies, the annual mean temperature is around –6 °C to –11 °C, and 60 to 85 per cent of the average annual precipitation of 500–800 mm at the snowline falls in summer. In the headwaters of the Yarkand River, in the Karakoram, seven of the nine glaciers observed retreated at a rate of 10–35 m per year between 1976 and 1985. Glacier No. 8 in the Tien Shan retreated an average of 6.7 m per year from 1962 to 1986. In the Nyinchintangla the annual velocity of the Zapu, recorded by the width of ogives, declined from an average of 125 m in 1978–89 to an average of 119 m in 1985–9, in line with recession of the Kaqin Glacier observed in the same area (Ageta *et al.* 1991). The Qiangtang Plateau, the western Qilian and the southern Kunlun, all with cold desert continental climates, have more stationary glaciers than other regions. Annual precipitation is only of the order

of 300–500 mm, and mean annual temperature as low as –13.4 °C, as on the Guozacuo Glacier on the south of the Kunlun. The margin of the Chongce Icecap and the termini of the Guozha Glacier and an unnamed glacier adjacent to it in the west Kunlun, changed little from 1970 to 1987, though the southern margin of the icecap withdrew 1–2 m between July 1987 and August 1988 (Chen *et al.* 1989). However the tongue of the Chongce Glacier was found to have retreated 420 m (±14) since 1971. On the north slope of the Gangdiseshan, on the southern edge of the Qiangtang Plateau, the glaciers retreated no more than a metre between 1907 and 1976 (Wang Zongtai 1989).

Over the last two decades mass balance studies have increased in China. A joint Chinese–Japanese group has worked on the Tibetan Plateau since 1989 (Seko *et al.* 1994, Pu and Yao 1994). Observations have been made on both the Da Dongkemadi and Xiao Dongkemadi in the Tanggula Mountains in the middle of the Plateau, and also on the Meikuang Glacier in the eastern Kunlun in the north. The mass balance values found up to 1992/3 were all positive. In 1993 the large seasonal amplitude of air temperature and strong solar radiation affected the Xiao Dongkemadi causing considerable internal accumulation, superimposed ice being formed during the main ablation period. In 1992/3 the mass balance on the Tanggula was higher than that in the east Kunlun (Pu and Yao 1994).

The mass balance of Glacier No. 1, in the Tien Shan (Figure 8.2), which was negative in the 1980s (Shi

Yafeng and Ren Jiawen 1990) and 1990/1 shifted to positive in 1991/2 and back to negative in 1992/3. The Kangwure Glacier (28°27'17"N, 85°45'E) on the north side of Mt Xixiabangma, in the Himalaya, had a large negative balance between 1991 and 1993, which is attributed to low precipitation (Liu *et al.* 1996).

A survey of recent climate change in Asia outside the tropics and east of 70°E found that between 1880 and 1900 temperatures rose by about 0.5 °C and precipitation declined to about 90 per cent of the 1950–80 mean. In the course of the twentieth century, temperatures rose unsteadily to peak around 1990, while precipitation was about 5 per cent higher in the second half of the century than in the first half (Hulme *et al.* 1994).

8.4 THE ASIAN AND ALPINE LITTLE ICE AGE RECORDS COMPARED

Throughout the vast extent of Asia, from the Caucasus to China, the ages of moraines, wherever they have been dated, indicate that glaciers reached advanced positions in the fourteenth century. The dates of succeeding intermittent advances vary from one region to another. In China, the Tien Shan and Caucasus, glaciers advanced in the late seventeenth century when, as will appear in Chapter 14, cooling was experienced progressively further east. In most of the mountain ranges there is evidence of glacial advances lasting a decade or two in the early, mid and again in the late nineteenth century. Recession was dominant in the twentieth century, except for a decade or two around 1970. It is always tempting to seek for universals and they are easier to promote when the basic data are imprecise, as they often are in Asia, than when they are firmly established. Nevertheless it should be recognized that the main features of the sequence of advances and retreats of the Asian glaciers, so far as they are known, correspond with those of Alpine glaciers in Europe and were presumably associated with variations in temperature and in the circulation of the atmosphere on a hemispheric scale.

9

NORTH AMERICA

Glaciers are scattered from 38°15′N in southern California to the arctic islands of northern Canada, from the Colorado Front Range and the mountains of Wyoming to the maritime coastal mountains of the northwestern United States, and from Alaska and the Yukon to the eastern Canadian Arctic (Figure 9.1). The many thousands of glaciers include representatives of nearly all the topographical and physical types. Tiny glaciers in the Sierra Nevada, none covering more than 2 km², nestle in deep cirques or beneath north-facing cliffs, well below the orographic snowline (Raub *et al.* 1980). In 1975 W. O. Field calculated that ice in the Pacific Mountain systems of Alaska and the Yukon covered over 100,000 km², some 18 per cent of the glacierized area of the northern hemisphere apart from Greenland.

The earliest recorded observations of North American glaciers were made along the southern part of the Alaskan coast where La Pérouse gave an account of Lituya Bay in 1787 and Captain George Vancouver described and mapped Glacier Bay in 1794. The longest documented records of terminal positions begin in the 1880s when glaciological studies were initiated by outstanding scientists such as Reid (e.g. 1897, 1915), Russell (1897, 1898) and Tarr and Martin (1914). Valuable records were left by government survey teams and boundary commissions as well as by individual observers. In 1907 A. O. Wheeler founded the Canadian Alpine Club, with glacier observations as one of its objects. His photographs of glaciers in Banff National Park and elsewhere are important sources for

tracing the fluctuations of the Drummond, Peyto and other glaciers over the last century (Wheeler 1931).

Apart from Cooper's studies in the Glacier Bay area, American glacial studies languished until Field decided in 1926 to take up Reid's investigations of the 1890s. He joined the staff of the American Geographical Society in 1940 and for three decades his office was an informal headquarters for glaciologists. It became more formal when a World Data Centre for Glaciology was established there.[1] Many key glaciological investigations were initiated by the Geographical Society and the Geological Survey.

Since the 1940s, research has proliferated and gained in sophistication, much of the impetus coming from universities and centres attached to them, notably the Quaternary Research Centre at Seattle, the Institute of Arctic and Alpine Research at Boulder, and the Institute of Polar Studies at Columbus, Ohio.

Wheeler's series of observations of the Yoho and other glaciers between 1901 and 1931 were exceptional. For the most part, regular observations did not begin until two or three decades later, as a perusal of the Unesco volumes on glacier fluctuations reveals, where the bulk of the American entries appear for the first time in the 1970s. Many of the glaciers listed in these entries were selected for their accessibility and are not necessarily representative. Only a small proportion of the North American glaciers have been named, let alone having their characteristics recorded scientifically.

In Canada, though the compilation of a glacier inventory was under way by 1968, no published

---

1 The Data Centre was eventually moved to the US Geological Survey at Tacoma.

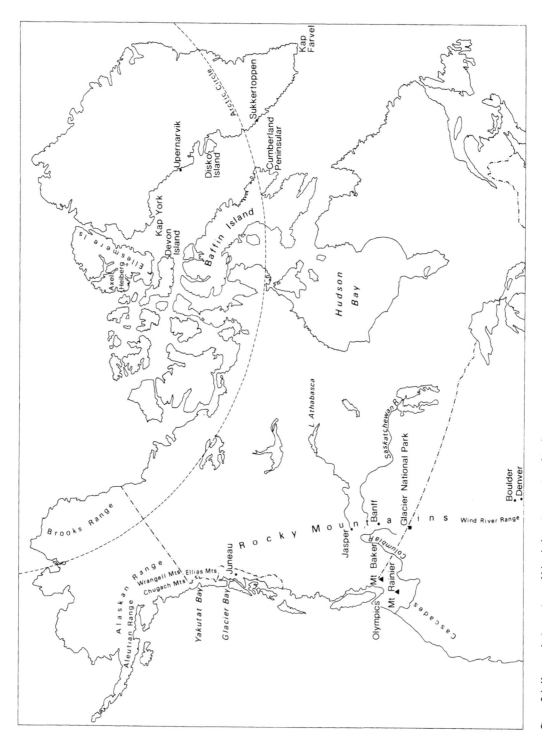

Figure 9.1 Key to glacier regions of North America mentioned in the text

inventories were available for many years (Luckman 1990). Topographical maps of the whole country are available on a scale of 250,000 but only a selection of the 1:50,000 sheets have been completed and the quality of the coverage varies considerably. By 1980, 33,000 glaciers had been identified out of an estimated total of about 100,000, and complete data as requested by Unesco were available for only 1500 (Ommanney 1980). The published inventory data for the USA in 1980 covered only the Sierra Nevada (Raub *et al.* 1980). Photographic collections, more especially that assembled by Austin Post at Tacoma, provide invaluable resources for glacier studies. The Canadian Water Resources board adopted methods of assessing volumetric changes of glaciers using ground-based photogrammetry as well as aerial survey. In recent decades detailed mass balance studies have been under way in both the USA and Canada (e.g. Müller 1962, Meier and Tangborn 1965, Østrem 1966, Letréguilly 1988, Letréguilly and Reynaud 1989, Brezel *et al.* 1992, McCabe and Fountain 1995, Haeberli *et al.* 1996, Luckman 1998).

Moraine sequences dated by radiocarbon, tree-ring and lichenometric methods, abound in North America (e.g. Bray and Struik 1963, Osborn and Taylor 1975, Luckman 1988). Methodologies were at first imprecise; Cooper (1937), for instance, counted the growth rings of two or three trees in the Glacier Bay area and on this basis made an estimate of the time of recession there. Lawrence's (1950) studies in southeast Alaska were influential in extending the use of dendrochronology to date moraines, so that by 1967 Viereck was able to assemble data on fifty-one glacier forefields. In the Canadian Rockies dates are now available from some forty glacier forefields (Luckman 1993).

Heusser (1956) had already come to the conclusion, on a basis of the North American data then available, that conditions had favoured glacial expansion in the mid sixteenth, early and mid seventeenth, early, mid and late eighteenth centuries, and in some regions in the early twentieth century. In each century advances were greater in some areas than others and not all areas were affected at any one time. It is now quite clear that glacial expansion took place within the last few centuries in both maritime and continental environments. An attempt is made here to present some of the data about the regional pattern of fluctuations over North America in recent centuries, concentrating on those areas for which most information is available (Figure 9.1).

9.1 THE CASCADES, OLYMPICS AND SOUTHERN BRITISH COLUMBIA

These mountains close to the Pacific coast receive a very high annual snowfall and support well over a thousand glaciers, most of them quite small but together constituting a very high proportion of the total ice in the conterminous USA. The glaciers of Mt Rainier (4392 m) are the largest. This ancient volcano, the focus of one of the oldest National Parks in the country, has been visited by naturalists and scientists since the mid nineteenth century. Records of the position of the terminus of the Nisqually (Figure 9.2), the largest of Rainier's glaciers, are the longest in the USA, going back to 1857; its moraines have also attracted much attention (e.g. Harrison 1956a, c, Johnson 1960, Meier 1963, Veatch 1969, Sigafoos and Hendricks 1961, 1972).

In order to date the Nisqually moraines, Porter (1981b) established lichen growth curves on both andesite and granodiorite surfaces of known age, taking care to avoid unusual local conditions near streams or waterfalls or where snow lies for long periods. Measurements of the diameters of lichen thalli were made on fully exposed surfaces of known age, the largest circular or near-circular thalli being measured to the nearest millimetre. Twin curves for the two rock types were obtained, spanning the previous 120 years, the growth rates being faster on the andesite than on the granodiorite. After an initial period of very rapid increase in size, the growth rate of *Rhizocarpon geographicum* thalli was found to diminish progressively. The ages of lichen-bearing surfaces older than 1857 had to be estimated by extrapolation of the curves on the assumption that the decreasing rates inferred from the data persisted into the past. Although this assumption could be invalid, Porter argued that data from the Italian Alps, where the growth rate of *Rhizocarpon geographicum* is similar, indicates that differences in age extrapolated from growth rates plotted as both linear and exponential functions are not significant and that for the first four or five hundred years probably do not exceed the error resulting from field measurement. If

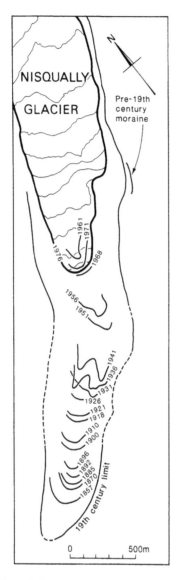

Figure 9.2 Nisqually Glacier frontal positions (*Source:* from Porter 1981b)

such measurements are taken to the nearest milli-metre, the accuracy of values obtained can be taken to range from three to five years for surfaces 120–200 years old, and ten to twenty years for surfaces 200–400 years old. This careful study, with its attention to the degree of uncertainty of the results, appears to be much more accurate than tree-ring dating of moraines or most other lichenometric work previously published (although there is always some uncertainty as to the

exact length of time required for lichen growth to be initiated after the retreat of the ice and the stabilization of the moraine).

Other glaciers on Mt Rainier also have well-marked moraines (Plate 9.1), a number of which have been dated using Porter's (1981b) lichenometric curve supplemented by tree-ring dating for the older moraines (Burbank 1981). Periods of moraine stabilization, when ice fronts receded after readvances, have been dated to 1519–28, 1552–76, 1613–24, 1640–66, 1691–5, 1720 and 1750. The sixteenth-century dates were obtained from tree-ring counts on one set of moraines only, whereas those from the seventeenth century were identified on two sets of moraines by lichenometry as well as tree rings. Burbank (1981, 1982) stressed that the interval between moraine stabilization and the es-tablishment of trees may have been underestimated and the ages of the moraines may consequently be greater than indicated by these dates; certainly the excesis pe-riod, known to have been as much as 100 years, has varied greatly from place to place, so that on these grounds the results might be regarded as less precise than those based on lichen growth. More reliably dated, more recent moraines gave ages of 1768–77, 1823–30, 1857–63, 1880–5, 1902–3, 1912–15 and 1924. The synchronicity of the fluctuations of the glaciers on Mt Rainier as they retreated from their late eighteenth- and early nineteenth-century maxima was taken to suggest a common response to summer ablation conditions. After 1915 the rate of recession more than doubled, the extent of withdrawal varying according to local conditions.

Calculated mass balance variations based on monthly temperature and precipitation records back to 1909 for Longmire on the southwest flank of Mt Rainier, and extended further back by correlation with other Washington stations with longer records, are in agree-ment with observed glacier behaviour since 1850. Com-parison of 3-year running means of net balance at Mt Rainier with dates of moraine stabilization reveals a clear correlation. Confirmation of these results is provided by observations of standstills of the Nisqually Glacier around 1875 and 1900 and a thickening of the ice for several years after 1932 and again after 1945 (e.g. Harrison 1956a, Veatch 1969).

Data from Longmire, as from California, Oregon and Washington, indicate that although temperature

Plate 9.1 Moraines left by the retreat of the Emmons Glacier on Mt Rainier, July 1970
Photo: A. T. Grove

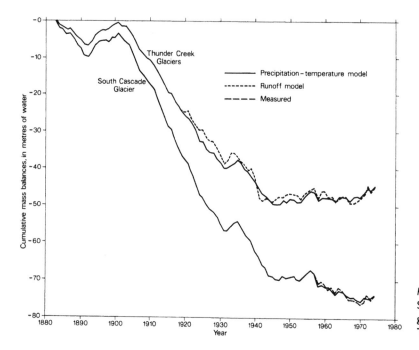

Figure 9.3 Cumulative mass balance for South Cascade and Thunder Creek glaciers, 1885–1975 (*Source*: from Tangborn 1980)

fluctuated, there was no long-term rise in mean annual temperature between 1850 and 1980. On the other hand, the lichenometric data presented by Porter (1981b) and Burbank (1981) suggest that a rise in mean annual temperature of the order of 1 °C took place earlier, between 1770 and 1850, and that this was responsible for initiating and sustaining the major recession of the succeeding 150 years. Dominantly negative mass balances calculated for the period 1850 to 1890 were associated with thinning of the glaciers; retreat of the termini at that time was not great, but the tongues were made susceptible to rapid retreat from 1910 to 1940 when ablation season temperatures rose and precipitation diminished.

The extreme sensitivity of the Olympic glaciers to small climatic changes was demonstrated by Tangborn (1980) who reconstructed mass balance values for the South Cascade Glacier and the neighbouring Thunder Creek Glacier from the late nineteenth century onwards. He calculated that mean summer temperature only 0.5 °C lower or winter accumulation 10 per cent greater

would have resulted in mass balances of the two glaciers remaining continuously positive during the twentieth century (Figure 9.3).

Four moraine sequences in the Dome Peak area (48°20′N, 1°21′W) were investigated by Miller (1969) using tree rings and ash layers for dating (Figure 9.4). He recognized that the true dates of moraine formation were likely to be several decades earlier than tree-ring counts. Glaciers were found to have come further forward in the last few centuries than at any time since the Fraser Glaciation of 10,000 years ago. The outermost moraine of the Chickamin Glacier he dated to the thirteenth century on a basis of trees which 'began growing more than 680 years ago', though the moraine could be older than this.[2] The outermost moraines of the South Cascade Glacier were dated to the sixteenth or seventeenth century and those of the Le Conte and probably the Dana to the sixteenth century. The ratios of the areas of these little glaciers in the 1960s to their areas at their Little Ice Age maximum extensions were similar (Table 9.1).

---

2   The accuracy of tree-ring dating of moraines has been criticized by Bradley (1999) because of the uncertainty of the excesis period and the possibility that the trees dated are not the oldest.

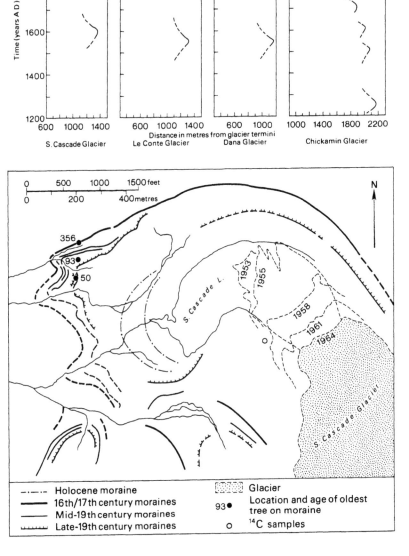

Figure 9.4 Fluctuations of termini of South Cascade, Le Conte, Dana and Chickamin glaciers and moraines and retreat stages of the South Cascade Glacier (Source: from Miller 1969)

Sixteenth- and seventeenth-century advances were followed by retreat and then by renewed but smaller advances in the nineteenth century, some of which were multiple. Volcanic ash from Mt Mazama helps to confirm the dating. For instance, the Mazama 'O' layer, dated by radiocarbon to 400 BP (Wilcox 1965), is present within the South Cascade moraine dated to the sixteenth century but is absent from within the moraines attributed to the nineteenth century. A seventeenth-century advance was also identified on Mt Olympus when Heusser (1957) made a reconnaissance of the Blue and Hoh glaciers and suggested that the earliest datable Little Ice Age advance of the Blue Glacier occurred about 1650; this estimate depended on

Table 9.1 Comparison of glacier area and terminal altitude of Dome Peak glaciers during the Little Ice Age maximum and in 1963/4

	South Cascade	Le Conte	Dana	Chickamin
Area (km) Little Ice Age max.	4.15	2.98	3.99	7.80
Area (km) 1963	2.72	1.58	2.46	4.87
Ratio 1963/max.	1/1.5	1/1.9	1/1.6	1/1.6
Altitude of terminus (m) at max.	1490	1340	1270	1100
Altitude of terminus (m) in 1964	1615	1829	1768	1525

Source: Miller 1969

the ages of only two trees, an Alpine fir (*Abies lasiocarpa*) and a hemlock (*Tsuga mertensiana*), both growing on a moraine remnant. An advance in about 1850 tilted trees on Blue Glacier moraines which, except in one place, obscure moraines dating to the seventeenth century. The Hoh also seems to have reached its greatest recent extent in the mid nineteenth century. By about 1890 the Blue Glacier had retreated 535 m from its nineteenth-century maximum to the base of an icefall. There it remained until between 1924 and 1953 it rapidly retreated 700 m (Spicer 1989). After a short halt, it retreated to reach a position between 1958 and 1960 further upvalley than had ever been recorded. Thereafter, in the cool and wet late 1960s and early 1970s, it began to thicken and by 1980 had advanced 164 m to regain its 1941–3 position, and there it remained for the rest of the decade.

The glaciers in the Coast Mountains of British Columbia advanced about 2500 years ago and again in the early Little Ice Age, reaching a maximum in the mid nineteenth century since when they have retreated (Desloges and Ryder 1990).

In the extreme northwest, at the Alaskan border, two of the largest valley glaciers in British Columbia, the Grand Pacific and Melbern, have lost half their volume in the last hundred years. Clague and Evans (1994) see the features such as kame terraces, kame deltas and ice-dammed lakes associated with their retreat as being similar in style and scale to those resulting from the decay of the Cordilleran ice sheet at the end of the Pleistocene. They were able to compare recent air photos with ground photos and other data from Boundary Commission surveys of a century ago. The Grand Pacific glacier, calving into the deep waters of Glacier Bay and Tarr Inlet, lost much of its volume before and during the nineteenth century. It grounded at a pinning point about 1912 and its retreat slowed. The retreat of the Grand Pacific shifted the ice divide with the Melbern, thereby abstracting some of the latter's supply and causing its front to retreat, mainly in the twentieth century.

9.2 MONTANA AND COLORADO

In the relatively dry, front ranges of the Rockies, small cirque glaciers, now fed almost entirely by wind-drifted snow (Alford 1974), are bordered by moraine fields. Benedict (1968, 1973, 1985, 1990, 1991) made careful and detailed lichenometric studies in the Indian Peaks region, northwest of Denver.[3] He suggested the likelihood that an interval of fifty years had elapsed before lichens became established on moraines in the Front Range and concluded that *Rhizocarpon geographicum* colonized the oldest of the moraines examined about AD 1750 and that they marked an ice advance around 1700 greater than any other in the Holocene history of the region (see Chapter 15). The Arapaho, one of the best-known cirques, has an outer moraine attributed to the seventeenth century and two inner moraines, probably deposited in quick succession and perhaps dating to about 1820 and 1850.

---

3 He did not produce different growth curves according to the rock substrates nor did he indicate the degree of uncertainty of his lichen ages.

The Agassiz and Jackson glaciers in Glacier National Park, 80 km apart on either side of the Continental Divide in northwest Montana, appear to have reached their most forward positions in the Little Ice Age about 1860 (Carrara and McGimsey 1981). No prominent moraines mark the limit of that advance, possibly because it was short-lived. Within the limit, trees were less than a century old and diminished in age towards the glacier; outside the limit they were older. It was concluded that the equilibrium line, as calculated from estimated accumulation area ratios at the nineteenth-century maximum, had been depressed by 180 m in the case of Jackson and 300 m in the case of the Agassiz glacier. The Agassiz, by 1979, was merely a patch of stagnant ice covering 0.75 km² as compared with 3.38 km² in the mid nineteenth century. Retreat, judging from tree-ring counts, was slow until 1910, for on a US Geological Survey map of 1906 the Agassiz Glacier was still 2.8 km long and projected outside its cirque into the forest. Between 1917 and 1926 it shrank back at a rate of more than 40 m a year and over the next sixteen years, in a period of increased summer temperatures and reduced precipitation, the rate of retreat more than doubled. Glacier shrinkage in these continental areas at this time was much more drastic than in the maritime west (Johnson 1980). Retreat then slowed down but did not cease and by 1990 the area of Grasshopper Glacier, for instance, in the Beartooth Range of Montana, was only 38 per cent of its 1956 extent (Lockwood *et al.* 1992).

9.3 THE CANADIAN ROCKIES

The largest glaciers in the Rockies are in the front ranges near the boundary between Alberta and British Columbia where peaks rise to over 3000 m and several extensive icefields extend below the treeline (Figure 9.5 and Plate 9.2). Features attributed to the Little Ice Age (referred to as the Cavell) advances in the Rockies, include well-developed, fresh moraines, some of them unvegetated, within one or two kilometres of present glacier tongues. Larger glaciers generally show nested sequences of moraines, whereas many of the steeper glaciers have only one complex moraine formed by repeated readvances (Kearney and Luckman 1981).

Historical accounts date back no further than the nineteenth century and refer to only a very few of the glaciers. Reconstructing their fluctuation history has to depend on dating moraines by dendrochronology, lichenometry and radiocarbon and, most precisely, by dating forest trees overwhelmed by the ice as it advanced. This last has been made possible by the careful assembly of ring-width chronologies, permitting calendar dating of *in situ* stumps and detrital wood (Luckman 1994, 1995, 1996a, b).[4] Dating moraines by dendrochronology and lichenometry gives minimum ages for the eighteenth and nineteenth centuries, but cannot be very precise; error terms are 10–50 years at best, and can be much larger. Radiocarbon dating also introduces errors of 60 years or more (Grove and Switsur 1994). Only a few moraines predating 1700, consisting of small, locally preserved fragments situated on the margins of nineteenth-century deposits or slightly downvalley of them, have been identified by these methods and the dating of these is poorly constrained. Because the nineteenth-century advances were the most extensive, the current retreat is uncovering progressively more fossil wood relating to earlier periods.

The presence of larch wood dated by radiocarbon to between AD 900 and 1100 at the margins of the Columbia Icefield, 90 km northwest of its present range, points to warmer conditions in the Rockies about a thousand years ago (Luckman 1986). Ice advances followed between AD 1200 and 1370 which terminated some distance upvalley of maximum Little Ice Age advances (Luckman 2000). There followed a period when trees with earliest ring dates of between AD 1315 and 1446 suggest treelines were again high between about AD 1350 and 1700 (Heusser 1956, Luckman 1988).

The first calendrical dating of an early Little Ice Age advance in North America came from the Robson

---

4   Several long chronologies have been developed by Luckman and his associates, including one from *Pinus albicaulis* from a treeline site near the Bennington Glacier with an earliest date of 1112, another from the Peyto Glacier (948–1991), and also at Peyto a *Picea engelmannii* covering the years 760–1992. At Grey Pass, Purcell Mountains they have a *Larix lyallii* chronology (801–1993) (Luckman 1996a, b).

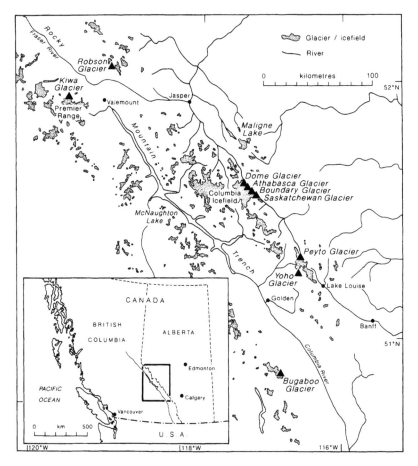

Figure 9.5 Location of dated Little Ice Age moraines in the Canadian Rockies and Premier Ranges

Glacier (Luckman 1995). Heuser (1956) had obtained a radiocarbon date of 450 ± 150 BP on fossil wood overridden by the Robson Glacier and inferred that the Little Ice Age had begun between AD 1300 and 1600.[5] On a basis of three more radiocarbon dates from logs, Luckman (1986) revised this dating to the twelfth to thirteenth centuries. About 500 to 700 m inside the terminal moraine, *in situ* tree stumps and detrital logs which had been sheared off and killed by the ice were exposed between 1925 and 1950 in channels cut by proglacial streams (Luckman 1996a, b). Dated stumps preserved in their growth position revealed both the date and location of the ice when the trees were killed. Ring widths and maximum density chronologies, de-

veloped by crossdating the stumps against living tree chronologies, showed that between about AD 1150 and 1250 the ice reached within 400 m of its Little Ice Age maximum and may have continued to advance into forest until 1350, but without extending beyond the outermost Little Ice Age Robson moraine.

Dates from the outer rings of detrital logs showing shearing marks provided not only kill dates, but also information about the duration of the period of advance. Distances between stumps at the upstream end of the 'southern channel' (R8309, R8306, R8708) suggested average advance rates of 3–4 m annually. Estimates over longer distances (R8713–R8723, R8713–R9207; Table 9.2) indicated rates of advance

---

5   This was an early radiocarbon date. Much more precise information became available after 1956 about the relationship between radiocarbon and calendar ages.

Plate 9.2 The Columbia icefield with its Holocene and Little Ice Age moraines, July 1970
Photo: A. T. Grove

of about 2.6–3.2 m annually. These are lower rates of ice advance than some recorded in recent decades in the Canadian Cordillera of about 6–20 m per year (Luckman *et al.* 1987).

The earliest ring date of AD 1142 from an isolated *in situ* stump in another channel was a minimum as the outer surface showed some loss of rings. This is surprisingly early, though corroborated by one of the detrital logs which has a similar kill date of 1153 (R814).[6] The stump was about 200 m up the channel from stumps dated at 1214, again suggesting only very slow or halting advance of the ice. It appears that the Robson began to invade the forest between about AD 1142 and 1150 and continued to advance, intermittently, until at least 1350. The ice advanced at about 3.8 m per year

between about 1214 and 1261, more quickly than it did between either 1142 and 1214 or 1261 and 1350 when the rates were about 2–2.5 m per year. No evidence has been found on the Robson forefield for any advances between 1315 and 1783.

Evidence from a tree overridden by the Kiwa in the Premier Range 50 km west of the Robson Glacier reveals a similar early advance. Stumps of trees killed by an early advance have also been retrieved from the Peyto forefield. Crossdating of stumps and detrital material seems to indicate advance periods between 1246 and about 1375, and again in the period 1721–1845 (Figures 9.7, 9.8 and 9.6) (Luckman 1996a, b). Periods of advance in the mid fourteenth century in the Peter Lougheed Provincial Park, Alberta, have been inferred

---

6   At the Greenland Summit accumulation measured at GISP2 was above average for the Holocene during a prolonged Medieval Warm Period which lasted till AD 1150, suggesting warmer conditions, while a cold period with low accumulation was placed around 1250 (Meese *et al.* 1994) (see Chapter 16, p. 552).

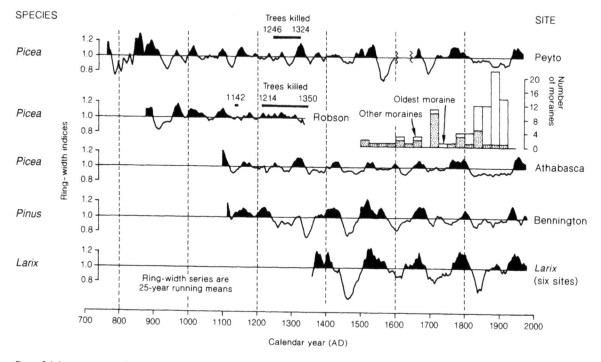

Figure 9.6 Long tree-ring chronologies and glacier fluctuations in the Canadian Rockies. The indexed ring-width chronologies are plotted on the same vertical scale and smoothed with a 25-year mean. Periods when tree-ring indices are lower than the mean for each record are shaded. The *larix* chronology is the mean of 6 *larix* chronologies. The long chronologies are based mainly on living trees growing on sites along some 400 km of the Continental Divide and on both sides of it. Note the differences in the nature and amplitude of the response between species (*Source*: after Luckman 1996a)

Table 9.2 Estimated rates of glacial advance across the Robson forefield

From	Date	To	Date	Distance (m)	Years	Rate (m/y^{-1})
Southern Channel						
R8713	1142	R8309	1214	175	72	2.4
R8309	1214	R8306	1246	113	32	3.5
R8306	1246	R8708	1261	65	15	4.3
R8708	1261	R9207	1300–20	147	39–59	2.5–3.8
R8713	**1142**	**R9207**	**1300–20**	**500**	**158–78**	**2.8–3.2**
R8309	1214	R8708	1261	178	47	3.8
Northern Channel						
R8713	1142	R8702	1214	220	72	3.1
R8702	1214	R8723[a]	1350	322	136	2.4
R8713	**1142**	**R8723**	**1350**	**542**	**208**	**2.6**

Source: from Luckman 1995
Notes: Rates are estimated between selected *in situ* stumps. Estimates in bold type are average values calculated using the maximum possible travel distance for each 'Channel'
a Although R8723 was not *in situ*, this figure provides a maximum rate for glacier advance over the entire exposed area of forest

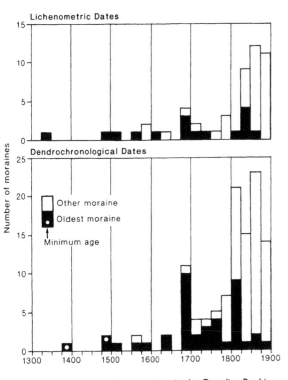

Lichenometric Dates

Dendrochronological Dates

Number of moraines

Other moraine

Oldest moraine

Minimum age

1300 1400 1500 1600 1700 1800 1900

Figure 9.7 Dated Little Ice Age moraines in the Canadian Rockies and Premier Ranges (*Source:* from Luckman 1996a)

from calibrated radiocarbon dates from wood and interpreting tree-ring width records from trees associated with moraines (Smith *et al.* 1995).

Tree rings from a series near the Bennington and Athabasca glaciers show synchronous declines in ring width during AD 1170–80, 1280–90 and 1330–50. Further periods of narrowed rings followed in 1430–50, 1530–40, 1690–1705 and 1810–25. The decadal-scale fluctuations which occurred in the later Little Ice Age were evidently a feature of the earlier part as well (Chapter 16 and Luckman 1993). The relationship between cool periods recorded by narrow tree rings and the timing of glacial advances in the Rockies is shown in Figure 9.6. Periods of suppressed growth

in AD 1140–60, 1230–1250s, 1330–1350s, 1440–1500, *c.*1580–1620, *c.*1690–1750 and during most of the nineteenth century suggest that there were probably periods of glacial advance between 1300 and 1700 for which well-dated evidence has not yet been recovered (Luckman 1994).[7] The strong similarities in the ring-width records from Athabasca, Peyto and Bennington in recent centuries is striking, and suggests that they were responding to regional forcing, probably temperature (Luckman 1996a, b). The lower amplitude and greater variability of the Waterton record, 200 km south of the *Larix* sites, may reflect regional differences in setting. The agreement between records during the thirteenth and fourteenth is less good, but this may be due in part to local effects from the glacier as it approached the trees. However all the chronologies display narrow rings coinciding with the killing of the last trees by the Robson and Peyto glaciers in the mid 1300s.

The times of moraine formation in the later part of the Little Ice Age are still being pieced together. The ages of trees growing on the moraines of 12 glaciers in Banff, Jasper and Yoho National Parks and the Robson Provincial Park were dated by Heusser (1956). He obtained the fullest record for the Robson Glacier, identifying readvance or recessional moraines of which the oldest was dated to 1782 and others to 1801, 1864, 1891 and 1907. Luckman (1977) used written records, photographs and tree rings to prepare a growth curve for lichens on the moraines of the Mt Edith Cavell and Penstock Creek glaciers in Jasper National Park. On the Angel Glacier moraines, where Heusser (1956) had cored 33 trees, Luckman cored an additional 75, many of them older. In a third of the trees his cores failed to reach the middle and these were not used for dating purposes. His dates were earlier than Heusser's, partly because he assumed that the interval between stabilization of the moraines and colonization by trees was 15–30 rather than 10 years, but he stressed that the ages he obtained were minima values (Table 9.3).[8]

---

7 Especially, one might have expected signs of glaciers advancing to show up in the years with narrow tree rings and, it has been postulated, low temperatures, in the mid to late fifteenth century (see Mann *et al.* 1998). There were such advances in Alaska, but there are no signs of them at that time in the Rockies (or indeed in most other parts of the world) in spite of the fact that the other cold periods in the Rockies find their correlative glacier advances elsewhere.

8 Excesis estimates from glacier forefields in the Canadian Rockies and adjacent areas have varied between 10–15 years and about 80 years (Luckman 1986, table 1).

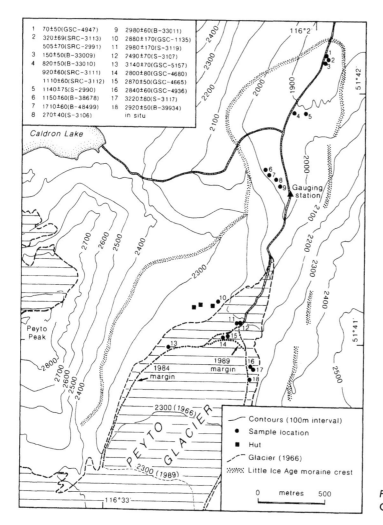

1	70±50(GSC-4947)	9	2980±60(B-33011)
2	320±69(SRC-3113)	10	2880±170(GSC-1135)
	505±70(SRC-2991)	11	2980±170(S-3119)
3	150±50(B-33009)	12	2490±70(S-3107)
4	820±50(B-33010)	13	3140±70(GSC-5157)
	920±60(SRC-3111)	14	2800±80(GSC-4680)
	1110±60(SRC-3112)	15	2870±50(GSC-4665)
5	1140±75(S-2990)	16	2840±60(GSC-4936)
6	1150±60(B-38678)	17	3220±80(S-3117)
7	1710±60(B-48499)	18	2920±50(B-39934)
8	270±40(S-3106)		in situ

Figure 9.8 Sampling sites and dated trees at Peyto Glacier (*Source*: from Luckman 1996b)

Table 9.3 Angel Glacier moraines retreat dates

Heusser (1956)	1723	1783	1871	1901
Luckman (1977)	1700–10	1715–25	1851–6	1881–9

Luckman used lichens to date the moraines of six other glaciers and emphasized that, because of the short calibration period and the shape of his growth curve, the ages he obtained by this method were also minima. All lichen dates earlier than 1700 depend upon extrapolation of growth curves which further reduces their precision.

The available data, referring to twenty-three glaciers from the Middle Canadian Rockies, were assembled by Luckman and Osborn (1979). By 1996 Luckman was able to present a table specifying the dates of moraines of some sixty-one glaciers, including several in the Premier Ranges, including details of the methods of dating used (Luckman 1996a, table 1). Many of the moraine dates were estimated from the ages of the oldest trees found growing on them. Excesis periods used varied between 10 and 20–40 years, depending on the area and, no doubt, the judgements of the individual workers concerned.

The closely spaced multiple moraines of the Columbia, Athabasca, Peyto and Bow glaciers mark nineteenth-century advances to positions comparable with those reached in preceding centuries, differences

Table 9.4 Little Ice Age moraines dated by ice-damaged trees in the Canadian Rockies

Glacier	Year	Evidence	Source
Athabasca	1714[a]	Tilted tree	Heuser 1956
Saskatchewan	1807, 1813	2 tilted trees	Heuser 1956
Small River	1838–9	Tree killed	Luckman 1995
Crowfoot	1839	Tilted tree	Leonard 1981
Peyto	1837–46[b]	7 trees	Luckman 1996b
Columbia	1842	Tilted tree	Heuser 1956
Athabasca	1843–6	Tilted tree	Luckman 1988
Dome	1846	Tilted tree	Luckman 1988
Bow	1847	Tilted tree	Heuser 1956

Source: from Luckman 1996a
Notes:
a The 1714 date from Athabasca is from a lateral moraine, the 1843–4 date from the terminal moraine
b Six trees were killed along the lateral moraine between 1837 and 1841. Several trees were killed close to the position of the terminal moraine; the last of these died in 1836

between the patterns of Little Ice Age moraines in front of the various glaciers probably depending on geomorphic setting and the extent to which more recent deposits have overwhelmed older ones. Evidence of an early eighteenth-century advance of the Athabasca has been preserved in only two places, just outside the mid nineteenth-century moraine (Luckman 1986). Since the late nineteenth century, recession has been the rule (Brunger *et al.* 1967). Only a few sixteenth-century moraines have been found (Figure 9.7), but that is almost certainly due to concealment of earlier by later deposits. Major advances seem to have occurred in the late seventeenth to early eighteenth, early to mid nineteenth and late nineteenth to early twentieth centuries. 'Regional glacier extent was probably greatest in the 1840s', and glaciers 'were still close to their maximum positions in the early 1900s' (Luckman 1996a: 363, 367).

While the moraine ages obtained from the ages of the trees growing on them or by lichenometry do not produce very closely grouped patterns, the close grouping of dates of trees damaged as glaciers advanced into forests growing along lateral trimlines or on older moraine surfaces is clear (Table 9.4). These dates are not minimal but specific; their coincidence suggests that the glaciers were responding almost synchronously to similar forcing, despite differences in size and geomorphic setting. Luckman (1996a) pointed to the coincidence of the two main periods of moraine formation in Figure 9.7 with the timing of 'coldest condi-

tions in the early and late 1500s, 1700 ± 20 and 1840 ± 15' obtained by Bradley and Jones (1993) from a variety of proxy data from North America.

Tree-ring densities have allowed temperature reconstructions to be made for the Canadian Rockies over several centuries (Luckman *et al.* 1997, Luckman 2000). They point to low April–August temperatures from 1180–1330 and again from 1430–1500, a brief but strong cooling in the 1690s, cool summers in the early eighteenth century and a prolonged cold period between 1800 and 1840, followed by warming until 1950. Precipitation, it would appear, was relatively high, at least in the Banff area, in the interval 1515–50, and in both Jasper and Banff and further southwest in the 1880s. Temperature variation seems to have had more influence than precipitation on glacier advances and retreats in this continental setting. Lake varves provide additional palaeoenvironmental chronologies but they are not readily interpreted in climatic terms (Luckman 2000, figures 10 and 11).

Documentary records for Canadian glaciers are short, beginning only in the nineteenth century. In 1858 Dr Hector, of the Palliser expedition, discovered the Lyell Glacier and described how 'after crossing shingle flats for about a mile, we reached a high moraine of perfectly loose and unconsolidated materials, which completely occupies the breadth of the valley, about 100 yards in advance of the glacier' (Thorington 1927). This, one of the earliest descriptions of a glacier in the Canadian Rockies, shows that modern recession from

Plate 9.3 William Vaux photographing Illecillewaet Glacier, which was already in retreat, in August 1898 from 'observation rock' (*Source:* from the Vaux family collection of the Whyte Museum of the Canadian Rockies)

the mid nineteenth-century maximum had already begun.

Measurement of glacier fronts started soon after the opening of the Canadian Pacific Railway to passenger traffic in 1886. The following year, George Vaux, a leading Quaker businessman from Philadelphia who was also a dedicated mineralogist and supporter of the Philadelphia Academy of Sciences, visited the Rockies with his three children, William, George and Mary. They stayed at the newly opened Glacier House Hotel and went to see the Great Glacier of the Illecillewaet. Later, William Vaux Jr was to comment in his monograph *Modern Glaciers* (1907), 'during railway construction the glacier was doubtless often visited by those stationed on the work but no records were made until July 17th 1887 when our party passing through, roughly mapped the tongue and made a photographic record of the conditions as they existed'. When the Vaux returned in 1894 and noticed that the glacier

front had retreated they realized the value of their 1887 photographs and became so interested in the study of glaciers that they initiated the first measurement of glacier front positions in Canada (Cavell 1983). The family was in a position to establish a routine which involved Mary and her father spending most of the summer in the Rockies, with George and William visiting as often as business would allow (Plates 9.3, 9.4 and 9.5). Inevitably they came into close contact with Wheeler, Dominion Surveyor working in the Selkirks. Already in 1899 he noted with approval that the Vaux surveys of the Illecillewaet and Asulkan had become 'properly systematised'.

The most detailed Vaux studies were of these two glaciers but they also photographed and observed the Yoho, Victoria and Wenkchemna glaciers, establishing standard photographic points and measurement procedures. Their observation points were later adopted by Sherzer (e.g. 1905) and Wheeler (e.g. 1910, 1933).

Plate 9.4 Mary Vaux at the foot of the Illecil ewaet Glacier, 17 August 1899 (*Source*: from the Vaux family collection of the Whyte Museum of the Canadian Rockies)

Plate 9.5 The smooth tongue of the retreating Illecillewaet Glacier in the summer of 1902 (*Source*: from the Vaux family collection of the Whyte Museum of the Canadian Rockies)

These simple reconnaissance studies are the only ones from Canada dating back to the time when the glaciers had yet to retreat far from their most forward positions in the nineteenth century. All the available evidence goes to show that recession has been predominant ever since. The withdrawal of the Athabasca and Columbia glaciers was accelerated when terminal lakes formed in front of them, the Athabasca losing about two-thirds of its 1870 volume in the subsequent century (Kite and Reid 1977).

Few glacier monitoring programmes were attempted in the twentieth century. The longest series lasted only thirty-six years, being discontinued in 1980 (Luckman 1990). The frontal recessions of a selection of glaciers with the longest records in British Columbia and Alberta are shown in Figure 9.9.

Comparison of 1970 and 1973[9] air photographs with a compilation of the Little Ice Age maximum extent of glaciers in the Premier Ranges, just west of the Rocky Mountain Trench, based on mapping of moraine positions, indicates a loss of about 23 per cent of the ice cover between the Little Ice Age maxima and 1970 (Luckman 1990). Average elevation of the snouts rose by 262 m over the period, the smallest glaciers suffering the greatest percentage loss of area. (It had to be assumed that moraine positions mapped from the air photographs could be taken as approximations for the downvalley extent of the glaciers in the mid nineteenth century). Ice cover loss of the North Saskatchewan–Mistaya catchment between 1948 and 1966 was estimated at 10 per cent (Henoch 1971). In the Kananaskis area, south of Banff, the average increase in the altitude of snouts between 1916 and 1988 was 137 m (McCarthy and Smith 1994). Kite and Reid (1977) estimated the loss of volume of the Athabasca tongue between its nineteenth-century maximum and the early 1970s at 32 per cent; few other attempts to estimate volumetric change have been made.

On average, about 25 per cent of the ice cover has disappeared since the Little Ice Age maximum, mainly during the twentieth century, a smaller proportion than the 35 per cent estimated for the Alps by Haeberli and Hoelze (1995).[10] Reduction has been substantially greater for small glaciers than for larger. In the Premier Ranges, glaciers of less than half a square kilometre have lost 46 per cent of their areas, and in the Columbia Icefield region, where the average loss has been 25 per cent, the smaller glaciers have lost 57 per cent of their maximum cover (Luckman 1996a).

9.4 COAST MOUNTAINS OF BRITISH COLUMBIA

In the northernmost part of the Pacific Ranges of the Coast Mountains near Bella Coola, early Little Ice Age advances have been identified on the forefields of two glaciers, both outlets of the 350 km² Monarch Icefield (Desloges and Ryder 1990). The Jacobson Glacier is judged to have advanced in 1440–1630 (S-2979), but this date is tentative. It came from a succession exposed in the margin of a lobe of the left lateral moraine, and was derived from a composite sample of discontinuous Ah horizons found within a cumulic regosol developed within gritty sand, burying a mature podzolic soil, itself superimposed on till (Desloges and Ryder 1990). A section through the main outer moraine of the Purgatory Glacier revealed three superimposed tills, separated by palaeosols. Leaf litter, 'clearly in situ', from the lower palaeosol (S-2977), records ice advanced over older till after 1210–90, while litter from the upper palaeosol (S-2978) indicates that the ice withdrew sufficiently for 2 cm of leaf litter to accumulate again before the glacier readvanced after 1290–1410.

An advance of the Klinaklini Glacier, an outlet of the Silverthrone Icefield further south in the Coast Ranges (51°N–51°30′N) was recorded by a date (WS-1567) from an *in situ* tree trunk, but this gave a calendar age range of 1040–1210, too wide to reveal whether it relates to an interim advance during the Medieval Warm Period or at the beginning of the Little Ice Age (Figure 9.10). The trunk was found above the current ice and 150 m below the crest of the right lateral moraine. An advance of the Franklin/Confederation Glacier occurred some time after 1160–1260, a date

---

9 Unfortunately recent reduction in routine air photography has curtailed this approach to glacier monitoring. In most cases only satellite data are available for the last 15–20 years (Luckman 1996a).

10 In both Rockies and Alps the percentage ice volume loss has been considerably greater than the percentage area loss (Luckman 1998).

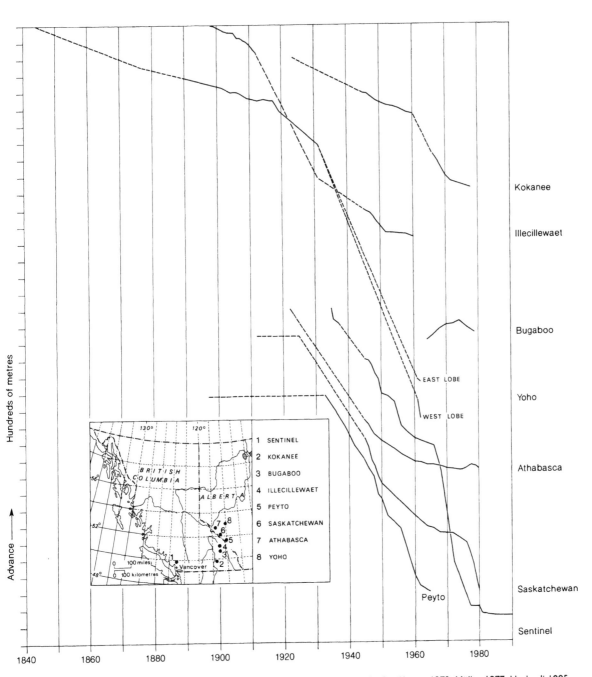

Figure 9.9 Glacier retreat in the Canadian Rockies (*Sources*: derived from figures published in Kasser 1973, Müller 1977, Haeberli 1985, Haeberli and Hoelzle 1993)

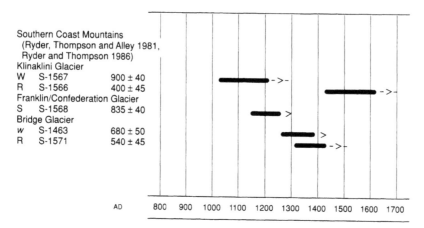

Southern Coast Mountains
  (Ryder, Thompson and Alley 1981,
  Ryder and Thompson 1986)
Klinaklini Glacier
W   S-1567   900 ± 40
R   S-1566   400 ± 45
Franklin/Confederation Glacier
S   S-1568   835 ± 40
Bridge Glacier
w   S-1463   680 ± 50
R   S-1571   540 ± 45

AD   800 900 1000 1100 1200 1300 1400 1500 1600 1700

Figure 9.10 Key radiocarbon dates relating to Little Ice Age advances in the coastal range of British Columbia *Notes*: The numbers in the margin give the laboratory identification number and the ^{14}C age. Horizontal lines represent the calibrated ages. - > - = Advance around or after indicated age; > = Advance after date; W = Wood *in situ*; w = Wood not *in situ*; S = Soil; R = Roots *in situ*

given by wood from a palaeosol (S-1568), but there is no direct evidence of the length of time involved. Recession of the Bridge Glacier, an outlet of the Lilloet Icefield, has exposed a nunatak carrying overriden soil and trees about 100 m above the present ice level (Ryder and Thompson 1986). Tree trunks and roots protrude from the highest moraine. Dates from a 1-m diameter tree trunk (S-1463) and a root (S-1571) indicate that the forest was overwhelmed some time after 1280–1390 and 1330–1430 respectively. Though the age range is wide, it is clear that the Bridge Glacier advanced in the Little Ice Age. Ryder and Thompson concluded from the history of a lake impounded by the glacier that the nunatak had not been overriden for a long period previously, perhaps as long as a thousand years.

Desloges and Ryder (1990) suggested that the Little Ice Age may have begun slightly later and ended generally later in the Coast Mountains of British Columbia than in the Rockies, noting that the mass balance of coastal glaciers is strongly controlled by precipitation, while temperature and solar radiation are important controls in inland regions.

9.5 ALASKA AND THE YUKON

Many of the earliest observations of glaciers in North America were made by early explorers and pioneer scientists seeking a seaway from the Pacific to the interior. Glacier Bay raised their hopes of success but when Vancouver's surveyors mapped its crenulated coast and icefields their hopes were dashed. Pre-twentieth-

century data are otherwise sparse in comparison with the size of the country and the range of environmental conditions. Much of the information that does exist was brought together by Field (1975) without being fully analysed. Serious attempts have been made in recent years to extract detailed chronologies of glacier behaviour for both land-based and tidewater tongues, using tree-ring dating in association with radiocarbon and lichenometry (e.g. Wiles and Calkin 1994, Wiles *et al.* 1995, Evison *et al.* 1996, Barclay *et al.* 2001, Calkin *et al.* 2001).

Donald Lawrence in 1950 was impressed by the synchronism of the glacier oscillations in the northwest of North America. The glaciers he had studied had advanced in the early or middle eighteenth century and had then begun to retreat about 1765. As they mainly emanated from the Juneau Icefield, their uniformity of behaviour, he wrote, would be expected.

But the synchronism with the Glacier Bay area, 65 miles to the northwest, those of Garibaldi Park (in British Columbia) 800 miles to the south-east, and of Eliot Glacier on Mt. Hood, 110 miles to the southeast, and even glaciers of Norway and Iceland, can hardly be assumed to be merely coincidental. . . . Even the morainal features of North Iliamna Volcano, 700 miles north-west of the Juneau Icefield, seem similar . . . to those just examined. Only some of the glaciers of the Prince William Sound region, 500 miles northwest of the Juneau Icefield, and of a few other areas on this continent, have behaved in a significantly different way. . . . Surely some widespread climatic event must have been the reason for the concordant behavior of glaciers so widely spaced geographically.
(D. B. Lawrence 1950: 213–14)

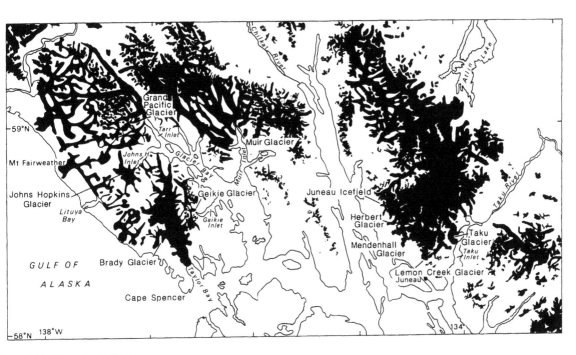

Figure 9.11 Glacier Bay in Alaska

Much of the Alaskan coastal region is subject to earthquakes. Rapid and seemingly anomalous glacier advances at the beginning of the century after a series of severe shocks centred near Yakutat Bay led to the notion that they had been caused by snow and ice being shaken onto the glaciers (Tarr and Martin 1914). The hypothesis was seriously questioned by Miller (1958) who pointed to the exceptionally heavy precipitation in the 1870s and 1880s and argued that diastrophism was a relatively minor factor. A major earthquake (8.4–8.6 on the Richter scale) in 1964, with an epicentre immediately south of the heavily glaciated Chugach Mountains, provided an excellent opportunity to test the earthquake advance theory (Post 1965). Extensive air surveys of the glaciers in Alaska and Canada were available for comparison with air photographs taken after the earthquake and little significant avalanching was found to have been caused. It can be safely concluded that earthquakes have not been responsible for the anomalous behaviour that characterizes some of the best-known Alaskan glaciers. Surging is a much more important factor, especially in the St Elias, Wrangell, Chugach, Alaska and Aleutian ranges (Post

1969). However, surging is not the whole explanation for the behaviour of the glaciers in, for example, the Taku, Glacier Bay and Yakutat Bay areas (Figure 9.11).

Tidewater glaciers are known to produce major anomalies (Field 1979). Reaching down to sea level they are generally nourished by firnfields situated well above the average elevation; a shift in the position of the snowline can trigger a rapid movement at the front (Mercer 1961). Wasting, affected by calving at the terminus as well as seawater ablation, is greatest when the calving cliff is longest and ends in deep water. The consequent instability allows many large icebergs to break away from the tongue, and sediments deposited from the ice are spread thinly in deep waters as swift recession proceeds. The rate of ice flow is maintained right to the terminus and may even increase towards it as there is little frontal resistance. With ice discharging faster than it can be replaced from the accumulation area, the glacier surface is rapidly lowered, the snowline migrates upglacier and the accumulation area shrinks, thereby diminishing the ice cover inland. Reduction of volume resulting from the feedback effect is largely independent of climate. On the other hand, when a terminus is advancing and reaches shallow water, deposition promotes further shallowing and material accumulates at the base of the ice cliff forming a barrier that protects the terminus from tidewater. As deposition continues, reduction in the rate of terminal ablation may initiate an advance. As it proceeds, a moraine-outwash barrier is pushed ahead of the ice until equilibrium is established between the flow of ice and the melting of the terminus. Conversely, very rapid recession occurs when the terminus recedes off a bar into deep water, as has occurred at the Columbia ice front (Meier *et al.* 1979, Weller 1980). The varying behaviour of the ice in the Glacier Bay, Yakutat Bay and Taku regions can be explained adequately by the operation of these controls. It is evident that when attempts are being made to discern the climatic sequence for the coastal and near-coastal ranges of northwestern America, records of advance and retreat of surging and tidewater glaciers have to be ignored or at least viewed circumspectly.

The glaciers of the coastal ranges of Alaska were generally in advanced positions in the eighteenth and nineteenth centuries and most of them have since receded. Recession was interrupted by minor advances between 1910 and 1914 and again around 1935. There were less widespread advances in the late 1940s and again in the 1970s and 1980s. Glaciers with higher névéfields have receded least and have been most apt to show small advances, while those with the firnline close to sources have receded most. A composite of mean annual temperature pieced together from available weather bureau records from 1895 onwards shows a net rise of 1–1.5 °C from the late 1900s to 1960. This was complicated by a series of fluctuations, with lows in 1895, 1920 and 1935, of which the first was much the most marked. The rise in temperature in the early twentieth century ceased in the 1940s and 1950s (Hamilton 1965) but has since resumed.

9.5.1 Glacier Bay and Lituya Bay

In 1794, Glacier Bay was 'terminated by a solid, compact mountain of ice, rising perpendicularly from the water's edge and bounded to the north by a continuation of the united, lofty, frozen mountains that extended eastwards from Mount Fairweather' (Vancouver 1798). From tree-ring evidence, Cooper

(1937) argued that by the time of Vancouver's visit the glacier had already retreated 10 km from a position it had earlier occupied near the mouth of Glacier Bay; and he placed the beginning of the retreat at between 1735 and 1785.

Since Vancouver saw it the trunk glacier has almost disappeared and Glacier Bay has been enlarged from less than 10 km in length to extend inland for over 100 km at the head of Tarr and Johns Hopkins inlets and 86 km at the head of Muir Inlet. The water area of the bay has been enlarged by about 130 km², i.e. by an amount comparable with the combined area of all the glaciers in the Swiss Alps. In Muir Inlet, where the ice was receding at about 400 m a year in the late 1970s, the terminus lay at the base of a cross-section that had contained ice over a kilometre thick in the 1890s. Recession was equally fast in the northwest arm of Glacier Bay, where the ice tongue split in the 1880s into the Grand Pacific Glacier in Tarr Inlet and the Johns Hopkins Glacier in the Johns Hopkins Inlet. The Grand Pacific shrank back to the head of its fjord between 1925 and 1935, after which it readvanced over 2 km. The Johns Hopkins, having retreated to the head of its fjord about 1930, readvanced through 1.8 km. The Hugh Miller and Geikie glaciers in the southwest part of Glacier Bay receded to land in the first half of the twentieth century, about 2 km and 3.5 km away from tidewater. Both were receding at about 60 m a year in the late 1970s (Field 1979).

The Brady Glacier, 52 km southeast of the Fairweather Range, discharges both north into Reid Inlet at the landward end of Glacier Bay, and southward into Taylor Bay about 30 km west of the entrance to Glacier Bay. When Vancouver's party visited Taylor Bay, Lieutenant Whidbey was ordered to survey it on 10 July 1794. Setting out from Cape Spencer, despite being much inconvenienced by 'immense numbers of pieces of floating ice', he reached the side of Taylor Bay which he found took

nearly a north direction for about three leagues to a low pebbly point; from which, five miles further, a small brook flowed into the sound and on its northside stood the ruins of a deserted Indian village. To reach this station the party had advanced up an arm about six miles wide at its entrance but had decreased to about half that width and there further progress was now stopped by an immense body of compact perpendicular ice, extending from shore

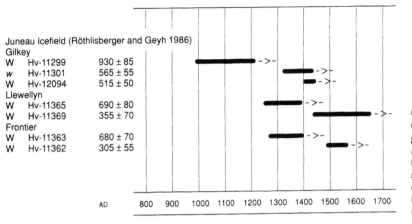

Juneau Icefield (Röthlisberger and Geyh 1986)
Gilkey
W   Hv-11299   930 ± 85
w   Hv-11301   565 ± 55
W   Hv-12094   515 ± 50
Llewellyn
W   Hv-11365   690 ± 80
W   Hv-11369   355 ± 70
Frontier
W   Hv-11363   680 ± 70
W   Hv-11362   305 ± 55

AD   800   900   1000   1100   1200   1300   1400   1500   1600   1700

Figure 9.12 Key radiocarbon dates relating to advances of Juneau Icefield glaciers (*Source:* from Grove 2001b)

Notes: The numbers in the margin give the laboratory identification number and the ^{14}C age. Horizontal lines represent the calibrated ages. - > - = Advance around or after indicated age; W = Wood *in situ*; w = Wood not *in situ*

to shore, and connected with a range of lofty mountains that formed the head of the arm, and, as it were, gave support to this body of ice.

Klotz (1899) compared Vancouver's chart with that of the 1894 Boundary Commission and concluded that the front of the Brady had advanced 5 miles (8 km) in the intervening period, at a time when the glaciers at Taku inlet and Glacier Bay were retreating. Derksen (1976) did not agree that the Brady glacier advanced at all at this time, arguing that Vancouver's chart, though it reproduces the regional coastline roughly, is distorted in detail. He identified the locations mentioned in the report in such a way as to suggest that in 1794 the glacier front was quite close to its 1970 position. This interpretation rests essentially on his identification of the 'low pebbly point', 5 miles SSE of the terminal position in 1794, with an undated and probably Holocene moraine 5 miles SSE of the present terminus. Derksen surveyed the area in detail, dating the climax of ice advance to 1876–88 on tree-ring evidence at trimlines, and accepted Bengtson's (1962) conclusion that a forest incorporated in till 24 km north of the present terminus of the Brady was destroyed by advancing ice about 685 ± 40 BP (UW-14, AD 1280–1390). He accordingly visualized the most recent

major advance of the Brady glacier as extending over the period from the mid thirteenth to the late nineteenth century.

The Juneau Icefield, covering about 1820 km² of the most heavily glaciated sector of the Coast Range (Figure 9.11), is drained by a dozen or so valley glaciers. It has been studied intensively and a chronology of glacier advance and retreat was worked out in some detail (e.g. Miller 1964, 1965, 1970, 1977, Field 1975). The Juneau area experienced a brief warm interval recorded by forest growth between cal AD 838 and 1300 (Calkin and Wiles 1991).[11] Using the same approach he had employed in the Alps and Himalaya, and avoiding those glaciers with floating tongues or known to surge, Röthlisberger and Geyh (1986) dated three sets of moraines made by Juneau Icefield outlet glaciers. Röthlisberger found that the Gilkey Glacier overran forest during the first millennium, and either during the Medieval Warm Period or early Little Ice Age left an *in situ* stump dated 1000–1210 (Hv-11299) (Figure 9.12). The Gilkey had advanced into forest again around 1330–1430 (Hv-11301) and 1400–40 (Hv-12094). The Llewellyn advanced about 1260–1390 (Hv-11365) and again about 1445–1650 (Hv-11369), on each of these separate occasions killing trees and leaving *in situ*

11   Calkin and Wiles (1991) presented radiocarbon ages in terms of their calendar equivalents, using the method of Stuiver and Reimer (1986). They did not quote the original radiocarbon dates or numbers. Calibration was found to have little effect on lichenometric curves and their derived ages were converted directly to calendar years. This procedure results in an appearance of greater accuracy than is warranted by some of the data. However Calkin and Wiles warn that 'no assumption of equal weight' can be placed on their chronologies.

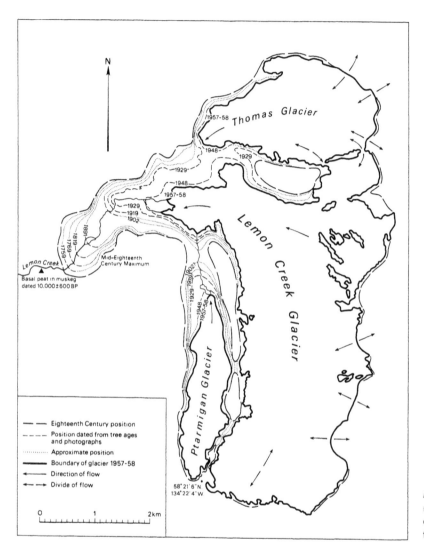

Figure 9.13 Lemon Creek Glacier retreat stages since the mid eighteenth-century maximum (*Source: from Heusser and Marcus 1964*)

stumps. The second tree killed by the Llewellyn had 120 rings; evidently the climate was sufficiently improved between the two advances for trees to grow at the site. The Frontier Glacier advanced and killed an *Abies lasiocarpa* tree, for which the stump *in situ* gave a date of 1280–1390 (Hv-11363), and advanced again two centuries later, 1495–1565 (Hv-11362), as well as during the later part of the Little Ice Age (Hv-11361). His results confirmed that in early advances between AD 1265 and 1401 the Davidson Glacier had sheared off tree stumps and in a subsequent advance reaching a culmination in 1752 had destroyed trees over 200 years

old (Field 1975). Nine other glaciers also reached their most forward Little Ice Age positions in the mid to late eighteenth century (Lawrence 1950). Ten of the Juneau outlet glaciers receded from 1870 maxima. Lawrence (1950, 1951, 1958) and others counted the rings of trees growing on the larger moraines and showed that there had been an important advance at just about the time when the glaciers that formed them were first seen by Europeans, as Cooper (1937) had also concluded.

As part of the Juneau Icefield Project, a detailed study was made of the forefield of the Lemon Creek Glacier (Heusser and Marcus 1964) (Figure 9.13,

Table 9.5 Changes in the area and terminal position of Lemon Creek Glacier, Alaska, *c.*1750–1958

Dated terminus Max c.1750	Cumulative area loss (%)	Recession rate (m/y)
1759	0.8	
1759–69	1.5	12.5
1769–1819	2.4	3.5
1819–91	3.3	3.8
1891–1902	8.0	61.4
1902–19	10.2	4.4
1919–29	11.7	7.5
1929–48	21.4	32.9
1948–58	25.1	37.5

Source: Heusser and Marcus 1964

Table 9.5). Cores were taken from Sitka spruce (*Picea sitchensia*) growing on the remnants of trimlines and moraines and it was assumed from the ages of the oldest spruce growing on a moraine (dated from air photographs) that the time taken for their establishment was ten years. It was found that all the Juneau outlet glaciers without exception were expanded in the mid to late eighteenth century. The pattern of halts and readvances in the long-term recession corresponded to that of the large Juneau outlet glaciers except the Taku and the Hole-in-the-Wall. The Herbert, for instance, retreated only very slowly between the mid eighteenth and the late nineteenth centuries and then withdrew rapidly before slowing down again in the early twentieth century. The most rapid retreat was in the middle of the twentieth century (Table 9.5).

The Taku, the major trunk glacier draining the southeast of the Juneau Icefield, was in an enlarged state, blocking the headward end of Taku inlet when Vancouver first saw it in 1794:

A compact body of ice extended some distance nearly all round . . . from the rugged gullies in the side were projected immense bodies of ice that reached perpendicularly to the surface of the ice in the basin, which admitted of no landing place for boats but exhibited as dreary and inhospitable an aspect as the imagination can possibly suggest.
(Vancouver 1798)

He also mentioned that 'the basin, was about 13 miles from the mouth of the inlet' indicating that the ice-front was not far from Taku Point. This agrees with local Thlingit accounts of an ice barrier which prevented access to the interior valley of the Taku river 'before the White Men came'. In the nineteenth century, however, the ice had withdrawn and Taku inlet and the valley at its head were being used by the Indians to cross the Coast Range to the Atlin area. In the 1870s to 1890s the route was also used by prospectors (Miller 1964). By the end of the century, Klotz (1899) found only the Foster Glacier discharging directly into the sea and noted that 'the gullies are not now so generally filled with ice'. Already around 1890 the Taku had begun to advance again. Between 1890 and 1952 its front advanced 5550 m and by 1961 a further 600 m. In the early 1960s, 80 per cent of the surface of the Taku was above the firnline compared with 60 to 70 per cent of the surfaces of the Norris and Mendenhall glaciers, which were retreating.[12] This circumstance may help to account for the strongly positive regime and the anomalous behaviour of the Hole-in-the-Wall, but it is probably much more to the point that these are tidewater glaciers. The Taku front was in water 100 m deep in 1890 and was calving actively until 1937. Deposition of sediment from the Taku and Norris glaciers and by the Taku river caused so much shoaling that by 1941 exposed sediments fronted most of the Taku terminus. The average advance rate of over 100 m a year of the 1890–1952 period dropped, the Taku's front grounded on its shoal moraine and outwash apron, advancing only about 75 m between 1964 and 1968 (Matyka and Begét 1996).

Lituya Bay, to the west of the Fairweather Range, was first visited and surveyed by La Pérouse (1787) in the course of his scientific expedition of 1786. He found it

disturbed by the fall of enormous masses of ice, which frequently separate from five different glaciers. . . . It was at the head of this Bay that we hoped to find channels by which we might penetrate to the interior of America. . . . At length, having rowed a league and a half-mile we found

---

12   The Mendenhall Glacier, which attracts about 300,000 sightseers a year, is close to Juneau, which has experienced a warming of 1.6 °C since 1943. Between 1930 and 1997 the glacier thinned by an average of about a metre annually and its front retreated nearly a kilometre. Exceptionally, its mass balance was positive in 1999/2000 owing to unusually heavy snowfall upglacier.

Table 9.6 Dates of the outermost moraines of the Juneau Icefield outlet glaciers

Glacier	Recession date	Source
Norris	1750s	Lawrence 1950
Taku	1750s	Lawrence 1950
Twin Glaciers	1775–7	Cooper 1942
Twin Glaciers	1777	Field 1932
Lemon Creek	1750	Heusser and Marcus 1964
Gilkey	1783	Heusser and Marcus 1964
Mendenhall (middle)	1765–9	Lawrence 1950
Mendenhall (east)	1786–8	Lawrence 1950
Herbert	1765	Lawrence 1950
Eagle	1785	Lawrence 1950

Source: Viereck 1967

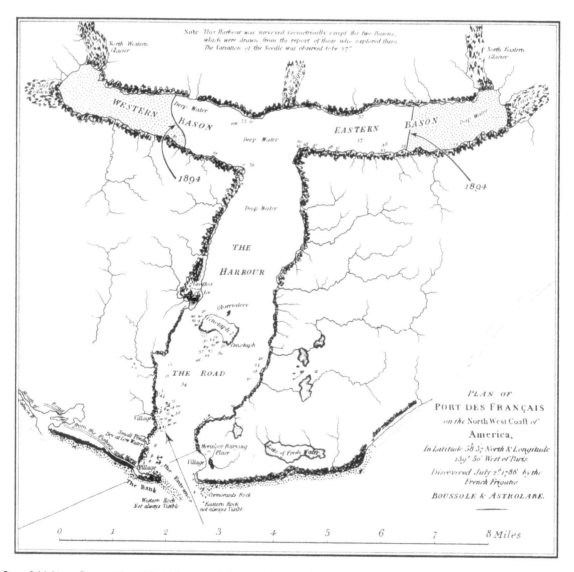

Figure 9.14 Lituya Bay: position of glacier fronts as depicted on the map of the Canadian–Alaskan Boundary Commission of 1894 superimposed on La Pérouse's map of 1799

the channel terminated by two vast glaciers. . . . Messrs Lelangle de Monte and Daglet, with several other officers attempted to ascend the glacier. With unspeakable fatigue they advanced two leagues, being obliged at the extreme risk of life to leap over clefts of great depth, but they could only perceive one continued mass of ice and snow, of which the summit of Mt Fairweather must have been the continuation. . . . I sent M. Momeron and M. Bernizet to explore the eastern channel which terminated like this at two glaciers. Both these channels were surveyed and laid down in the mouth of the Bay.

Klotz (1899) compared La Pérouse's chart with the map of the area made by the Canadian–Alaskan Boundary Commission of 1894 (Figure 9.14) and demonstrated that a substantial advance of the Lituya Glacier had taken place in the course of the nineteenth century.

9.5.2 St Elias, Wrangell and Chugach Mountains

Moraine sets in the White River valley on the northern flanks of the St Elias and Wrangell Mountains were dated by Denton and Karlén (1977), using lichenometry, to between 1500 and the early twentieth century. The work was based on measurements of the largest thalli of *Rhizocarpon geographicum* (possibly with some *Rhizocarpon superficiale* included as they are difficult to differentiate). The growth curve was based on tree-ring dating of surfaces together with some radiocarbon dating. Full details were given of the control surfaces. The growth curve was rated as only 'reasonably accurate' but 'affording a valuable reconnaissance tool'. Omitting data from glaciers known to surge, four sets of moraines were identified with lichens having maximum diameters of 37–45 mm, 27–29 mm, 15–16 mm and 6–10 mm. The geomorphological descriptions seem to make the attribution of these moraines to the last four centuries acceptable. Denton and Karlén suggested that there had been fluctuating glacial expansion between 1500 and the twentieth century but they did not attempt to itemize these expansions chronologically, and their growth curve was insufficiently accurate either to do so or even to confirm that the outer moraines were caused by glacier expansion as early as 1500.

On the seaward side of the St Elias Mountains, annual precipitation of 3500 mm at the town of Yakutat, on the coast south of Yakutat Bay, supports dense spruce and hemlock forests. In the Holocene the Hubbard Glacier, fed by heavy snowfall on the mountains 100–150 km inland, several times advanced into Disenchantment Bay, its eastern lobe penetrating the northwest end of Russell Fjord and blocking it to form a lake that overflows south. The advance of the Hubbard is slow as it pushes forward and travels over its outwash and morainic debris; the retreat is speedy as the terminus retreats and calves in deep water. Variations in the Hubbard's frontal position are consequently out of step with smaller land-based glaciers nearby (Barclay *et al.* 2001).

Tree stumps on moraines of the land-based Beare Glacier, west of Icy Bay, have been dated from their rings and by radiocarbon and point to it having reached an advanced position between 1646 and 1788. The Guyot Glacier, which reached down to the Bay about the same time, has retreated 40 km from its terminal moraine since 1904 (Calkin *et al.* 2001).

Westwards, at the head of the Gulf of Alaska on the coastal side of the Chugach Mountains, the firn limit is below 1000 m, rising inland to 1500 m in interior valleys. Ice covers 10,400 km², an area second in Alaska only to that of the adjacent St Elias Mountains. The whole crest of the range is covered with interconnected névéfields, the largest of which is the Bagley Icefield. A first appraisal of the voluminous and scattered evidence indicated that with few exceptions non-tidal and non-moraine covered valley glaciers had receded 1–2 km from the positions they had occupied within the preceding two and a half centuries (Field 1975).[13] Many have been receding from nineteenth-century maxima, but not continuously. The glaciers of the lower Copper River and adjacent basins, for example, are all retreating from moraines formed when they were up to 5 km longer. Some of these moraines have been dated. West of the Copper River, trees overrun by ice and minimum ages of terminal moraines record culminations of Little Ice Age advances of the Sheridan Glacier around AD 1284, 1747 and 1897 (Calkin *et al.* 2001). East of the river, the outer moraine of the Martin River

---

13   The advances since the 1960s of the Ahtna, Center and South MacKeith on Mount Wrangell, an active volcano, may be due to increased volcanic heating changing their basal conditions (Sturm *et al.* 1991).

Glacier seems to have been formed by 1650 (Reid 1970); the Miles was in an advanced position in 1880 (Tarr and Martin 1914); the Grinell advanced in 1904 and many of the glaciers in the area, including the Childs, Allen, Henry and Grinell, advanced about 1910.[14]

The early twentieth-century advances were of crucial interest to the operators of the Copper River and Northwestern Railway, put through the valley between 1906 and 1910 and routed between the Miles and Childs glaciers. During its construction both glaciers advanced; the distance between their termini diminished until the Childs was less than 1500 m from the bridge carrying the railway over the river. The track then crossed the tongue of the Grinell Glacier, 400 m of stagnant ice thinly covered by ablation moraine overgrown by alder thickets and cottonwood trees. An advance of the Grinell was then in progress which, it was feared, might continue long enough to break up the stagnant ice in front of its active tongue. Worse still, the tongue might push forward over the 500 m separating it from the river and form an ice-dammed lake. The railway also passed over the Allen Glacier for 8 km; fortunately the advance of the Allen in 1912 'was short-lived and did not communicate appreciable motion to the stagnant ice on which the track was laid' (Field 1932). Between 1911 and 1931 the Grinell retreated and by the time it was seen by a party from the American Geographical Society in 1966 a further retreat of several hundred metres had taken place. There is known to have been a minor advance of the Allen in 1966 and of the Childs in 1968, but there do not seem to have been any more observations of the Grinell (Field 1975).

9.5.3 Prince William Sound

The advances and retreats of the numerous glaciers flowing down to Prince William Sound, to the east of Seward, and also those of the Kenai Peninsula to the west, have not always conformed with expectations.

Trees overrun by the iceberg-calving Columbia Glacier and still in growth position, show that about AD 1215 it pushed into Terentiev Bay (Calkin *et al.* 2001).

Tree-ring dates from 21 subfossil trees, exposed by the glacier's recent retreat, show that it may have continued to advance from the fifteenth century into the nineteenth, with the creation of a series of ice-marginal lakes that drowned extensive forests. Having reached its terminal moraine in AD 1850, the front of the glacier oscillated behind the moraine until the lower reaches disintegrated. Since the early 1980s retreat has been rapid.

Six glaciers terminating in or near College Fjord were monitored after 1931, initially by surveying and photogrammetry and latterly using satellite imagery. Amongst them, the Yale and Harvard glaciers issue from the same snowfield into the same fjord. While the Harvard advanced at an average rate of nearly 20 m per year after 1931, the Yale retreated about 50 m yearly (Sturm *et al.* 1991). Terminal fluctuations here and, as we have noted already, in many places elsewhere, are determined by a variety of factors: the topography and dynamics of the individual glacier, the temperature of the ice, the presence or absence of a lake or fjord at the tongue, by aspect in relation to mobile weather systems, as well as by the regional climate.

Evidence from the Sargent Icefield, on the west side of Prince William Sound between Seward and Whittier, for three phases of glacier advance in the Little Ice Age is particularly precise, being based on radiocarbon ages and tree-ring crossdating of 130 subfossil logs at eight glacier forefields and tree-ring ages of twenty-eight moraines at fifteen glaciers. Thirteenth-century wood in the end moraine of the Nellie Juan, a northerly outlet of the Sargent Ice Field, provides evidence of an early Little Ice Age advance (Field 1975). Wood extracted from till of the Princeton, Tebenkof and Billings glaciers, has been crossdated with living trees, and it confirms that these glaciers, like the Nellie Juan, were burying forest from AD 1190 through 1300 (Wiles *et al.* 1999, Calkin *et al.* 2001). Trees buried by six glaciers show that the most extensive Little Ice Age advances were under way by 1540, with the Ultramarine and Nellie Juan glaciers advancing at 16–18 m per year. This second phase continued into the seventeenth century and gave terminal moraines which were stabilized by the mid eighteenth century. The third Little Ice Age phase is marked by the advances of ten

---

14   The Martin River, Guyot, Malaspina and Hubbard glaciers are all known to surge.

Table 9.7 Dates AD when glaciers extended beyond their present margins on the eastern, maritime flank and western, continental flank of the Kenai Mountains, and around Prince William Sound, Alaska

Kenai Mountains		Prince William Sound
Maritime	Continental	
400–660	490–630	620–700
890–970	1125	1200–1330
1420–60	1440–60	
1625–70	1690–1710	1630–50
c.1750	1720–60	1713–42
1880–1910	1830–60	1880–95

glaciers that built moraines dating to between 1874 and 1895 (see Tables 9.6 and 9.7, based on Calkin *et al.* 2001, figure 2).

A tree-ring chronology from hemlock, which is temperature sensitive, records several cooler than average, multidecadal periods, the first between AD 1100 and 1200 and others centred on 1400, 1660 and 1870. All of these periods were accompanied and followed by glacial advances. It therefore appears that, on a century timescale, cooling was the main factor responsible for phases of glacier expansion within the Little Ice Age in this coastal region of Alaska.

9.5.4 The Kenai Peninsula

On the Kenai Mountains which form the backbone of the Kenai Peninsula west of Seward, glaciers of a wide range of morphological types occupy a total area of 200 km². Intensive field surveys of the forefields of 16 of the land-based glaciers have allowed their Little Ice Age history to be reconstructed in considerable detail (Wiles and Calkin 1994). Radiocarbon dates of subfossil wood in moraines have been crossdated against tree-ring chronologies to obtain more precise dates than would have been possible using radiocarbon alone. Photographic records were used to determine the excesis (colonization) interval; it was found to be 15 years. Where trees were lacking, lichens were used to

obtain minimum ages for moraine formation, though in the boreal maritime climate of the southern mountains, the growth of lichens is so rapid as to limit the timespan over which they are useful age indicators.[15] All radiocarbon ages were calibrated (Stuiver and Reimer 1986) to give a range of calendar dates; Wiles and Calkin reported the mean ages of these ranges, their interpretations being based on the 1-sigma ranges.

The results indicate that some of the calving glaciers in the region fluctuate over centuries in concert with neighbouring land-based glaciers, only occasionally exhibiting divergent behaviour; others are quite out of phase (Wiles and Calkin 1990, 1993, 1994). Four glaciers, all of them southeastern outlets of the Harding Icefield, can be taken as examples. The Northwestern and McCarty glaciers, which terminate in the waters of fjords with those names, advanced strongly in the sixth century AD and retreated in the following century, as did the land-based Dinglestadt Glacier nearby and the Grewingk Glacier on the other side of the mountains (see Chapter 15). Both the Northwestern and McCarty advanced again as early as the tenth century, when there is no sign of a Grewingk advance. They also advanced in the fifteenth century and retreated in the late nineteenth century in phase with the surrounding land-terminating glaciers. The Aialik and Holgate Glaciers at the head of Aialik Bay were more discordant, both advancing during the Medieval Warm Period about AD 1100–1250 when their land-terminating neighbours were receding, and both showing relatively minor advances in the Little Ice Age.

The land-based glaciers on both the western and eastern sides of the Kenai Mountains are much more conformist. On both sides land-terminating tongues advanced during three succeeding phases between AD 1300 and 1850. While there were expansions on both flanks starting about 1300, the majority advanced in the fifteenth century and again in the eighteenth century. On both sides of the peninsula, glaciers advanced to positions close to their maxima about 1814–28, 1847–58, 1872–88, 1903–9, 1913–15 and 1920–30. There was then a general retreat until an advance around 1951.

---

15 A lichen curve was developed showing an initial growth rate of 0.37 mm per year, almost three times that in the Wrangell–St Elias Mountains (Denton and Karlén 1977) and more than four times that in the Brooks Range (Calkin and Ellis 1980). No estimate of probable accuracy was given.

The climate of the Kenai Mountains is dominated by the Aleutian Low giving a counter-clockwise circulation over the Gulf of Alaska. Mean annual precipitation and temperature are consequently higher on the east side (with 1690 mm at Seward, and an annual temperature of 4.1 °C) than on the west (where the precipitation at Homer is only 572 mm, and the mean annual temperature 3.0 °C). Most of the valley glaciers are accessible and much of the area has been mapped on a scale of 1:63,360. All the glaciers have receded from their nineteenth-century positions over distances varying from about 100 m to 2.3 km. Recent moraines abound. 'Pre-1750' moraines occur within a kilometre of the Sargent Icefield snout. Inside them are moraines dated to 1830 and smaller ones dated to 1890 and 1950. The Tebenkof Glacier, the largest in the Blackstone area, has an outer moraine, dated by Viereck (1968) to 1875–85, from which it had receded 250–350 m, according to Tarr and Martin (1914), and a further 300 m by 1935 (Field 1975).

Distinct differences have been found in the fluctuation histories of glaciers on the eastern more maritime flank of the Kenai Mountains and the more continental western flank though the time of onset of Little Ice Age advances seems to have been much the same on both.

On the west side of the Kenai Mountains, the land-terminating Grewingk Glacier had extended several kilometres outside its present position by 1442, and 100 k to the northeast, the Tustemena was approaching its Little Ice Age maximum by 1460. The Grewingk and Exit, according to the dates of wood on lateral moraines, reached their Holocene maxima by 1650 (Calkin and Wiles 1991). Advance phases of the western glaciers have been recognized from 1440 to 1460 and 1650 to 1710. They were retreating from their maxima from the early 1700s to late 1800s and re-advanced between 1830 and 1860. Dates of recessional moraines show that minor pulses of glacial activity occurred around 1880, from 1900 to 1920 and around 1950. Reconstructions suggest that the volume of the Grewingk–Yalik Ice Field at the nineteenth-century maxima may have been as much as 30 per cent greater than at the present day, and with equilibrium line altitudes 260–320 m lower. Present glacier fronts are between one and five kilometres behind their furthest extent in the nineteenth century.

On the east side of the Kenai Mountains, the Bear, Exit and Yalik glaciers, were advancing by 1460, 1587 and 1650 respectively. The eastern glaciers experienced advance phases about 1420–60, 1640–70, around 1750, and between 1880 and 1910. Retreat from the outermost end moraines took place from the late nineteenth century to the early twentieth century when equilibrium line altitudes increased by 100–150 m.

Differences in the time at which eastern and western glaciers reached and retreated from their most extended positions might have been caused, at least in part, by westward migration of the icefield divide, leading to enlargement of the eastern accumulation areas and progressively greater glacier extension through time (Wiles and Calkin 1994). The western, more continental glaciers advance as a result of cooler summers, the eastern maritime glaciers mainly as a result of high winter precipitation (Table 9.7) (Calkin and Wiles 1991, Wiles and Calkin 1994). Periods of low mean summer temperature were registered at Seward in 1908–10, from the early 1930s to mid 1940s, and during the late 1970s through the 1980s. During these periods the eastern glaciers did not expand as might have been expected. Around 1950 and after 1980, when moraine dating is relatively accurate and precise, summer temperatures were up to 0.5 °C above the mean of the preceding decades yet several glaciers advanced. Positive mass budgets during these periods appear to be due to increased winter precipitation accompanying the higher temperatures. This explanation is supported by mass balance data from Wolverine Glacier, 40 km to the east of the Kenai Peninsula (Mayo and March 1990, Haeberli et al. 1996). By contrast periods of moraine formation on the west side of the mountains correspond with times of low tree growth, and low summer temperature as shown by a tree-ring chronology from the Kenai Lowlands near Homer.

9.5.5 The Brooks Range

The Brooks Range forms a barrier roughly 1000 km long and 200 km wide which arcs east to west across northern Alaska between latitudes 67° and 70°N. In the centre of this Arctic range, mountain peaks reach 1200 to 1800 m in the south and 2100 to 2400 m in the north. In the Franklin and Romanzof mountains, with a combined ice cover of 260 km², most of the glaciers, including the largest in arctic Alaska, lie above 1500 m and include over 300 small subpolar icecaps, cirque glaciers, and valley glaciers up to 8 km long. Precipitation comes mainly from the Bering Sea to the west and varies between 500 and 1000 mm a year, about half falling as snow. At Anaktuvuk Pass in the north-centre of the range the mean annual temperature is about −10.1 °C and mean annual precipitation about 280 mm, of which 75 per cent falls as snow. Chronology depends to a great extent on lichenometry as the glaciers lie above the treeline. A lichenometric curve for the central Brooks Range was set up using measurements of the largest diameter of *Rhizocarpon geographicum sp.* and the faster-growing *Alectoria miniscula/pubescens* (Calkin and Ellis 1980, Evison *et al.* 1996). Calkin and Ellis estimated their curve to be accurate to within 20 per cent,[16] and identified clusters of advances around 1200, 1570 and 1860, displayed at 45, 85 and 42 glacier forefields respectively (Calkin and Wiles 1991). The 1570 advance has been supported by radiocarbon dates from moss and wood overrun by the ice. Ice margins may have remained close to those of 1570 until about 1780. Moraines from about 1860–90, formed immediately behind those of the sixteenth century advances, are most prominent in the eastern part of the range (Evison *et al.* 1996).

Major periods of moraine stabilization for fifteen cirque glaciers in the Killik river area were dated at 750, 320 and 70 BP (presumably AD 1200, 1630 and 1880) and this work was later extended to other areas in the Brooks Range where similar results were obtained (Haworth *et al.* 1983). More than half of the glaciers investigated in the west and central parts of the range had especially well-defined moraines dated to 475–325 BP (AD 1475–1625). These dates cannot be taken as accurate as the growth curve was calibrated by radiocarbon dates of which two were less than 400 BP, as well as dendrochronologically dated control points and dated surfaces from an Inuit settlement abandoned between 1873 and the 1880s, and gold placer mines believed to have been abandoned soon after 1901. It seems likely that glaciers in the western part of the

Brooks Range, where precipitation is heavier, may have advanced earlier in the Little Ice Age, as did those further east, but the evidence would probably have been buried by later advances. In the northeast, moraines dated about 450 lichen years BP (AD 1500) represent the most extensive Little Ice Age (and Holocene) glacial advances. They are similar in size and distance from the present ice fronts to those formed during the event dated to about 1890 (Evison *et al.* 1996).

On Seward Peninsula in westernmost Alaska, moraines of three cirque glaciers on the Kighaik Mountains mark advances which, according to a 'preliminary' lichen curve, culminated around 1645, 1675 and by 1825, associated with equilibrium line altitudes being depressed about 170 m (Calkin *et al.* 2001).

There is no doubt that the little glaciers in the Brooks Range and other interior mountains receded in the twentieth century. Comparison of present conditions with those displayed in photographs from 1901 and 1911 show general recession and thinning throughout the area, which still continues. Maxima were probably associated with equilibrium line altitudes 100–200 m lower and July temperatures 2° to 3 °C lower than the mean 1978–83 values (Calkin *et al.* 1985).

Calkin and Wiles (1991) concluded that the clear records of Little Ice Age expansion in Alaska were separated by an interval, correlative with the Medieval Warm Period, from earlier Holocene episodes of expansion, during which ice margins were at or behind their current positions. Advances of both land-terminating and calving glaciers occurred in the thirteenth century, but major advances in the highland area may have been delayed until the fifteenth century. Many mountain glaciers reached their maxima by the sixteenth century and, especially in the southern maritime areas, in the middle to late eighteenth century. Little evidence is found in arctic and continental interior areas of moraine building during the eighteenth century. Glaciers expanded throughout Alaska in the nineteenth century, those not blocked by older moraines expanding further than in earlier centuries, and in some cases reaching Holocene maximum positions. A small number of glaciers in areas of high precipitation have remained in

---

16 Evison *et al.* (1996) redrafted the Calkin and Ellis curve using calibrated radiocarbon ages (Stuiver and Reimer 1993) and judged that the original curve may overestimate the ages of lichens more than 20 mm in diameter by 20 per cent, but decided not to adjust the curve because it is only partly based on radiocarbon.

these positions, but retreat was dominant in the late nineteenth and early twentieth centuries. Equilibrium lines by the 1980s were 100–200 m higher in the arctic and up to 400 m higher in the maritime regions, possibly reflecting a rise in mean summer temperature of the order of about 2–3 °C and/or changes in precipitation (Williams and Wigley 1983).

Records from a thermal borehole in the Ogoturuk Creek area in northwest Alaska indicate that mean ground surface temperature had risen by 2–2.5 °C during the preceding 100 years, while similar records from the Barrow area indicated a temperature rise of 4 °C since 1850, with about half the increase occurring after 1830, although there were indications from both sites that recent cooling had penetrated the upper parts of the profiles (Péwé 1975).

9.6 MID TO LATE TWENTIETH-CENTURY GLACIER FLUCTUATIONS IN NORTH AMERICA

In North America there was a change from marked glacial recession towards stability or advance in the 1940s. At Tatoosh Island, near the Straits of Juan de Fuca, a trend towards decreasing temperature and increasing precipitation was under way by 1943 and glacier growth in the Cascades and Olympics followed (Figure 9.1) (Hubley 1956). The upper Nisqually Glacier began to thicken in 1944 and the Palisade Glacier in the Sierra Nevada in 1947. By 1953–5, of seventy-seven glaciers investigated in this region, fifty were advancing at rates of 3–100 m a year. Of the remainder, all but one was showing clear signs of increased thickness. The first glacier known to advance was the Coleman on Mt Baker in the North Cascades in 1949 (Harrison 1956b). The tongue of the Nisqually advanced 100 m between 1961 and 1971, and the Blue Glacier advanced 210 m between 1965 and 1981 and then remained relatively stable for the remainder of the decade (Spicer 1989). The response of glaciers in the Olympics was less marked than those in the North Cascades: only the Blue, Hoh and Black on Mt Olympus, where precipitation is greatest, advanced after about 1960. Inland, the glaciers in the Wind River Range were swelling at high levels in the early 1950s, although the fronts themselves were still retreating (Harrison 1956a).

Further north in Banff National Park, the Athabasca and Saskatchewan slowed down their precipitate re-

treats after 1950 as did the smaller Drummond Glacier (Nelson et al. 1966). The expansion of the 1950s observed in the Rockies, Cascade and Olympic mountains did not show up in the Juneau area, although there was a decrease in the rate of retreat of the Lemon Creek Glacier between 1948 and 1958.

In 1961, the state of 475 glaciers in western North America was surveyed from the air (Meier and Post 1962) (Figure 9.15). Detailed studies in North America and Europe have shown that the accumulation area ratio of glaciers, that is the ratio of the area above the equilibrium line to the total area, is closely related to specific net budget or mass balance. For a glacier with a linear increase in net budget with altitude and a symmetrical distribution of area about median altitude, an accumulation area ratio (AAR) of 0.5 would indicate that the glacier was in equilibrium and neither advancing nor retreating. AARs can be readily calculated for mid latitude glaciers from air photographs taken towards the end of the summer season. It is also often possible to judge from air photographs whether glaciers are advancing, retreating or stationary (La Chapelle 1962). Advancing glaciers are characterized by convex tongues with many crevasses; retreating glaciers commonly terminate in a fine feather edge and have fewer crevasses; stagnant tongues often have concave surfaces and are frequently more or less covered with ablation moraine.

Meier and Post examined 960 glacier tongues appearing on oblique air photographs taken in 1961. A consistent pattern was revealed (Figure 9.15). The area along the Pacific coast from the Kenai Peninsula to latitude 53°N in British Columbia had AARs of 0.6 and the glaciers appeared healthy. Over a large area of eastern Alaska, the southern half of the Coast Range of British Columbia and the Monashee Mountains, the AARs ranged from 0.5 to 0.6. It was believed that values in the eastern part of the Alaska Range and the Wrangell Mountains were probably similar but little data were available because of weather conditions. The glaciers of the Canadian Rockies, the Cascade Range and Montana had AARs of 0.25 to 0.5. In the western part of the Alaska Range, the Rocky Mountains in Wyoming, and the remnant glaciers of Idaho, AARs were less than 0.2. The consistency of AAR patterns over such wide areas must reflect the homogeneity of mass balance fluctuation histories (Letréguilly and

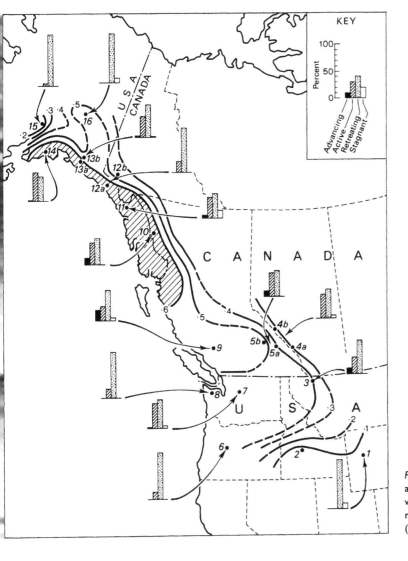

Figure 9.15 Distribution of values of accumulation area ratio (AAR) over western North America in 1962 in relation to glacier advance and retreat (*Source:* from Meier and Post 1962)

Reynaud 1989) and ultimately be associated with circulation patterns and synoptic situations.

Where AARs were more than 0.5, more than half the glaciers appeared to be active or advancing; where they were less than 0.2, all were retreating or stagnant. At the time of the 1961 survey, glaciers in the Cascades such as the South Cascade had negative mass balances once more, but further small increases of mass balance were to take place in the 1970s. Small glaciers like the avalanche-fed Vesper in the northern Cascades are much more sensitive to climate fluctuations than the more

massive glaciers in the Rockies (Dethier and Frederick 1981).

Recession was rapid in the Canadian Rockies between 1920 and 1950, but then slowed down markedly. Tree-ring widths near the treeline reached a twentieth-century maximum during the 1950s and 1960s and, in the neighbourhood of the Columbia icefield during the late 1960s and early 1970s, seedlings established themselves above the treeline (Luckman 1998). In the Premier Range, a study of some thirty-one glacier fronts between 1973 and 1976 found 90 per cent of them advancing or stable (Luckman *et al.* 1987). Between 1970 and 1985 fourteen of the glaciers advanced and built small terminal or lateral moraines. The advances seem to have resulted from positive mass balances caused by the greater winter precipitation between 1951 and 1968 and more especially lower summer temperatures during 1954–68, as recorded at Valemount (Luckman *et al.* 1987, Osborn and Luckman 1988). They had some positive years between 1965 and 1976 but a small negative balance overall. Between 1977 and 1995, winter precipitation and accumulation diminished and the Peyto's balance was consistently negative (Luckman 1998).

There were advances between 1950 and 1980 at many localities in Alberta and southern British Columbia, though none of the glaciers for which mass balance data are available advanced. In the Coast Mountains, the Sentinel was stationary for some years (Figure 9.9) (Osborn and Luckman 1988) and there were advances in the Clarendon Ranges between 1970 and 1979 (Ricker 1980). From the mid 1960s the Bugaboo advanced for a decade (Figure 9.9); in a maritime climate, its mass balance variations, at least in the short term, appear to be controlled by winter precipitation rather than by summer temperature (Lettréguilly 1988).

In the Franklin and Romanzof Mountains of the northeastern Brooks Range, changes have been traced from photographs of nine sample glaciers (Evison *et al.* 1996). Since the late nineteenth century their fronts have retreated by between 150 and 1330 m, and their total areas have diminished by 2 to 15 per cent. Thinning has been more important than recession, the altitude of trimlines showing that surface lowering may account for as much as 9 per cent of loss of volume. The available data is insufficient to trace minor interruptions of the general trend.

The subpolar Grizzly, Marmot and Buffalo glaciers in the Atigun Pass area of the Brooks Range were monitored from 1977 to 1982 using pit studies and stake networks (Calkin *et al.* 1985). Mass balances were negative in all years except 1981, when below normal summer temperatures over most of Alaska were associated with an unusually deep low pressure system over the Arctic Ocean. No significant variations in winter precipitation occurred over the period 1977–82, but equilibrium line altitudes fell progressively from 1977 to 1981. Variation in summer temperature appears to control changes in contemporary mass balance in this region of Alaska.

The terminal positions of six glaciers on Mount Baker, in the north Cascades, have been mapped using photogrammetry for the period 1940 to 1990. Rapid retreat until the early 1950s was followed by thirty years of advance, until retreat resumed in the early 1980s and continued at least until 1990 (Harper 1993). The termini of these maritime glaciers, with a high percentage of their areas at high altitudes, respond rapidly and sharply to climatic changes, especially to variations in precipitation.

Nine glaciers, representing a range of geographic and topographic characteristics, were selected for mass balance monitoring in the North Cascade Glacier Climate Project which was designed to identify the nature of the response of glaciers to regional climate change (Pelto 1996).[17] The mean annual net balance between 1984 and 1994 was negative at 0.38 m per year, resulting in a mean loss of 4.2 m of ice thickness, a substantial fraction of the total as the 1971 depth of the North Cascade glaciers was estimated to be only 30–50 m (Post *et al.* 1971). Positive balances, especially in 1984 and 1991, have not been sufficient to compensate for such losses. In 1985, thirty-eight of the forty-seven glaciers observed were retreating, and by 1994 all but one were doing so. Recession rates increased during the interval and some small glaciers disappeared. Comparison of records from eight weather stations in the Cascades showed that during 1984–94 the mean ablation season temperature was 1.1 °C above the 1950–80 mean, which implied that retreat would continue for several more years.

Pelto and Hedlund (2001) extended their studies to thirty-eight North Cascade glaciers and concluded they could distinguish three different kinds of glaciers so far as terminus behaviour and response times are concerned. Type 1 glaciers, steep, extensively crevassed and with fast-moving ice near the tongue responded to

---

17 See also correspondence in *Journal of Glaciology* (1997) 43: 192–6.

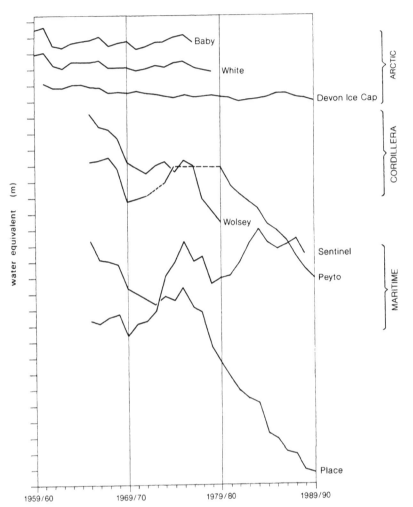

Figure 9.16 Cumulative mass balance of selected glaciers in arctic, cordilleran and maritime North America (*Sources*: derived from figures published in Kasser 1967, 1973, Müller 1977, Haeberli 1985, Haeberli and Hoelzle 1993)

climatic change in 20–30 years; type 3 glaciers, gently sloping, not much crevassed and slow-moving near the tongue took between 60 and 100 years to respond. Type 2 glaciers are intermediate in all respects.

The 32-year mass balance record of the South Cascade Glacier (Haeberli *et al.* 1996: fig. 9),[18] is long enough to allow its relation to variations in atmospheric circulation pattern to be considered (McCabe and Fountain 1995). Since the mid 1970s winter mean 700 mb heights over western Canada and the northwestern contiguous United States have increased and winter mean 700 mb heights in the eastern North Pacific Ocean, centred near the Aleutian Islands, have fallen. This change in pressure distribution has reduced the movement of storms and moisture into the region, and has also caused an increase in subsidence, resulting in warming and drying of the air, helping to reduce precipitation and also to increase the ratio of rain to snow during the winter. The decrease in winter precipitation, with increased winter temperatures, raised the ratio of rain to snow and led to diminished winter accumulation balances. Primarily because of this, the yearly net

18 South Cascade Glacier (perversely sited at the crest of the North Cascade Mountains), was selected as a benchmark glacier for the US Geological Survey's *Glacier Monitoring Program*.

balance of the South Cascade has been decreasing since the mid 1970s. The glacier has consequently suffered a dramatic decrease in volume during the last two decades. It may be taken that the correlations between the mass balance variations of the South Cascade and atmospheric circulation also apply to other glaciers in the region.

The mass balance of glaciers in Washington and southern British Columbia has at times been inversely correlated with that of glaciers in Alaska (Meier *et al.* 1980). The sudden decrease in mass balance of the South Cascade in the mid 1970s was mirrored by a corresponding increase in the Wolverine Glacier in Alaska, associated with rises in both accumulation and winter air temperature. This is explicable in terms of increase in 700-mb heights over western Canada and decrease in heights over the eastern north Pacific since the mid 1970s having led to enhanced southerly air flow from the Gulf of Alaska into Alaska. This promoted the intrusion of moist, warm air into Alaska, causing increased winter precipitation and positive mass balances.

The poverty of modern measurements in Canada stands in strong contrast to the quality of data on early Little Ice Age events which is emerging from investigations of moraines and related tree-ring sequences. It is to be hoped that a deliberately planned system of monitoring, preferably including mass balance observations of glaciers selected to represent the widely varying environments in which Canadian glaciers persist, may be initiated as part of the palaeoclimatic data base identified by the IGCP Scientific Steering Committee as a major goal (Luckman 1990).

Some fifteen glaciers on Mount Wrangell in Alaska have been monitored for the last three decades by surveying, photogrammetry and satellite observations. Most termini have been stationary or have retreated slightly, but glaciers on the active North Crater of this volcano began to advance in the early 1960s and have advanced steadily between 5 and 18 m since then. This can reasonably be attributed to changes in basal conditions caused by volcanic meltwater (Sturm *et al.* 1991).

The longest twentieth-century glaciological record from the US arctic comes from the McCall Glacier (69°18′N, 143°48′W) in the northernmost part of the Romanzof Mountains, northeast of Brooks Range

(Rabus *et al.* 1995). This 8-km long glacier has an average width of 640 m, the ice emerging from three cirque heads and extending from above 2700 m on the north face of Mount Hubley to 1359 m at the terminus. The McCall was observed during the International Geophysical Year in 1957/8, and from 1969 to 1975 as part of the activity of the Hydrological Decade. Mass balance maps were constructed each year from 1968/9 to 1971/2. The mass-exchange rate was found to be low; mean annual accumulation was only +0.16 m and ablation −0.3 m in 1969–72. The ice front rose from about 1320 m in 1958, to 1327 m in 1970, and 1354 m in 1993. Mean mass balance became roughly twice as negative in 1972–93 as it had been in the period 1958–71. The ice surface was consequently lowered everywhere; by about 3 m in the cirques and as much as 42 m near the present terminus, the rate of lowering doubling around 1975. These changes reflect the renewed warming of the Arctic in the 1970s which started in the Barents and Kara Seas and, spreading westwards, was most marked in the Alaskan region where annual temperatures rose by 1 °C (Kelly *et al.* 1982). Warming had started after 1890 and culminated in the 1930s, with winters about 2.5 °C and summers 1.3 °C warmer than in the 1880s. The Arctic cooled in the 1950s and remained cool in the 1960s with annual temperatures 0.85 °C below those of the 1930s. Glaciers in Arctic Alaska withdrew by 150–700 m from their Little Ice Age moraines as a result of the overall warming (Calkin 1988). The McCall Glacier appears to have a surprisingly fast reaction time, less than a decade. It will not be known how representative the changes in mass balance are of other glaciers in the region until further data have been collected.

9.7 A SUMMARY OF THE LITTLE ICE AGE GLACIER ADVANCES IN NORTH AMERICA

The first Little Ice Age glacier advances in North America, in the Canadian Rockies, were somewhat earlier than most of those in Eurasia. On the mountains of the Pacific coast of Alaska the first advances were at a similar time to those in Europe. Fifteenth-century advances, rarely recognized in Eurasia, have been described from the Kenai Peninsula and the Brooks Range of northern Alaska, and early sixteenth-century advances from the Cascades. Little Ice Age glacial maxima were reached at various times; in the thirteenth century in the Cascades and Chickamin mountains, in the late fifteenth century in the northeast Brooks Range, the mid seventeenth century in the western Kenai Mountains, the mid to late eighteenth and mid nineteenth centuries by glaciers in the Canadian Rockies, the Juneau Icefield and Prince William Sound. Advances in the nineteenth century appear to have been at similar but not identical times to those commonly identified in Eurasia, namely around 1820 in the Cascades, 1850 in the Cascades, Juneau Icefield, western Kenai and the Brooks Ranges, 1885 in the Cascades, Canadian Rockies and around Prince William Sound. Twentieth-century recession was commonly interrupted about 1940, somewhat earlier than in Eurasia.

10
ARCTIC ISLANDS

The Arctic islands consist of Greenland, with Svalbard (Spitsbergen, Franz Josef and other islands nearby), Jan Mayen and Novaya Zemlya to the east, and the Canadian Archipelago to the west (Figure 10.1). All these islands support extensive and massive glaciers and icecaps. Together they make up about 12 per cent of the globe's ice cover, some two million square kilometres, 90 per cent of it in Greenland.

For much of the year most of the islands are surrounded by sea ice which maintains coastal air temperatures near freezing into the early summer. Where open water persists through the winter, temperatures of coastal areas are prevented from falling far below zero, evaporation takes place from the sea surface early in the summer, and precipitation is heavier than elsewhere. Consequently the extent and persistence of sea ice, which depends in large part on the ocean currents, has an important bearing on the economy of the glaciers and their Little Ice Age histories.

On all the islands the summits of mountain ranges emerge from the lower courses of ice streams flowing from central ice domes down ancient valley systems towards the coast. Many of the ice streams terminate in fjords. Coastal plains and land-based ice termini are most extensive in southwest and southeast Greenland, well outside the Arctic Circle. Other ice-free coastal areas are far to the north of the Arctic Circle in western Spitsbergen and southern Novaya Zemlya.

The coastal margins of Greenland and the Canadian Archipelago have long been homes to the Inuit. Their ancient settlement sites are beginning to yield palaeoenvironmental information to archaeologists. More valuable for our purposes have been the settlement histories and archaeological remains of the Norse settlers of Greenland, dating from the Medieval Warm Period.

As in Scandinavia and Alpine Europe, some arctic glaciers are believed to have advanced in the fourteenth century and to have reached a maximum extent either at this time or a few centuries later. The history of arctic exploration suggests there may have been a warming phase in the sixteenth and early seventeenth centuries which allowed Russian seamen to sail beyond North Cape in 1496, Willoughby to discover Spitsbergen and winter on the Kola Peninsula in 1553, Borough to reach the Kara Sea in 1556, and Barents to discover Novaya Zemlya in 1594 (Armstrong 1984). Norwegians visited Greenland again in 1577 while, to the west of Greenland, Sebastian Cabot in 1508–9, Frobisher in 1576, Davis in 1585–7 and Baffin in 1616 penetrated ever further north. Then, for a century and a half, the effort to find Arctic passages to the Pacific stalled. Warming towards the end of the nineteenth century and in the first half of the twentieth century, which Overpeck *et al.* (1997) would put at 2 °C[1] has been associated with a general retreat of ice fronts (Figure 10.2).

Greater greenhouse warming in the Arctic than elsewhere is predicted by most general circulation models of the atmosphere on account of positive feedbacks between air temperature, ice extent and surface albedo. However, weather observations, ice sheet

---

1 Rather more than some would take it to be.

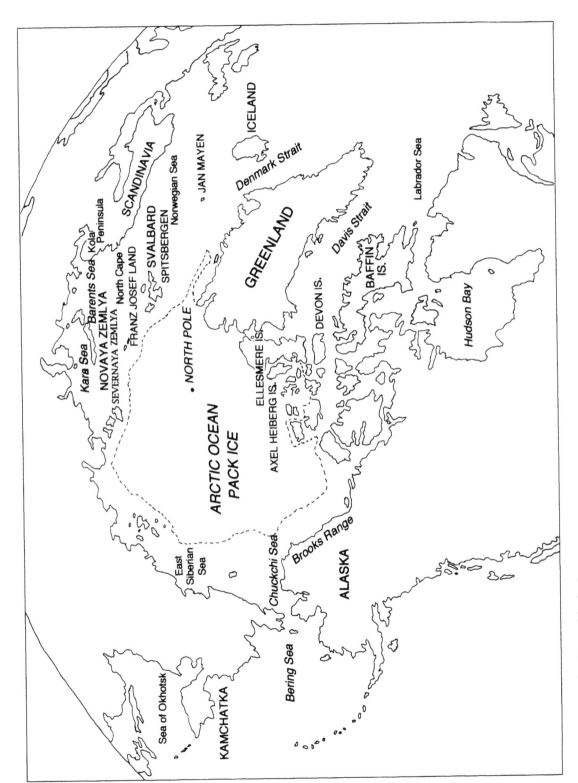

Figure 10.1 Location sketch map of the Arctic

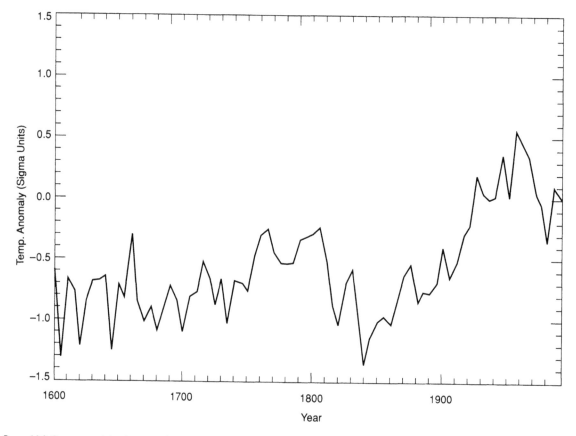

Figure 10.2 Summer-weighted means of Arctic temperature for 1600–1990 based on proxy records. The results are 5-year averages plotted as normalized deviations from observed temperature means (*Sources*: from Overpeck *et al.* 1997 and Serreze *et al.* 2000)

accumulation layers and glacier mass balance measurements all indicate widespread cooling on land in the Arctic for a decade or two in the latter half of the twentieth century (Braithwaite *et al.* 1992, Kahl *et al.* 1993). As in most other parts of the world, temperatures fell in the 1960s and 1970s, and then there was renewed warming at the end of the millennium (Lefauconnier and Hagen 1990, Serreze *et al.* 2000). Overpeck *et al.* (1997) concluded that Arctic temperatures in the twentieth century were the highest in the last four centuries, with an overall warming of 2 °C concentrated in the period 1840–1960. Arctic glaciers at low altitudes have responded to twentieth-century warming by retreating almost everywhere.[2]

10.1 SEA ICE INFLUENCES

Arctic pack ice covers an average of 14 million km² at its greatest extent in February[3] and is about half this area in the autumn (as shown by pack ice in Figure 10.1). With its high reflectivity of incident solar radiation, this great ice mirror is a major driver of global weather systems and greatly influences the circulatory systems of the ocean.

---

2 Inevitably the warming has also had a powerful effect on Arctic wildlife. See http://www.greenpeace.org/climate/arctic99/reports/seaice3.html.

3 This is about equal to the total area of the globe covered by ice.

A reduction in sea ice offshore causes a step-up in temperature inland, as we have seen in Iceland (Chapter 2, p. 25). In the Arctic, sea ice cover fluctuated about a mean in the first half of the twentieth century and then in the second half decreased by about 15 per cent in summer and by 8 per cent in spring (IPPC 2001: 125). Most of the sea ice decline so far has been in the Barents, Kara and East Siberian seas north of Russia, as well as in the Sea of Okhotsk near Kamchatka. In sharp contrast, an increasing amount of sea ice developed in the Davis Strait and the Labrador Sea between Canada and Greenland and persisted into the 1990s. The major ice domes have not been much affected directly by the warming as temperatures above 2000 m have remained below freezing, but retreating outlet tongues must eventually cause rates of inland ice discharge to increase.

Associated with sea ice reduction has been a decline in average sea level pressure over the central Arctic Ocean, allowing cyclonic depressions to penetrate far to the north (Chapman and Walsh 1993). Warm air masses arriving in spring as well as in summer caused the melt season to lengthen from fifty-five days in 1979 to seventy-five days in 1996. Open water 'leads', which reflect less sunlight than the ice, widened, thereby accelerating the warming process. Submarine reconnaissances beneath the ice which began in 1958 have found that ice thickness has diminished by about 40 per cent (IPPC 2001: 126), though some other estimates are lower.

Warm salty surface water brought north by the Gulf Stream and the North Atlantic Drift is cooled in the Labrador Sea and Norwegian Sea. Losing its heat to the atmosphere and becoming more salty and dense as ice forms on the surface, it sinks to deeper layers in the ocean and moves south. This formation of deep water is the main force driving the Ocean Conveyor which eventually brings the dense cooled water back to the surface in the Indian and Pacific Oceans (see Chapter 17, pp. 568–9). If the Conveyor were to cease to function, because of an excess of reduced ice formation, the thermohaline circulation would slow down or cease and the Gulf Stream and North Atlantic Drift which warm northwest Europe might fail to penetrate so far north (Broecker 1997, Clark et al. 2002).

In 1968, a large area of unusually cold, fresh water, which seems to have been derived from melting Arctic pack ice, appeared off the west coast of Greenland. Now known as the Great Salinity Anomaly, it significantly reduced deep water formation in the North Atlantic and could conceivably be responsible for the subsequent increase in sea ice in the Davis Strait and Labrador Sea (Mauritzen and Häkkinen 1997). Another threat to the Conveyor comes as the climate warms, cyclonic depressions reach further north, and more rain and snow fall on the Arctic Ocean and the North Atlantic reducing the saltiness and density of the water. Over the last forty years, increases in annual precipitation of 15–30 per cent have been reported from western Alaska and elsewhere in the Arctic. The analysis of long hydrographic records has shown that the flow of cold dense water over the sills of the Denmark Strait and the Faroe–Shetland channel into the North Atlantic has steadily diminished over the last four decades and has already led to sustained and widespread freshening of the deep ocean (Dickson et al. 2002).

10.2 GREENLAND

10.2.1 Norse settlement and medieval climate change

On the western and eastern coasts of southern Greenland the climate was relatively benign when early voyagers, including Erik the Red, succeeded in establishing permanent settlements towards the end of the tenth century AD. An early Medieval Warm Epoch has been detected in ice cores extracted east of the central Greenland ice divide, between Summit and Northice, but no pronounced Little Ice Age (Friedmann et al. 1995). Archaeological evidence indicates that Inuit may also have been influenced by the climatic amelioration, for about AD 1100 they moved into Greenland from arctic Canada and eventually, by about AD 1300, had set up shelters in the outer fjords of the Norse west coast settlements (Figure 10.3).

Norse farmsteads persisted in Greenland for nearly 500 years, surviving there, isolated, for some decades after contact with Scandinavia had broken. As Lamb (1977) postulated, it appears that although economic and political factors played their parts, the isolation of the Greenland settlements and their eventual demise were primarily the outcome of climatic deterioration. Drift ice discouraged navigation at a time when the attention of the authorities in Scandinavia was being

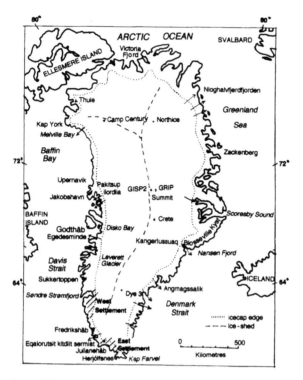

Figure 10.3 Location sketch map of Greenland

increasingly diverted to continental affairs. The Norse farmsteads, based on rearing livestock, dwindled and were eventually deserted, probably in the course of extremely hard winters, and Greenland was left to the Inuit for the rest of the Little Ice Age.

Much effort is currently being directed towards throwing more light on the relative importance of the parts played in the end of the settlement by environmental and human factors and their interrelationships. For instance, Fricke *et al.* (1995) have argued that the proportion of $\partial^{18}O$ in mammal teeth reflects the proportion of that isotope in the water their users have ingested and should therefore indicate the temperature in their lifetimes. From an examination of the oxygen isotope composition of the enamel in mammal teeth from various archaeological sites they concluded that temperatures in Greenland declined between about AD 1400 and 1700.

The climatic significance of foraminifera and mud layers in sediment cores from Nansen Fjord on that part of the east coast of Greenland closest to Iceland

has been investigated by Jennings and Weiner (1996). They note that sea ice today affects the Greenland coast throughout the year, whereas it usually lies 200 km off the west coast of Iceland because of the Irminger Current moving clockwise around the island. Sediment has accumulated on the floor of Nansen Fjord at a rate of a millimetre or two annually. The assemblages of foraminifera species in the sediment are regarded as sensitive to the relative strengths of the East Greenland Current at different times: colder, somewhat fresher Polar water, alternating with somewhat less cold and slightly saltier Atlantic Intermediate water. In addition, the thicknesses of layers of mud from icebergs calving off the Christian IV glacier into the head of the Fjord are interpreted as indicating periods of low temperature. Sediment cores dated by radiocarbon, assuming a reservoir age of 140 ± 50 years, have provided a surprisingly detailed record. The regional climate around Nansen Fjord, it has been concluded, was warmer and more stable than today between about AD 730 and 1100, then colder in the decades around 1150 and again towards 1370. The foraminiferal assemblages point to short-lived, warmer conditions culminating around 1190 and 1470, temperatures then declining toward a severe but variable cold period lasting from 1630 to 1905.

Isotopic signals in Greenland ice cores suggest particularly low annual temperatures in AD 1308–18, 1324–9, 1343–62 and 1380–4, with summer temperatures especially low between 1349 and 1356 (see Chapter 16, p. 550). These dates for cooling correspond extraordinarily well with times of climatic stress that have been identified in western Europe (Chapter 18).

Christiansen (1998) has found sedimentary evidence in nivation hollows in the Zackenberg area that the periglacial landscape of northeast Greenland responded quickly to Little Ice Age climatic deterioration. Greater winter wind speeds between AD 1240 and 1420 are indicated by increased nivation activity. Aeolian sedimentation followed between 1420 and 1500/1580, and then there was much snow and renewed wind action between 1500/1580 and 1690.

Documentary evidence relating to Norse Greenland, collected together in two scholarly works, *Grönslands Historiske Mindesmærker* by Finnur Magnússon (1838 and 1845) and *Grænland Í Miðaldaritum* by Ólafur Haldórsson (1978), point to cold years around 1320 and the late 1340s when the old route from Iceland to Greenland was no longer used, at least in part because of the presence of sea ice (see Chapter 2, p. 21).

A variety of archaeological evidence, including well-preserved fossil beetle and insect faunas, especially flies, indicates that Greenland's West settlement ended some time between 1341 and 1363 (Barlow *et al.* 1997, Buckland and Wagner 2001).

Archaeologists have thought that the Norse failed to adopt the harpoon technology of their Inuit neighbours and may therefore have been less successful as seal hunters and survivors (Barlow *et al.* 1997). However, measurements of the stable isotope ^{13}C in the bone collagen of Norse settlers, living at times dated by radiocarbon, are claimed to indicate an increase in the proportion of food of marine origin in Norse diets from about 20 per cent in the eleventh century to 80 per cent in the fourteenth. It seems that seals could well have become the main source of food for the Norse Greenlanders, possibly in response to a deteriorating climate (Arneborg *et al.* 1999), as they had long been for the Inuit.

The East settlement, with winters somewhat less severe than those of the West settlement (both are on the west coast, with the East settlement to the southeast of the West, Figure 10.3), managed to survive for another century or so. It is known from archaeological evidence that Herjolfsnes remained inhabited until after 1480 but the climate had evidently deteriorated several generations earlier, for costumes on bodies buried before the 1350s were penetrated by tree roots, while those of subsequent decades remained undisturbed and in a remarkably good state of preservation, indicating that not long after the interments the ground had become frozen a metre or two deep throughout the year (Nörlund 1924, Hovgaard 1925, Gad 1970).

10.2.2 Advances and retreats of Greenland outlet glaciers in the eighteenth to twentieth centuries

The positions of the ice edge and the extremities of the glaciers remained unrecorded until the eighteenth century when Paars, the Danish Governor of Greenland, wrote of a visit he had paid in 1729 to the edge of the Inland Ice in the Godthåb district. He mentioned that the ice edge was near to a waterfall; the only possible location for which can be identified well forward of the present ice front, near a trimline marking the extreme limit of ice advance in recent centuries.

Weidick's (1959, 1963) careful studies of the historical evidence show that the ice in southern Greenland was everywhere more extensive in the eighteenth than in the twentieth century. For instance a trader, Peder Olsen Walloe, who set off towards Julianehåb in an *umiak* (a large Inuit canoe) in the summer of 1751 and made a reconnaissance of southernmost west Greenland, wrote in his diary

> the Greenlanders say that the ice increases every year, which is mostly recognizable from the fact that tracks where the Greenlanders used to go hunting are now quite overridden and covered by ice, and, as far as may be concluded from their simple chronometry, the change that has taken place in a score of years is very considerable.
>
> (Weidick 1959)

Detailed confirmation comes from Otho Fabricius, the incumbent missionary at Fredrikshåb from 1768 to 1773:

> the ice spreads out more and more every year . . . the experiment has been tried of erecting a post on the bare ground a good distance from the ice, and the next year it was found to be overtaken by it. So swift is this growth that present day Greenlanders speak of places where their parents hunted reindeer among naked hills that are now all ice. I myself have seen paths running up towards the interior of the country and worn in bygone days but now broken off at the ice, which confirms the Greenlanders' statement. The ice advances especially in the valleys and where these reach the sea and the heads of fjords (I mean the inner ends of the fjords) it becomes so dominating as to have great floes hanging over the water.
>
> (Weidick 1959)

It was not only the Greenlanders' reindeer hunting paths that were disappearing under the ice. In 1765 an Icelandic pastor, Egill Thorhallesen, was sent to Godthåb and made summer expeditions to both north and south for two years before joining the Godthåb mission. He wrote a brilliant description of the region (Gad 1970, 1973), in which he mentioned that the Greenlanders 'also speak of other ruins, some of which are still to be seen and some are already under the ice which has laid itself over the entire hinterland, indeed over the highest mountains and filled the valleys between them'. At the same time, he does not seem to have been over-impressed by the severity of the climate, remarking in his report that he had no doubt Icelandic farmers could still make a livelihood for themselves if they settled on the fjords of southwest Greenland.

From his studies of the moraines of the two Tasiussaq glaciers in the Sukkertoppen area, Beschel (1961) inferred glacier advances about 1600 and 1750. His pioneering study, though influential in initiating the use of lichenometric dating, was not sufficiently well based to establish that there really was an advance about 1600. The historical evidence makes it clear that the glaciers of West Greenland advanced in the mid eighteenth century, as did those in Iceland and Norway, but it has not yet been ascertained whether or not there was an earlier advance.

For the interval between the mid eighteenth and mid nineteenth centuries, little information about fluctuations in the position of the ice is available except from one glacier, the Sermitsiaq in Tasermiut, which was reported to be advancing in 1833 (Weidick 1959).

For glaciological purposes the best of the early charts and maps are those drawn by reindeer hunters and compiled by Rink in the mid nineteenth century. Though often distorted in outline, they reveal good knowledge of detail. Like Jan Møller's pictures, they are now safely housed in the Royal Library collection in Copenhagen (Weidick 1959). Additional pictorial records survive from the latter half of the nineteenth century. In 1877 the Danish government appointed a commission to promote geographical and geological investigations in Greenland (Kommissionen for Ledelsen af de geografiske og geologiske Undersøgelser i Grønland) and the journal *Meddelelser om Grønland* began to appear.

Weidick (1968) has reviewed the evidence for the fluctuations of both the Inland Ice and the local glaciers in western Greenland. Compiling data on some 500 ice lobes, he plotted the fluctuations of 135 lobes between Kap Farvel in the south and Upernavik in the north for which information is available over more than three decades, providing detailed lists of all his sources. The results show a very marked consistency of behaviour although a few anomalies are revealed, especially amongst calving glaciers. The oscillations of 94 per cent of the Inland Ice lobes were in phase, despite their wide geographical distribution and differences in subglacial topography, size and dynamics.

Predominant retreat of the ice margins since the mid nineteenth century was interrupted by halts or readvances around 1880 and 1920. In some places in the south the readvances of 1880 reached a maximum extent, covering nearly all the 'trim zone', the area covered by previous late Holocene advances. In Disko Bay, glaciers generally reached their historic maxima in the mid nineteenth century. The front of Jakobshavn Isbrae at the southern end of the Bay receded 25 km between 1850 and 1950 leaving fresh moraines as evidence and a trimline 250–300 m above the surface of the tongue. Since then, this outlet glacier has been relatively stable though its front fluctuates seasonally through about 2 km (Weidick 1992).

The Pakitsup ilordlia ice lobe in Qingua kujatdleg was exceptional in advancing beyond the earlier maximum, over a vegetated area. Following a recession, a readvance in the 1920s on a smaller scale tended to be more marked in the north than the south (Davies and Krinsley 1962). Subsequently a general retreat set in. Since the historic maximum, most of the Inland Ice lobes have retreated on average by about a kilometre. Most of those fluctuating strongly, and all of those that have been known to oscillate through more than 5 km, are calving lobes.

Around 1950, major sections of the ice margin were thinning and retreating, though a few cases of growth were recognized in highland areas. Then, in the 1980s, a widespread advance commenced with marginal expansion spreading from the highlands to parts of adjacent lowlands, around Eqalorutsit kitdlit sermiat and Sarqap sermerssua for instance (Weidick 1991). The advance in the highland areas may reflect long-term changes in the accumulation of the interior of the ice sheet with a time lag of a century or more; alternatively, it could result from changes in the mass balance of the lower, marginal parts of the ice sheet, more vulnerable to short-term fluctuations in meteorological conditions.

Detailed observations of a part of the lowland ice margins near the coast have been made of the Leverett Glacier, 20 km west of Kangerlussuaq (Figure 10.3) (Tatenhove *et al.* 1995). The fluctuations in the position and surface altitude of this small outlet glacier of the main ice sheet have been traced from air photographs for the period 1943–92. From 1943 to 1968 the ice surface dropped at a rate of −0.6 m per year.

Then it rose at a mean rate of +1.2 m annually from 1968 to 1985, and at a rate of +0.4 m from 1985 to 1992. The average summer (May–September) temperature at Kangerlussuaq was relatively high during 1943–68, relatively low 1969–85 and then warmer again 1986–92. For the period before 1985 the observed change in elevation might be explained in terms of reduced ablation rates caused by the lower summer air temperature. Thickening in the years around 1990 seems to require either increased inland ice discharge or increased snowfall at the ice margins. Changes in the ice sheet margin must depend upon the relationship between meteorological conditions near the coast and changes in outflow from the inland ice sheet which is probably affected by climate variations on a longer timescale.

Small glaciers, sensitive to short-term meteorological changes, are clearer indicators of Little Ice Age climatic fluctuations than those emerging from the Inland Ice, but only a few of the many in Greenland have been investigated.[4] The most numerous records of the behaviour of local glaciers come from the Sukkertoppen area and Disko Island.[5] Like the lobes of the Inland Ice, most of these local glaciers reached their maximum recent extents before the nineteenth century, perhaps as early as 1750. The majority remained close to their most advanced positions until the mid nineteenth century, although a few receded earlier leaving large areas of dead ice. A general reactivation of the fronts between 1880 and 1890 brought advances that generally failed to reach the previous maxima. Subsequent retreat was interrupted by minor advances between 1915 and the mid 1920s that in many cases sufficed only to stabilize moraines. A general retreat in the twentieth century, fastest between the 1920s and 1940s, coincided with a period of increasing temperatures at coastal stations that culminated in the 1940s and 1950s.

According to Weidick et al. (1992) the timing of the retreat of valley and cirque glaciers since 1850 has coincided, although the distances involved have differed. However, this generalization was based on quite

a small number of cases. While some small glaciers appear to have thinned very little since the mid nineteenth century,[6] the rise in the altitude of the snowline during the twentieth century has separated some of the tongues of outlet glaciers from their sources in ice-covered plateaus inland, leaving dead ice in the lower valleys (Weidick et al. 1992). Cairns in Godthåbsfjord, erected by deer hunters, melted out of semiperennial snow cover on the plateaus in the 1970s (Rosing 1988). Radiocarbon dates on moss indicate that the cairns were built between AD 1290 and 1400 (Weidick et al. 1992). This suggests the possibility that snow and ice were extending in West Greenland early in the Little Ice Age, at about the same time as glaciers were advancing in the Canadian Rockies and the European Alps.

Air photographs on a scale of 1:10,000, taken by the Danish Geodetic Institute in 1968–9, provided a base on which ice fronts were remapped in 1978. Seven glacier forefields in the Ikamiut kangerdluarssuat mountains, north of Sukkertoppen (65°24′N, 52°52′W), were mapped and moraine ridges were dated by lichenometry (Gordon 1980, 1981). In addition, investigations of the environment of two outlet glaciers of a small icecap were made. The results must be regarded as tentative as the linear growth rates of the lichens had to be estimated from the maximum size of thalli plotted against distance from the ice margin as it appeared on 1942 air photographs. The estimates of ice retreat from moraines dated by the lichens to about 1745, 1850, 1885, 1930 and 1944 are in agreement with documentary and other data tabulated by Weidick (1963). Evidence of a more extensive glaciation before 1745 was found on two of the forelands. The most prominent end moraines and trimlines were formed about 1850. Between 1945 and 1968/9 the nine glaciers receded over distances of between 180 and 570 m, three of them continuing to recede until 1978. No lichens or plants had started to grow near the fronts of three of them in 1978 and they are presumed to have continued to recede until that year. Patches of moss were growing right up to the margins of six glaciers

---

4   Of the world's local glaciers, 10 to 20 per cent are in Greenland (Weidick et al. 1991).

5   Several of the land-based glaciers on Disko are known to have experienced large fluctuations in the last few centuries which may have been surges, but they have shown no recent tendency to surge.

6   The snow and ice cover on the northeast side of Hjortetakken (Kobbefjord, near Godthåb) is instanced on the basis of comparison of a 1984 air photograph and a painting by H. Rink in 1853.

which had readvanced 20–158 m between 1968 and 1969.

Elsewhere in Greenland in the last half of the twentieth century, some glaciers were stationary and some advanced. Weidick's (1968) graphs show stability or a reduced rate of retreat after 1940. By 1960 there were signs of readvances around Disko Bay, and the valley glacier Sermikavsak on Upernavik Island was advancing in the 1970s (Gribbon 1979).

An important question is whether the oscillations of the Inland Ice and the local glaciers are synchronous. The major trends in the movements of the ice margins of south and west Greenland distinguished by Weidick appear also to have affected north and east Greenland. The historical data for northern Greenland between Kap York and Nioghalvfjerdfjorden were examined by Davies and Krinsley (1962), who compared their observations with the accounts by Chamberlain and Salisbury written after the Peary auxiliary expeditions of 1894 and 1895. All the glaciers had retreated markedly, though by 1960, out of the 203 glaciers considered, 160 appeared to be stationary. Much less analysis has been attempted of the scattered data available for the east coast.

Four ice cores drilled to the north of Summit have provided evidence of the Little Ice Age in central Greenland. Values of $\partial^{18}O$ were found to be distinctly low during the seventeenth and the first half of the nineteenth centuries, indicating a fall in temperature of about 1 °C, while the salt aerosol load increased by 20–30 per cent, suggesting a more active circulation.

10.2.3 The question of current Greenland inland ice equilibrium

The uniformity of trend displayed in Weidick's (1968) graphs strongly suggested that temperature dominated ice marginal behaviour. For instance, minor readvances of the Sukkertoppen glaciers are explicable in terms of periods of low summer and autumn temperatures in the 1880s, low summer temperatures in the 1890s, and low temperatures at all seasons in the 1920s, as recorded at the meteorological stations at Godthåb (64°10′N, 51°33′W) and Sukkertoppen. Between 1875 and 1920, periods of climatic cooling were short and the general recession was broken by only short periods of enlargement. Rapid frontal retreats between 1920 and 1942 were associated with higher spring, summer and autumn temperatures giving long ablation seasons. A decrease of mean summer temperature which shortened ablation seasons in the Arctic between 1942 and 1960 was accentuated during the 1960s and 1970s. In West Greenland, lower summer temperatures were accompanied by colder and drier autumns. The mean temperature at Godthåb of 3.4 °C for 1968–73 was the lowest 5-year mean since records began in 1875. Winter and summer precipitation increased sharply in the 1950s and 1960s so that accumulation benefited from higher precipitation as well as cooling.

Glaciological and hydrological investigations have been stimulated in recent years by the opportunities for developing hydroelectric power resources. By 1981, mass balance measurements were being made on eight glaciers between Søndre Strømfjord and Kap Farvel,[7] and the Geological Survey of Greenland was mapping and making annual measurements of glacier margins. Air photograph cover of the whole of Greenland was complete by 1985 and work had started on compiling information for an atlas showing all West Greenland glaciers on a scale of 1:250,000. By 1991 nearly the whole of Greenland was covered by 1:250,000 maps and an inventory was published in 1992, based on natural hydrological basins, of all the glacier units in West Greenland between 59°30′ and 71°00′N, with a 1:300,000 map section (Weidick et al. 1992).

In general, small cirque glaciers in Greenland have lost 15–30 per cent of their volumes since their Little Ice Age maxima 150 years ago. Whereas the response time of valley glaciers is 20–30 years and of the inland ice margins very much longer, small glaciers respond rapidly to temperature change and with a time lag of about a decade to precipitation change (Weideck 1990).

Efforts to acquire more knowledge about the past and present economy of the Greenland ice sheet have

---

7   Weidick et al. (1992) provide details of the dates and locations where mass balance measurements have been made both on mountain glaciers and on outlets of the Inland Ice. Most continued only for short periods; none were comparable in length with the longer series from North America.

been driven by concern about global warming, with the possible release of meltwater and icebergs in large enough volumes to cause a significant rise in sea level. Greenland ice covers 1,785,000 km². The equivalent of a metre thickness melted from the ice sheet would raise the ocean surface by about 4 mm (not taking into account hydro-isostasy); this compares with a current observed global sea level rise of 2 ± 0.4 mm/y.

Current estimates of the mass balance of the ice sheet vary considerably because of the spatial and temporal variability of the inputs and outputs to the system, and the difficulty of acquiring adequate data from such a large, remote and inhospitable region. Monitoring the bulk of ice loss resulting from icebergs calving at ice cliffs at the heads of the numerous fjords is a complicated matter, though greatly simplified by satellite surveillance. On the inland ice, wind drifting, which greatly affects snow accumulation, varies spatially with the relief of the ice sheet surface. Snow accumulation and melt rates vary annually and decadally at rates believed to be nearly a factor of ten larger than the long-term elevation changes (Dahl-Jensen 2000). Precipitation values measured by gauges at weather stations have to be corrected according to the design of the gauge and, so far as possible, the wind strength at the time snow was falling. Ablation losses caused by sublimation and by evaporation of meltwater are not readily established. Satellite radar sensing and laser altimetry of the changing level of the ice surface are beginning to provide more accurate information.

Although fluctuations of outlet glaciers are primarily controlled by ablation and thus by climatic forcing, frontal behaviour of an individual glacier is commonly a function of that particular glacier's dynamics. Surging glaciers have been identified from repeated air photography in many parts of Greenland, around Disko island and Nûgssuaq, and also around the Blosseville Kyst in East Greenland, notably between Scorseby Sound and Kangerdlussuaq (Weideck 1988) (Figure 10.3). Other examples occur in the far north, for instance at the head of Victoria Fjord (Higgins and

Weideck 1988). Calving glaciers are especially prone to dynamic decoupling from climate (Warren and Glasser 1992). During surging, ice fronts in fjords can advance several kilometres in a year or two, such an active phase being followed by recession at rates of 10–100 m per year that may last several decades. Surges in the twentieth century on the whole failed to reach Neoglacial maxima (Weideck 1988).

Ohmura and his colleagues (1999) concluded that mean annual precipitation on the Greenland ice sheet in recent decades has been equivalent to 340 mm of water, this figure being based on regular observations at a limited number of meteorological stations, most of them not located on the ice sheet itself. Measured values were between 500 and 1000 mm at the southeastern margins of the ice sheet, declining to less than 100 mm in the northeastern interior. Precipitation in individual years, as in many other parts of the world, can differ from the mean by ± 50 per cent and there may well be decadal or longer-term variations.[8] Average evaporation losses are put (with considerable uncertainty) at 35 mm, varying from 2–300 mm at the southern margins to less than 5 mm on the central watershed. It is suspected that an additional 8 mm are lost as runoff from marginal areas. Average annual accumulation on the icecap, the difference between precipitation and evaporation, was then calculated to be about 300 mm water equivalent.

The mean mass of annual precipitation remaining on the ice sheet can also be assessed from the thicknesses of annual accumulation layers visible in shallow ice cores, of which a large number have been extracted over the years. In the study by Ohmura et al., values from fifty-two accumulation layers were utilized. The mean values of the thicknesses of the annual accumulation layers in shallow cores varies from 400 mm to 800 mm at the southeastern edge of the ice to less than 100 mm in the northeast interior.

Of all losses from the ice sheet, calving at the margins is the most important and contributes immediately to sea level rise. At the time of the study by Ohmura

---

8   At Egedesminde, with a mean of about 350 mm of precipitation between 1951 and 1997, annual totals varied from about 250 mm in the mid 1960s and 300 mm in the early 1990s to nearly 600 mm in 1979. In Greenland generally, and especially in southern Greenland, annual precipitation is inversely related to the North Atlantic Oscillation Index (NAOI), the correlation coefficient being −0.8 over southern Greenland and −0.75 for Greenland as a whole. In northern Scandinavia, in contrast, precipitation varies with the NAOI.

et al. (1999), calving rates were available only from Weideck's study of glacier fronts on the west coast and at the southern tip of Greenland. The southwestern portion of the ice sheet from which this calving takes place constitutes only 29 per cent of the total ice sheet area. For this portion, the calving rate was estimated to be equivalent to a surface lowering of 250 ± 40 mm per year, giving a net loss of mass from the ice sheet equivalent to a surface lowering of around 100 mm annually.

The rate of calving of a glacier is believed to depend, in part at least, on the depth of water at its seaward end in relation to the thickness of the ice. Such a relationship can usefully supplement observations made from maps, air photos and satellite images. The values for ice mass lost by calving obtained by Reeh (1989) were less than the difference between the values he had adopted for precipitation and evaporation on the ice sheet surface. He therefore concluded that Greenland was experiencing net accumulation. Bigg (1999) added to Reeh's data and, applying various flux equations to the calving observations, came to the conclusion that net mass balance was zero in southeast Greenland, but there was net accumulation in all other areas.

R. Thomas *et al.* (2000) adopted a different approach to avoid the troublesome problem of measuring calving. They compared the thickness of annual accumulation layers in shallow ice cores with the rate at which the ice is removed by flow away from the central N–S ice-shed. This ice-shed is nearly 2000 km long and lies at a height of 2–3000 m above sea level (Figure 10.3); ice flows away from it in the direction of steepest slope towards coastal inlets, 500 to 600 km away north of Disko Bay (between 70° and 80°N), and a diminishing distance away from the ice-shed south of the Bay (between 70° and 60°N). The rate of flow of the surface ice was measured by making repeated surveys to discover the spatial displacement of ice stations using the GPS (Global Positioning System). Ice discharge was assessed by summing movement downslope through cross sections between measuring stations which were spaced about 30 km apart near the 2000-m contour, taking into account ice thickness as determined by radar sounding, and a calculated velocity gradient from surface to bed. The removal values for several inland ice regions, each region comprising flowlines through a few tens of survey stations, were compared with snow

accumulation rates obtained from cores and weather stations in order to obtain values for ice sheet thickening or thinning over the last few decades. As Dahl-Jensen (2000) commented, it involves subtracting one large figure from another large figure, so that the uncertainty of the estimate is of the same order of magnitude as the expected long-term trend from climate and dynamic history. The results were believed to be least reliable in the extreme southeast where precipitation is highest, about 800 mm of water equivalent per year, and spatial variability is greatest, with *sastrugi* (wind-formed sharp ridges and hollows) often present. South of 70°N a negative balance of 300 mm per year was found to the east of the ice-shed, and a positive balance of about 200 mm per year to the west (where Ohmura *et al.* 1999 had found a negative balance). As the western slope is about 50 per cent longer than the eastern slope, gain and loss south of 70°N would seem, according to R. Thomas *et al.* (2000), to be roughly in balance. North of latitude 70°N, the surface of the eastern ice was found to be rising at a rate of about 20 mm annually, while the surface of the western ice was falling at about twice that speed. The net loss from the entire 1,785,000 km^2 area of the ice sheet was put at between 4 and 18 mm annually, enough to give a sea level rise in a century of the order of 5 mm.

Airborne laser altimetry in 1993 and 1999 suggested that the surface of the Greenland ice thinned by 11 ± 7 mm/y in the south but that for the entire region there was a rise of 5 ± 5 mm/y. After subtracting 5 mm for bedrock uplift (part of which, at least, is caused by recovery from the weight of the Pleistocene ice sheet), the change in ice thickness is zero. Below the 2000-m contour thinning predominated and was estimated to amount to 51 km^3/y. Such a thinning would give a mean annual sea level rise of 0.13 mm for the 1980s and early 1990s, at a time when Greenland temperatures were half a degree cooler than earlier in the twentieth century (Krabill *et al.* 2000).

McConnell *et al.* (2000) made use of ten ice cores taken from above the 1700-m contour around the southern ice sheet, each core penetrating at least 20 annual accumulation layers. Radar sounding showed that altitude changes of the ice sheet surface corresponded with accumulation values and the Crete record suggested that this has been typical of the last few centuries. Rates of elevation change were found to vary spatially by as much as 500 mm over distances of 200 km. It was concluded that, south of latitude 73°N, there was thinning to the east of the ice-shed, as Thomas *et al.* (2000) had concluded, and strong thickening to the west. This result could be explained in terms of greater snow accumulation in the west in recent years.

Outlet glaciers seem to respond to temperature change within a decade or two, in much the same way as do Alpine and other European glaciers. A change in temperature of a degree or two is unlikely to have much effect on the mass balance and rate of flow of the inland ice, but a change in precipitation might have an almost immediate effect. Presumably, a removal of ice mass from the outlet glaciers steepens ice surface gradients and increases rates of discharge from the ice inland in the longer term. Further studies are needed to establish the current balance of the inland ice and the nature of its fluctuations over time. It may turn out that the fluctuations of the outlet glaciers are in response to changes over decades in temperature and precipitation in the coastal zone, superimposed on variation in the net balance of the inland ice responding to changes in temperature and precipitation over centuries.

10.3 SVALBARD

The Svalbard archipelago lies between 73° and 81°N, far above the Arctic Circle in the same latitude as northern Greenland and in the same longitude as northern Scandinavia. The West Spitsbergen current, a northerly branch of the North Atlantic current, crosses the Iceland–Faroes Rise and brings relatively warm, saline water into the Norwegian Sea (Figure 2.2). Svalbard temperatures are consequently anomalously high for the latitude, currently varying near sea level from means of −15 °C in February and March to 6 °C in July.

In the early decades of the twentieth century, the seas to the south of Svalbard, instead of being covered with pack ice throughout the winter, became free of ice through much of it. In consequence, temperatures in central-west Spitsbergen rose steadily between 1895 and 1925, by 6 °C according to Brázdil (1988). In western Svalbard, the warming between 1912 and 1920

was 2 °C (Hagen and Liestøl 1990, Lefaucconier and Hagen 1990).

With strong contrasts between cold polar air in an anticyclonic circulation to the north, and warm Atlantic air in depressions originating around Iceland and approaching from the south, the weather fluctuates violently. Precipitation, brought mainly by southerly air masses, diminishes from 1200 mm on the southern mountains to 400 mm at the north coast. Mean annual temperatures at sea level are well below zero so that bare ground is permanently frozen to depths of a few metres. Thin ice margins are frozen to bedrock in the ablation zone whereas the base of thick ice at higher altitudes in the accumulation zone may be at the melting point on account of heat flow from the earth and the pressure and insulating effect of the ice mass itself. Measurements of the change of temperature with depth in Scott Turnerbreen show that the depth of ice required for pressure-melting is about 100 m (Lefauconnier *et al.* 1999).

Out of Svalbard's total area of 63,000 km², about 60 per cent is covered by ice.[9] The ice sheets consist of huge continuous masses streaming between outcropping mountain ridges and nunataks festooned by cirque glaciers. Most of the ice is of the subpolar type, partly frozen to the underlying rock, with its margins on land below freezing. Radar sensing of Kongsvegen, which is situated on the west coast of northwest Spitsbergen in the Kongsfjorden area, showed an 80–100-m thick cold upper layer in the ablation zone and a temperate accumulation area. Most of the cirque glaciers are entirely below freezing. Equilibrium altitudes rise from 200 m near the south coast where snowfall is greatest, to 800 m in the north where it is least. In the Little Ice Age the glaciers were close to their late Holocene maximum extent and they remained there until early in the twentieth century.

Most of the glaciers have very low velocities of only a few metres per year most of the time, especially in the lower parts of ablation areas. On Austfonna, in Nordaustland, Dowdeswell and Drewry (1989) measured a maximum mean velocity of 0.13 m/d. On Austre Brøggerbreen, the maximum velocity close to the equilibium line is less than 0.01 m/d, and on Midre

9   Much of the information given here is from *The Glacier Atlas of Svalbard and Jan Mayen* (Hagen *et al.* 1993).

Lovénbreen about 0.02 m/d (Liestøl 1988). Both of these are small glaciers with areas of only 5–6 km², near the coast of western Spitsbergen. On such slow-moving ice, firn and ice build up in the accumulation zone and surface gradients gradually increase until the shear stress reaches a critical value; then a surge takes place. The velocity suddenly increases by as much as a hundred times and remains high for a year or two, occasionally for as much as ten years. Between surges, the glacier remains sluggish for some decades and its front retreats. About 80–90 per cent of Svalbard glaciers have been regarded as subject to surging, though this is now being questioned.

In the cases of glaciers terminating in the sea, tributary glaciers as well as the main ice body are liable to be affected by the surging motion. In glaciers terminating on land, only the surging tongue itself is likely to be affected immediately. Hambergbreen, which had last surged towards the end of the nineteenth century and had since been retreating by 240–50 m per year, pushed forward more than 5 km between 1961 and 1970 at a rate of roughly 0.5 km per year (Ziaja 2001). Between 1958 and 1961, the front of Fridtjuvbreen advanced 6 km, filling Fridtjou Harbour and pushing up the sea-floor and banks of shell-bearing clay. Nyribreen surged forwards 12 km along a 15 km front in 1955–6; Bråsvellbreen advanced 20 km along a 30-km wide front in 1937–8. Such ice front movements are responses to long-term changes in mass balance. Their timing is determined by the dynamics of the particular ice body and the morphology of its floor. Each glacier is characterized by its own unique time interval between surges, a time usually of a few decades when feed to the lower glaciers is very slow, accumulation areas steepen, and ice fronts retreat.

Evidently changes in the position of a single Svalbard glacier front is not a good indicator of its economy and of the climatic trend. Of eighty-six Svalbard surges reported since 1860, it may (or may not) be significant that 24 per cent took place in the 1930s and 19 per cent in the 1960s. (The relatively small number reported for the intervening decades could be related to war-time and early post-war paucity of observations.)

There is good evidence that the ice in southern Svalbard was well back from its present limits in early medieval times. Drillings by Russian teams on Svalbard ice domes indicate that at altitudes of several hundreds

of metres accumulation rates have been fairly constant for the last 500–600 years (Hagen et al. 1993).

A Medieval Warm Period could conceivably be represented by certain peaty accumulations. About 450 m in front of Werenskioldbreen, which is known to have retreated 1.5 km between 1925 and 1975, the Polish Scientific Spitsbergen Expedition found a bed of mossy vegetation 5 cm thick sandwiched between two till layers. The bottom of the vegetation layer (St-5068) yielded a radiocarbon date of 1565 ± 235 BP, calibrating to AD 240–690, and the top (Gd-264) gave a date of 760 ± 145 BP, calibrating to AD 1157–1327; a sample (St-4695) from the whole thickness of the stratum dated to 1080 ± 105 BP, which calibrates to AD 810–1040 (Baranowski and Karlén 1976). Olsson and Blake (1962) had found peat in moraine on the northern shores of Hornsund, at the foot of Rotjesflellet, which dated to 1390 ± 85 BP, calibrating to AD 600–700. However, radiocarbon dates from peat are to be viewed with some suspicion as they depend so much on the mode of sample preparation.

Svalbard ice core $\partial^{18}O$ values point to the second half of the sixteenth century and the beginning of the seventeenth as a cold period, with mean summer temperature approximately 1.5 °C lower than the twentieth-century mean. This was followed by a short warmer period from 1625–45 with mean summer temperatures approximately 0.5 °C higher than in the preceding one. It was a time when the whales spread out across the Greenland Sea and were consequently difficult to kill. Then mean summer temperatures dropped again, pack ice increased, whales congregated at the edge of the ice, and kills greatly increased (Hamilton and Haedrich 1999, Hacquebord 1999). Greater warmth between 1715 and 1750 saved the whales, for a time. Then, closed pack ice returned and in the course of the early nineteenth century Spitsbergen whales were exterminated by hunting.

Dutch and English whalers who operated around Svalbard in the seventeenth and eighteenth centuries plotted glacier fronts on their maps. Though imprecise, they suggest that many glaciers at that time were not much more extensive than they are today. Since about 1920, ice retreat has been general. In Sørkappland, southernmost Spitsbergen, the areal extent of glaciers diminished by 18 per cent between 1936 and 1991 and further north in Nordenskiøldland between

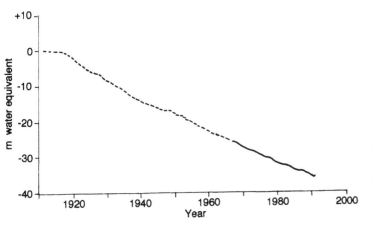

Figure 10.4 Cumulative net balance of Brøggerbreen, 1912–91; 1966–91 values observed, 1912–65 values estimated from July to September temperature at Longyearbyen (*Source:* from Lefauconnier and Hagen 1990, and Hagen *et al.* 1993)

1936 and 1995 by 44 per cent. Since 1936 equilibrium line altitudes have risen by about 150 m, according to Ziaja (2001), who notes that the ice connecting Sørkappland with the rest of Spitsbergen was narrower in 1995 than it had ever been since it was first seen in 1596. Thinning continues but has not been as rapid in recent decades as it was in the first thirty years of the twentieth century; nevertheless, some glaciers which used to surge are now cold throughout and relatively quiescent.

In 1976 Russian scientists extracted a 200-m core from the cold firn zone on the Lomonosonov Plateau, at an elevation of 1000 m above sea level. The upper part of the core's isotopic profile, which was characterized and age-calibrated by the effects of nuclear testing between 1951 and 1976, indicated an average annual accumulation between those years of 820 mm water equivalent, annual values varying from 653 mm in 1958/9 to 1270 mm in 1964/5. The Chernobyl layer of 1986 was detected in several ice cores drilled in 1997 on Lomonosovfonna (Nordenskiöldbreen). The mean annual net balance between 1986 and 1996 of 760 mm water equivalent is not significantly different from that measured in 1976 for the period since 1951 (Tarussov 1992).

Mass balance measurements were instituted by the Norwegian Polar Institute on Finsterwaldbreen in south Spitsbergen in 1950, and continued to be made every other year until 1968. Measurements have been made annually on two small, low altitude glaciers with a north-east aspect, situated 5 km apart, Austre Brøggerbreen (Figure 10.4) measured since 1966, and Midre Lovénbreen since 1967/8, and also the larger Kongsvegen tidewater glacier since 1987.

On Finsterwaldbreen, between 1950 and 1968, positive mass balances were recorded only for the years 1952/4 and 1956/8. Mean annual surface lowering over the whole period amounted to 0.25 m water equivalent, a value which because of the very low velocity of the ice is almost the same as that of the mean annual net balance.

The surface of Austre Brøggerbreen, though it is a cold glacier, has been entirely an ablation area in recent decades; it was lowered by a mean value of 0.43 m annually from 1967–88, net balances in individual years varying from +0.13 to −0.93 m. Winter accumulation exceeded summer ablation losses only in 1990/1 (+0.13 m). (Lefauconnier *et al.* 1999, table 1).[10] There was a good correlation between the net mass balance, on the one hand, and summer and autumn positive degree-days and winter precipitation on the other. Extending the sequence of ice losses from this small cold glacier back to 1912 on a basis of the temperature record gave an accumulated loss by 1988 equivalent to 30 per cent of its volume. Significantly, summer temperatures at Longyearbyen are believed to have risen by nearly 2 °C between 1912 and 1916, presumably with the disappearance of sea ice in the vicinity of

10   Or also in 1987 (+0.22 m) according to Lefauconnier and Hagen (1990).

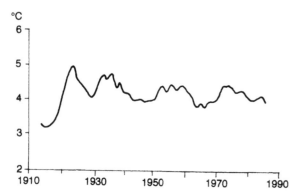

Figure 10.5 Reconstructed running 5-year means of summer temperature at Longyearbyen (*Source:* from Hagen and Liestøl 1990)

western Svalbard (Hagen and Liestøl 1990) (Figure 10.5). Wastage continued into the 1990s, with the balance year 1992/3 the most negative since measurements began as a result of low winter accumulation and extremely high ablation (Haeberli *et al.* 1994).

On Lovénbreen, a partly temperate glacier which is believed to have surged at the end of the nineteenth century, the net balance over the period 1967–88 was –0.36 m/y, a little less negative than on Austre Brøggerbreen, probably because its higher mean elevation gives greater winter accumulation and less summer ablation. Here too, 1990/1 was distinguished as one of four years with positive balances; this was the year of the Pinatubo eruption (see Chapter 17, p. 579).

Kongsvegen, located 25 km southwest of Ny-Alesund where a meteorological station started records in 1969, is 25 km long, much larger than Brøggerbreen and Lovénbreen. It last surged in 1948 and is building up for another such effort. The cold layer, somewhat paradoxically, extends from the surface to the bed in the ablation area. With an equilibrium altitude 100 m higher, Kongsvegen's accumulation area is at a higher elevation than that of the two smaller glaciers, and since 1987 it has had a mean net balance of +0.11 m (Lefauconnier *et al.* 1999). The year of the largest negative net balance on record at Kongsvegen was 1992/3, as was also the case at Brøggerbreen and Lovénbreen. That year, the year of a great El Niño (see Chapter 17), mass balances were strongly negative elsewhere in the Arctic islands, for instance on Ellesmere Island and Axel Heiberg Island (Wolfe and English 1995).

Soviet mass balance investigations on relatively small glaciers near the coast have been made over a shorter period than those of the Norwegians. The sets of results obtained by the two countries correspond quite well, the Russian results from Vöringbreen giving correlation coefficients of 0.81 with Brøggerbreen and 0.88 with Bertilbreen.

Hodgkins *et al.* (1999) compared a recent topographic survey with earlier maps of Scott Turnerbreen, a glacier only about half the size of Lovénbreen and Austre Brøggerbreen and in the same area. They concluded that since 1936 the surface had been lowered on average by 36 m, equivalent to a negative net mass balance of –0.58 m/y. Scott Turnerbreen is known to have surged around 1930 and so much of its bed at that time must have been at the pressure melting point. Hodgkins *et al.* (1999) argue that as it is only 76 m thick at the deepest point, it must now be entirely frozen to its bed. The thinning since the end of the Little Ice Age has resulted in it no longer being a surging glacier.

10.3.1 Franz Josef Land

Airborne radio-echo sounding was used by Russian scientists in 1994 to estimate the change in area and volume of twenty-six icecaps of various shapes and sizes in the Franz Josef Land archipelago, eastern Svalbard, since air photographs were taken of them in 1953 (Macheret *et al.* 1999). The total area of ice involved was about 13,500 km². On a basis of the relationship found between ice area loss and volume change, it was concluded that over the 40-year period a reduction in the area of the icecaps of 210 km² (about 1.5 per cent) was equivalent to a decrease in volume of 42 km³ (about 2 per cent) and a surface lowering of 0.6 m. Although the mass balance has been negative in the last forty years, it is believed that mean annual losses were about three times greater over the preceding thirty years, 1930 to 1960. This result agrees with observations that mean summer temperature on Franz Josef Land decreased slightly in the 1980s.

10.4 NOVAYA ZEMLYA

Novaya Zemlya, a narrow island some 800 km long and 100 km wide, with an average altitude of 1000 m, extends north and northeast off the Russian mainland

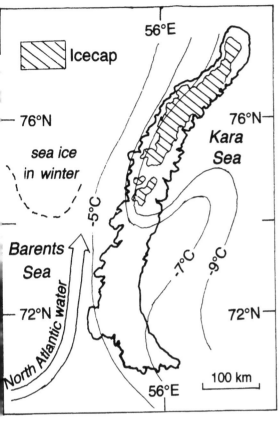

Figure 10.6 Novaya Zemlya showing inferred isotherms
(*Source:* after Zeeberg and Forman 2001)

between 71° and 77°N. Glaciers cover the northern part above 600 m. To the east, the Kara Sea is usually ice-covered whereas the Barents Sea to the west is ice-free, being open to North Atlantic water which reaches Novaya Zemlya when the North Atlantic Index has been positive for a few years. The glaciers are nourished by eastward moving cyclonic depressions which increase in frequency when the NAO index is high.

The glaciers were scarcely observed until the twentieth century when some of them were mapped by Russian expeditions.[11] Ice-cored moraines, being free of lichens, are suspected to have been deposited in the Little Ice Age. An *Astarte borealis* shell from marine sediments incorporated in the outermost lateral mor-

aine of the Shokal'ski glacier has given a radiocarbon date of 660 ± 45 BP which suggests that the Little Ice Age maximum occurred some time around or after AD 1300 (Zeeberg and Forman 2001).

In the first half of the twentieth century, tidewater glaciers retreated rapidly, typically by about 300 m annually. Then, after 1961, temperatures measured at weather stations, contrary to model predictions, were lower in the winter. Summer temperatures remained low for some years, responding only slowly to the positive NAO index values, and over the next three decades, with increased precipitation from more frequent cyclonic depressions, glacier fronts scarcely receded. Eventually, towards the end of the century, rising summer temperatures increased ablation, and mass balances became negative.

10.5 THE CANADIAN ARCTIC ARCHIPELAGO

The region between the Mackenzie delta and Baffin Bay is extremely inhospitable, with mean July temperatures generally no more than about 4 °C. While on the east side of Baffin Bay the frigidity of the west Greenland coast is ameliorated by the relatively warm West Greenland current, the east coast of Baffin Island is chilled by the southward flowing Arctic water.

In spite of the cold, Inuit are thought to have been active on the islands of northern Canada, possibly even in the latter stages of the Last Glaciation. Archaeologist Patricia Sutherland believes she has found traces of Norse shore stations on northern Baffin Island dating back to the fourteenth century, or possibly much earlier, in the form of textiles made from arctic hare fur and goat hair found at Inuit archaeological sites (Pringle 2000).

Most of the icecaps and glaciers in Arctic Canada have been retreating for several decades (Koerner 1989). In the Queen Elizabeth Islands of the Canadian Archipelago, ground temperatures appear to have increased by 2–5 °C (Taylor 1991). On westernmost Banks Island, the Inuvialuit at Sachs Harbour say freeze-ups now come three or four weeks later and the sea ice breaks up earlier than used to be the case. There are more cloudy days, more open water in winter, and a

11   This brief account is based on Zeeberg and Forman (2001).

lack of sea ice in summer. Meat stored in the permafrost has been spoiling (Riedlinger 1999).

10.5.1 Baffin Island

Baffin Island has been regarded as one of the most likely places for a glaciation to gather momentum. The present-day glaciation limit for that part of Baffin Island around the north of the Barnes icecap is between 0 and 300 m above the surface of the plateau on which the icecap lies. Using an observed mean lapse rate of 0.5 °C per 100 m, Bradley (1973) argued that a mean summer cooling of 1.5 °C would lower the limit below a substantial part of the land surface. Ives (1962) had suggested that extensive lichen-free areas represented the ice cover of the Little Ice Age and that while only 2 per cent of the region is now glacierized, it has recently been as much as 70 per cent ice-covered. Undisturbed, vegetated patterned ground features were revealed by rapid recession of a thin body of the Tiger Icecap in northern Baffin Island (Falconer 1966).

The reliability of lichen-kill zones as indicators of Little Ice Age snow cover has been questioned by Koerner (1980) but there is no doubt that the topography and geographical position of Baffin make it peculiarly sensitive to changing climate (Williams 1979) and it may be accepted that glacial extension here at high levels in recent centuries was on a much larger scale than in other parts of North America discussed in Chapter 9. It seems probable that the area mapped by Andrews et al. (1976) may be taken as giving a reasonable indication of the extent of snow cover in recent centuries, whether the zone represents that in which lichen was killed by glacial extension or that in which lichen sparsity is due to high incidence of extensive seasonal snow cover.

Recent moraines in the Cumberland Peninsula, the easternmost part of Baffin Island surmounted by the Penny Icecap, have been dated by lichenometry (Miller 1973). The forefields of forty-six glaciers were examined and the lichens on 100 moraine crests. Advances were found to have culminated around 750, 380 and 65 years BP, but the lichen growth curve, dependent on radiocarbon data, was regarded as a first approximation only and its reliability was rated at about 20 per cent.

In the Watts Bay area of southwest Baffin Island, relative age units have been defined using both lichenometry and weathering-rind data (Dowdeswell 1984), the classification of the data being confirmed by discriminant analysis (Dowdeswell and Morris 1983). Moraines of three age groups were identified at the margins of the Grinell icecap and a nearby cirque glacier. The two youngest sets of recessional moraines, formed during temporary slowing of more general retreat, were judged to have been formed not more than 130 years ago and the older ridges some sixty years earlier.

Moraines about 100 years old were also identified near the Terra Nivea icecap, about 40 miles east of Watts Bay (Müller 1980), in northern Baffin near the Barnes icecap (Andrews and Barnett 1979), and also in the Pangnirtung area (Davies 1980). Twentieth-century recession, so marked in the northwest USA, western Canada and Alaska, has also taken its toll of the glaciers in the eastern Canadian Arctic. Results from a climate change monitoring station near the Lewis Glacier, in operation since 1989, indicate that the Lewis front has been retreating at about 25 m per year, and other areas of the northwest margin by 10–30 m per year. It seems that lower regional summer temperatures between 1960 and 1990 did not significantly slow the recession which has been under way in the northwest section of the Barnes Icecap for the last three centuries (Jacobs et al. 1993).

The lowering of temperature which affected western North America in the second half of the twentieth century was also felt in the eastern Canadian Arctic (Bradley and Miller 1972, Bradley 1973). Seasonal means of temperature on Baffin Island showed a decrease of as much as 2.1 °C in the June–August ablation season for the period 1960–9 and an increase of as much as 2 °C for the accumulation season September–May. Winter precipitation increased markedly everywhere, with surprising consistency considering that local topographic features cause total precipitation amounts to vary widely from place to place. The lower summer temperatures were apparently caused by increasing advection of cool air from the north and east and a concurrent decrease in warmer air from the west and southwest.

A single decade was sufficient for the snow cover to respond to the new conditions. Comparison of air photographs showed that while snowbanks had been diminishing between 1949 and 1960, in the next ten years permanent snow cover increased markedly. Snow patches less than 10 m in diameter doubled in size; larger snow patches, more than 50 m in diameter, expanded less than the smaller ones. In the early 1970s, permanent snow was lying in many areas that had been snow-free at the same seasons in the 1960s. Extension of snow over lichen-covered surfaces suggested that a decade of climatic deterioration had been sufficient for the snow cover to become as extensive as it had been forty years earlier. Glaciers, because of their longer response times, continued to retreat between 1960 and 1971.

A 335-m long core from the Penny Ice Cap has revealed a statistically significant inverse relationship from 1901 to 1990 between changes in Baffin Bay spring sea ice extent and sea salt concentrations in the icecap.

Sea salt concentrations in the uppermost ice core provide a record of Baffin Bay sea ice and hence temperature extending back over the last 700 years. Low sea salt concentrations and strongly negative $\partial^{18}O$ values in the ice from AD 1500 to 1600 and again from 1650 to 1710 point to low temperatures in the region in the midst of the Little Ice Age. Over the last thirty or forty years, summertime sea ice extent in most parts of the Arctic has decreased in tune with northern hemisphere warming, but in parts of the western Arctic including the Baffin Bay region it increased. The explanation would seem to be that, with a strongly positive North Atlantic Oscillation Index, stronger westerly winds gave lower surface temperatures in the northwestern North Atlantic and eastern Canadian Arctic.

10.5.2 Devon Island

An abrupt transition to cooler summers took place on Devon Island from 1962–4. As annual degree-day totals here are highly correlated with mass balance values, it has been possible to reconstruct the mass balance for the northwest of the icecap back to 1947–8 (Bradley and England 1978). It emerges that the mass of ice lost in the period 1947/8 to 1962/3 was six or seven times greater than in the succeeding decade 1963/73. Icecap growth here is, however, limited by low values for precipitation so that an occasional warm summer can wipe out the cumulative gain of a succession of years with positive mass balances. A fall in

temperature in this high Arctic environment has therefore a much less noticeable effect on glacier expansion than it has in regions with higher precipitation.

10.5.3 Ellesmere Island

Mass balance measurements began on the Gilman Glacier in northern Ellesmere Island, 82°N, in 1957 (Hattersley-Smith 1960). They have since been extended to more than a dozen different glaciers and icecaps (Wolfe and English 1995).

Snow beds and small glaciers on northern Ellesmere Island were reported by Hattersley-Smith and Serson (1973) to be expanding but Bradley (personal communication, November 1984) observed that the northern Ellesmere Island icecap shrank by 11–14 per cent between 1972 and 1983 and mass balance measurements showed a net loss of about 0.4 m in thickness. Differences in nourishment on nearby glaciers have caused noticeable variations in response. Carl Troll Glacier in the Borup Fjord area of northwest Ellesmere, which had advanced 20 m between 1950 and 1959, had retreated again 50–70 m by 1978. The Webber glacier advanced 60–150 m during the period 1959–78. Its accumulation comes from a large icefield nourished by superimposed ice and drifting snow, whereas the Carl Troll depends mainly on accumulation in small cirques (King 1983).

Quviagivaa Glacier is in a shallow valley at an elevation of between 560 and 1250 m, on the Fosheim Peninsula of west-central Ellesmere Island. Mountains to the east block disturbances from the Arctic Ocean and Baffin Bay, and July mean temperatures (1988–93) are 9.8 °C, remarkably high considering the latitude of 79°56′N. Wolfe and English (1995) measured Quviagivaa's 1992/3 mass balance and found it to be −532 mm, strongly negative on account of the summer melt being about double the normal in this El Niño year. Much the same was the case with Baby and White glaciers on Axel Heiberg Island, the Meighan and Agassiz icecaps, and other glaciers in the Arctic and elsewhere in the world that year. Quviagivaa's ice cover retreated between 1959 and 1993 by about 100 m, almost as far as it had retreated by 1959 from its Little Ice Age maximum, which is marked by a lichen boundary trimline 40 m above the current ice surface (Wolfe and English 1995, figures 2 and 7).

In the 1960s and 1970s, as the front of Twin glacier retreated at a rate of about 4 m per year, a well-preserved high arctic plant community was revealed (Bergsma *et al*. 1984). Radiocarbon dates of the plants, (WAT 778) 400 ± 140, (WAT 789) 430 ± 90 and (SI 6033) 410 ± 45, calibrate to AD 1300–1655, 1419–1627 and 1438–1619, which suggest that the ice advanced over the plants in the fifteenth and sixteenth centuries.

10.5.4 Axel Heiberg Island

Mass balance studies have been made on White and Baby glaciers, 79°5′N on Axel Heiberg Island, since 1960 (Wolfe and English 1995). White Glacier, 80°N, which occupies 38.7 km² in the Expedition Fiord area, extends from 1782 m, where annual accumulation is about 300–400 mm, to 56 m above sea level, where annual ablation is typically 2000–4000 mm. The mass balance switched from strongly negative in 1960 to slightly positive in 1961 (Müller 1962). Over the period of observation 1960 to 1994 the net mass balance varied from −780 mm to +350 mm, averaging −100 mm (Haeberli *et al*. 1996).

10.6 ARCTIC ISLANDS SUMMARY

The climate of the Arctic islands is much influenced by the extent and distribution of sea ice which in turn depends on ocean currents and on the Arctic and North Atlantic Oscillations. As a consequence, the direction of temperature change on the decadal scale may differ between regions east and west of Greenland. The fragmentary evidence of glacier advance and retreat does not suggest that the climate record in the Arctic as a whole in recent centuries differs greatly from that elsewhere.

Dates from moraines and isotopic indications of cooling from ice cores suggest that the Little Ice Age began there around AD 1300 and isotopic signals in Greenland cores indicate several episodes of strong cooling in the fourteenth century. Archaeological findings and documentary accounts relating to the Norse settlements in Greenland point in the same direction. A slight amelioration in the fifteenth to seventeenth centuries was interrupted in the decades around 1600 and was followed by increasing severity of the climate in the mid eighteenth century. In the Canadian Archipelago, cooling may have come some decades earlier than in the east. Generally in the Arctic islands, as in Iceland and much of North America, it remained cold until late in the nineteenth century with advances of glaciers in Greenland around 1850 and 1885. Warming in the twentieth century was interrupted by cooling around the middle of the century with temperatures rising again in the final decades of the millennium.

11

LOW LATITUDES

Tropical Latin America, East Africa and New Guinea

The tropical glaciers are amongst the least known in the world. All of them are believed to have retreated in the course of the twentieth century. Documentary evidence of the timing of their fluctuations in earlier centuries is sparse. There are very few historical records concerning the glaciers of Latin America, though it is conceivable that additional information is hidden away in Iberian, Vatican or other archives. For Africa and New Guinea no such hopes can be entertained (Figure 11.1); only the Lewis Glacier on Mount Kenya has been the subject of detailed and sustained monitoring in recent decades (Hastenrath 1989, Kaser 1995). New information from ice cores is summarized here and additional detail can be found in Chapter 16.

The regimes of equatorial and tropical glaciers differ in many respects from those of middle and high latitudes (Kaser and Osmaston 2002). There are no appreciable seasonal temperature variations. Precipitation near the equator usually occurs in two seasons of the year, at the passage of the Intertropical Convergence Zone. At higher latitudes, towards the tropics of Cancer and Capricorn, precipitation occurs during a single season and is usually sparse. At the tropics, ice equilibrium lines and snowlines are at very high altitudes, especially in the lee of high mountain ranges, because of the aridity. Ablation, which may persist throughout the year, mainly involves sublimation, direct transition from ice to water vapour as a result of strong solar radiation inputs throughout the year. In more humid areas, cloud build-up in the afternoon results in glaciers being sited more commonly on western than east-facing slopes. With the position of the 0 °C level and the snowline almost constant throughout the year, any increase in temperature puts the lower

parts of intertropical glaciers into the ablation zone in every season (Kaser 1996a). The total area of the intertropical glaciers, most of it in Peru (Kaser 1995), is now only about 2500 km², constituting only about 0.16 per cent of the global ice cover and shrinking rapidly.

11.1 INTERTROPICAL LATIN AMERICA

Glaciers are distributed along the eastern and western ranges of the Andes in Venezuela, Colombia, Ecuador, Peru, Bolivia and northern Chile. In northern Colombia a reconnaissance expedition to the Sierra Nevada de Santa Marta in 1939 found a striking sparsity of glaciers and a profusion of glacial deposits indicating rapid recent wasting (Cabot 1939). By 1969 approximately a third of the ice present in 1939 had gone (Wood 1970). Since the 1970s, warm events associated with El Niños have become more frequent and more intense and ice recession greatly accelerated in the last decade of the twentieth century (Francou *et al.* 2000).

11.1.1 Mexico

Small glaciers survive on the peaks of two active volcanoes in Mexico. Following eruptions in the sixteenth and seventeenth centuries, glaciers on the Pico de Orizaba at the eastern end of the Trans Mexican Volcanic Belt extended down the northern and western slopes on top of the new lava. They reached their most advanced positions marked by moraines at an altitude of about 4400 m in 1850, according to Heine (1994). Since the formation of these moraines the front of the ice had retreated about 800 m to an altitude of 4720 m

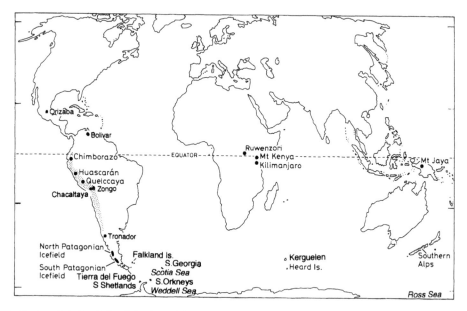

Figure 11.1 Distribution of glaciers in low latitudes and the southern hemisphere

by 1994 and its thickness had been greatly reduced (Palacios and Vázquez-Selem 1996). Popocatépetl to the southwest, only 60 km from Mexico City, supports two small glaciers at an altitude of about 5000 m with a total area of about 50 ha. They were in retreat throughout the twentieth century except for a short-lived advance between 1968 and 1978.

11.1.2 Venezuela

Pollen analysis of a peat-bog at 4080 m in the Páramo de Piedras Blancas in Venezuela, dated by a single radiocarbon analysis of organic matter from the lowest part of the core and assuming a constant rate of sediment accumulation, indicated a climate similar to that of today from the seventh to the eleventh centuries AD, followed by warmer and moister conditions. From the mid thirteenth century onwards, forest retreated to a level 200 m lower on account, it is hypothesized, of a cooler and drier climate. Analysis of an undated core taken from the edge of Laguna Victoria, about 20 km away, also provided evidence of cooling at a time not long before the appearance of *Rumex* pollen, a weed of cultivation, associated with the arrival of European agriculturalists. Over the last four centuries warmer and wetter conditions have returned (Rull *et al.* 1987).

Maps and photographs show that the glaciers on the Pico Bolívar in the Sierra Nevada de Mérida of Venezuela retreated dramatically during the twentieth century, some thinning by as much as 100–50 m, and the ice-covered area diminishing by as much as 80 per cent (Schubert 1972). Since 1972 the rate of deglaciation in the region appears to have accelerated; the Timoncito Glacier on the southeastern face of the Pico Bolívar massif practically disappeared between 1972 and 1991, as did the Espejo and the El Encierro on the north face (Schubert 1992).

11.1.3 Ecuador

Many of the glaciers in Ecuador are situated on great volcanoes aligned along fault zones on the western flank of the Cordillera Oriental and the eastern flank of the Cordillera Occidental. As the Atlantic is the main source of precipitation here, snowlines are lower on the eastern mountains than the west. Mercer (1967) and Hastenrath (1981) gathered together the scattered information available. It includes a few documentary descriptions from the Quito archives dating from the early years of Spanish rule which suggest that conditions in the sixteenth century were probably somewhat more severe than at present, with permanent snow on the crests of Pichincha, Corazon and other mountains now free of perennial snow (Figure 18.1, p. 593). No information has been found for the period between 1580 and 1730, but records of the French Academy's geodetic mission of 1735–43 provide reliable data. Velasco, who lived in Ecuador from his birth about 1727 until 1767, and von Humboldt, who travelled through Ecuador in 1802, left useful accounts. The German geologists Reiss and Stubel in their studies (1886) of volcanoes between 1870 and 1874 made careful observations of contemporary glaciers. They were accompanied by the Ecuadorian artist Troya, some of whose pictures were used to illustrate the geological publications which eventually emerged; the originals, which were once in Leipzig, do not seem to have survived. Whymper (1892), who was climbing in Ecuador in 1880, published several sketches of the glaciers. Photographs by Meyer (1907) are an important source of information; forty-three of them were published with an explanatory text in English (1908). They show that the largest glaciers on the northeast side of Chimborazo (6272 m, 1°25′S) had all receded by the beginning of the twentieth century from fresh terminal moraines girdling the mountains at heights between 4000 and 5200 m. Air photographs taken in 1960 and 1977 of the Reschreiter Glacier on Chimborazo show the ice bulging above the crests of its lateral moraines but by 1983 the surface of the ice was lower (Clapperton 1983). Topographical maps of Ecuador based on air photographs have been produced in recent years by the Instituto Geografico Militar. The maps and photographs were used by Hastenrath (1981) in his 1974, 1975 and 1978 field studies and for assembling glacier records for individual mountains. He concluded that the ice-covered area is an order of magnitude smaller now than it was at the Little Ice Age maximum and that the regional snowline has risen intermittently over the last 250 years by about 300 m. He estimated that the volume of the Caldera Glacier on El Altar, 5319 m, diminished by 50 million m³ between 1870 and 1900, and by a similar amount since then. Such a shrinkage would have required a diminution of mean annual precipitation by about 500 mm, a change of albedo of about 15 per cent, or a reduction of cloudiness by one- or two-tenths.

Antizana Glacier 15 (0°28′S), at between 4800 and 5760 m, is an icecap sitting on a volcano 40 km east of Quito. Since 1956 the low part has separated into two. Air photos from 1956, 1965, 1993 and 1997 indicate a sudden acceleration in retreat after 1993, associated with ENSO warm events in 1994–5 and 1997–8. The ENSO of 1992–3 was relatively ineffective, probably because it followed soon after the eruption of Pinatubo when the tropical troposphere at 500 hPa was unusually cold (Francou *et al.* 2000).

11.1.4 Peru

Many of the Peruvian glaciers are in remote, unmapped places. The Cordillera Blanca[1] is the most heavily ice-covered mountain range, not only in Peru, but also in the tropics as a whole. An initial inventory of glaciers there made by three Osterreichischer Alpenverein expeditions between 1936 and 1939 led to the publication of several good topographical maps and a sequence of scientific papers. The glaciers proved to be strikingly different from those in the Alps, the majority being thin, fissured slope glaciers, characterized by gigantic, paper-thin cornices and abundant icicles; the valley glaciers are generally heavily covered in debris. In 1988 a Peruvian Glacier Inventory was completed, based on air photographs from 1970, which listed 722 glaciers covering 723 km² in the Cordillera Blanca (Ames 1998).

The Peruvian glaciers were in advanced positions in the 1870s (Spann 1948). Subsequent retreat was especially rapid between 1890 and 1910; then came a readvance. The ice limit on Huascarán rose 430 m in altitude between 1886 and 1909, descended 150 m between 1909 and 1932, and then rose again by 500 m over the next ten years (Broggi 1943). The glaciers near the head of the Laguna Perron in the Huandoy group shrank greatly between 1932 and 1947. The Atlante glacier in the northeast of the group is near to a mine-working and more is known about its variations than most others in the range. Sievers (1914) had

---

1 The Cordillera Blanca (8°08′–10°02′S, 77°53′–77°09′W) is the largest of the twenty mountain groups carrying glaciers in Peru (Kaser *et al.* 1996).

been told in 1909 that the firnline in the Cordillera Blanca had risen about 50 m since 1895. In 1932, Kinzl (1949) heard that a small advance had taken place about 1920, closing the entrance to one of the mines.[2] The ice advanced in 1923/4 for 12–15 m and began to retreat again in 1927. In 1932 it had already retired about 45 m (Kinzl 1942). The recession that followed between 1939 and 1948 averaged about 10 m a year but slowed between 1948 and 1957 to 2 m a year (Smith 1957).

The ice-covered area in the Santa Cruz–Pucahirca group in the north of the region diminished by 10 per cent between 1930 and 1950, while in the Huascarán–Chopicalqui massif in the central Cordillera Blanca, ice cover in 1950 was only 84 per cent of that in 1930, dropping much more slowly to 82 per cent by 1970[3] (Kaser 1996b) and then more rapidly (Ames 1998). An upward shift in equilibrium line altitudes of about 95 m, as it was determined on the Huascarán for the interval 1920 to 1970, is probably to be explained mainly by a rise in temperature and effective global radiation accompanying reduced cloudiness and precipitation.

Information about the Yanamarey Glacier, including descriptions and maps made in the 1930s by Kinzl (1942), and the results of surveys made by Unidad de Glaciología e Hidrología Electroperú, have proved sufficiently detailed for changes in volume, area, length and terminus elevation to be assessed quantitatively for a series of twentieth-century dates (Hastenrath and Ames 1995). The length of this glacier decreased from 1600 m in 1948 to 1250 m in 1988, its estimated volume diminishing from 29 to 25 million m^3.[4] The surface of the nearby Uruashraju Glacier lowered by 1040 mm per year from 1977 to 1987; nearly half the present water discharge coming from ice melt rather than precipitation (Ames and Hastenrath 1996). Retreat of the Yanamarey, Uruashraju and Broggi, about which there

is most data, continued into the 1990s and may be taken as representative.

Systematic monitoring of glaciers in the Cordillera Blanca has been prompted by disasters with serious loss of life resulting from avalanching and the overflowing of moraine-dammed lakes (Kinzl 1970, Lliboutry et al. 1977). Valley glaciers in the Cordillera Blanca are typically fronted by well-marked systems of Little Ice Age moraines, in many cases set apart from older moraines. Their number and size vary from glacier to glacier, some forming long crescentic loops 50 m high (Clapperton 1972), many of them holding back proglacial lakes. Lliboutry et al. (1977) noted signs that several glaciers had advanced towards these moraines again in the early twentieth century.

Recession after 1932 involved the appearance of pools of water that united and enlarged to form supraglacial lakes, which themselves expanded until they reached frontal moraines. Kinzl's maps and photographs of 1948–50 disclosed that ten big lakes had appeared between Quebrada Ulta, above Carhuas, and Quebrada Shallap, above Huaraz. By 1970 there were at least 137 lakes, most of which had come into existence in the twentieth century (Ames 1998). Some of them enlarged swiftly: the Laguna de Safuna Alta was still only a pool in 1950, but by 1969 a body of water 4.85 million m^3 in volume had collected and was threatening a small town 40 km downstream. Glacier recession and the consequent formation of proglacial lakes behind morainic dams caused a great increase in large and sudden floods of liquid mud (aluviones), not all resulting from the bursting of moraine-dammed lakes, but the largest from huge rock and ice avalanches, notably the 1962 aluvion which overwhelmed Ranrahivca, drowning 4000 people, and that of 1970, which again devastated Ranrahivca and also the neighbouring town of Yungaya, causing 15,000 deaths. A similar disaster had occurred in 1725[5] but no other had been reported

---

2 Kinzl (1942) had found the sequence of moraines in front of many valley glaciers included a compound moraine or bastion with fresh younger moraines on the proximal side, which he attributed to a short advance phase occurring not long before the first Alpenverein expedition. The Atlante data appear to confirm this view, especially as other Peruvian mines had trouble with advancing tongues at the same time (Oppenheim 1945).

3 Ice cover extent was measured from available air photographs and from 1:25,000 maps of the Dirección Generál de Reforma Agraria y Asentamiento Rurál (1972).

4 Although Ames (1998) gives the volume loss, since some date not stated, as 64 million m^3.

5 An ice avalanche which totally destroyed the town of Ancash in January 1725 was described by Cosme Bueno in his Descripción de las Provincias del Arzobispado de Lima (Ames and Francou 1995).

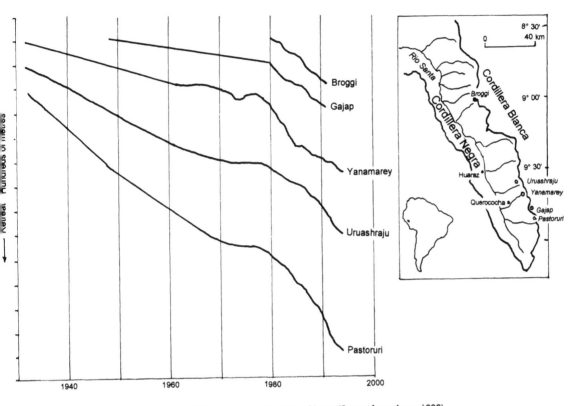

Figure 11.2 Variations in the frontal positions of glaciers on the Cordillera Blanca (*Source*: from Ames 1998)

until 1938, when the mid twentieth century retreat was in full swing. A commission intended to control the lakes of the Cordillera Blanca, set up in 1941, succeeded in gradually lowering the levels of thirty-five of the most dangerous lakes by means of pipes and concrete-lined trenches, so effectively that an earthquake in 1971 failed to cause any outbursts from lakes that had been treated. After a pause in the retreat of the glacier fronts following lower temperatures and higher precipitation as recorded at Querecocha, recession accelerated in the 1980s (Figure 11.2).

The glaciers in the Cordillera Huayhuash, a spectacular range 30 km long, east of the Cordillera Blanca, are steeper, thinner and more crevassed than those in the Alps. The largest are about 4 km long. In 1909 Sievers (1914) found that many had retreated about 150 m from massive, fresh moraines. They were still in much the same positions in 1939 and Kinzl (1955) considered that, like others in the Cordillera Blanca, they had probably advanced in the 1920s. Between

1936 and 1954, although the larger glaciers showed little change, the smaller ones shrank noticeably (Mercer 1967).

The glaciers in the Andes Centrales to the north and south of the trans-Andean railway have receded since the middle of the nineteenth century in a spectacular fashion. When Hauthol (1911) visited the area in 1908 he found the glaciers receding but inferred they had advanced briefly but sharply in 1886–7. Some peaks such as Paragate had small glaciers in 1917, but these had shrunk by 1923 and disappeared completely by 1942. The snowline is estimated to have risen from about 4600 m in 1862, to 4900 m in 1923 and 5100 m in 1940 (Mercer 1967).

The Quelccaya Icecap (13°56′S, 70°50′W) offered a peculiarly good opportunity to obtain palaeoclimatic information from ice cores. With an area of 55 km² and a central summit at 5650 m rising 400 m above the snowline, temperatures during the 1970s and 1980s were so low, less than –3 °C, that percolation was inhibited. The relief of the ice surface is low enough to allow accumulation layers to remain little distorted. As net radiation was 'for all practical purposes zero', so that no energy was available for evaporation or melting, the mean annual net balance was about the same as mean annual accumulation (Thompson 1980). Though the annual temperature range is small, precipitation is seasonal, with the Atlantic, 1800 km away, the main original source of moisture, and thunderstorms recycling the water en route. The ice is well stratified on account of the seasonal occurrence of accumulation; 80–90 per cent of the snow falling in summer and autumn, between November and May. Annual values for the last 1500 years were obtained from the cores. It was fortunate that they were extracted in the 1970s and 1980s, for in 1991 it was found that the margins of the icecap had receded drastically and, what is more, the oxygen isotope profile had been smoothed as a result of recent warming and percolation so that annual variations in $\partial^{18}O$ could no longer be traced (Thompson et al. 1993).

Preliminary cores 15 m long had revealed a periodic variation of microparticle content, $\partial^{18}O$, and radioactivity which appeared to be annual (Thompson 1980). Two cores 155 m and 163 m long penetrating to bedrock were therefore extracted from the Quelccaya summit area in 1983, in the hope that they would provide a key to regional atmospheric conditions over a long time period and possibly register the occurrence of El Niño conditions in the eastern Pacific, 400 km away.

The Quelccaya cores proved to be extremely informative (Figure 11.3). Dating, depended not only on a combination of annual dust layers, microparticulate concentrations and conductivity values but also on the presence of an ash layer attributable to a great eruption of Huaynaputina (16°35′S, 70°52′W) known to have taken place in February and March 1600. The resultant chronology is rated as accurate to ±2 years back to AD 1500. Decadal temperatures, inferred from $\partial^{18}O$ values, were generally low between 1530 and 1880.[6] Values of conductivity and microparticle concentration throughout most of the 1530–1880 period were higher

than those obtaining for the fourteenth, fifteenth and twentieth centuries. Greater dust deposition in recent centuries was probably due to higher wind velocities across the high altiplano, where westerly and northwesterly winds dominate in the dry season. The annual layering indicated that in the last thousand years the highest precipitation occurred between 1500 and 1720, whereas the period from 1720 to 1860 was very dry.

The Little Ice Age stands out in the Quelccaya record as an important climatic event that affected this part of the southern hemisphere as clearly as the northern hemisphere (Thompson et al. 1986). Thompson (1992) visualized the Little Ice Age as spanning the period between AD 1490 and 1880 when microparticles and conductivities were 20–30 per cent above the averages for the entire core.[7] A particularly noticeable dust event occurred around 1490, though an even larger microparticle level was recorded in the first decade of the fifteenth century. This is the only direct evidence presently available about the timing of the Little Ice Age in the Cordillera Blanca.

Elsewhere, a radiocarbon date on a basal sample of peat and fossil soil (WIS-1202) from Laguna Huatacocha, closely associated with the end moraine of Glaciar Huallcacocha, near Huascarán (9°25′45″S, 77°26′40″W) (Röthlisberger and Geyh 1986) has given a calendar span of AD 1220–90 which could relate to the onset of the Little Ice Age (Wright 1984). A compound moraine, consisting of 'bulldozed' peat on the forefield of a glacier at the head of the Upismayo Valley, in the Cordillera Vilcanota (near 13°45′S, adjacent to the Quelccaya Icecap) gave a date of 630 ± 65 BP (DIC-678), indicating glacial activity after cal AD 1290–1400. Mercer (1977: 604) commented that 'the most recent glacial expansion in the Cordillera Vilcanota-Quelccaya icecap area that culminated between about 600 and 300 BP, was followed by a smaller readvance which coincides broadly with the globally recognised "Little Ice Age"'.

Abnormally high precipitation in the northern coastal desert regions of Peru and Ecuador, and drought in the southern highlands of Peru and northern Bolivia, are associated with strong negative values of the El

---

6   The details of temperature fluctuations at Quelccaya, plotted as departures from the 1881–1975 mean, compare remarkably closely with those for the northern hemisphere compiled by Groveman and Landsberg (1979).
7   The Little Ice Age was initially dated at AD 1500–1900 (Thompson et al. 1985, 1986).

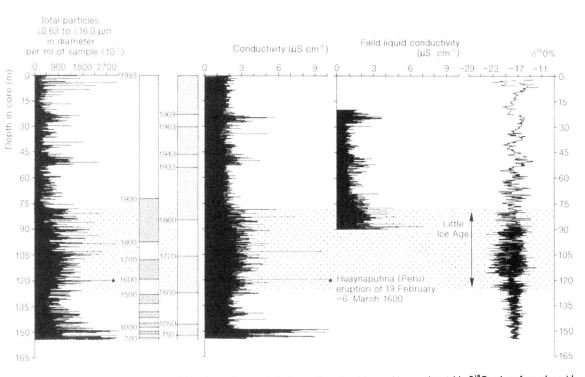

Figure 11.3 The Quelccaya summit core, 153 m long, showing high dust and conductivity, and low and variable ∂¹⁸O values from the mid sixteenth to the mid nineteenth centuries. Total microparticle content, and liquid conductivity were measured in the field. The second column shows the stratigraphic timescale and illustrates the compression of time with depth. The third column shows the dry and wet (darker shading) periods reported in Thompson *et al.* (1985). The hatched area indicates the Little Ice Age (*Source:* from Thompson *et al.* 1986)

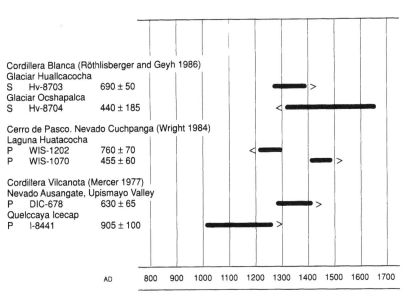

Figure 11.4 Key radiocarbon dates relating to Little Ice Age advances in tropical South America (*Source:* from Grove 2001b)

Notes: The numbers in the margin give the laboratory identification number and the ¹⁴C age. Horizontal lines represent the calibrated ages. > = Advance after date; < = Advance before date; P = Peat; S = Soil

Niño/Southern Oscillation Index.[8] During El Niño events in 1976 and 1982–3 net accumulation on Quelccaya in the south was reduced and, with sparser cloud, net radiation increased (Thompson *et al.* 1992). The smaller accumulation caused concentrations of soluble and insoluble particles to increase at the same time as the dust flux across the altiplano of southern Peru intensified because of the drought. This suggested the possibility of using the ice core record to extract an ENSO chronology, but other climatic conditions, such as the severe drought in 1935–45, produce similar variations in the ice core parameters. Detection of short-term ENSO events affecting wide areas requires evidence from cores extracted from more than one icecap.

In 1993 Thompson obtained two further cores from the col of Huascarán (9°06′41″S, 77°36′53″W), 970 km to the north of Quelccaya (Thompson *et al.* 1995a, Davis *et al.* 1995). This site is much closer to the zone of maximum response to ENSO events, and at 6048 m altitude the oxygen isotope profile had not been affected by recent warming; neither surface melting nor meltwater percolation have been observed, and the palaeoclimate record is still intact.[9] It appears that long-term Holocene cooling culminated in the seventeenth and eighteenth centuries, and that the $\partial^{18}O$ decrease was similar to that recorded at Quelccaya.

11.1.5 Bolivia and northern Chile

The glaciers in the widest part of the Andes in Bolivia and northern Chile received relatively little attention until recently, except for mapping carried out by the Osterreichischer Alpenverein in 1928. The Illampu–Anconhuma–Casiri area of the Cordillera Real, north-west and southeast of the La Paz railway, between 15°50′ and 16°60′S, was surveyed in 1928 and the altitudes of most of the terminal moraines were determined. The glaciers were then quite close to moraines but there was evidence of some shrinkage. These surveys were repeated in 1963, 1975 and 1983. The resulting 1:50,000 maps indicate that retreat continued through the twentieth century, but not so fast as in the Austrian Alps. Advances similar to those which began in Austria in the 1980s were not observed in the Cordillera Real (Finsterwalder 1987).

Parts of the Bolivian section were little known until the 1950s when a British expedition mapped the snouts of fifty-one glaciers. Both retreating and advancing tongues were found; the fronts of many glaciers were then reaching past old terminal moraines and covering plants that were at least ten years old (Melbourne 1960). An accurate base from which future change can be measured now exists; 1:70,000 air photographs taken in July/August 1975 have been used to map some 1775 glaciers in sixteen mountain ranges of the Eastern Cordillera of Bolivia (Jordan 1985).

Annual mass balance data had been collected from two glaciers in the Cordillera Blanca in the 1980s (Kaser *et al.* 1990). Because accumulation on tropical glaciers takes place in the wet season and coincides with the warm period when ablation is also at a maximum, it was judged necessary to make monthly measurements. In 1991–2, an El Niño year with high temperature and low precipitation, mass balances were found to be negative. The following year was normal and the mass balance positive. Observations were subsequently extended to two glaciers in the Cordillera Real, the Chacaltaya, a tiny glacier 20 km northeast of La Paz at 16°S, and the Zongo on the Huayna Potosí massif, also at 16°S, and a little to the east in the Amazon basin (Haeberli *et al.* 1994, Francou *et al.* 1995). Between 1940 and 1983 the Chacaltaya lost 62 per cent of its mass. In the next two decades recession increased dramatically and by

---

8 Sea level pressure across the tropical Pacific see saws in such a way that when it is high over Tahiti it is low over Darwin. When pressure is high over the eastern Pacific cold water upwells along the coast of Peru, sea surface temperatures are anomalously low for the latitude, and the climate of the coastal areas is hyperarid. When, every few years, pressure is low over the eastern Pacific upwelling ceases along the Peruvian coast, sea surface temperatures rise and a warm 'El Niño' event, marked by heavy rainfall occurs. The El Niño/Southern Oscillation (ENSO) is a global-scale phenomenon, with many long distance teleconnections. Apart from the annual cycle, it is the dominant signal in the climatic system on timescales of months to a few years (Diaz and Markgraf 1992).

9 Glaciers on Huascarán–Chopicalqui massif retreated in the twentieth century. Equilibrium levels fell 95 ± 5 m between 1920 and 1970 (Kaser *et al.* 1996) and annual temperatures from nearby meteorological stations and the isotopic records from 1950 to 1993 show warming trends.

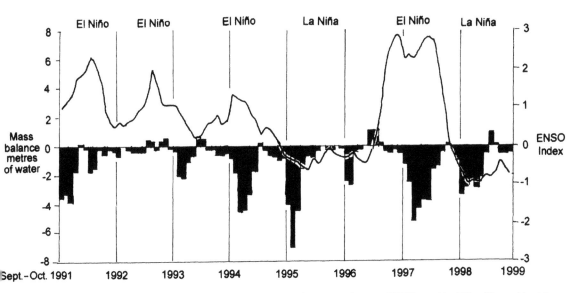

Figure 11.5 The changing mass balance (monthly bars) of the Chacaltaya Glacier in relation to El Niños and La Niñas. The multivariate ENSO index line expresses bimonthly standardized departures from a mean (*Source:* from Francou *et al.* 2000)

1998 Chacaltaya's volume was only 7 per cent of its volume in 1940 largely as a result of the El Niños of 1991-2, 1994-5 and 1997-8 (Figure 11.5). Very soon it is likely to disappear completely (Francou *et al.* 2000). The Zongo, which extends from 6000 to 4900 m a.s.l. was found to be much affected by a vigorous El Niño event in 1997/8 as a result of the reduced precipitation and higher temperatures giving a large extent of low-albedo bare ice. The resultant decrease in albedo was responsible for a sharp increase in all-wave radiation causing much greater melting than in the preceding La Niña year (Wagnon *et al.* 2001).

11.2 EAST AFRICA

Of the three mountains with glaciers in East Africa, Kilimanjaro (5895 m) and Mt Kenya (5199 m) are both volcanoes. Ruwenzori, on the border between Uganda and Zaire, is a horst, an uplifted block of crystalline rock, a mountain range overlooking the western arm of the Great Rift Valley. Kilimanjaro rises from rather arid plains and has a higher snowline than either of the other two. On both Kenya and Ruwenzori, accumulation dominates in March–May and September–November, ablation in the intervening months. The four seasons are reflected in the firn profiles on

Ruwenzori, though their impress is limited by their short duration and the lack of strong seasonal contrasts (Temple 1968).

11.2.1 Mount Kenya

The East African glaciers, especially those on Mt Kenya, were the subject of a book by Hastenrath (1984). He provided a remarkably detailed presentation of the evidence of the changing state of all the glaciers since the nineteenth century, with full photographic and cartographic as well as documentary inventories. Most remarkable are his photographs of the glaciers that he took from a hot air balloon in 1974.

Sets of moraines, only very sparsely vegetated, in front of all the glaciers mark more advanced positions of the ice in recent centuries. The moraines in front of the Lewis Glacier on Mt Kenya, which is the most accessible and closely studied of all the East African glaciers, are multiple, with two or more crests spaced up to about 50 m apart (Charnley 1959). They lie at a height of between 4270 and 4420 m (Figure 11.6). Between 1899 and 1974 the area of the glacier diminished from 0.63 to 0.31 km² and over the same period the elevation of the snout rose by 130 m (Hastenrath 1975). Shrinkage was at a maximum in the early part

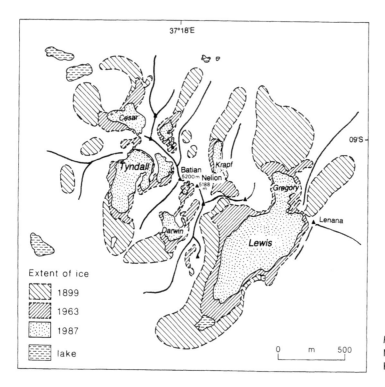

Figure 11.6 Changing extent of glaciers on Mt Kenya, 1899, 1963 and 1987 (*Source:* from Hastenrath and Kruss 1992)

of the century, slowed down in 1934, and almost ceased in the 1960s. After 1963, though the tongue continued to retreat, there were signs of thickening of the upper glacier. This came to an end in 1974, by which time the terminus was 445 m back from its 1899 position and the maximum velocity of flow had slowed from 15 to 5 m per year (Hastenrath 1989). Over the next four years the volume of the Lewis Glacier is estimated to have diminished by 1.2 million m³, i.e. by one-sixth (Bhatt *et al.* 1980). Between 1978–82 and 1982–6 ice flow slowed down progressively (Hastenrath 1987). Between 1974 and 1986 the ice volume shrank by nearly 50 per cent and the area diminished from 0.31 to 0.25 km². The snout rose by 18 m and retreated by a further 73 m. Simultaneously the maximum flow velocity fell from 5 to below 3 m per year as the glacier became progressively smaller, thinner and more sluggish.

The net balance of the Lewis Glacier during the period 1986–90 was overall less negative than it had been in 1978–86. Measurement of the change in thickness in 1990, based on repeat air photography, was affected by fresh snow cover which reduced its accur-acy (Hastenrath and Rostom 1990, Hastenrath 1992). Changes over the period 1986–90 are summarized in Table 11.1, the uncertainty being estimated as 0.2 m thickness equivalent for the entire glacier. Changes in flow velocity were determined to within 0.1 m per year. The decrease in the extent to which net balance was negative, compared with the period 1978–86, caused the decay of the Lewis to be less drastic than had been expected. Though the flow of the ice became slower, it was still capable of redistributing mass. Hastenrath forecast substantial further decay in the period 1990–4 on the basis of modelling projections, with ice thinning less than 1 m in the upper parts of the glacier, but more than 7 m in the lower part.

11.2.2 Ruwenzori

The story of the Ruwenzori glaciers is very similar, although their alimentation seems to depend more on moisture derived from the Atlantic than is the case on Mt Kenya. Measurements of precipitation at six stations in the early 1950s suggest that it reaches a

Table 11.1 Changes in the thickness, width and surface velocity of Lewis Glacier, 1986–90

Longitudinal distance	Thickness (m)		Width (m)		Surface velocity (m/y)		Volume flux (100 m³/y)	
1900	1986	1990	1986	1990	1986	1990	1986	1990
100	12	12	142	142	0.6	0.6	7	5
200	24	22	155	155	0.9	1.2	39	29
300	28	25	275	265	1.3	1.4	70	65
400	20	19	272	267	2.0	1.5	76	56
500	28	27	435	410	2.3	2.2	131	115
600	33	29	373	358	2.2	2.3	127	110
700	28	24	308	288	2.5	2.2	101	68
800	24	18	208	255	1.8	1.8	57	38

Source: after Hastenrath 1992, table I
Notes: Ice thickness and width were measured in metres, at the longitudinal distances specified, as were centre-line surface velocity in metres per year and volume flux in hundreds of cubic metres per year

maximum at around 3500 m (Osmaston 1989) It is clear that in the late nineteenth century the ice cover was substantially greater than now. Since then the larger ice masses have become fragmented and the smaller ones have disappeared. Some glaciers, including the Elena and Speke, advanced slightly in 1952–6, leaving small moraines (Kaser and Noggler 1991), while others receded only slowly (Whittow *et al.* 1963, Temple 1968). Temple (1968) suggested that the small advances had been caused by increased precipitation.

The Speke, the largest of the Ruwenzori glaciers, receded 35–45 m between 1958 and 1977, and a further 150 m by 1990 (Kaser and Noggler 1991). Between 1967 and 1977 the western side of the Speke serac zone had thinned at around 4700 m and by 1990 almost half of the western side of the tongue had disappeared, but the seracs on the eastern side were still thick and active. That the eastern tongue alone remained extended and active was attributed to it being protected against incident shortwave radiation by topographical shading until midday, while the western side has no such protection and so is more vulnerable to radiation during the morning hours. By the early 1990s the central Ruwenzori mountains had lost about three-quarters of the ice cover they had carried in 1906, let alone the much greater area which they supported

at the Little Ice Age maximum, which has not been quantified. If the climatic trends of recent decades persist there will soon be little ice left on Ruwenzori.[10]

11.2.3 Kilimanjaro

The glaciers of Kilimanjaro, all of which are either on the western Kibo cone or within the crater, have shrunk continuously since the end of the nineteenth century, those on the eastern crater rim disappearing entirely and the rest likely to do so within the next few years (Figure 11.4, Plates 11.1a and 11.1b).

11.2.4 Twentieth-century shrinking of the East African glaciers

The shrinkage of the East African glaciers in the course of the twentieth century seems to reflect important changes in the climate of the region. It is known that Lake Victoria and other lakes in East Africa were much enlarged in the early 1870s as a result of higher rainfall than the present. Between 1920 and 1949, according to Sansom (1952), rainfall decreased by about 150 mm on Mt Kenya, though in the neighbourhood of Kilimanjaro and Ruwenzori it increased by 75 mm. Mean annual temperatures oscillated between 1915 and 1963

---

10   Details of the history of observations of the Ruwenzori glaciers are given in Kaser (1993) and Kaser and Noggler (1996).

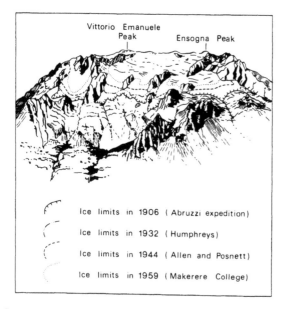

Figure 11.7a The retreat of the ice on the eastern side of Mt Speke, Ruwenzori, Uganda, 1906–59 (*Source:* from Whittow *et al.* 1963)

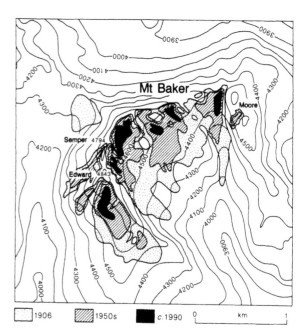

Figure 11.7c The reduction of ice cover on Mt Baker, Ruwenzori during the twentieth century (*Source:* after Kaser and Noggler 1996)

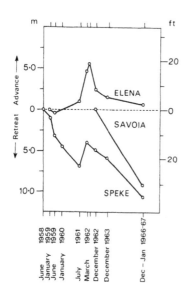

Figure 11.7b Fluctuations of the termini of the Speke, Savoia and Elena glaciers between June 1958 and January 1967 (*Source:* from Temple 1968)

with interannual amplitudes of about 2 °C and a slight cooling between 1933 and 1963 (Whittow *et al.* 1963). More important, about the time in the early 1960s when the glaciers began to advance or retreated more slowly for a few years, lake levels rose sharply as a result of higher rainfall totals, especially in November 1961, and subsequently remained at high levels for several years. Subsequently the glaciers recommenced their rapid retreats. Lake levels rose again in the late 1990s as a result of heavy November rainfall but any associated changes in the glaciers have yet to be reported.

Looking back to the last century, Hastenrath (1985) interpreted numerical modelling of the Lewis Glacier by Kruss (1984) as indicating that the retreat since 1890 was caused by annual precipitation diminishing in the two decades after 1883 by something of the order of 150 mm, with an accompanying decrease in cloudiness. He also found evidence that temperatures in East Africa had been a fraction of a degree higher in the twentieth than in the latter part of the nineteenth century and that there was a particularly marked rise of temperature in the 1920s.

Table 11.2

(a) Changes in the surface areas of Ruwenzori glaciers in km²

	Elena	Speke	Moore	Savoia
1906	0.315	0.453	0.166	0.121
c.1955	0.315	0.372	0.088	0.043
1990–3	0.113	0.369	0.016	0.043

(b) Areas of Ruwenzori glaciers as percentages of those of 1906

	Elena	Speke	Moore	Savoia
1906	100	100	100	100
1950–60	63	82	53	64
1990–3	36	44	10	36

(c) Glacier cover on Mt Stanley, Mt Speke, Mt Baker and Central Ruwenzori in km² and as a percentage of the area in 1906

	1906		c.1955		c.1990	
	km²	%	km²	%	km²	%
Mt Stanley	2.854	100	1.885	66	0.999	35
Mt Speke	2.187	100	1.306	60	0.555	25
Mt Baker	1.468	100	0.617	42	0.160	8
Ruwenzori	6.509	100	3.808	58	1.674	26

Source: after Kaser 1997: 176
Notes: The values have been extracted from maps, photographs, air photographs, ground surveys, expedition reports and published papers. The 1:40,000 map of the Ruwenzori Range produced in association with the Duke of Abruzzi's expedion of 1906 (De Filippi 1909) provided coverage of all the glaciers for that year. Full details of sources are given in Kaser (1997)

The moraines bordering the East African glaciers still offer opportunities to learn more about the fluctuations of the ice in recent and earlier centuries. Some attention has been directed to the record provided by the ice itself. A core 11 m long, penetrating to bedrock, extracted from the Lewis glacier lacked annual banding but was characterized by peaks in the concentration of particulate material (Davies *et al.* 1977). These were interpreted as representing periods of ablation, each lasting some years or even decades. A layer of ice and firn above the highest dirt band probably represented the last period of net accumulation

starting in the early 1960s. The dirt band beneath, it was thought, might represent ablation and the concentration of particulate material over the period since about 1865. Other cores of a similar length have since been obtained (Hastenrath and Patnaik 1980, Thompson 1981, Thompson and Hastenrath 1981) but the information provided in the form of ice and firn horizons, oxygen isotope profiles and microparticle content is difficult to interpret.

11.3 NEW GUINEA

In the western half of the great Cordillera, 2000 km long, which forms the backbone of New Guinea, several peaks of the Merauke range rise above 4500 m. Three in Irian Jaya, Mt Mandala (Juliana 4640 m), Ngga Pilimsit (Idenburg 4717 m) and Puncak Jaya (Carstenz 5030 m) carry glaciers. An icecap on Puncak Trikora (Wilhelmina 4730 m) disappeared entirely at some time between 1939 and 1962, and all the little glaciers

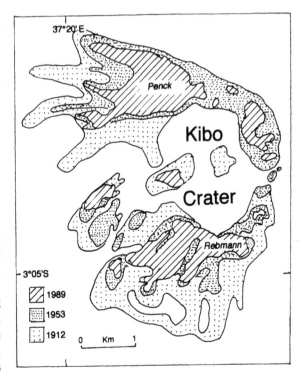

Figure 11.8 The shrinkage of the glaciers around the Kibo summit of Kilimanjaro, East Africa (*Source*: from Hastenrath and Greischer 1997)

Plate 11.1a The Kibo glaciers as seen from about 4330 m in the 1880s (*Source*: from Meyer 1890)

Plate 11.1b The Kibo glaciers as seen from about 4330 m in 1986: the Ratzel Glacier has almost disappeared
Photo: William Grove

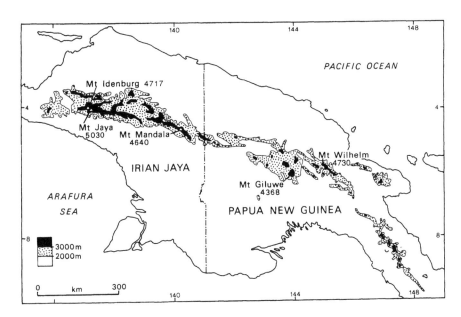

Figure 11.9 The mountains of New Guinea

scattered along the central cordillera have greatly diminished (Peterson *et al.* 1973). Puncak Jaya (4°5′S, 137°10′E), part of a huge block of Miocene limestone, carries the largest area of ice; five separate glaciers on its horseshoe-shaped margins had a total area in 1972 of 6.9 km². Amongst these glaciers, the Meren and the Carstenz (discovered as late as 1913) were the object of concentrated attention by Australian expeditions in the early 1970s.

The glacier snouts reach down into valleys that appear to have been abandoned by ice in quite recent times, leaving behind a series of moraines. On the youngest ones, with sharp crests and unstable slopes, plant colonization is at a very early stage, for no organic matter has accumulated between valley walls and lateral moraines which presumably formed within the last century or two. The retreat on Puncak Jaya over the last century or so has been traced from air photographs, satellite imagery, dated cairns and expedition records (Figure 11.10 and Table 11.3). Extrapolation of retreat rates from the period 1936 to 1974 suggests that recession from the most recent moraines began in the mid nineteenth century. This conclusion is supported by numerical modelling of the retreating glaciers (Allison and Peterson 1976). Such observations as are available from elsewhere are sufficient to confirm

that the very marked recession of the Carstenz glacier is typical of all the glaciers in New Guinea.

Puncak Jaya is only 80 km from the Arafura Sea. The outstanding feature of its climate is its constancy through the year and the high humidity. The mean diurnal temperature range for much of 1972 was only 2.7 °C at 4250 m and 3.4 °C at 3600 m, while the range of temperature for 10-day means at 3600 m was less than 1.5 °C. On most days cloud builds up in the middle of the day and then, in the afternoon and evening, rain falls at low altitudes and snow on the peaks (Allison and Bennett 1976). Total precipitation measured in the Meren area in 1972, an unusually dry (El Niño) year, was about 2440 mm, and it is thought that values between 2800 and 3300 mm are probably typical of the area as a whole, with the highest monthly totals between December and March. Hardly surprisingly, no significant stratigraphy was noted in 10-m cores taken from the firnfields of the two glaciers.

The observed retreat of the Carstenz glacier could be accounted for by a warming at a rate of 0.6 °C per century starting in 1830–50, with a 30-year lag before significant retreat began (Allison and Peterson 1976). This corresponds to the warming of the tropics calculated by Mitchell (1961). Retreat might also be associated with increased cloud cover and reduced radiation

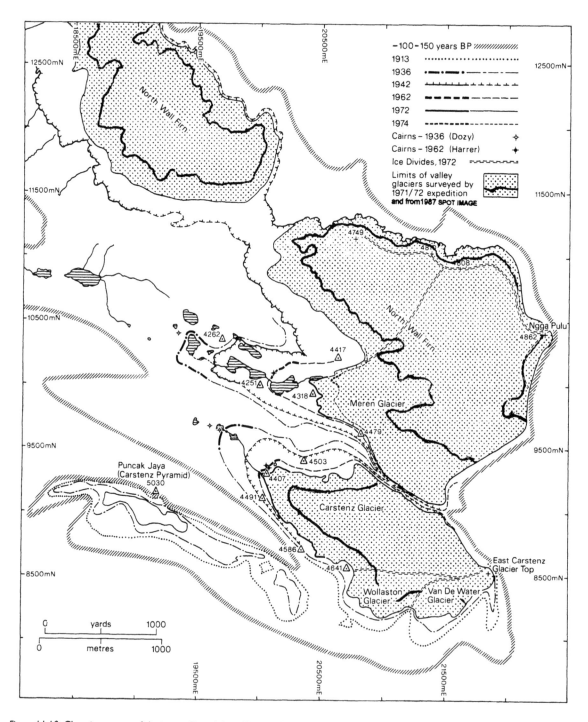

Figure 11.10 Changing extent of the ice on Puncak Jaya (Carstenz), New Guinea. Since 1974 the margins of the main ice masses have retreated some 200–300 m (*Sources*: from Allison and Peterson 1976, Peterson and Peterson 1994)

Table 11.3 Rate of retreat of Carstenz Glacier, New Guinea, 1850–1974

	m/y
1850–1936	24
1936–42	54
1942–62	24
1962–72	37
1972–3	15
1973–4	100

Source: Allison and Peterson 1976

inputs. A change in precipitation alone is unlikely to have been responsible for the recession; it has been calculated that a decrease in mean annual precipitation of about one metre would have been required to account for it (Allison and Kruss 1977). A minor but significant and unusual role may have been played by decreased albedo of the ice caused by increased colonization by 'cryovegetation', mainly algae, some black, some highly coloured, which is unusually prominent on these glaciers (Kol and Peterson 1976).

Comparison of older air photographs with a SPOT satellite image of 1987 and oblique air photographs taken in 1991 shows that retreat has continued. Less than 3 km² of ice now remain of about 19 km² at the Holocene maximum (Peterson and Peterson 1994). The Meren, which had become dead ice by 1991, must soon disappear altogether. The Carstenz Glacier has retreated less, despite having thinned greatly since 1973.

11.4 THE LITTLE ICE AGE IN LOW LATITUDES

Mercer's (1977) studies in the Cordillera Vilcanota suggested that the Little Ice Age in the northern Andes began between AD 1290 and 1400. The Quelccaya cores, with the ash of the Huaynaputina eruption providing an AD 1600 datum, show some negative $\partial^{18}O$ value in the late thirteenth and early fourteenth century (Thompson *et al.* 1993). Thompson (1992) pointed to 1490, the time when microparticles and conductivities exceeded 20 per cent of the average for the core, as marking the beginning of the Little Ice Age. The Huascarán record indicates that long-term Holocene cooling culminated in the seventeenth and eighteenth centuries.

Many glaciers in low latitudes disappeared in the twentieth century, especially in the last decade or two, and yet satellites and radiosondes do not indicate any warming of the lower troposphere in the tropics since 1979 (IPPC 2001).

Strong El Niño events have been felt particularly severely in the tropical Andes where high temperatures in 1997–8 coincided with low precipitation. Under such conditions, there is little snow to protect the ice from the sun's rays and, with the low albedo, shortwave radiation is absorbed abundantly causing high rates of melting. Streams fed from the glaciers may currently receive an adequate supply of water from the melting glaciers, but in the longer term the storage reservoirs constituted by the ice accumulated in the course of the Little Ice Age are being depleted and will soon be exhausted.

12

SOUTHERN HEMISPHERE MID LATITUDES

The southern Andes and New Zealand

The Southern Alps of New Zealand and the southern Andes are the southern hemisphere equivalents of Norway and the Western Cordillera of North America. In both of the southern regions, mountain ranges face snow-bearing westerly winds; the Southern Alps standing athwart the Roaring Forties, the Andes of southern Chile projecting into the Furious Fifties.

12.1 THE SOUTHERN ANDES

The elevation of the Andes decreases from 6000–7000 m in northern Chile and Argentina to less than 2000 m in the far south. The snowline lowers southwards at a similar rate. The arid zone reaches to the northern limit of winter cyclonic depressions, about 30°S, and some of the intertropical Andean peaks are high enough to carry ice. In northern Chile between 18°S and 32°S, 88 glaciers cover an area of 116 km². Some are very small, little more than stationary sheets of ice, lacking crevasses, accumulating in one year, ablating in another over the whole of their surfaces. As they have no moraines, early fluctuations are difficult to detect, though old photographs can help in tracing more recent history. In the 1940s and 1950s they showed no signs of recession (Lliboutry et al. 1958).

In central Chile 1402 glaciers have a combined area of 934 km². Most of the ice lies north and south of Santiago, between 32°30′ and 35°S, where glaciers on the slopes of Aconcagua (7040 m), for instance, are up to 12 km long (Corte and Espizua 1981). Those best known are Nevado Universidad and Juncal Sur, both of which surge. About their earlier fluctuations little seems to be known, apart from Röthlisberger and Geyh's (1986) dating of moraines of the Glaciar Cipreses (34°33′15″S, 70°23′15″W) in the Cordillera Central where a radiocarbon date of a fossil soil superposed on bedrock gave a minimum age of 625 ± 155 BP (Hv-10915) for one of four lateral moraines which calibrates to AD 1260–1440. An advanced stand of the ice was dated to about AD 1858.

In 1945, many of the glaciers in the Mendoza–Santiago region were retreating rapidly (Mercer 1962, Corte and Espizua 1981). As glaciers are of critical importance for Chilean water supplies, the Dirección General de Aguas started to monitor them in the 1970s (Casassa 1995). The work has included the compilation of glacier inventories covering large parts of the country, most of them unpublished, plus mass balance measurements of Glaciar Echaurren Norte, near Santiago. The results of the measurements for the period 1975 to 1983 showed large variations from year to year, with no trend apparent.

On the Argentinian side of the Andes, when river discharges diminished in the San Juan Cordillera in the mid twentieth century, alpinists were requested to donate photographs taken before 1962, the year in which the area was covered by air photography, to facilitate study of the timing of volumetric changes in the ice cover (Bader 1973). Argentinian glaciers from 28° to 35°S were inventoried by the Institute of Snow and Ice at Mendoza (Corte and Espizua 1981, Cobos and Boninsegna 1984, Aguado 1984) and south of 35°S by Rabassa (Rabassa et al. 1978). In the Cajo del Rubio, the headwaters region of the Las Cuevas River, mass balance measurements of the Piloto and Alma Blanca glaciers began in 1979. Although measurements are lacking for some years, it is already clear that heavy snow precipitation in the central Andes is correlated with ENSO events, in contrast to the situation further north where El Niño years are liable to be dry (Leiva and Cabrera 1996).

In the Mendoza region, an accurate photogrammetric survey of the Plomo glaciers was carried out by Helbling in 1909. His photographs were subsequently used as a basis for tracing the movements of glacier fronts. Corte and Leiva (1977) were able to compare Helbling's photos with air photographs taken in 1974. More recently, Llorens and Leiva (1995) have utilized 1992 satellite images for this purpose. Several surging glaciers, including the Grande del Nevado del Plumo, Juncal I and Innominad, all occur in this part of the central Andes. Since 1909, the glaciers in the Plomo basin have retreated, some of them by as much as 5 km. Recession has not been unbroken, several glaciers advancing between 1974 and 1991.

A wooded site at 33°S in central Chile was sampled by Villalba *et al.* (1990). The resultant ring thickness record, which extends back over eight centuries, appears to be indicative of variations in precipitation. The longest period of high, sustained growth was found to be the interval between 1820 and 1906, with interruptions around 1860 and 1890. Nineteenth-century amelioration of the prevailing aridity probably accounts for the glaciers north of latitude 32°S reaching their maximum extent at the beginning of the twentieth century.

On the western, Chilean slopes at about 34°S, the glaciers seem to have reached their maximum extent in the mid nineteenth century. The Los Cipreses Glacier advanced in the fourteenth century and again in the 1860s (Röthlisberger and Geyh 1986). The Ada Glacier was observed in 1858 to reach down to 1797 m; by 1882 the front had risen to 1929 m, and by 1980 to 2500 m above sea level.

The glaciers on the Argentinian side, between 34°20'S and 35°20'S, depend on redistribution of snow by southwesterly winds. In the past, ice from several tributary valleys reached down into the valley of the Atuel. The situation there in 1895 was observed by Hauthol (1895) who reported that 'the glaciers clearly indicate not only that until a short time ago they were of much greater extent but also that they are now los-

Table 12.1 Recession of the Humo Glacier, 1914–82

	m	m/y
1914–47	−3200	97
1948–55	−300	43
1955–63	−160	20
1963–70	−300	43
1970–82	−150	12

Source: Cobos and Boninsegna 1983

ing volume and receding at a great rate'. At that time the lowermost kilometre of the Humo tongue consisted of dead ice covered in moraine. It was observed in 1914 and again in 1937 by Groeber (1947, 1954). Between his visits the glacier front receded 3200 m, involving the loss of 0.5 km³ of ice. An analysis of the fluctuations of glaciers in the upper Atuel was made by Cobos and Boninsegna (1983) who found the ice mainly confined to cliff glaciers in sheltered ravines with only the Corto and Humo glaciers entering the Atuel valley river basin.

The Cordon Limite between Argentina and Chile, rising above 4000 m, supports glaciers on both sides. Striking changes over the last century or more have been reasonably well documented by travellers in the region. A base map published in 1947 was updated in 1961 and air photographs were flown in 1948, 1955, 1963 and 1970. The retreat of the Humo and its neighbours continued, at a diminishing rate, at least until 1982 (Table 12.1). All the other glaciers surveyed by Cobos and Boninsegna also dwindled rapidly during the twentieth century, with the single exception of a glacier in the Laguna valley which advanced 1400 m between 1970 and 1982.

Further south in Chile, between 37°S and 41°30'S, 82 glaciers cover 266 km², the largest being Tronador (41°10'S, 71°53'W) on the mountain of the same name which rises to 3003 m on the border beteeen Chile and Argentina (Figure 12.1) Air photographs flown in 1945 revealed much more ice in Argentina south of 40°S than had been indicated on earlier maps (Mercer 1967). High avalanche cones and reconstituted glaciers were found to be common between 43° and 45°20'S, signs of recent recession being widespread

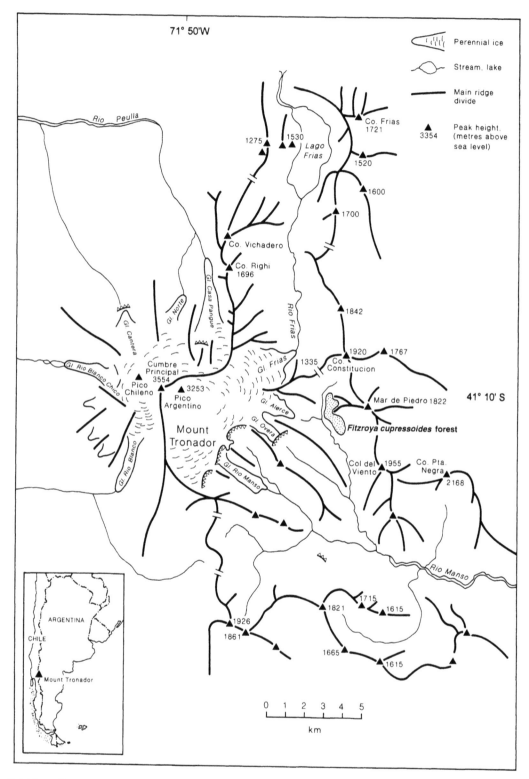

Figure 12.1 Mount Tronador. Four of the glaciers descend into Argentina and six into Chile. Cores for the Rio Alerce tree-ring chronology were collected in the *Fitzroya cupressoides* forest (*Source*: from Villalba *et al.* 1990)

and comparable in extent with retreat in the Rockies. Most information is available for the glaciers of Monte Tronador. The gently sloping accumulation area surrounding the steep summit ends in cliffs over which the ice avalanches thunderously, giving its name to the mountain. Some of the reconstituted glaciers below are easily accessible. Little Ice Age advances brought the tongue of Rio Manso on the southeast side to its Holocene moraine (see Chapter 15). Lawrence and Lawrence (1959) found evidence from tree rings suggesting that the latest Rio Manso moraines were formed on occasions when the glacier reached advanced positions between the early eighteenth century and 1795, and again in 1809–21, 1832–4 and about 1847. Tables presented by Rabassa et al. (1978) show that over the periods 1942–53 and 1953–70, three of the main Tronador glaciers were retreating at mean rates of something like 10 m annually. However, an advance of the Rio Manso Glacier, which the Lawrences thought was probably still continuing, had brought the front forward in the late 1950s into forests dating from the 1920s, and figures given by Rabassa and his colleagues indicate readvances of the Tronador glaciers in the 1970s (see Table 12.2).

The recent moraines of the Frias Glacier on the northeast side of Tronador have been dated to about 1638, before 1722, 1747, 1839, about 1881, 1914, 1942 and 1977 (Villalba et al. 1990). These dates, which are consistent with information from old drawings and photographs, depend on ring counts of the oldest trees found on each moraine, together with precise ages for ice positions that caused damage to living trees. Glacial retreat rates suggested were slow until about 1850, increased to over 7 m per year between 1850 and 1900, and about 10 m per year between 1910 and 1940. The terminal position of the Frias has been measured since 1976. In 1977 it advanced 32 m to a position 835 m above sea level (Table 12.2). It was then about 800 m upvalley of the position shown in a photograph taken by Steffen in 1896.

On a basis of the similarity found between the timing of the Frias fluctuations and tree-ring indices from *Fitzroya cupressoides* in the Rio Alerce valley, Villalba et al. (1990) suggest that Little Ice Age expansion of the

Table 12.2 Fluctuations of Mt Tronador glaciers, Argentina, 1942–77 (m/y)

Castano Overa		Alerce		Frias	
1942–53	0	1942–53	−11.1	1942–53	−16.9
1953–70	−1.8	1953–69	−9.9	1953–70	−9.8
1970–4	+8.2	1969–70	−16		
1974–5	−15	1970–5	−6.8	1970–6	+47.4
1975–6	−12	1975–6	+18		
1976–7	+9	1976–7	+6	1976–7	+32

Source: Rabassa *et al.* 1978

Frias probably began between AD 1270 and 1380. The lack of dated moraines from this period they attributed to subsequent overriding by glacier advances between AD 1520 and 1670 during a longer interval of cool, wet summers that were particularly marked between 1630 and 1650. The suggested correlations seem neat, but moraine dating based on the number of rings in the oldest trees can only be approximate at best, and may be misleading. Furthermore, the relationship found between climate and tree-ring indices depends, in large part, on data from weather stations outside the immediate Tronador region. However identification of the start of Little Ice Age expansion as occurring much earlier than the sixteenth century is strengthened by radiocarbon dating by Röthlisberger and Geyh (1986) of *in situ* tree trunks in the lateral moraine of the Ventisquero Negro (Hv-11800) to cal AD 1000–1220.[1]

A great many independent glaciers and two extensive icefields lie polewards of 46 °S. Present knowledge of these icefields, the largest in the mid latitudes, was reviewed in detail by Warren and Sugden (1993). It is commonly accepted that both icefields consist almost entirely of ice at pressure melting point (Lliboutry 1956, Aniya 1988, Warren and Sugden 1993). Despite concentrated efforts by Japanese investigators during recent decades (e.g. Aniya 1988, 1992, Wada and Aniya 1995, Naruse and Aniya 1995, Aniya and Wakao 1997), not even preliminary estimates of total ice volume have yet been made, little is known about mass balances, and the dynamics of the glaciers are inadequately

1   Calibration according to Stuiver and Reimer (1993).

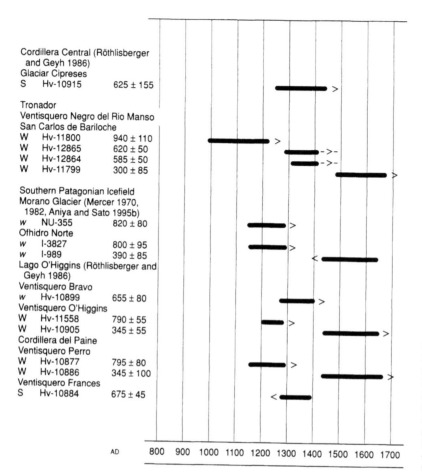

Cordillera Central (Röthlisberger
  and Geyh 1986)
Glaciar Cipreses
S   Hv-10915   625 ± 155

Tronador
Ventisquero Negro del Rio Manso
San Carlos de Bariloche
W   Hv-11800   940 ± 110
W   Hv-12865   620 ± 50
W   Hv-12864   585 ± 50
W   Hv-11799   300 ± 85

Southern Patagonian Icefield
Morano Glacier (Mercer 1970,
  1982, Aniya and Sato 1995b)
w   NU-355   820 ± 80
Ofhidro Norte
w   I-3827   800 ± 95
w   I-989   390 ± 85
Lago O'Higgins (Röthlisberger and
  Geyh 1986)
Ventisquero Bravo
w   Hv-10899   655 ± 80
Ventisquero O'Higgins
W   Hv-11558   790 ± 55
W   Hv-10905   345 ± 55
Cordillera del Paine
Ventisquero Perro
W   Hv-10877   795 ± 80
W   Hv-10886   345 ± 100
Ventisquero Frances
S   Hv-10884   675 ± 45

AD   800 900 1000 1100 1200 1300 1400 1500 1600 1700

Figure 12.2 Key radiocarbon dates relating to Little Ice Age advances in temperate South America (30° to 51°S) (*Source:* from Grove 2001b)

Notes: The numbers in the margin give the laboratory identification number and the ^{14}C age. Horizontal lines represent the calibrated ages. - > - = Advance around or after indicated age; > = Advance after date; < = Advance before date; W = Wood *in situ*; w = Wood not *in situ*; S = Soil

understood. Although water resources in Patagonia are largely glacial in origin, annual mean discharges and water balances are almost unknown (Escobar *et al.* 1992).

12.1.1 The Northern Patagonian Icefield

The Northern Patagonian Icefield (Hielo Patagónico Norte), extending from 46°30'S to 47°30'S and 40–5 km wide, has an area, according to Landsat imagery, exceeding 4200 km². Most of it lies between altitudes of 1000 m on the west and 1500 m on the east. Gravity anomalies along an east to west traverse between the Soler and San Quintin glaciers have shown that the bedrock topography is irregular, descending below sea level in places, with ice thickness as much as 1400 m in troughs (Casassa 1987). Similar investigations gave a maximum depth of 575 ± 75 m for the Soler Glacier (Aniya *et al.* 1988). Almost all the twenty-eight outlet glaciers with areas greater than 5 km² terminate on land (Aniya 1988). Twelve have proglacial lakes but calving does not appear to form a dominant part of the total ablation.[2] The exception is the San Rafael Glacier

---

2 However, drastic increases in calving into proglacial lakes producing apparently large recession of a few glaciers were noted by Aniya and Wakao (1997). The processes involved have been the subject of more detailed observations in New Zealand, mentioned later in this chapter (pp. 361–3).

Plate 12.1 The calving front of the San Rafael Glacier in 1986
Photo: Charles Harpum

(Plate 12.1). One of the world's fastest flowing glaciers, draining some 18 per cent of the total area of the icecap (Aniya 1988), it ends in an ice cliff 3 km long, producing huge icebergs (Harrison 1992). The accumulation area of the San Rafael is by far the largest of all the outlet glaciers, although its total area of 760 km^2 is practically the same as that of the less active San Quintin further south. The piedmont lobe of the San Quintin (San Tadeo), which is separated from the Pacific by 5 km of outwash plain, has an ablation area of 402 km^2, 26 per cent of the ablation area of the entire icefield.

The climate of the Northern Patagonian Icefield is dominated by eastward moving cyclonic depressions which bring very heavy precipitation to the coast and to the high western side of the icefield. Mean annual precipitation over the centre of the icefield has been estimated to be 8000 mm. East–west precipitation gradients are exceptionally steep, typically of the order of several metres over tens of kilometres (Escobar *et al.*

1992). Mean annual cloudiness is over 85 per cent on the western side, whereas the east is sunny. Annual accumulation on the San Rafael has been estimated to be 3450 mm of water equivalent (Yamada 1987). Surface ablation, which has been studied on the San Rafael and Soler, occurs at almost any altitude in summer; snow can fall at low altitudes at any time of year (Warren and Sugden 1993, Aniya 2001). Equilibrium line altitudes probably fluctuate widely from year to year.

Dates around 1600 and 1880 have recently been obtained for pieces of wood brought to the surface by thrusting of the Soler Glacier into underlying sediments (Aniya and Shibata 2001).

In the course of the twentieth century the outlet tongues of the Northern Patagonia Icefield retreated from their nineteenth-century maximum positions. Those originating on or near Cerro San Valentin (3970 m), the highest peak in the southern Andes, have heavy debris covers and their shrinkage has been horizontal rather than vertical. The outlets on the western, more

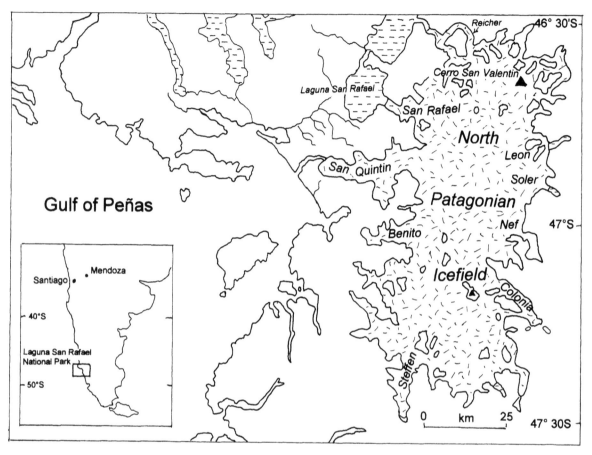

Figure 12.3 The Northern Patagonian Icefield

maritime sides of the icefield are almost free of surface debris and the 1945 air photographs showed them to be much healthier than those on the east. Both the San Rafael and San Quintin advanced between 1935 and 1950 and the Glaciar Steffen in the far south was still at its historic maximum in 1945 (Lliboutry 1956), whereas the easternmost outlets had retreated leaving well-marked trimlines. The most rapid retreats took place between the early 1930s and the early 1960s (Clapperton and Sugden 1988). Since 1944/5 fluctuations have been traced by Aniya and Enomoto (1986), Aniya (1988, 1992, 1999, 2001), Wada and Aniya (1995) and Aniya and Wakao (1997), using various ground surveys, 1:50,000 topographical maps published by the Instituto Geográfico Militar of Chile in 1982, plus vertical air photographs taken in 1974/5

and oblique air photos of 1985, 1990, 1993/4, 1995/6 and 1998/9.

Since 1944 the western outlets have been retreating faster than the eastern. The annual average rate of recession between 1944/5 and 1985/6 was 68 m per year, though the Reicher tongue made a small net advance. Between 1985/6 and 1990/1 the rate of retreat on the western side accelerated to as much as 300 m per year and then declined between 1990/1 and 1993/4. Retreat was still faster on the west side than the east, but the difference was less than in previous years. An exception was again the Reicher, which retreated at a rate of 1183 m per year when its snout disintegrated in a proglacial lake. Then, between 1991 and 1994, the two snouts of the Reicher scarcely receded and it appeared to have stabilized, but both snouts resumed their recession in 1999 (Aniya 2001). Between 1993/4 and 1995/6 the recession rate of most eastern glaciers continued to decline and two, the Piscis and León, actually advanced while at the same time recession of the western outlets speeded up slightly.

The total area of ice lost from the Northern Patagonian Icefield between 1945 and 1986 has been put at 41 km²; the volume loss by recession of the glaciers amounting to between 6.4 and 19.2 km³, and by thinning to between 212 and 291 km³ (Aniya and Wakao 1997, Aniya 1999). Over the same period, surface temperatures south of 46°S rose by between 0.4 and 1.4 °C (Rosenblüth et al. 1995). Meteorological stations are few and far between, and the steepness of climatic gradients in the region casts doubt on the relevance of remote climatic records to climatic trends around the icefields themselves (Warren and Sugden 1993). General recession after the 1940s was attributed by Aniya (1988) to decreased accumulation, with local variations in the general pattern due to topographic and orographic factors, the perimeter length of tongues, and the ratio of accumulation to ablation areas of individual glaciers. According to Philander (1990) the fluctuations do not appear to have been influenced by El Niño–Southern Oscillation events.

The San Rafael Glacier, a western outlet of the Northern Patagonian Icefield with an area of 760 km², calving into the 140-m deep Laguna San Rafael (Plate 12.1), has attracted particular attention (e.g. Warren 1993, Warren et al. 1995a, Winchester and Harrison 1996). It continues to arouse interest partly because its behaviour contrasts so much with that of the neighbouring San Quintin (Escobar et al. 1992). The distribution of lateral moraines indicates that the front of the San Rafael has in the past been much wider and more extensive than it is now. The outermost moraines of recent centuries were originally dated by the Lawrences (1959) on tree-ring evidence to 1882 and this was taken to show that the San Rafael was more extensive at that date than it had been for more than 500 years. If an excesis time of six years is taken into account it seems that retreat probably began about 1876 (Winchester and Harrison 1996). When the first bathymetric survey was made in 1875, the tongue extended 8 km across the Laguna. In 1776, Father José García Alsue observed it was touching the edge of the Laguna with large icebergs on the water. John Byron during his voyage of 1742 had made no mention of seeing icebergs, and neither had Antonio de Vea in 1675 when he wrote of the glacier 'which extends from the beach to the land inshore' (Müller 1959). The Lawrences inferred from the scatter of information that a large advance had taken place between 1742 and 1776; such an advance would have obliterated evidence of any less extensive Little Ice Age event. This observation is not otherwise substantiated. Retreat since 1882 has not been continuous. In 1905 the glacier extended 7 km into the Laguna, and was still 4 km from the western shore according to a survey of the front made by a ship's officer of the *Pilcomayo* (Winchester and Harrison 1996). An inner moraine was stabilized by 1910, according to ring counts on trees growing on it. Retreat may have been rapid between 1910 and 1935, when the front had reverted to about its 1776 position, but this is uncertain. According to Lliboutry (1956), both the San Rafael and the neighbouring San Quintin advanced between 1935 and 1950. Thereafter the front of the San Rafael was more or less stationary until 1959, when a sharp readvance was observed and trees established in the 1950s were overthrown (Müller 1959). The glacier front then retreated very rapidly, at an average rate of 200 m per year from 1974 to 1984, and 190–300 m per year from 1985 to 1990. Recession of the southern side of the front slowed to 20 m per year between 1990/1 and 1993/4, while the northern margin advanced 10 m between 1991 and 1992 before halting for the next year (Warren 1993). A strong advance from 1996–9 regained an area of 0.86 km² that was lost again by rapid retreat in 1999/2000. The San Quintin, having advanced 1200 m from 1991 to 1993, then retreated faster than has ever been recorded, losing an area of 7.55 km² between 1996 and 2000, the extensive snout being circumscribed by a huge proglacial lake into which it was disintegrating (Aniya 2001).

Analyses of the San Rafael frontal oscillations published by Warren (1993) and Winchester and Harrison (1996) differ in some respects, concerning the position of the front in 1905 and 1935 for instance, but agree on the general trend of recession during the last century or more. Both glaciers probably reached their maximum Little Ice Age positions sometime before 1876, and then began to retreat. While the San Rafael has oscillated over more than 10 km, with changes in frontal position over two or three decades interspersed with small intermediate fluctuations, the San Quintin has oscillated over about 2 km at most, with smaller scale intermediate fluctuations over a distance of a kilometre or so continuing for a longer period. The contrasting behaviour can probably be attributed to effective wasting of the San Rafael being restricted to a front in the Laguna San Rafael only 3 km wide, while the front of the San Quintin extends over 60 km so that even a large loss of ice volume has resulted, until recently, in only a small amount of frontal recession (Aniya 1988, 2001).

The fluctuations of a floating tongue calving into water cannot normally be taken as directly reflecting climate. Nonetheless, it may be noted that advances or standstills have followed the three wettest periods this century and recent assessments have been in agreement in concluding that temperature changes appear to have had no significant influence: precipitation rather than temperature variations has been the dominant climatic factor controlling fluctuations of both the San Rafael and San Quintin (Warren 1993, Winchester and Harrison 1996).[3] It has been argued that although oscillations of San Rafael on a timescale of years or decades may not be related to climatic change, on a longer timescale they might well provide a proxy climate record, primarily reflecting precipitation variations. Warren (1993) was inclined to attribute the slackening in the retreat of the glaciers between 1996 and 1999 to a precipitation increase that started in 1971 and lasted at least until 1989 but the validity of the precipitation record is uncertain (Aniya 2001). More mass balance and meteorological data are needed to evaluate the hypothesis.

In the course of the twentieth century all the independent glaciers between the northern and southern icefields retreated as well as many of those in the area to the east, between Lago Buenos Aires and Lago San Martin. A glacier flowing east towards Rio Meyer, when it was photographed in 1897, had receded only a short distance from its end-moraines, although it had obviously thinned considerably (Hatcher 1903). By 1945, the snout was 2 or 3 km back from the end-moraine (Mercer 1967). On the other hand, a small glacier on Cerro Hermoso that showed signs of recent retreat on the 1945 air photographs, was reported in 1961 by Magnani to be up against its terminal moraine.

12.1.2 The Southern Patagonian Icefield

The Southern Patagonian Icefield (Hielo Patagónico Sur, or Hielo Continental Patagónico), stretches 350 km along the southern Andes from 48°20′S to 51°30′S. With an average width of 30–40 km, it has an area of about 13,000 km^2. The ice surface, which is broken by many nunataks, has a mean elevation of 1500–2000 m. Until very recently neither a comprehensive glacier inventory nor even estimates of the thickness of the main icefield were available, though Casassa (1992), using ice radar sounding, found 616 m of ice occupying the trough of the Tyndall Glacier. Bertone (1960) had produced a simple inventory using preliminary maps produced from air photographs taken in 1944/5, but listed only the Argentinian outlets, and did not provide a location map. Topographical maps, even on a scale of 1:250,000, still cover only a part of the icefield. Aniya *et al.* (1996) made an inventory of all 48 outlets of the icecap using as base map a 1986 Landsat TM mosaic (Figure 12.4).[4] Glacier divides and accumulation and ablation areas were delineated by using stereoscopic interpretation of air photographs. Four glaciers were found to have accumulation area ratios of more than 0.9. The average value, 0.75, compares with a mean value of 0.63 for the Northern Patagonian Icefield. The equilibrium line altitude rises from 650 m in the south to 1500 m in the northeast, probably on

---

3   The nearest meteorological station is at Punta Arenas, about 250 km to the south. Mean annual temperature there dipped a degree or two below the twentieth-century mean between 1906 and 1910, increased until the mid 1920s and was low again between 1963 and 1972; precipitation was above the mean in the 1940s and 1950s and below in the 1960s.

4   Only Landsat TM images taken on 14 January 1986 were found sufficiently cloud-free for remote sensing of the entire icefield because of the extremely difficult weather conditions of the region. The parameters compiled were: latitude and longitude of snout location, length of outlet, total area, accumulation area and aspect, ablation area, AAR, ELA, positive or negative evidence of calving, highest elevation, lowest elevation, relief difference between lowest and highest elevation. The results were listed in detail.

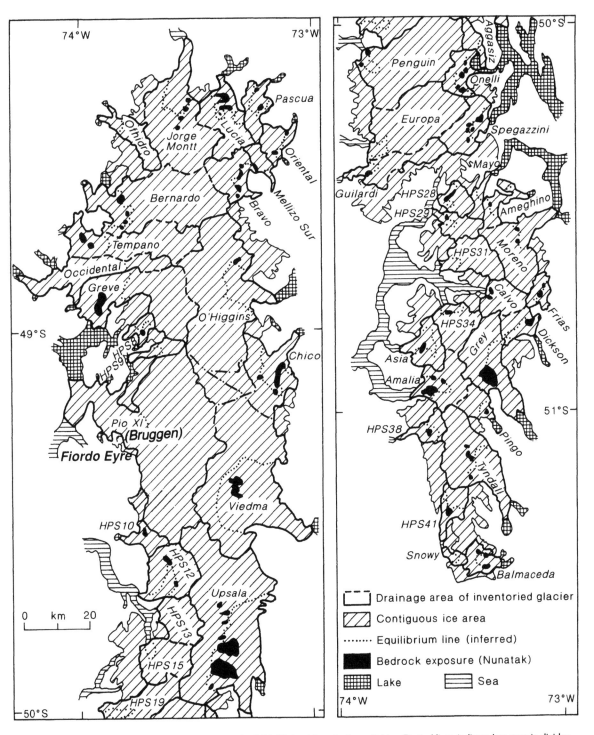

Figure 12.4 Outlet glaciers of the Southern Patagonian Icefield. Chained lines indicate divides. Dotted lines indicate less certain divides inferred from air photographs (*Source*: adapted from Aniya 1988)

account of the combined influence of latitudinal difference and the east–west contrast in climate. Except for the Frias Glacier (50°45′S) on the southeast side and the Bravo (48°35′S) on the northeast, all the outlets are calving, those on the west mostly into fjords, those on the east into proglacial lakes. Calving is evidently a dominant feature of the icecap's net mass balance and short-term fluctuations of glacier fronts are consequently only indirectly related to climate change (Warren and Rivera 1994, Warren et al. 1995b).

The magnificent fjords of southern Chile are often choked with floating ice. At Evangelistas (52°24′S), south of the icefield, mean monthly temperatures range between 4.4 °C and 8.7 °C. Mean annual precipitation here at the coast is 2620 mm; on the icefield it has been estimated to be about 7000 mm. Blizzards in summer can cause drifting in winds up to 200 kph (Shipton 1962). It is wild country, with a climate at high levels ideal for nourishing glaciers and, only a few kilometres away at lower levels, rain forest so dense as to be impenetrable without a machette (Heusser 1960). Many of the western glaciers were scarcely known at all before the first air photographs were taken in 1949. South of 50°20′S, bedrock altitude diminishes and the icefield becomes fragmented. Its remote southwestern margins are still scarcely visited (Warren and Sugden 1993).

In 1968, all the glaciers on the western side of the southern icefield were receding but were much closer to their recent maxima than those on the eastern side (Mercer 1970). The Brüggen or Pio XI glacier, ending in Fiordo Eyre, has attracted particular attention. In 1557, Ladrillero, in command of a Spanish ship, saw it calving into the fjord (Prieto and Herrera 1998). When it was seen from the *Beagle* in 1830, it was the source of a river flowing down to the sea across a narrow lowland plain. By 1926 the ice had advanced to reach the opposite side of the fjord (Agostini 1941). It had retreated 3 km by 1945 to a fjord head position (Lliboutry 1956), but by 1962 had readvanced 6.5 km (Mercer 1967) and by 1986, 8.5 km (Naruse and Aniya 1995). One of its two tongues dammed the Greve valley

creating a lake which caused calving of the front of the Greve Glacier (Warren and Sugden 1993). By 1993 the Brüggen had reached its Holocene maximum and was still advancing (Warren et al. 1997). Comparison of Landsat images of 1976 and 1986 showed that, while the northern tongue had continued to advance by 120 m a year, the southern tongue had retreated by about 60 m annually (Aniya et al. 1992).

Five western tongues in the north, Ofhidro Norte, Ofhidro Sur, Bernardo, Tempano and Hammick, were investigated in 1967–8 by Mercer (1970). An excesis period of seventy years was suggested by the discrepancy between the age of a tilted but living cypress (*Pilgerodendron uvifera*) and the ages of trees on the outermost recent moraine of the Tempano. All the tilted *Nothofagus* (Southern Beech) trees proved to have rotten centres and so were useless for dating by tree-ring counts. Radiocarbon dating of Holocene moraines shows that the two Ofhidro tongues have not entered the Fiordo Ofhidro for several thousand years. *In situ* tree trunks, rooted in peat and buried in outwash, occur outside the recent moraine sequence. The centre part of a 100-year-old tree has been dated to 800 ± 95 years BP (I-3827), which gives the time when the trees were overwhelmed by a glacial advance as having been during the period cal AD 1170–1290,[5] pointing to an early onset of the Little Ice Age. When Mercer visited it, the Ofhidro Norte ended in a proglacial lake bounded by a massive, multiple end-moraine. Outside, but within 200 m of the ice, were three older moraine ridges. Mercer examined sections of the trees on each moraine and, assuming the growth rings to be annual, found the ages to be 45, 50 and 105 years. Allowing 70 years[6] for the establishment of trees, he concluded that the moraines were formed between 1790 and 1850. Air photographs of 1945 show the ice still (or again) reaching the base of the multiple, undated end-moraine; by 1958 the terminus had retreated as much as a kilometre. The main tongue of the Hammick ends 12 km from the Fiordo Tempano, less than 50 m above sea level. Moraines of this glacier dated in the same way, suggest that the glacier reached its maximum recent

---

5   Calibrated according to CALIB 3 at 68 per cent probability. At 95 per cent probability the range is AD 1030–1300.
6   Winchester and Harrison (1996) estimated the excesis period for *Nothofagus nitida* on the forefield of the San Quintin at about 6 years on the basis of previous descriptions and local information. It is very probable that excesis times vary in different localities around the Patagonian icefields, and may differ for different species.

Plate 12.2 The Southern Patagonian Icefield from the west, with Lago O'Higgins, Lago Viedma and Lago Argentino seen from left to right on the far side, 10 March 1973
Photo: G. M. Grechko and Y. V. Romanenko, from Salyut-6

extent about 1750 and readvanced about 1840. The terminus retreated 50–60 m between 1945 and 1968, when the ice margin was only 50–200 m from an afforested end-moraine, with trees 150 years old on the outer face and 60 years old on a smaller moraine superimposed on the inner face. The calving fronts of the Tempano and Bernardo were also seen to have reached their greatest extents of recent centuries in the eighteenth century, with moraines formed between 1750 and 1800.

To the east, most of the very large number of smaller outlet glaciers are 'inaccessible, unnamed and unknown' (Warren and Sugden 1993: 318). However there are also several very large outlet glaciers including the Upsala, the largest glacier in the southern hemisphere outside Antarctica, calving into large lakes such as Argentino and Viedma in westernmost Argentina (Warren *et al.* 1995b). There are more travellers' reports

about these glaciers (Plates 12.2 and 12.3), than about those flowing west towards the Pacific.

The Perito Moreno Glacier, with an area of 257 km² and calving into Lake Argentino, is notorious for its anomalous behaviour (Nichols and Miller 1952, Lliboutry 1953, Liss 1970). In 1981 it was advancing into forest on the north side, though the southern arm of the lake was not then dammed. Until the late 1980s, unlike most of the eastern outlets, it showed no sign of major recession. Widening trimlines and a break in the recent periodicity of damming episodes may indicate that retreat is now imminent (Warren and Sugden 1993). Since 1917 the glacier has dammed a southern arm of Lake Roca at least seventeen times, causing floods in the Santa Cruz River – the sole outlet of Lake Argentino – whenever the dam collapses in response to the water level rising behind it. Depetris and Pasquini (2000) have shown that, at least since 1955, most

Plate 12.3 The Southern Patagonian Icefield seen from the east with Lago Argentino. The Moreno Glacier was almost closing the southern arm of Lago Argentino when this handheld photograph was taken from the Shuttle on 31 October 1985

ruptures of the dam have accompanied or followed shortly after El Niño events. During these events snowfall in the accumulation area of the Perito Moreno and rainfall on Lake Argentino greatly increase, causing the ice to advance and the water level behind it to increase until the ice dam gives way.

In the early part of the twentieth century the eastern outlet glaciers of the Southern Patagonian Icefield retreated rapidly from extended positions reached in the mid to late nineteenth century, leaving clear trimlines in the vegetation (Lliboutry 1956, Clapperton 1983). While some had retreated markedly by the 1920s, the most rapid recession seems to have taken place between the 1930s and the early 1960s (Clapperton and Sugden 1988). By 1960 the O'Higgins had retreated 8 km from its 1933 position and had thinned by over 300 m vertically (Shipton 1962), possibly because the terminus had been dislodged from a morainic shoal (Aniya 1999). The Upsala

thinned and retreated 1.6 km between 1981 and 1984 and a similar distance between 1990 and 1994.

In all, the area of the Southern Patagonian Icefield contracted by about 202 km² between 1945 and 1986, probably on account of rising temperatures. The contrasting fluctuation trends of the eastern and western outlets were first noticed by Mercer (1968, 1970). In 1945 the western outlets were close to recent moraines or vegetation trimlines, but all the eastern outlets were shrunken, except the Moreno. A summary of the known fluctuations of the outlets between specified dates was published by Warren and Sugden (1993, table 2). In many cases data were available only for two intervals; most information was available for the Upsala East and Dickson glaciers. On the northeast of the icefield all the glaciers between the Upsala and the Lucia, except Glaciar Viedma, have been retreating at rates of over 100 m per year. The Viedma remained stable with only a small advance in the middle of the century (Lliboutry

Table 12.3 Fluctuations of selected large outlet glaciers of the Southern Patagonian Icefield

Glacier	Observation period	Frontal change (m)	Mean rate of change (m/y)
Tyndall	1945–75	–3500	–120
	1975–86	–700	–64
Dickson	1897–1943	–780	–17
	1945–67	–1100	–50
	1967–75	–230	–29
	1975–92	–1400	–82
Moreno	1899–1917	+1000	+55
	1917–90	0	0
Upsala East	1945–67	–2100	–95
	1967–75	–100	–13
	1975–86	–2100	–191
	1986–92	–500	–83
Tempano	1945–68	–1000	–43
	1968–75	–1000	–143
	1975–85	–3000	–300
Pio XI:			
West arm	1945–76	+10,000	+323
	1976–86	–600	–55
	1986–92	+700	+117
North arm	1945–76	+4100	+132
	1976–86	+1200	+120
Amalia	1945–86	–6000	–146

Source: after Warren and Sugden 1993

1953). Its stability has been attributed to the terminus being lodged on a shoal (Aniya 1999). On the west and southwest sides of the icefield only the Amalia has retreated substantially. All the other major outlets have retained their positions, advanced strongly or retreated only slightly. The known fluctuations of some of the glaciers, such as the Brüggen and Moreno, have been more complex than the summary figures show (Table 12.3).

Major investigations of the eastern glaciers were initiated by the American Geographical Society together with the Museo Argentino de Ciencias Naturales 'Bernadino Rivadavia' in 1948. A series of expeditions starting in 1949 (Nichols and Miller 1951) and ending in the late 1960s (Mercer 1965, 1968, 1970) visited major outlet glaciers and studied a substantial selection of moraines. The data collected provide some insight into the fluctuation history of the Patagonian glaciers but the work was essentially exploratory. Very few of the 350 outlets of the eastern margin were carefully examined (Bertone 1960). In general, there is no doubt that the eastern glaciers, which all appear to be less well nourished than those to the west, have taken part in the global shrinkage of the twentieth century (Bertone 1960, Magnani 1961). Some of the small glaciers, for example those between Cerro Mellizo Sur and Brazo Norte Occidente of the Lago San Martin, had already lost a third to a half of their areas by the 1960s (Mercer 1965). The meteorological station at Lago Argentino, the nearest to the eastern outlet glaciers, registered a decrease in precipitation and a slight rise in air temperature during the last half of the twentieth century (Ibárzabal unpublished, cited Naruse and Aniya 1995).

Detailed glaciological, meteorological and geomorphological investigations have been carried out by a series of Japanese research teams which have also included members from Chile and Argentina (Naruse and Aniya 1995, Aniya and Naruse 2001). Attention has been directed particularly to the Upsala, Moreno, Tyndall and Ameghino outlets of the Southern Icefield (e.g. Aniya and Skvarca 1992, Malagnino and Strelin 1992, Skvarca *et al.* 1995, Nishida *et al.* 1995, Aniya and Sato 1995a, b, Aniya 1999). As a result, more precise data are building up rapidly, but the ice-covered area is extremely large, glaciers for which most data are available have floating tongues, variations in air temperature and precipitation are much affected by local conditions (Takeuchi *et al.* 1995), and many uncertainties remain.

Mercer (1976) attempted to summarize the Little Ice Age history of the Southern Patagonian Icefield, pointing to glacial advances culminating in the seventeenth, eighteenth and nineteenth centuries. Much of his dating was dependent on dendrochronology. He generally assumed an excesis time of about a hundred years, though he was aware that in Patagonia the interval can vary over quite short distances. Near the Tempano Glacier he used 70 years as the time taken by *Nothofagus* species to colonize bare ground, whereas Heusser (1964), for the San Rafael area, suggested 25 years. According to Innes (1992) *N. dombeyi* can colonize newly exposed surfaces rapidly, in as little as two years according to Veblen *et al.* (1989). Seedlings of

N. butuloides and *N. nitida* were found in the early 1990s growing on surfaces known to have been under the ice in the 1980s (Warren 1993). On the other hand, Nichols and Miller (1951) reported that the Ameghino moraines were still without vegetation after 80 years. Presumably microclimate, dominant wind direction, lithology and distance away of possible colonizing species must all influence excesis times. Many more detailed case studies are needed than are so far available; meanwhile, calculations involving excesis estimates cannot be taken as altogether reliable.

Radiocarbon dates from the Patagonia icefields region are still too scarce to trace early Little Ice Age culminations adequately, especially as the topographical influences and glacier dynamics which have caused recent fluctuations to vary in both timing and extent must also have operated in the past. However evidence is emerging that the Little Ice Age here may have begun in the thirteenth century. Aniya and Sato (1995b) dated wood from a trimline of the Moreno (NU-355, 820 ± 80) to the period cal AD 1160–1270,[7] commenting that this indicates 'the glacier reached its recent maximum around the 12th century'. They noted 'this date is close to AD 1300 ± 90 obtained for the initial stage of the Little Ice Age advance at Ofhidro Norte Glacier (Mercer 1970), a tongue known to have been land based for several thousand years'.

Röthlisberger and Geyh (1986: 218–26)[8] obtained radiocarbon dates from the forefields of three glaciers in the Cordillera del Paine (around 51°S). A date of 795 ± 80 BP (Hv-10887) from the Ventisquero Perro, a small glacier separate from the main icefield, provides particularly clear evidence of an advance around cal AD 1163–1288 as it came from *in situ* fossil trees growing near bedrock and overwhelmed by ice. The implication of similar dates from two other small glaciers, Ventisquero de las Torres and Ventisquero Francés (Hv-11553 and Hv-10884), indicating advances after cal AD 1023–1188 and 1285–1388 respectively, are less clear cut as they come from wood which was not *in situ* and from fossil soil in a lateral moraine (see Figure 12.2). In the Lago O'Higgins area, around 49°S, Röthlisberger and Geyh (1986) dated wood from a

fossil forest, growing from bedrock above the Ventisquero Bravo, a land-based outlet of the Southern Icecap. Sample Hv-10899 gave a calibrated date of AD 1281–1400 indicating that the forest was overwhelmed by ice in the late thirteenth century. Röthlisberger was also able to show that the minimum age for a left lateral moraine of the Ventisquero O'Higgins was probably cal AD 1213–84 (Hv-11558); this is a less significant date as the O'Higgins tongue is floating.

Though Röthlisberger recognized in his summary that there were two phases of ice extension between the eleventh and sixteenth centuries, he nevertheless placed the start of the Little Ice Age at about AD 1600 (e.g. 1986: 249, 239). He had taken part in the Swiss research effort which established that glaciers in the western European Alps had advanced well before the sixteenth century (Röthlisberger *et al.* 1980), but he seems to have been unprepared to accept thirteenth-century advances as having been part of the Little Ice Age (see Chapter 6). In the Southern Patagonian Icefield, considerable evidence of a culmination around AD 1600–1700 is available, but it does not seem that investigations have yet been sufficiently dense to establish the full Little Ice Age sequence. Röthlisberger himself made important headway by examining the forefields and moraines of small glaciers in the region and, as we have seen above, produced evidence of early Little Ice Age advances, but further intensive investigations of this sort are still required.

12.1.3 The Straits of Magellan and Tierra del Fuego

From a careful study of shipping records between 1520 and 1620, Prieto and Herrera (1998) have discovered that only in the second half of the sixteenth century did Spanish navigators make direct references to icebergs originating from glaciers in the Magellan Strait. Ladrillero in 1557/8 and Sarmiento de Gamboa in 1579/80 both encountered icebergs calved from the snouts of glaciers on islands in the western entrance to the Strait. Today these glaciers no longer reach the sea.

---

7 Calibrated at 68 per cent probability according to CALIB 3.
8 Röthlisberger and Geyh's book has been sadly neglected, even in recent publications.

The journals of Thomas Cavendish, who went through the Straits in 1587, and of later navigators make no mention of such sightings. The iceberg sightings follow ship reports of unusual cold off the southern Pacific coast in 1520/1, 1539/40 and 1553/4, and northerly winds in the summer of 1557/8. Prieto and Herrera (1998) note that, according to Quinn and Neal (1992), the years 1578 and 1579 experienced the first recorded severe El Niño events and suggest that the possibility of a link with the presence of the icebergs should not be disregarded.

In the heavily glaciated Darwin Cordillera (54–5°S, 69–71°W) of southwestern Tierra del Fuego, ten peaks rise to over 2000 m. Changes in glacier extent have been traced from air photographs taken in 1943–4, 1960 and 1984–5 and from field investigations in 1993 and 1994 (Holmlund and Fuenzalida 1995). Lliboutry in 1956 noted no signs of major change during the first half of the twentieth century, but photographic evidence indicates a slight retreat of the northern and northwestern glaciers before 1943. Glaciers on the northern and eastern sides have retreated in the course of the twentieth century, but those on the southern side are either stable or advancing, with some fronts close to their twentieth-century maximum positions. Deep fjords cut into the southern rim are occupied by calving glaciers. Some glaciers to the north also calve. The particularly rapid recession of Ventisquero Marinelli stands out as typical of the behaviour of calving tongues, and is in part attributed to non-climatic factors. Holmlund and Fuenzalida (1995) consider all the southern glaciers as one set, whether calving or land-based, as they all exhibit the same recent behaviour pattern. There is no evidence concerning the longer-term history of the calving glaciers which form the majority of those discussed, and so it seems premature to conclude that they may be used as climatic indicators. The state of the glaciers terminating on land requires explanation. As no significant trend in winter precipitation has been registered and mean monthly temperatures increased by about 0.5 °C in the twenti-

eth century, Rosenbluth and Fuenzalida (1991) speculated that the prosperous state of the southern glaciers might have been caused by southward migration of the subtropical convergence zone, with winds bringing increased orographic precipitation on the windward side of the mountains and dry föhn winds on the leeward side. Climatic data is currently too sparse to evaluate this view. Although ENSO events are generally considered not to influence the behaviour of the Patagonian glaciers, they certainly influence those of the Southern Alps of New Zealand.

12.2 NEW ZEALAND

The most substantial data on southern hemisphere fluctuations during the nineteenth century come from New Zealand. The two main islands, situated between 33° to 47°S and 167° to 178°30′E, stretch northeast across the latitudinal belt influenced by the high pressure cells of the subtropics and the low pressure zones of the Southern Ocean. Weather and climate are governed by a procession of anticyclones separated by troughs of low pressure and depressions, nearly always moving from west to east (Trenberth 1976). Latitudinal shifts in the southern margin of the high pressure zone are important controls of the synoptic sequence. Regional weather patterns of the South Island are much affected by the Southern Oscillation and the impact of El Niño events (Fitzharris *et al.* 1992).

Though North Island has many peaks above 1200 m, only Ruapehu, a massive volcano near the centre of the island rising to almost 2800 m, is high enough to support glaciers. On South Island, over 3000 glaciers are scattered between the tiny outlying ice remnants on Mt Tapuaenuku on the inland Kaikoura Range in the northeast and Caroline Peak above Lake Hauroko in the southwest (Figure 12.5).[9] The largest are in the Mt Cook region, around 43°35′S.[10] Of the ten main trunk glaciers, the Franz Josef, Fox and Volta/Therma descend to the west of the main divide, the Tasman, Murchison, Hooker and Mueller, the Godley, Bonar

---

9   An inventory of the glaciers in New Zealand carried out by air survey gave a total of 3144 (Chinn 1991). The inventory map, annotated with catchment names, was published by Chinn (1996) and without annotations by Chinn (1999).

10 Gellatly *et al.* (1988) reviewed the main glacial regions in New Zealand, together with maps. The first published map of the glaciers of the Southern Alps appeared in a paper by Anderton (1973) concerned with their significance for water resources.

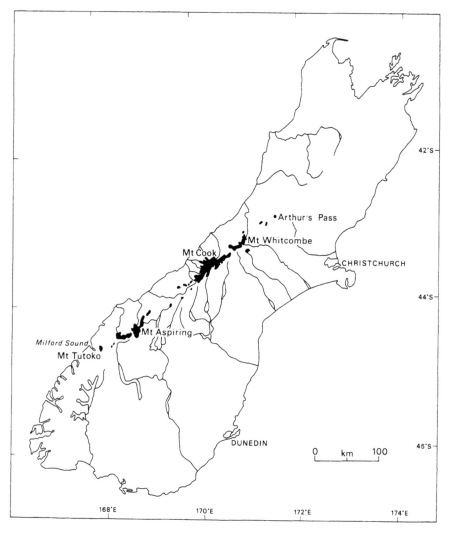

Figure 12.5 Distribution of glaciers and permanent snowfields in South Island, New Zealand

and Douglas flow down to the east. The Tasman, on the flanks of Mt Cook, is the largest by far, containing nearly one-third of the ice in New Zealand.[11]

Exploration of the larger glaciers began in 1862 with the remarkably detailed investigations of Von Haast, which were supported by planimetric diagrams of the fronts of the Godley, Classen and Tasman glaciers (Figure 12.6) and many field sketches, some of which were later interpreted in water colour by John Gully (Paul 1974, Gellatly 1985a) (Plates 12.4 and 12.5). The first photographs were taken in 1867 when 'Mr A. P. Sealey made some long expeditions with his camera' (Harper 1896). Serious surveying and mapping made good progress in the 1890s with the explorations of C. E. Douglas with A. P. Harper (1896) on the western side of the Southern Alps, and of Harper and T. N. Broderick

---

11 The Tasman has an estimated volume of 155 million m^3 and the Murchison, next largest, 36 million m^3 (T. Chinn personal communication, January 2000).

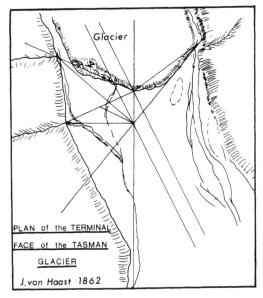

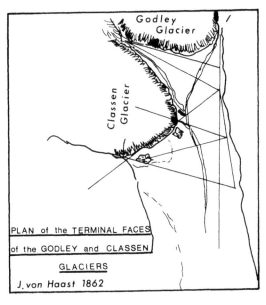

Figure 12.6 Sketchplans of the tongues of the Tasman, Godley and Classen glaciers made by J. Von Haast in 1862. (Copy by Dr A. Gellatly)

on both sides (Figure 12.7). Douglas and Harper's (1895) fieldbooks form part of an unusually informative repository of written records and illustrations, allowing the fluctuations of the New Zealand glaciers to be traced through the late nineteenth and into the early twentieth centuries (Gellatly 1985a, Chinn 1994). Identification of phases of glacier expansion earlier than the second half of the nineteenth century has to be based on moraine dating.

In 1893 Harper noted that

The great glaciers on the east side of the watershed are chiefly flat, hummocky ice ... covered roughly speaking for a quarter of their length with a considerable quantity of surface moraine, formed of detached masses of rock and debris of all kinds, with which the Swiss moraines cannot be compared for roughness and extent: it is lifted in heaps or hillocks of 50 feet or more above the general level of the glaciers [Plate 12.6]. The old lateral moraines too are most marked, especially on the Hooker [Figures 12.8 and 12.9].

Harper mistakenly concluded that the glaciers east of the watershed differ essentially from those to the west. This assessment was evidently based on comparison of the Fox and Franz Josef in the west with the Tasman and other glaciers to the east (see Plate 12.6) (Harper 1893: 37).

All the large, low gradient valley glaciers on both east and west sides of the divide have morainic covers; debris supply being high because the greywacke rocks are highly fractured, the whole area is very active tectonically, and slopes are correspondingly steep. Surveys from the air have shown that debris covers 19.3 per cent of the area of the western glaciers and 29.7 per cent of those to the east (Chinn 1996). Most of the largest glaciers are on the east and are readily accessible, while debris-covered glaciers on the west such as the Balfour and La Perouse are rarely visited. The steep tongues of the Fox and Franz Josef, both unobscured by debris, are unique among New Zealand glaciers both morphologically and in rate of ice turnover.

Harper noticed that the western tongues reached lower altitudes than the eastern ones, the Fox 'being the lowest of all the glaciers' and 'like the others on this side having tree-ferns and bushes growing almost on the moraine, and in some cases overhanging the ice'.

The plentiful moraines east of the continental divide provide evidence of repeated fluctuations. Their interpretation has not been straightforward. An initial linear lichen growth curve (Burrows and Lucus 1967, Burrows and Orwin 1971) was employed to date several sequences (Burrows 1973, 1975, 1977) and it was

Plate 12.4 View from Mt Cook range to the Tasman and Murchison glaciers. A watercolour by Julius Von Haast (1822–87), made on 12 April 1861. This workmanlike sketch by the Provincial Geologist was used as a basis for the painting in Plate 12.5 (*Source:* from Alexander Turnbull Library Collection, Wellington, New Zealand. Ref. no. 52096)

Plate 12.5 Watercolour of the Tasman and Murchison glaciers painted by John Gully on the basis of Julius Von Haast's sketch (*Source:* from Alexander Turnbull Library Collection. Ref. no. 51360)

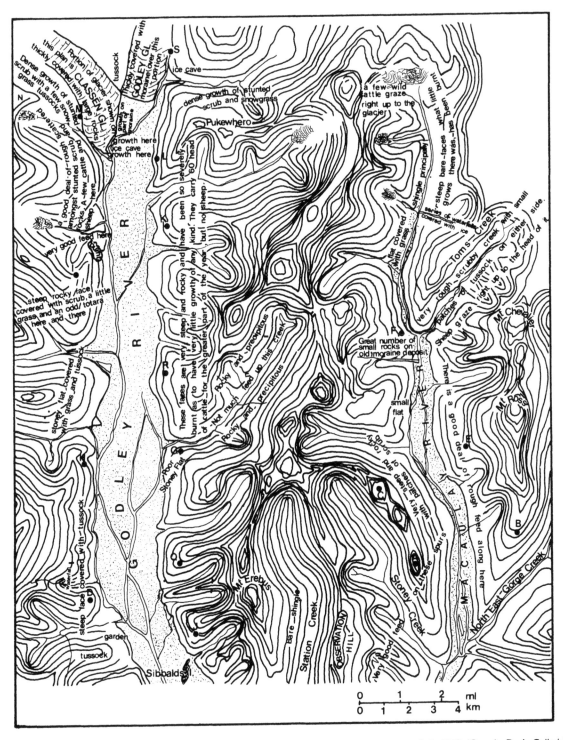

Figure 12.7 Broderick's map of the tongues of the Godley and Classen glaciers and their environs made in 1888. (Copy by Dr A. Gellatly)

Plate 12.6 The moraine-covered surface of the Mueller Glacier, typical of the glaciers east of the watershed. The Stocking Glacier can be seen hanging on the mountainside on the far side of the Mueller in the centre of this 1928 photograph (*Source:* from Alexander Turnbull Library, Radcliffe Collection. Ref. no. 7550)

concluded that there had been regular, well-marked periods of glacier activity occurring at least once every hundred years since the twelfth century (Burrows and Greenland 1979). Differences in soil development on moraines dated only forty years apart according to this curve, together with the unusual steepness of the curve itself, led to suspicions of its inaccuracy (Burrows 1980, Birkeland 1981, Burrows and Gellatly 1982).

Three important control points on the Mueller forefield having been redated, further control points were identified, and a new curve set up for the Mt Cook area (Gellatly 1982, 1983) (Figure 12.10). This non-linear curve was based on a combination of five historically dated surfaces, all post-1860, and nine surfaces approximately dated by weathering-rind thickness. Rates of rind growth on fine-grained sandstone were calibrated by Chinn (1981) using surface rocks on radiocarbon-dated landslides. At the time, he considered his dates to be within 20 per cent of actual values, though it may be noted that three of the ten

radiocarbon dates on which he based his curve were more recent than 400 BP, within the time range when such values are not unambiguous (Klein *et al.* 1982).

The forefields of the Mueller, Hooker, Tasman, Murchison, Godley and Classen glaciers have been remapped and the moraine-dating substantially revised. This has resulted in the recognition of separate peaks of glacial expansion occurring about AD 1140, 1350, 1640 and 1845. In view of the dependence of the lichen growth curve on weathering-rind thickness, these dates must be regarded as approximate. However, the convergence of results from different lines of relative dating is striking, with lichen size, weathering-rind thickness and degree of soil development varying accordingly.

The earlier part of the chronology is substantiated independently by radiocarbon dating (Figure 12.11). All these dates refer to organic material providing a minimum age for an earlier advance and a maximum age for a later one. Size frequency histograms of lichen

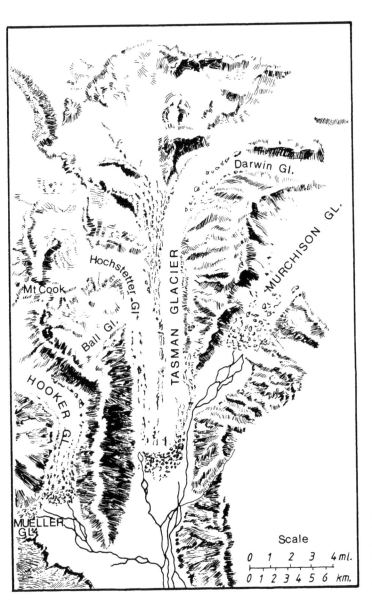

Figure 12.8 Green's map of the Great Tasman Glacier together with its tributaries and part of the Hooker Glacier made in 1884, with additions from Dr von Lendenfeld's map. (Copy by Dr A. Gellatly) (*Source:* first published in the *Proceedings of the Royal Geographical Society,* 1884)

populations growing on the various dated surfaces and differences in modal rind thicknesses suggest that the recognition of four main phases of advance in the Little Ice Age is justified.[12]

Records obtained in 2000 from a speleothem from the northwest of South Island, between Farewell Spit and Punakaiki, provide additional support for there

having been four main phases of advance in the Little Ice Age, with low $\partial^{18}O$ values indicating four cool periods at AD 1050–1100, 1360–1400, 1570–1660 and 1830–1910 (Figure 12.12).[13]

For the present, it seems reasonable to accept the chronology proposed by Gellatly (1982, 1983, 1984, 1985a, b) as being the most accurate available for

12   Discriminant analysis of the data might well be used to test its statistical validity (Dowdeswell and Morris 1983).
13   Data kindly provided by Professor Paul Williams, December 2001.

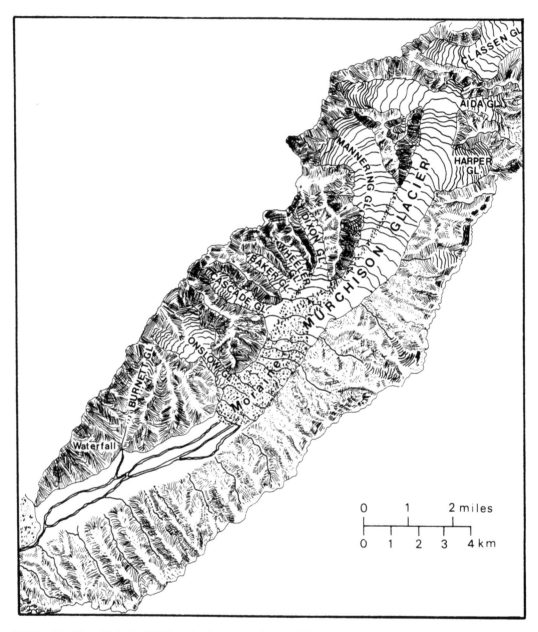

Figure 12.9 The Murchison Glacier in 1892 from the Government Survey of the central portion of the Southern Alps of New Zealand, with additions by A. P. Harper *et al.* (Copy by Dr A. Gellatly) (*Source*: first published by the Royal Geographical Society, 1893)

glacier fluctuations east of the continental divide. The earliest of the Little Ice Age advances she recognized was a well-marked event dating from about the fourteenth century. When more moraines have been dated it should be possible to obtain a more detailed picture of the several advanced phases which have occurred within the last millennium.

The fluctuations of eighteen glaciers on the densely vegetated and steep western face of the Southern Alps were investigated by Wardle (1973) using both

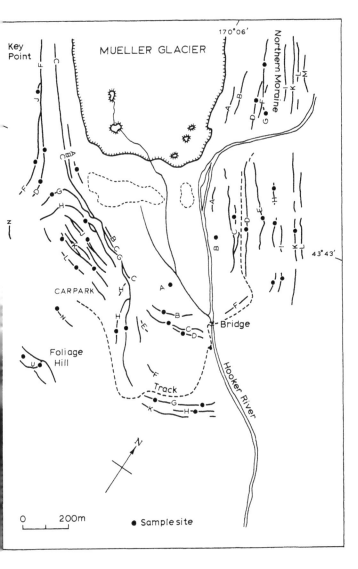

Figure 12.10 Forefield of the Mueller Glacier, South Island, New Zealand. Age of moraine rock surfaces from lichens and weathering rinds; A 100, B 135 ± 35, C 340 ± 88, D 580 ± 150, E 840 ± 280, F 1150 ± 300, G 1490 ± 385, H 1830 ± 476, I 2160 ± 562, J 2540 ± 660, K 2940 ± 765, L 3350 ± 870, M 3790 ± 960, N 4200 ± 1090, U 7200 ± 1870 (*Source:* from Gellatly 1984)

tree rings and lichens to supplement early historical sources. Difficulties arise with both these techniques. *Rhizocarpon geographicum* occurs only locally. It is excluded from moraines associated with the Franz Josef and Fox by more rapidly growing lichens and by mosses and has not been found on surfaces known to have been exposed for less than thirty-five years. Moreover, it includes in addition to the usual yellow form, a dull greenish-yellow more diffuse type which seems to be faster growing. Wardle (1973) used maximum lichen diameters to estimate age in the absence of other evidence. He plotted the diameters of the largest thalli of

Rhizocarpon geographicum that he found against the approximate ages of moraines derived from other evidence, but did not have a properly constructed growth curve at his disposal and so could not use lichen size as an independent dating tool. The number of woody plants with well-defined annual rings growing on the Westland moraines is limited. *Dracophyllum traversii*, with the clearest rings, is uncommon on moraines and does not appear in the early stages of plant succession. Some species of *Olearia* have clear rings as young shrubs but not as older trees. The main trees at low altitudes are *Weinmannia racemosa*, in which the central rings

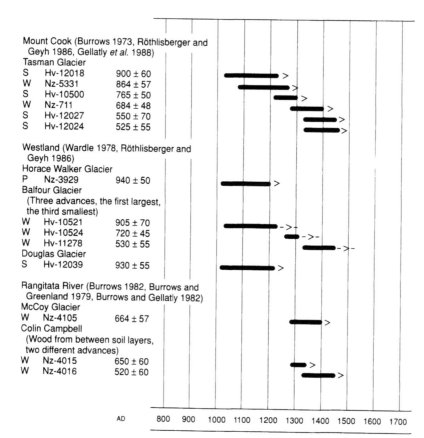

Figure 12.11 Key radiocarbon dates relating to glacial advances in New Zealand

Notes: The numbers in the margin give the laboratory identification number and the ^{14}C age. Horizontal lines represent the calibrated ages. - > - = Advance around or after indicated age; > = Advance after indicated date; S = Soil; W = Wood; P = Peat

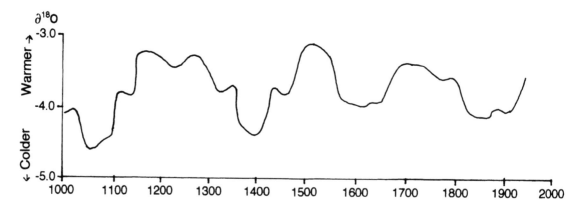

Figure 12.12 Raw ∂^{18}O records from a speleothem at North Westland, New Zealand (at the northwest of South Island), plotted as a 5-point running mean

are invariably decayed in trees over 300 years old, and *Metrosideros umbellata* which has indistinct rings and wood too hard to penetrate with an increment corer. Two podocarps, *Dacrydium cupressium* and *Podocarpus ferrugineus,* grow on the moraines of the Fox and Franz Josef but the evidence that growth rings are annual is not conclusive. Fortunately *Nothofagus menziesii,* which colonizes the moraines of the Hooker Range, has a lifespan of 600 years, tolerably distinct rings and is easily cored. The growth rings of subalpine trees and plants can reasonably be regarded as annual (Wardle 1973) but the seeding period is extremely uneven. Very coarse moraine can remain bare for a century, while seedlings can be found on finer moraine still underlain by ice. Wardle concluded that seedlings of *Olearia, Dracophyllum, Nothofagus, Weinmannia* and *Metrosideros* are likely to appear in favourable sites within five to ten years but *Podocarpus ferrugineus* and *Dacrydium cupressinum* do not enter the succession on wetland moraines during the first century of exposure.

Despite all the difficulties, Wardle was able to obtain thirty incremental cores providing minimum dates for moraines and produced a chronology for the Westland glaciers. He attributed the multiple moraine sequences that he found to a succession of glacial advances occurring during the last four centuries, of which the largest were those in the seventeenth and eighteenth centuries. The number of moraines formed in recent centuries differs from one forefield to another and compound moraines with crests of two or more ages are not uncommon. Wardle found few moraines that he could not confidently date as either younger or much older than 350 years. He recognized advanced phases of the Westland glaciers in the mid to late eighteenth century, early to mid nineteenth century and late nineteenth century. His results from the Westland glaciers conform reasonably well with those of Gellatly from the east of the divide.

The recorded history of glacier fluctuations in Westland National Park was summarized by Wardle (1973) and in the Mt Cook National Park by Gellatly (1985a). The present whereabouts of manuscripts, early maps and photographs were specified and some account provided of the observations made by early scientists and travellers. The early observers found nearly all the glaciers in advanced positions (Figure 12.7, Plate 12.7), although J. B. Ackland noted as early as 1866 that the Colin Campbell Glacier was retreating and he described it as looking in 1880 like 'an old deserted quarry heap' (cited Chinn 1994). Von Haast, the Canterbury Provincial Geologist who explored the Mt Cook area in 1861 and 1862, wrote of the Godley Glacier: 'from the fact that several older moraines densely clothed in sub-alpine vegetation were already half buried in the present terminal moraine of the glacier, it was clear to me that the glacier, after a period of retreat, was now advancing' (Von Haast 1879). Between Von Haast's discovery of the Tasman in 1862 and his second visit in 1869 the snout advanced half a mile. It was clearly still enlarging in 1882 when it was recorded that

> the bare boulders were piled up into a rampart about sixty feet high, over which the ice of the glacier rose in a vertical wall of from twenty to thirty feet. By continually dropping stones from its upper surface on to the top of the moraine it was thus daily, before our eyes, building up the higher rampart which may be, in the years to come, the only record left of the existence of the great Tasman Glacier.
>
> (Green 1883: 199)

The Westland glaciers were also advancing; early photographs show the Franz Josef further forward in 1885 than in 1875. Evidently small advances were widespread in the final decades of the nineteenth century. Since then retreat has predominated, though this has been broken by a number of small readvances and the six main trunk glaciers on the east side of the divide were all within 1 or 2 km of their late nineteenth-century moraines in 1981 except for the Godley which had withdrawn about 4 km (Gellatly 1985a).

Rapid twentieth-century wastage, swiftest since 1930, has caused not only the retreat of the fronts of trunk glaciers but also the complete disappearance of some small ice bodies. The fluctuations of the Franz Josef are the best known for the historic period (Plates 12.7, 12.8, 12.9 and 12.10, Figures 12.12 and 12.13). Small advances, each insufficient to redress previous retreat, took place in 1907–9, 1929–34, 1946–51, 1965–7 (Sara 1970), and 1983–2000.[14] By 1982 the

---

14 Many earlier accounts include an advance in 1921–34, but this was queried by Peter Tyson. On checking, a map interpretation error was found in the early publications (Tyson *et al.* 1998).

Plate 12.7 The Franz Josef Glacier as portrayed by Sir William Fox in 1872. The terminus was then about 1000 metres wide and the ice was probably advancing. The rock on the right of the picture is the 'Sentinel Rock', later used as a convenient observation point (*Source*: from Alexander Turnbull Library Collection. Ref. no. 69165)

Plate 12.8 The Franz Josef Glacier in about 1905. The front had withdrawn considerably since 1872, though the ice was still not very far from Sentinel Rock (*Source*: Macmillan Brown Library, Canterbury University)
Photo: A. C. Graham

Plate 12.9 By the end of the 1930s the tongue of the Franz Josef Glacier had thinned and a lake occupied the low ground near Sentinel Rock which was in part ice-covered in 1905. Note the rapid growth of vegetation on moraines exposed during the twentieth century (*Source*: Franz Josef Waiau Area Office, New Zealand)
Photo: M. C. Lyson, *c.*1940

Plate 12.10 The Franz Josef Glacier during its advance in 1985
Photo: David Norton

Plate 12.11 The Franz Josef Glacier in January 2002 as seen from Sentinel Rock
Photo: William Grove

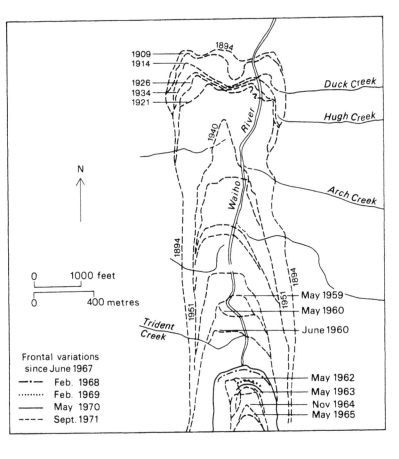

N

0   1000 feet

0   400 metres

Frontal variations
since June 1967
—·— Feb. 1968
········ Feb. 1969
———— May 1970
- - - - Sept. 1971

Figure 12.13 Frontal variations of the
Franz Josef Glacier, 1894–1971 (*Source:*
from Kasser 1973)

front had receded further than at any other time since observations started (Chinn 1999), but renewed expansion had begun by the end of the year. Between early 1984 and mid 1985 'the terminus ... advanced more than 200 m' (*New Zealand Herald*, 5 July 1985); this advance, it was suggested, could be attributed to an increase in precipitation associated with more frequent cool southwesterlies in the early 1980s (Plate 12.10). The trend continued; a brief standstill in March 1998 was followed by further expansion.

Unfortunately, regular monitoring of terminal positions of the New Zealand glaciers ceased after 1992,[15] although the behaviour of the western tongues, especially that of the Franz Josef, continued to attract some public interest. In 1996 the fronts of both the Franz

Josef and Fox were precipitous, showing all the characteristics of rapid advance (Plate 12.11). When observed from Sentinel Rock in 1996, the Franz Josef appeared to have regained a position approximately equivalent to the one it occupied in 1961, when a photograph from the same viewpoint by Sara (1970) shows a 2-m high moraine immediately in front of the ice.

During the twentieth century the ice-covered area in New Zealand was reduced by 23–32 per cent, a sample of 127 glaciers shortening by an average of 38 per cent, and losing a quarter of their areas (Chinn 1996). An upward shift of the mean elevation of the glaciers has been approximately equivalent to a temperature rise of 0.6 °C. The response of individual glaciers to climatic changes has varied greatly, according to size,

15 It could however be resumed without difficulty because early workers left accurate information about their observation stations, some of which are still clearly marked.

sensitivity and location (Chinn 1999). Swiftest wastage in the middle decades of the twentieth century coincided with a period when the subtropical high shifted poleward through about 4° (Fitzharris *et al.* 1992) and when El Niño events were relatively few and far between.

In the absence of high altitude meteorological data and with only limited mass balance studies (e.g. Thompson and Kells 1973, Anderton and Chinn 1978), climatic interpretation of glacier behaviour has been controversial. It has been suggested that precipitation plays the dominant role in controlling the fluctuations of the Franz Josef (Suggate 1950, Hessell 1983). Salinger and Gunn (1975) pointed out that mean temperatures in New Zealand were low between 1900 and 1935 and that warming was experienced from 1935 to 1970. This analysis was disputed by Hessell (1983), but the evidence of warming is convincing (Salinger 1979, 1982). Instrumental records show overall warming in New Zealand between 1860 and 1990. The increase was small between the 1860s and 1940s, but mean annual temperatures then rose sharply by 0.5 °C. The West Coast and Southland experienced a mean temperature rise of 0.9 °C from 1841–1990 (Salinger *et al.* 1995). It seems that predominant twentieth-century retreat of the glaciers took place against a background of rising temperature and longer ablation seasons (Gellatly and Norton 1984).

Full explanation of glacier oscillations must include the influence of precipitation as well as temperature. Changes in the extent of the Te Wae Wae (or Stocking) Glacier, one of the debris-free glaciers on the east side of the divide (Plate 12.6) were related by Salinger *et al.* (1983) to climatic variables represented by smoothed monthly temperature values from Hokitika and precipitation data from Otira, both on the west coast. Principal component analysis showed variance of these values could account for 83 per cent of the changes in glacier front position; the relationship with temperature was found to be closer than with precipitation.

Since 1982 however the part played by precipitation has been much more prominent, probably because of the greater number and strength of ENSO events.

The mean elevation of snowlines at the end of summer, which corresponds to the equilibrium line altitude (ELA), rises from about 1500 m on glaciers on the northwest side of the Southern Alps to about 2200 m on the southeast side. The gradient, which reflects the greater snowfall on the northwest, is steep in years with high ELAs and less in years with low ELAs (Lamont *et al.* 1999). Aerial surveys of snowline heights at the end of the ablation season have been made for forty-seven index glaciers in the Southern Alps since 1977 and used as surrogates for annual mass balance values (Chinn 1995). In retreat phases the 'end of summer snowline' has been about 50 m higher than average, and in advance phases 40 m lower. Snowline altitudes have fallen in most years since 1977, revealing the predominance of cumulative mass balance increases (Chinn 1999).[16] There are some difficulties with this method of monitoring; insufficient mass balance data are available with which to construct mass budget curves, the last day before the first winter snowfall is difficult to forecast, and difficult weather conditions limit observations.[17] Despite such shortcomings, the survey results correlate well with major climatic events and conform satisfactorily with known fluctuations of termini when reaction times are taken into account (Figure 12.14).

The cumulative mass balances of all the index glaciers, including those with the most extensive debris covers, increased every year from 1982 to 1994 except in 1989–90. Reconstructed cumulative mass balance variations for the Franz Josef for the period 1913–89 correspond neatly with the known history of advance and retreat (Woo and Fitzharris 1992). Its reaction time of about five years is short, because it has a high, gently sloping and extensive accumulation basin, and a steep tongue (Soons 1971, Chinn 1995). Following the onset of positive cumulative mass balances, many South

---

16 Snowlines rise when net annual accumulation diminishes and mass balances are negative. They fall when net accumulation increases and mass balances are positive. The mean values quoted here are for 1977–97. Since the middle of the nineteenth century they are calculated to have risen by about 100 m, equivalent to a rise in regional temperature of about 0.6 °C, mainly in the middle decades of the twentieth century (Chinn 1996).

17 In practice it has not been possible to survey all the glaciers every year; the work was made impractical for administrative reasons in 1990 and by weather conditions in 1991.

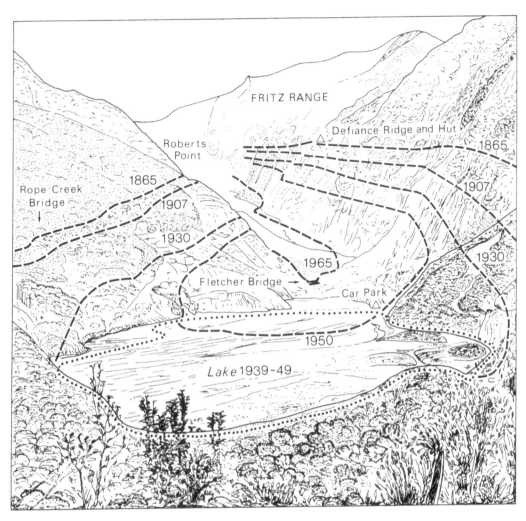

Figure 12.14 The position of the front of the Franz Josef Glacier between 1865 and 1965 as seen from Sentinel Rock. In 1865 the whole of the foreground would have been covered by ice. (Drawn from annotated photograph in Westland National Park Collection)

Island glaciers in addition to the Franz Josef and Fox were characterized in the late 1990s by the steep fronts and convex tongues typical of advancing glaciers. The cumulative mass balance of the Tasman, monitored for longer than that of any other index glacier, declined overall from 1959 to 1973, but increased during 1973–93, despite fluctuations within both periods (Figure 12.14).

The debris-covered fronts of the glaciers east of the watershed do not respond sensitively to climate. Their debris mantles form effective insulators, reducing the melt rates of the ice beneath by up to 90 per cent and

slowing down the response of the tongues to negative balances though still allowing rapid responses to positive changes, for instance the half mile advance of the Tasman between 1862 and 1869 (Kirkbride 1989). Eventually even slow ablation leads to disruption of insulating mantles, allowing the rapid growth of thermokarst lakes, their shape and location depending on the surface gradient and marginal topography of the ice. Despite its debris cover, the lower Tasman decreased in thickness by about 82 m between 1891 and 1963 with little or no shortening, about half the losses resulting from regional warming and much of the rest

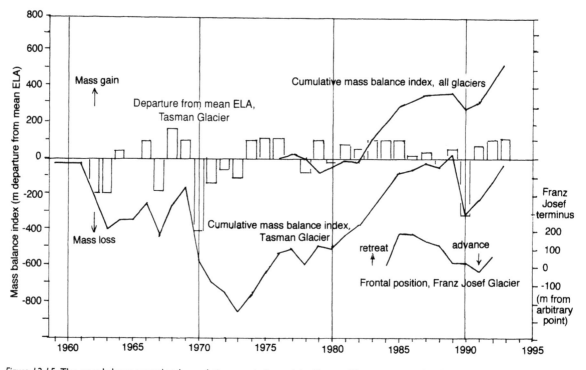

Figure 12.15 The mass balance record and cumulative mass indices of the Tasman Glacier, compared with the cumulative mass balance indices of all the index glaciers 1977–93, and with measured positions of the Franz Josef terminus (1984–92) (*Source*: after Chinn 1995)

from the feedback effect on the temperature at the surface of the ice of its decreasing elevation. After 1972 downwasting increased and disintegration of the tongue began (Kirkbride 1993, 1995a, b, Hochstein *et al.* 1995). Thermokarst ponds were first recorded on the Tasman as small isolated features in 1957. Subsequently they increased in number and size, especially in the 1970s. By 1986, they were encroaching on the glacier margin, coalescing around 1990 to form a large melt lake where ice had been 150–200 m thick. Calving began soon afterwards, and more lakes appeared further up the tongue. It is expected that rapid calving of the ice cliffs at the proglacial lake margins will soon lead to accelerated retreat of the glacier tongue when the ice begins to float. Imminent retreat of the Tasman front by as much as 8 km has been predicted (Kirkbride 1995b).[18]

The timing and extent of frontal retreat of glaciers with proglacial lakes varies according to local circumstances. The most dramatic retreat so far has been that of the Godley Valley glaciers, which were confluent when Von Haast saw them in 1862, and still confluent in 1917 but have since fragmented into the Maud, Grey, and Godley glaciers (Kirkbride 1993). The Godley had separated by 1950, while the Grey and Maud parted at the head of a 2-km long lake in 1990. No further major retreat is to be expected here; the system has now changed sufficiently for these glaciers to have regained equilibrium with the current cumulative mass balance situation (Chinn 1994). With the development of lakes at their termini, the Ivory cirque glacier began to calve in 1972, as did the Wilkinson Glacier about 1980. By the end of the twentieth century, the Mueller was still at the thermokarst stage, but the Hooker had

---

18 The timing and nature of the processes involved have been followed in particular detail with observations of both ice and hydrological characteristics.

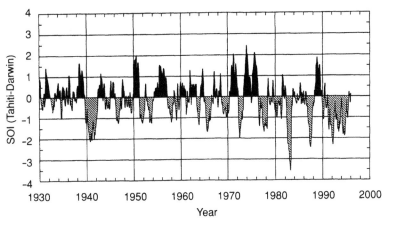

Figure 12.16 Variations of the incidence of negative values of the ENSO index, 1930–95 (*Source:* data supplied by Salinger 1996)

begun to retreat rapidly by calving. Chinn (1996) estimates the twentieth-century loss of ice in the Southern Alps at 23–32 per cent, which is somewhat less than the loss in the European Alps of 30–40 per cent, but he foresees melting accelerating as more lakes form at the tongues of the large debris-mantled glaciers on the eastern side.

Both temperature and precipitation are strongly influenced by the El Niño/Southern Oscillation (ENSO) system (Figure 12.16). Low temperature anomalies occur west of the Southern Alps during El Niño events (when the Southern Oscillation index is low) and conversely warm temperature anomalies during La Niña events (Gordon 1985).[19] Though temperatures have risen overall, there have been significant anomalies superimposed on this trend, more particularly in 1982–3, 1992 and 1993, and 1997, all of which were El Niño years and cool in southern New Zealand. In the whole New Zealand temperature record from 1871 to 1993, annual mean temperatures in 1992 were the lowest since 1930 (with a deviation of −1.07 °C from the 1951–80 average) and 1993 was the second coldest (with a deviation of −0.97 °C) (Folland and Salinger 1995). These anomalies are believed to have been the

result of the low index Southern Oscillation coinciding with the Mt Pinatubo eruption (Basher and Thompson 1996, Salinger 1995, Folland and Salinger 1995).

High pressure in the southwest Pacific accompanying an El Niño promotes frequent westerlies which bring plentiful precipitation as they rise over the Southern Alps, especially at high altitudes.[20] Between 1951 and 1975 precipitation measured at Hokitika was close to average, but after 1980 more frequent westerlies over the Southern Alps, associated with a succession of low index ENSO values (Figure 12.16) made the west of South Island wetter, and areas in the lee of the mountains drier (Salinger *et al.* 1995). The predominance of low index values since the late 1970s has been associated with many heavy precipitation events affecting the west, for example in 1979, 1982, March 1989 and December 1995. On 2 March 1996 we noted that 175.5 mm of rain had fallen that day by mid afternoon at Haast on the west coast although the total for the year, up to that point, was a modest 919.0 mm. Such heavy falls registered at the coast imply much more precipitation at higher altitudes. Maximum values of as much as 18,000 mm are believed to occur at high altitudes to the west of the Main Divide. Falls at such

19 When pressure is high over Darwin and low over Tahiti during El Niño events, the ENSO index is negative. When pressure is high over Tahiti and low over Darwin during La Niña events, the ENSO index is positive.
20 Precipitation in the Southern Alps was not effectively measured until the 1970s. Snow surveys began in the 1960s but it has proved difficult to separate snowfall from rainfall. Very high values of precipitation indicated by frequent floods in the region were confirmed by measurements of as much as 12 m of water equivalent per year. It is suspected that the maximum may be as high as 16 m per year (Henderson 1993).

altitudes must commonly be in the form of snow. Precipitation on the Franz Josef during its advance phase in the 1980s and 1990s varied greatly from year to year but on average is estimated to have been about half a metre greater in the accumulation season than in the preceding retreat phase, and temperatures in the ablation season 0.2 °C lower (Hooker and Fitzharris 1999).

The cumulative positive mass balances towards the end of the twentieth century may be attributed in part to a northward shift of the subtropical high pressure zone affecting the synoptic sequences in the Tasman Sea region. At the same time, more frequent El Niño episodes have given more persistent south to southwest air flows to the South Island, bringing more snow to the Southern Alps and causing valley glaciers to advance five to ten years after each episode. The New Zealand glaciers, like the maritime glaciers of western Norway, have consequently refused to conform in recent years with the dominant global trend of glacier retreat (Fitzharris *et al.* 1992, 1997, Hooker and Fitzharris 1999).

12.3 THE LITTLE ICE AGE IN THE ANDES AND THE SOUTHERN ALPS COMPARED

On the centennial timescale, the sequence of Little Ice Age events in the southern Andes and in New Zealand is similar. The first ice advances in the second millennium AD occurred on both sides of the Pacific in the thirteenth century or even earlier. In the southern Andes and Tierra del Fuego there is good documentary evidence of major advances in the latter half of the sixteenth century. Most glaciers in both regions reached their furthest forward positions in the mid eighteenth or mid nineteenth centuries and then retreated in the twentieth century to positions similar to those at the beginning of the second millennium.

The record of the glaciers in coastal New Zealand suggests the possibility of a twenty-year recurrence interval of advances over the last 150 years. On the decadal timescale, snow precipitation in both the central Andes, about 35°S, and the Southern Alps of New Zealand, about 45°S, is correlated with ENSO events, heavy falls accompanying El Niños. The response of the glaciers in the Southern Alps appears to be stronger than in the central Andes. The response to El Niños of the outlet glaciers from the Patagonian icefields is not readily apparent because of the various time lags, but there are signs that it also is positive.

ANTARCTICA AND THE SUB-ANTARCTIC ISLANDS

13.1 ANTARCTICA

The response to climate change of the East and West Antarctic ice sheets differs markedly from that of ice masses in lower latitudes. It is so cold that a change in temperature of a few degrees directly affects only the marginal ice shelves. These are generally floating but are grounded here and there. If melting should reduce the extent or thickness of the shelves, it has been argued, they would be liable to lift off the seabed and become more mobile, allowing continental ice from inland to move seawards more easily (Morgan *et al.* 1991). An increase in the velocity and calving of ice fronts might then cause icestreams issuing from the icecaps to accelerate, the effect progressing upstream into the interior. The whole deglacial process might then be accentuated by the feedback effect of the accompanying rise in sea level but the time lag between climatic change and marginal shifts would necessarily be lengthy.

Climatic fluctuations of the last millennium, though they are registered in Antarctic ice cores, have had scarcely any discernible effects on the ice sheet margins (see Chapter 16). Air and ice temperatures are so low that there appears to have been no melt of any significance. On the contrary, warming is liable to result in an increase in snowfall causing glaciers to advance possibly for several decades, centuries or even millennia, rather than retreat.

A close relation might be expected between accumulation on the West Antarctic ice sheet and the Southern Oscillation Index. Bromwich and Rogers (2001) found accumulation and index values to be in phase from the early 1980s to 1990, an interval that included the strong El Niño event of 1982/3. However, during the succeeding 1990–8 interval with its strong 1997/8 El Niño, accumulation and index abruptly shifted to antiphase, a conflict that has yet to be explained. Recent studies suggest that the West Antarctic ice sheet is slowly thickening but the situation is a very complex one and not yet well understood (Alley 2002).

13.1.1 The Dry Valleys of McMurdo Sound

Alpine-type glaciers such as those in the Dry Valleys of the Ross Sea area are believed to be useful as regional climatic indicators but only when they are used to integrate climatic change over periods of some thousands of years (Chinn 1981 and see Chapter 15, pp. 494–6).

Small glaciers near the coast of Antarctica are more sensitive indicators of climatic history and some evidence from them is beginning to be obtained, though dating remains a difficulty. Edmondson Point Glacier, in the Terra Nova Bay area of Victoria Land is believed to have withdrawn between the tenth and fourteenth centuries, advanced at some later stage, and is currently retreating (Baroni and Orombelli 1994).

Lakes in the Dry Valleys are more sensitive than the glaciers to changes in temperature but the link between summer energy change and lake level is very complex, and disparities between lake level rises are caused by differing shapes and area-altitude distributions of the glacier tongues supplying meltwater.[1] Systematic

---

1   The water levels and permanent ice covers of the Dry Valley lakes, and others in the McMurdo Sound area, were investigated by the New Zealand Antarctic Research Programme for some twenty years until 1987 when they were incorporated within the divisions of the New Zealand Department of Scientific and Industrial Research (Chinn 1993).

monitoring of lake levels in the Dry Valleys began in 1968–9. Levels are known to have risen by 14 m between 1903 and 1990 (Chinn 1993), Lake Vanda, in Wright Valley, rose by about a metre per year probably as a result of greater inflow from the Onyx River (Howard-Williams 1993–4). According to weather station records, temperatures in the Dry Valleys decreased by about 1 °C between 1986 and 2000 and, with less meltwater, lake levels fell. The cooling, especially marked in summer and autumn, was associated with clearer skies and lower wind speeds. Doran *et al.* (2002) found that as a result of the lower summer temperatures productivity in the lakes and the numbers of invertebrates in the soil both declined.

13.2 THE ANTARCTIC PENINSULA AND SUB-ANTARCTIC ISLANDS

Indicators of environmental changes in the course of the last millennium are more abundant in the Antarctic Peninsula and the sub-Antarctic islands than on the continent itself. The Antarctic Peninsula projects north towards Tierra del Fuego and Patagonia with

the sub-Antarctic islands lying either side of it in the Scotia Sea (Figures 12.1 and 13.1). South of the Antarctic Circle (66°30′S) organic material for dating is lacking; northwards there are moss and lichens in the South Orkney and South Shetland islands, and grass and moss communities in South Georgia. Tephra layers, probably originating from a volcano on Deception Island and abundant in the Peninsula and also on some of the sub-Antarctic islands, are beginning to provide a regional tephra chronology (Björck *et al.* 1991).

All the islands experience cyclonic depressions. Moving eastwards across the Scotia Sea, they develop mainly along the Polar Front Jet Stream which deviates south to follow a track between the mountains of southernmost South America and the Antarctic Peninsula. Being constrained by the topography and ocean circulation, the position of the jet stream is quasi-stable, varying slightly with changes in the strength of the Weddell Sea Low Pressure Cell and the extent of sea ice. Reviewing the previous literature, Clapperton (1990a) concluded that glaciers in South Georgia, the South Orkneys, South Shetlands and the Antarctic

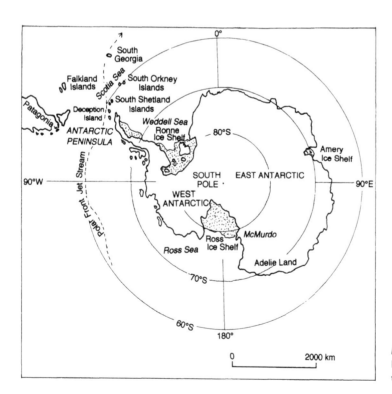

Figure 13.1 The location of the Antarctic Peninsula and sub-Antarctic Islands in relation to Antarctica and South America

Peninsula had all experienced at least two or three advance phases since the twelfth century AD.

While temperatures in the McMurdo Dry Valleys were falling towards the end of the twentieth century they were rising on the Antarctic Peninsula. Wherever temperatures on the Peninsula and neighbouring islands were high enough for melting to occur in summer, the surface of the ice was lowered. At Rothera Point, for example, the surface of an ice ramp fell by an average 0.32 m per year between 1989 and 1997 (Smith et al. 1998). On Signy Island (60°43'S), at the confluence of the ice-bound Weddell Sea and the warmer Scotia Sea, summer air temperatures rose by 1 °C between 1980 and 1999. Since 1950 the ice cover has receded by about 45 per cent and the nutrient content of lakes on the island has markedly increased (Quayle et al. 2002).

13.2.1 South Georgia

South Georgia, lying south of the Antarctic Convergence at 54–55°S, is about 300 km long by 5–40 km wide. Individual peaks along its axial mountain chain rise to well over 2000 m. The climate is cool and wet with a mean annual temperature of +2.0 °C and mean annual precipitation at Grytviken, where observations began in 1902, of about 1600 mm (Gordon and Timmis 1992). More than half of the island is covered by glaciers, many of them ending in the sea (Clapperton et al. 1989a, b), with coastal areas on the northeast side, below 70–100 m, occupied by grass and moss communities.

Mercer (1962) found that every glacier had 'a series of moraine ridges close to its snout, where the hummocky relief and scarcity of vegetation contrast with the older ground beyond'. Many glaciers had evidently reached more advanced positions in the last few centuries and had left behind multiple lateral and terminal moraines more than 300 m outside the margins of the land-based glaciers. The outermost moraines are more weathered and more densely covered with vegetation than those nearer the ice. The only glacier not calving into the sea for which details were available to Mercer (1962) had thinned by 60 m and retreated by 300 m between 1912 and 1958.

The environmental succession recorded in peat bogs and lakes on South Georgia suggested to Clapperton et al. (1989a, b) that Little Ice Age advances began after the late thirteenth century and peaked in the eighteenth, nineteenth and twentieth centuries. They estimated a Little Ice Age lowering of equilibrium line altitudes by an average of 50 m on the northeast of the island, corresponding with a cooling of 0.5 °C.

Difficulties associated with the uncertainties of ocean reservoir times arise when attempts are made to acquire Little Ice Age radiocarbon dates from marine material. Grasses and other vegetation provide a better possibility of reconstructing glacial history from organic fragments in moraines. Behind the terminal moraine of the Heaney glacier in St Andrews Bay, till buries a layer of peat which has given a radiocarbon date of 155 ± 45 (SRR 738) (Clapperton et al. 1978). This supports a Little Ice Age origin for the till, although it cannot be taken as providing a date more precise than somewhere in the age range AD 1670–1950. An attribution of Little Ice Age initiation to the thirteenth century AD depends upon a radiocarbon date from whalebone which, again, cannot be precise. A radiocarbon date of 1540 ± 70 BP (SRR-49) from a vegetated outer moraine of the Nordenskjöld Glacier (Gordon 1987) gives an approximate calendar age within the period AD 1300–1430, but only if a reservoir correction of 1000 years is employed (Harkness 1979). Uncertainties involved in estimating the magnitude of the appropriate reservoir correction for dates on marine samples prevent attempts to correlate results with the ages of deposits elsewhere from being anything more than guesses or, at any rate, preliminary estimates.

Since the first landing by Cook in 1775, thirty-eight glaciers on South Georgia have been mapped or photographed on more than one occasion. Thirteen terminating on beaches have shown no change in frontal position, nor has any evidence of surging been found (Hayward 1983). Since 1882 observations have been made sporadically, covering different periods from one glacier to another. The scatter of evidence suggests that an advance phase in the first decade of the twentieth century, giving a maximum about 1910, was followed by a slight retreat and then a rapid and longer advance. Cirque and small valley glaciers reached their maximum twentieth-century extensions in the mid 1930s. Slow retreat till the late 1940s, was terminated by a short, rapid advance till the early 1950s. Most glaciers reached their maxima by 1956. A slow withdrawal of the land-based glaciers followed, but the large valley and tide-water glaciers briefly advanced again in the 1970s (Gordon and Timmis 1992). In the late 1970s and early 1980s all the glaciers receded and thinned. Timmis (1986) compared areal changes of groups of small glaciers of similar size (0.1–4.0 km²), type and morphology in the northeast and southwest coastal areas, and found that both sets had well-defined Little Ice Age moraines up to several hundred metres outside their current fronts, with differences in the relative magnitude of change. He attributed these mainly to localized differences in weather patterns due to topographical effects.

The synoptic background to the advances culminating in the 1930s was associated with a seasonal net cooling of 0.5 °C in winter and 0.7 °C in summer. The marked recession of recent decades corresponds with an overall warming of 0.5 °C in both seasons, the effects on mass balances being slightly offset by a small increase in winter precipitation.

13.2.2 Deception Island

A detailed and lengthy record of mass balances from the sub-Antarctic was produced when volcanic eruptions in 1969 and 1970 ruptured the ice cover on Deception Island, exposing annual ice layers that had accumulated since 1680 (Orheim 1972, 1977). The layers were clearly delineated by summer dust surfaces and the stratigraphic interpretation was made relatively easy by the great width of the sections exposed; one of them was 50 m and the other 100 m high. As the sections were in the accumulation area, there is no reason to suppose that the mass balance record they provided has been affected by earlier volcanic events. The main effect of volcanism here is the deposition of pyroclastic material that is buried by snow the following winter and then ceases to affect surface albedo. The record provides a valuable indicator of regional climatic trends, as mean summer temperatures for 1944–67 for Deception Island show a strong positive correlation with temperatures at Orcadas in the South Orkneys, Argentine Island, Port Stanley, Punta Arenas and Grytviken.

Orheim (1977) compared the Deception Island accumulation series with mass balance series then available from Scandinavia (Liestøl 1967), the Alps, the

Polar Urals and northwest USA (Kasser 1973) and also with measurements from cores taken from Antarctic sites including the South Pole and Byrd Station. No relationship of any kind was found with the Antarctic series, probably because the mass balance values for Antarctica are almost entirely controlled by variations in accumulation while the Deception Island values are determined largely by summer warmth. Comparison of the Deception Island series with series from the northern hemisphere revealed that short-term variations in the mass balance were inversely correlated during the last few decades and, Orheim considered, probably over at least the last 150 years. Spectral analysis suggested that this relationship was caused by an anti-phase cycle with a period of about ten years. The cycles also occurred in the oldest part of the Deception Island series but were then apparently in phase with those of the northern hemisphere. This effect could be due to a gradual error in dating caused by missing years in the Deception record.

Long-term variations in mass balance at Deception Island, so far as they can be safely identified from the data, appear to be in phase with variations in most other parts of the world. There seems to have been a sharp decrease in mass balance on Deception Island between 1870 and 1880. Although this was both preceded and followed by long periods of more positive mass balance, the mass balances of the century after 1870–80 were significantly smaller than for the preceding hundred years. This pattern parallels the overall trend in the northern hemisphere and in low latitudes and also in New Zealand and the southern Andes, with glacier shrinkage predominating after 1850.

13.2.3 Kerguelen Island

The scattered information available from other sub-Antarctic islands was surveyed by Mercer (1962). The glaciers on Kerguelen were reported to be retreating when the gunship *Gazelle* called there on a voyage of 1874–6. Many glacier fronts were mapped during the voyage of the *Curieuse*, 1912–14, but this work is not known to have been repeated. In 1931, the outlets of the main icefields had evidently been retreating. At Port-aux-Français, 49°12′S, 70°12′E, temperatures increased by 1.5 °C between the mid 1960s and mid 1980s (Allison and Keage 1986).

13.2.4 Heard Island

More detailed information is forthcoming from Heard Island, 4000 km southwest of Perth (53°05'S, 73°30'E). The climate is cool and wet throughout the year and 80 per cent of the island is ice-covered. Heard Island is made up of two volcanic cones. The larger and still active one, Big Ben, rising to 2750 m, supports fifteen rapidly moving glaciers. Typically, they widen and steepen as they approach the sea and terminate in ice cliffs. A few piedmont glaciers on the eastern and southern sides, notably Brown and Stephenson glaciers, are now land-based with lobate tongues. The smaller cone, on the northwest side of the main mass of the island, forms the Laurens Peninsula which rises to 706 m. The ice here is thinner and does not reach the coast but terminates at about 200–350 m above sea level.

Little or no change in the general appearance of the Heard Island glaciers seems to have been reported between the visits of the Challenger expedition in 1874 and the Gauss expedition of 1902, nor again between 1902 and the Banzare expedition of 1929. This is probably not just the result of the inspections having been cursory but of the circumstance that many of the glaciers are calving into the sea. It was evident from trimlines that glaciers had been 30–90 m thicker not many decades before (Allison and Keage 1986).

Rather detailed knowledge of more recent fluctuations is to be attributed to the enthusiasm of Dr Budd, appointed medical officer to Heard Island in 1954. In spite of snow or rain falling on 200 days in the year, he took every opportunity to return to the island and make rapid surveys of the glaciers and wildlife (Budd 1970, Budd and Stephenson 1970, Radok and Watts 1975). It seems that a slight recession of the ice between the setting up of a station on Heard Island by the Australian Antarctic Research Expedition in 1947 and its abandonment in 1955, was accentuated between 1955 and 1963. About this time the Brown and Stephenson glaciers lost their coastal cliffs and became land-based. The Winston glacier, a little to the south, had retreated a mile up its lagoon between 1947 and 1963 when Budd surveyed the glaciers encircling Big Ben. In 1965 Budd returned again and found some indications that the glaciers were enlarging. Two years later on his next visit he found that the Winston and Vahsel in particular had expanded well outside the area they had occupied in 1954. The Little Challenger glacier had spread laterally and had developed sea cliffs 20 m high.

The readvance of the 1960s, which was quite marked, though it did not affect all the glaciers on the island, was still continuing in 1971, according to Radok and Watts (1975). They suggested that glacier fluctuations on the island were controlled by temperatures varying in accordance with the latitude of depression tracks affecting the region (Parkinson and Cavalieri 1982). Comparison of air photographs taken in 1980 with those of 1947 shows that recession, which had been general over the period as a whole, had been most marked since about 1970. Tidewater glaciers thinned; the steep glaciers on the north side of Big Ben had not retreated noticeably, but those on the east had receded several hundreds of metres. Small icecaps and glaciers on the Laurens Peninsula had shrunk so much by 1980 it was evident they would disappear entirely if the trend were to continue for long. The shrinkage seems to have been in response to warming since the mid 1960s, which has been recorded on Kerguelen and which appears to have been associated with a northerly shift of depression tracks (Allison and Keage 1986).

13.3 RECENT WARMING IN ANTARCTICA

Over the period from 1959 to 1996 the mean temperature trend for all stations in the Antarctic was +1.2 °C which, as Vaughan and Doake (1996) point out, is well above the global mean. There is considerable variation from place to place. Air temperatures at the Pole itself fell in the second half of the century whereas the Antarctic Peninsula, especially the northwest, got much warmer, reaching levels, according to ice core values, probably never experienced in the last 500 years. Sea ice in the Bellingshausen Sea to the west of the Peninsula has diminished and, as the seasonal snow cover has shrunk, the ranges of flowering plants have expanded. Surface temperatures off Heard Island, for example, rose by as much as 1 °C, and the number of king penguins on the island exploded from only three breeding pairs in 1947 to 25,000 (Pockley 2001). The warming has reached down from the ocean surface but to a much lesser degree; between the 1950s and the end of the century the upper 1000 m of the Southern Ocean warmed by 0.1 °C (Gille 2002).

13.4 ANTARCTIC SUMMARY

Evidence of the Little Ice Age in Antarctica comes not from the continent itself but from the Antarctic Peninsula and the islands to the north where glacier economies depend on local climates and ice cover. When glaciers began to advance in the last millennium is still unknown. As a result of mean air temperatures rising by 0.5° to 1.5 °C since 1850, glaciers retreated in the course of the late nineteenth and twentieth centuries from the advanced positions they had reached in the preceding few centuries.

14

LITTLE ICE AGE CLIMATE

Little Ice Age glacier variations provide useful information about past climates including changes in the atmospheric circulation. No full treatment is attempted here but a few themes are considered. First, the development by historians and climatologists of a methodology to elucidate the climate history of Europe is outlined. Then, an account is given of recent progress in unravelling the climatic history of Asia, especially China, where exceptionally long documentary records can be set alongside the results of analysing Tibetan ice cores. Little Ice Age climate in southern Africa is included mainly to illustrate the problems and possibilities of tracing the course of events in a part of the world for which there are fewer documents and very few ice cores. Finally, the glacial evidence of the Little Ice Age is evaluated in the context of climatic reconstructions by other means.

14.1 ADVANCES IN THE METHODOLOGY OF CLIMATIC RECONSTRUCTION

Tracing the vicissitudes of climate during the last thousand years or so has been greatly improved by advances in the methods used to reconstruct past environmental conditions.[1] New sources of proxy data, including isotopic and other characteristics of ice cores, have also emerged (Bradley 1999). Dendroclimatology has continued to develop (e.g. Fritts 1976, Schweingruber

1988, Briffa *et al.* 1988, 1990, 1999, Esper *et al.* 2002), and tree-ring series have been obtained from many localities scattered over the globe. A multitude of new documentary sources has been unearthed, the need for careful verification of data is now widely accepted, and greater insight has been obtained into the ways in which it may be used.[2] The importance of cross-dating different kinds of proxy data with descriptive evidence to produce a combined, more reliable record is widely understood though not always pursued. Computer listing and analysis of data would have been impossible until quite recently (Pfister 1992a, Schüle 1994).

Most proxy series relate to climate in a particular season of the year. Tree-ring indices, for instance, are mainly affected by conditions during the growing season, usually the summer months. The isotopic characteristics of ice cores naturally refer to periods during which snow was falling; extremely cold periods during which there was no snowfall leave no trace. On the other hand, ice core melt records are valuable pointers to summer temperatures having been above zero. Documentary sources, even if verified, may be biased towards recording unusual or extreme situations. It is therefore essential to use as many independent lines of evidence as possible.

A satisfactory understanding of past weather and climate requires the reconstruction of long series of temperature and precipitation values at many locations, ·

---

1   Historical climatology emerged as a discipline in the late nineteenth and early twentieth centuries, the first major work of synthesis being published by C. E. P. Brooks in 1949 (Ingram *et al.* 1978). Since the 1940s the methodology and approaches required to achieve reliable reconstructions of past climates have been increasingly refined, Manley (1974) and Lamb (1977) playing key roles.

2   The use of content analysis is an example of one of the methodological advances involved (e.g. Baron 1982).

together with maps of past pressure distributions. It is important to discover the extent to which means, interannual variability and the incidence of anomalies may have changed over time. The ultimate aim is to discern sequences of regional synoptic situations, and to discover the nature of the teleconnections between them. Such knowledge can assist in identifying the forcing factors responsible for climatic changes.

The distribution of data over the global is very uneven. Ice cores come from high latitudes in both hemispheres, and also from high altitude sites near the tropical latitudes, notably the Quelccaya icecap in Peru and the Dunde and Guliya icecaps in China (Thompson 1992). Tree-ring series have been obtained from subtropical, temperate and high latitude regions of both hemispheres, but the greater number come from the northern hemisphere, especially regions bordering the Arctic Ocean and the north Pacific. Dendroclimatic data are more difficult to obtain from the tropics because most tropical species do not form annual rings, though a few exceptions have been found.[3] Coral is confined to the tropics. Stalagmites and stalactites in caves in limestone areas from the tropics to high latitudes can provide a $\partial^{18}O$ record datable by Th-U.[4] Archival sources are richest and stretch back furthest in Eurasia, because literate societies have been recording climatic events and conditions for longer there than elsewhere. Arabic documents might conceivably disclose much interesting information for the Near East and parts of Africa. It is probable that detailed exploration of Spanish, Portuguese, Dutch and other colonial records could greatly extend knowledge of climate history in Latin America, southern Africa and the Far East.

14.2 EUROPEAN EVIDENCE FOR LITTLE ICE AGE CLIMATE

The Swiss archives hold a wealth of resources for paleoclimatic studies (Chapter 6) as do those of many other parts of central and northern Europe (Bradley and Jones 1992c).[5] Documents from medieval times, extending into the opening stages of the Little Ice Age, are often difficult to interpret, not only linguistically but also in distinguishing fact from fiction. Jones and Bradley (1992, table 13.1) listed forty-nine locations for which long, homogeneous temperature series were then available. Of this total, seventeen are European, the longest being Gordon Manley's (1974) reconstruction of monthly mean temperatures for central England.

Manley's study stands as a model of meticulous scholarship, providing full explanations of the ways in which values were derived, and the extent to which estimates should be considered reliable (Figure 14.1). In the course of his work he examined a great number of manuscript sources, all carefully listed, as well as many English weather diaries. Subsequent searches have discovered many more; Kington (1994), for instance, specifies forty-six sources covering various parts of the 40-year interval 1675–1715.

English documents predating the later Little Ice Age, allowed Lamb (1972) to construct a graph showing southwesterly wind frequencies for England from the fourteenth to the twentieth centuries (Figure 14.2). A high frequency of southwest winds over England can be taken to indicate zonal flow, a low frequency blocked or meridional circulation when North Atlantic Polar waters spread south to the latitude of the Shetland Islands, displacing warmer surface waters and shifting cyclonic storm tracks further south. In the course of the Little Ice Age these conditions alternated at intervals of a few or several decades.

Lamb led the way in attempting to construct indices of summer dryness/wetness and winter mildness/ severity in England from 1100 to 1969, and reached towards a more regional approach, first in constructing diagrams illustrating decadal excess of wet or dry summer months in different longitudes of Europe around 50°N for the same period, and then in producing maps

---

3 Publication of an extended tree-ring sequence from Java (Palmer and Murphy 1993) indicates the possibility of more paleoclimatological information being forthcoming from the tropics in future.

4 Such a record from Oman has been interpreted as indicating that the variations in the northward shift of the intertropical convergence zone and the summer monsoon rainfall belt between 9600 and 6100 years ago were correlated with changes in atmospheric radiocarbon and hence with solar activity. The results point to considerable fluctuations of climate from decade to decade and from one century to another (Neff *et al.* 2001).

5 See also Special Issue of *Climatic Change* 43 (September 1999) 'Climatic Variability in Sixteenth-Century Europe and its Social Dimension'.

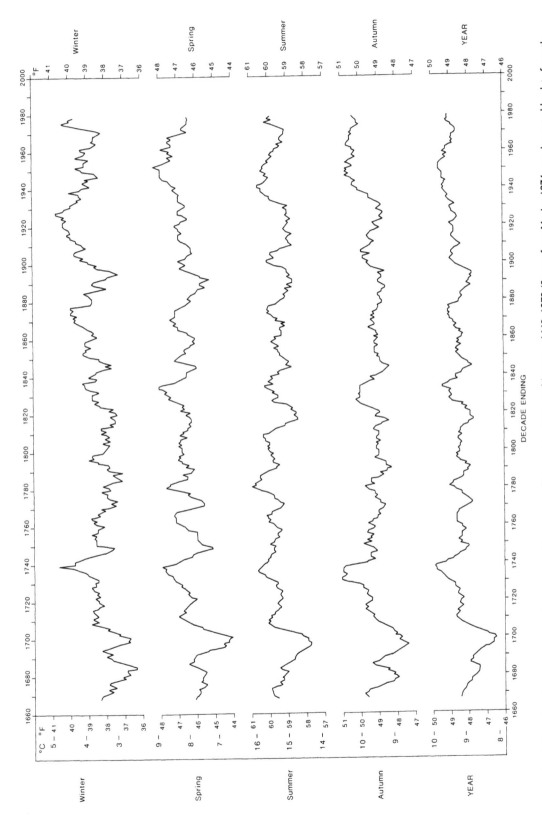

Figure 14.1 Central England temperatures by seasons and for the year. Ten-year running monthly means 1669–1979 (Source: from Manley 1974, supplemented by data from the graph which he kept in his study)

Figure 14.2 The frequency of southwesterly surface winds over England since AD 1340 (*Source:* from Kington 1994)

for specified months of particular years, showing both area weather reports and speculative mean sea level pressure and wind patterns.[6] (Figure 14.3).

A good demonstration of the profit to be gained by intensive archival searches is the extraordinary quantity of verifiable data relating to parts of western Europe found by Alexandre (1987). His book, *Le Climat en Europe au Moyen-Age*, is a major contribution to knowledge of climatic variations between AD 1000 and 1425, a period for which there are in general less data than for more recent centuries. Doubtless much more material remains to be found, and new documents are still being uncovered. An important discovery is the weather diary of Louis Morin kept in Paris between 1665 and 1713, which contains three readings daily of air temperature and pressure, as well as daily observations of

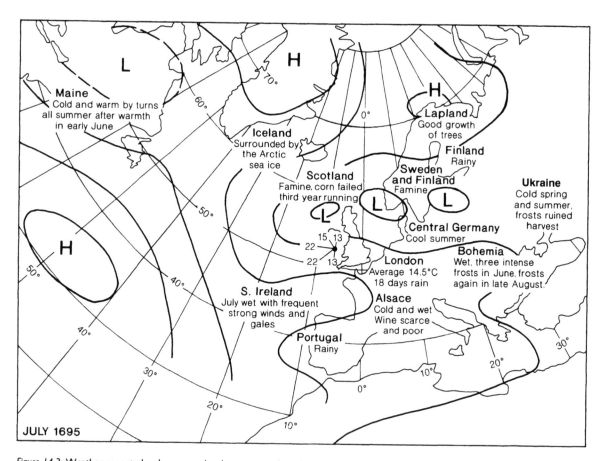

Figure 14.3 Weather reported and mean sea level pressure and wind pattern for July 1695 (*Source:* after Lamb 1977)

---

6 Although some of the details of his reconstructions have turned out to need some correction or additions in the light of subsequent investigations, it was Lamb's example, energy and enthusiasm that caused a great deal of the more recent paleoclimatic research carried out in England to take place.

wind speed and direction, duration and intensity of precipitation (Pfister and Bariess 1994).

Jacobeit *et al.* (1999) have constructed maps of monthly surface pressure and surface atmospheric circulation over Europe for months with outstanding climatic anomalies, according to written records, during the sixteenth century. Slonosky *et al.* (2000) have constructed such maps for all the months from the 1770s to 1995, making use of monthly instrumental pressure series from fifty-one locations some of them starting as early as 1755. They found that a positive annual NAO index, with pressure low over Iceland and high over Gibraltar as in the periods 1855–65, 1885–1930 and 1950–70, was associated with strong westerlies. A negative annual index as in the periods 1821–30, 1870–85, 1930–50 and 1970–90 was accompanied by reversals of zonal flow and frequent blocking in the eastern Atlantic and western Europe. Winter NAO index values were strongly positive from 1900 to 1930, even stronger than between 1980 and 1995. Negative winter values were strong in the 1960s and 1970s. Generally, index values were more extreme in the late eighteenth and early nineteenth centuries than in the twentieth, anomalies being most marked before 1830, with high positive values from 1775 to 1790 and 1800 to 1815 and low negative values from 1790 to 1800 and 1815 to 1830.

The longest instrumental pressure time series is the Paris–London circulation index, downstream from the North Atlantic, a useful indicator at least in winter of the zonal circulation over Europe. Slonosky *et al.* (2001) have constructed a circulation index between 1696 and 1708 by making use of Louis Morin's long series of daily pressure and temperature readings for Paris in conjunction with the similar readings of William Derham, Rector of Upminster in Essex.[7] Usually pressure is higher at Paris than at London and westerlies predominate. Over the 9-year period covered by both observers, which is known to have included some of the most severe winters of recent centuries, it was found that reversals of the north–south pressure gradient and easterly winds were more prevalent than during 1994–7 and it can be inferred that the circulation was more meridional and variable.

The NAO is now seen as part of a wider sub-Arctic system called the northern hemisphere Annular Mode (NHAM) affecting the climate of North America and the Arctic basin as well as Europe (Thompson and Wallace 2001). Daily mean data for the winters of the forty years 1958–97 show that high index conditions are characterized by westerly winds from Russia to Canada along 55°N, with temperatures on average about 5 °C warmer over much of central North America and Europe and also warmer over the Barents and Kara seas. Low index conditions are marked by cold anticyclones over central Canada and Russia, an anticyclonic surface circulation throughout the Arctic basin, and a much greater frequency of cold events over Europe, Russia and North America especially the Pacific Northwest.

Certain parts of Europe have specialized types of data of great significance; descriptions of arctic drift ice off the coast of Iceland (see Chapter 2) and the extent of ice cover in the Baltic are examples (Tarand 1992). Grapevine yields in Austria, Hungary, Germany and Switzerland underline the sudden shift in climate in the 1580s (Landsteiner 1999). Spanish archives contain records of *rogativas*, prayers of supplication for rain or for deluges to cease, which were performed with various degrees of solemnity according to the perceived seriousness of the situation (Alvarez Vázquez 1986). Unpublished collections of weather data by José Maria Fontana Tarrats (1977), which formed the basis of a general account by Font Tullot (1988) of climatic conditions in Spain during the last few centuries, demonstrate the existence of much material hidden away in the archives. Unfortunately Tarrats gave few details of the sources he used. A certain number relating to Andalusia in the first half of the seventeenth century have been verified by Rodrigo *et al.* (1994) but many remain to be critically examined.

Dendroclimatic data have been accumulated from northern Fennoscandinavia (Briffa 1994), central Europe (e.g. Briffa *et al.* 1999) and the Mediterranean. Mean annual (January to December), and summer (April to September) temperatures have been reconstructed from AD 1500 to the twentieth century for 10°W at 50°, 45°, 40° and 35°N (Serre-Bachet 1994).

---

7   Derham's temperature readings were used by Manley in his construction of central England's temperature record.

A large data bank of tree-ring values collected by Kuniholm and his associates, mainly for archaeological purposes, has the potential to provide information on past climates in the eastern Mediterranean (Kuniholm and Striker 1987), one of the regions within Europe which has been neglected by climatic historians.

14.3 THE EASTERN MEDITERRANEAN

The periods of glacier enlargement in the Alps and Pyrenees during the Little Ice Age can be attributed to a greater frequency of meridional circulation associated with blocking anticyclones. Synoptic conditions different from those of the twentieth century in northern and central Europe must have been accompanied by climates in Mediterranean Europe somewhat different from those we regard as normal. A fuller understanding of the history of European climate entails investigating the course of events in the south and especially the southeast, about which information has been deficient.

Carol Griggs has recently assembled over 500 of Kuniholm's oak tree-growth records from Greece and Turkey over the last 800 years and plotted the average ring widths for each year.[8] The resultant graph shows tree rings just as wide in the early thirteenth century as in the twentieth (Figure 14.4). Narrow tree rings (less than 75 per cent of the mean width) cluster in the intervals 1250–1320, 1484–1500, 1570–1610, 1700–60, 1810–20, 1840–50 and 1950–60. The narrow rings are attributed to low temperature, which is likely to be the explanation for narrow rings in trees near the upper treeline, but not necessarily at lower levels where lack of rain could be a more significant factor.

Some information has been acquired about the sixteenth- and seventeenth-century climate of Crete from documents in the Venetian archives (Figure 14.5). Lying midway between the Greek mainland and the Libyan coast, the island is affected by air masses from the Atlantic and the Sahara, from eastern Europe and western Asia. Inter-annual variability is considerable. Summers are hot and dry; winters mild and wet. Snow rarely falls on low ground and almost never lies near

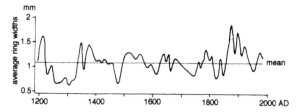

Figure 14.4 Mean ring widths of over 500 oak samples from the eastern Mediterranean region (a freehand smoothing of the original which was by Carol Griggs 2001) (Source: in Cornell University's Aegean Dendrochronology Project Progress Report, December 2001)

the coast for more than 24 hours. The situation is very different in the interior mountains rising to 2500 m.

Crete was ruled by the Venetian Republic from 1204 until it was conquered by the Ottoman Turks between 1645 and 1669. It was subsequently ruled by Turkey until 1898. The Turkish records have so far attracted little attention;[9] the Venetian archives have been found to contain a great deal of weather information (Grove et al. 1992, Grove and Conterio 1994, 1995). Administrators reporting back to Venice dated and signed their letters and named the places where they were writing them, thereby providing satisfactory sources for climatic reconstruction. They describe the weather over short and long periods; and are concerned about its effects on crops; they often give details of wind direction and comment on the effects of droughts, floods and other extremes. Such archival information can be supplemented by travellers' accounts and by the reports of French consuls, but no weather diaries have been found, as yet. The cessation of serious cereal growing in Crete in recent decades, the absence of records of vintage dates, and modern dependence on irrigation have so far prevented a phenological approach to the quantitative evaluation of the historical data.

For the century 1548 to 1648, the information extracted from Venetian records reveals that in some years and groups of years weather conditions were anomalous by twentieth-century standards. The most important anomalies were winter and spring droughts,

8   In the Malcolm and Carolyn Wiener Laboratory's (Cornell University) *Aegean Dendrochronology Project Progress Report*, December 2001, http:www.arts.cornell.edu/dendro.

9   But it is clear from the selection of documents translated into Greek by Staurinides (1975, 1976a, b, 1984) that much climatic information is to be found there.

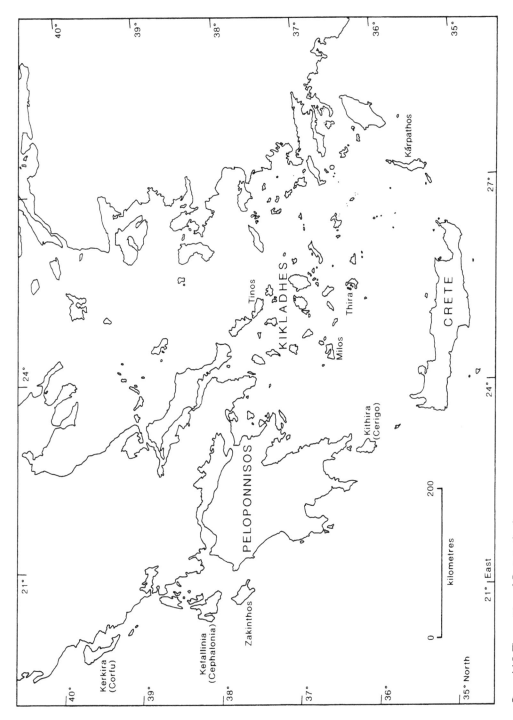

Figure 14.5 The position of Crete in the Aegean

Table 14.1 Classification of data relating to weather anomalies in Crete, 1548–1648

Winter	
Normal	No information suggesting unusual conditions
Severe	Buildings damaged by prolonged rain, difficulties with transport, or many mentions of 'cold'
Very severe	Exceptional falls of snow, prolonged periods of abnormal cold, or rain so excessive as to delay sowing of cereals till late spring instead of autumn
Drought	Prolonged drought after October
Spring	
Normal	No contrary indications
Good	Crops promise particularly well, good spring rains, or barley harvest early
Unfavourable	Crop forecasts poor, harvests expected late
Drought	Evidence of long spells without rain
Summer	
Normal	No unusual circumstances
Unfavourable	Crops damaged by south winds or heavy rain

and exceptionally severe winters. The method of data classification used is shown in Table 14.1.

The most serious weather anomalies afflicting Venetian Crete were winter and spring droughts (Figure 14.6). In February 1555, for example, it had not rained for three months.[10] Rain had still not come by April and it was recognized that there was no hope of a cereal crop that year. Heraklion was suffering badly from lack of water and the countryside around was 'burnt and parched. . . . the land so dry it is hardly possible to find water for men or animals to drink'.[11] The 1555 drought held for six months. Figure 14.6 shows a noticeable clustering of drought years in 1561–6 and 1600–7, spring droughts being more common than winter.

The precipitation records of Heraklion for 1910–92 include a number of dry years. In several the onset of the rains was delayed until October, or even November, as in 1987 and 1992. But since records began there have not been so many as three months of complete drought in winter. The incidence of dry years between 1547 and 1645 is not in itself remarkable, but in no twentieth-century drought years were there such long rainless periods in winter. Some of the Turkish records also mention droughts in Crete which appear to have been more severe than in recent times; for instance, 1690 was a very dry year 'with very few beneficial rains and as a result the seeds and cereals did not flourish'.[12] In 1690 and again in 1695 the authorities found it impossible to collect taxes based on crop production because of drought losses.

Between 1547 and 1645 there were twelve droughts in winter, twenty-one in spring and two at an unspecified season. Often the whole of Crete was affected by drought, sometimes much wider regions. Shortage of water in 1595–6 forced people to emigrate from Milos, Tinos and Cerigo (Figure 14.5), and in 1620–1 drought afflicted all the islands of the Greek archipelago. Saharan air extended far to the north in 1643 when protracted south winds in May caused 'great harm to the countryside throughout the realm'. 'Furious and burning winds from the south' continued into June.[13]

Eight very severe winters were experienced in Crete between 1547 and 1648 (Figure 14.7).[14] In 1595

---

10 Dando in a letter written 6 February 1555. ASV.PTM,Filza 728.
11 Michele, in a letter written 20 April 1555. ASV.PTM,Filza 728.
12 Staurinides (1978: 165).
13 Dust veils are suspected to have been especially heavy between 1640 and 1643, as a result of numerous eruptions on the Pacific rim. Many parts of the world experienced severe droughts or other anomalous weather conditions, notably the Ukraine and southeast Asia (Atwell 2001). In 1641 northern China may have experienced its driest year between 1470 and the present day, and in 1642 central China its second driest (Zhang and Lin 1992).
14 This might be compared with the ice core record which includes strong sulphuric acid spikes in two or more northern hemisphere locations in 1585–6, 1588–90, 1596–7, 1600–2 and 1605–8 (Atwell 2001: 56).

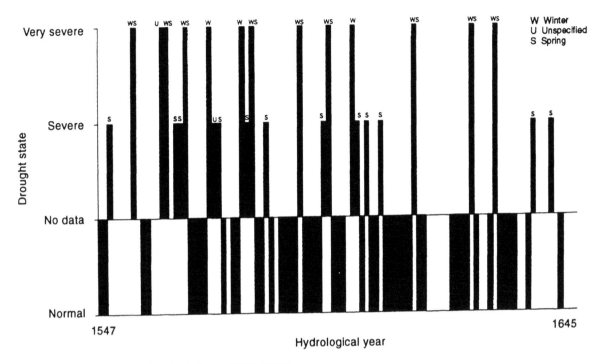

Figure 14.6 Winter and spring droughts in Crete, 1547/8–1644/5

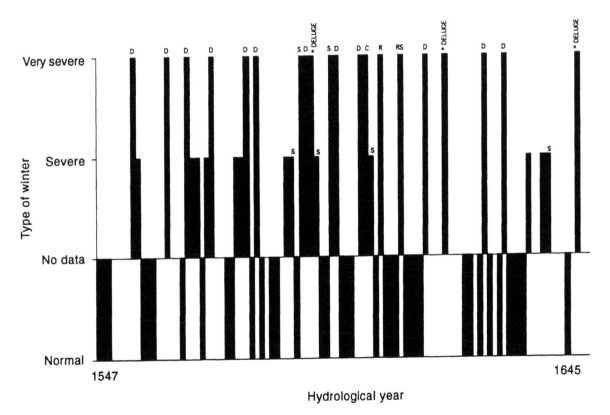

Figure 14.7 Severe winters in Crete, 1547/8–1644/5

excessive and long-lasting snow caused great suffering. In 1601–2, 'very bitter cold' caused many deaths among the crews of the Galleys of the Guard,[15] and a large number of animals died because of continuous rain and extreme cold.[16] Winter at high elevations seems to have been generally more snowy than nowadays and in 1659 snow fell 'up to the coast' for two months.[17]

Documentary and instrumental data from the eastern Mediterranean and southern Balkans, mainly Greece, have been compiled by Xoplaki *et al.* (2001) for the periods 1675–1715 (the Late Maunder Minimum) and for 1780 to 1830. Especially in the earlier of the two periods, variability of the climate was greater than in recent decades. Winters and springs in the decade 1675–85 were cold and wet, especially in 1676. Droughts were severe between 1685 and 1695. In the winter of 1682–3 snowfalls and continuing frost were reported in Serbia and all over Greece. In Crete that year, the Greek population complained that the Turks had forced them to carry 6300 loads of snow down from the mountains in the heat of summer (Staurinides 1976a: 192, Grove and Conterio 1994). The severe winter of 1686/7, when the ice on Lake Ioannina lasted for more than three months, was followed by famine in the Peloponnese and Thessaly. The winter of 1699–1700 was very cold everywhere in central Europe with long-lasting snow in northern Greece and Crete (Xoplaki *et al.* 2001). Tournefort, a botanist well known for the accuracy of his observations, visited Crete in 1700 and noted that the tops of the mountains were covered by snow all the year round (1715: 41–2).[18] After an interval of eight years there are reports of severe cold in Serbia and eastern Europe in 1709 and southern France experienced one of its coldest winters on record. Crete seems to have escaped lightly. Then there was a run of drought years in the Balkans from 1710 to 1714 concluding with plague and a great famine in Greece. The winter of 1713/14 was a European-wide phenomenon, the driest winter of the last 500 years in Switzerland, France and Great Britain (Xoplaki *et al.* 2001).

Information is then sparse for the eighteenth century until instrumental records begin to appear. In Milan, the winters of 1766 and 1767 were among the coldest ever recorded there; January means of −2.5 °C and −4.7 °C compare with a 1930–60 January mean of +1.7 °C. (Grove and Rackham 2001). In 1780 heavy rain and floods are reported in northwest Greece, and famine and plague in Crete. The winter of 1782 was harsh and cold in Greece, and in the winter of 1789/90 snow cover was excessive in Serbia (Xoplaki *et al.* 2001). About this time Swiss glaciers were advancing but Austrian glaciers were beginning to retreat. In the early nineteenth century, when Alpine glaciers were again advancing, northern Greece experienced heavy rainfall in the winter and spring of 1805 and severe cold in 1807/8 which also affected Mytilene, Lesbos, close to the Turkish coast. In both Milan and Rome 1845–54 was the coldest decade since instrumental records began, with mean temperatures 0.9 °C and 1.1 °C below the 1930–60 means. Minimum temperatures were especially low: the mean daily minimum temperature for the decade in Rome was 9.8 °C (8.5 °C in 1844 and 1845) compared to 11.6 °C in 1930–60. There were exceptional snowfalls in Crete in 1849 and 1858, which caused damage to trees and livestock (Parlamas 1949).

14.4 THE SYNOPTIC HISTORY OF EUROPE IN THE LITTLE ICE AGE

To gain an appreciation of the past climate, long series of values of temperature and precipitation are not by themselves sufficient; an impression of the spatial arrangement of atmospheric conditions is also needed. Several years ago, Hubert Lamb, among others, began to reconstruct weather and pressure system maps for individual years (Lamb and Johnson 1959, 1961, 1966,

---

15 Priuli in a letter written 10 March 1602. ASV.PTM,Filza 770. The summer of 1601, the year when Peru's Huaynaputina exploded, is thought to have been the coldest in the northern hemisphere in the past six centuries (Atwell 2001).
16 Ca'Taiapiero in a letter written 24 March 1602. ASV.PTM.Filza 770.
17 Provveditor Generale Barbaro in a letter written 25 April 1659. ASV.PTM,Filza 808.
18 By 1851–2, when Captain Spratt made his surveys of the island, this was no longer the case. The summits of the White Mountains 'after midsummer, appear bald and grey, but in winter and spring are covered with snow' (Spratt 1865: 149). The summits were not known to remain snow-covered in twentieth-century summers.

LITTLE ICE AGE CLIMATE 381

Lamb 1972, 1977). The methods used by different individuals and groups varied. In recent years many European researchers have followed the lead given by Pfister and have contributed to a common database which allows 'multi-proxy mapping' (Kington 1994).[19] It was first directed towards an exploration of the nature of European climate during the Late Maunder Minimum, widely suspected to have been the coldest phase of the Little Ice Age. It made use of more than 220,000 items of information mostly relating to northern, central and western Europe to present the spatial pattern of anomalies between 1675 and 1715 (Luterbacher *et al.* 2001). Monthly temperature and precipitation data from England, Switzerland and Hungary[20] form the backbone of the reconstruction, together with more sporadic observations from adjacent regions to the north and south.

Knowledge of the detailed spatial patterns of climatic change in Europe is now more complete for this period, 1675–1715, than for any other part of the Little Ice Age (Pfister 1994b). Whereas Hungary and England were relatively wet, Switzerland and France were generally drier than in the twentieth century (Pfister 1994a) which may account for the modest response of Alpine glaciers to the low temperatures. Trends frequently seem to have originated in the west and extended east. In the mid 1670s, for instance, cooling in the west spread east, with maximum winter severity in England in the 1680s, in central Europe in the 1690s and in the Carpathian basin in the 1700s. Subsequently, warming was observed first in England from 1697, around 1703 in Switzerland and after 1705 in Hungary. Northern hemisphere summers from 1691 to 1700 are believed to have been the coolest of the millennium, the coldest in west-central Europe being 1692 (Jones *et al.* 1998). A strong anticyclone centred about 60°N between the Norwegian Sea and the Baltic repeatedly brought warm southwesterlies to Iceland, and in northernmost Europe warm winters were commoner than cold ones. On the other hand, cold northeast winds in central Europe caused long winter droughts and in the Mediterranean extreme cold and long periods with snow in sixteen of the winters between 1675 and 1699. In the winters of 1683 and 1697 cold was experienced everywhere in the Mediterranean basin. Italy and Spain both experienced winter cold in 1681, 1691, 1693 and 1694. Cold in Spain in 1689 and 1695 coincided with long rainy periods in Italy. In other years, notably 1679, 1693 and 1698, cold in Spain coincided with rain in the eastern Mediterranean.[21]

The winter of 1680 was very dry in Castile, rather warm in Poland, but cold and snowy in Russia and wet in Greece. In central Europe, the winter of 1694/5 was the coldest in the last 500 years. No spring season between 1695 and 1703 exceeded the 1901–60 average temperature in either England or Switzerland. In France the lowest minimum temperatures in five centuries came in 1708/9 when a cold air mass extended from Finland and Russia right across Europe to the Pyrenees. The eastern Mediterranean was possibly less affected: while olive trees were killed by frost in Provence, the oil harvest in Crete was quite good. An extreme drought in the eastern Mediterranean, lasted over the winter and spring of 1713–14 (Grove and Conterio 1994) and in Switzerland that year the winter was one of the ten driest in the last 500 years, with river flows and lakes falling to extraordinarily low levels (Pfister 1994).

The analysis of seasonal and regional weather in the Maunder Minimum based on the Bern data bank has gone a long way towards substantiating the ideas of Manley and Lamb about the period, supplementing and also modifying them to some degree. Over the British Isles during this period (Figure 14.2), southwest wind frequencies showed a maximum in the 1670s (Figure 14.8a), decreased sharply to a minimum around 1690 (Figure 14.8b) and then increased again to reach another zonal circulation peak in the 1730s. The minimum of 1690 provides an explanation for the remarkably low temperatures (Figure 14.1) diagnosed by Manley (1974). For January in the 1690s, Lamb had

---

19 The EURO–CLIMHIST data bank (Schüle 1994).
20 The pre-1914 frontiers of Hungary were used.
21 All this came within the Maunder Minimum (see Chapter 17) and the cooling possibly due to the low solar activity was certainly reinforced by volcanic eruptions in various parts of the world in 1673, 1693 and 1694 (Luterbacher *et al.* 2001).

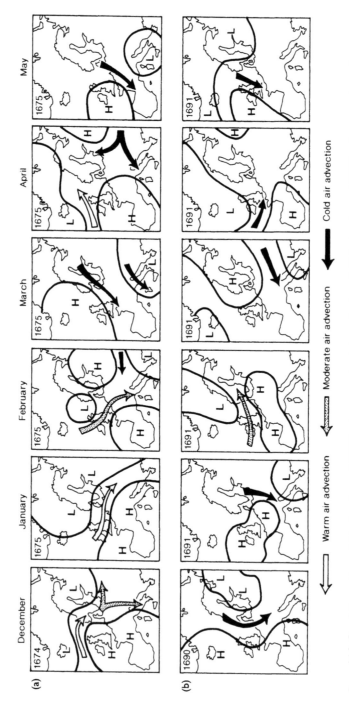

Figure 14.8 Circulation patterns: (a) in the winter of 1674/5, when westerly winds predominated, (b) in the winter of 1690/1, when westerly winds were rare (*Source*: from Pfister *et al.* 1994b)

identified a centre of high pressure over Scandinavia on his weather charts, with cold continental air dominating Europe, northeasterly air flows, and frequent outbreaks of cold air into the Mediterranean. This kind of pattern was found by the Maunder Minimum study to characterize certain individual years of the decade, including 1692 and 1694, but did not appear in others such as 1693. Lamb identified a southward displacement of the Icelandic low and truncation of the Azores high in July. This accords nicely with evidence of the predominance of northwest and north winds in summer in Denmark during 1675–1715 (Frich and Frydendahl 1994) and the complete absence of warm, dry summers in western and central Europe from 1685 to 1704.

The synoptic sequences during the Maunder Minimum could be defined more exactly given better data coverage especially from the eastern Mediterranean region. The Turkish archives may yet yield interesting data. The same methodology could be applied to a much longer period than the Maunder Minimum, and might eventually be extended outside Europe (Bradley *et al.* 1991). From what is already known it appears that in Europe at any rate, the decade 1690 to 1700, the deepest portion of the Maunder Minimum, was one of the coldest, if not the coldest decade, on record (see Chapter 17).

14.5 ASIAN EVIDENCE FOR CLIMATIC CHANGE

A great wealth of documents with references to weather and climate exists in Asia, especially in China and Japan, some older than any in European archives. So far no deliberate examination has been made of Indian language archives, though a few European colonial officials made notes on climate, especially regarding drought.[22]

Over the past twenty years there has been a surge of active interest in climatic reconstruction in Japan

(Mikami 1992). A 'Historical Weather Data Base' (Yoshimura and Yoshino 1988) is exploiting such sources as records of the time of appearance of cherry blossom going back to the ninth century (Arakawa 1956). Full use can be made of such occurrences by reference to phenological observations going back to 1927 (Kawamura 1992). Data on the incidence of precipitation are available from a wide range of sources dating from the late eighteenth and nineteenth centuries (Mikami 1992). Dates of the ice break-up on Lake Suwa in central Japan are known from 1443 to 1953 (Fukaishi and Tagami 1992).

Although in China instrumental weather records more than a century long exist for only a few places, such as Beijing and Shanghai, documentary sources of climatic proxy data are extraordinarily long and plentiful. Decadal dustfall frequencies have been assembled for the last 1700 years (Yan Zhongwei 1994), and records of the occurrence of thunder for the last 2200 years (Wang 1980). Official documents from as far back as the Tang Dynasty (AD 618–907) contain phenological data located in both space and time (Zhang De'er 1994). Daily weather records include official reports, such as the 'Clear and Rain records' from the Quing Dynasty (1636–1910),[23] describing sky conditions, wind direction and precipitation type (Zhang De'er 1992). Diaries containing daily accounts of weather, also date from this period (Wang and Zhang 1992).

Dendroclimatic studies began in China in the 1930s and 1940s and greatly increased after 1970 (Wu 1992). There are still wide gaps in the coverage and only a few records go back more than 400 years (Figure 14.9).

Cores from the Dunde icecap at 5325 m and the Guliya icecap at 6710 m on the Tibetan Plateau (Figure 14.10) have added a very important element to the range of data available for paleoclimatic reconstruction (Thompson 1992, Thompson *et al.* 1989, 1990, 1995b). Because of the ice core evidence attention is focused here on Chinese rather than Japanese climate history.

---

22 See R. H. Grove 1995: 304ff. An East India Company surgeon, Roxburgh published two papers on late seventeenth-century climate in Madras Presidency (1778, 1790).

23 The earliest Clear and Rain record for Beijing dates back to 1672. The system became countrywide in 1685. Unfortunately the records from many cities have been lost. The four cities with the longest and most complete records are Beijing (1724–1904), Nanjing (1722–98), Suzhou (1723–1810) and Hangzhou (1723–73) (Wang *et al.* 1992).

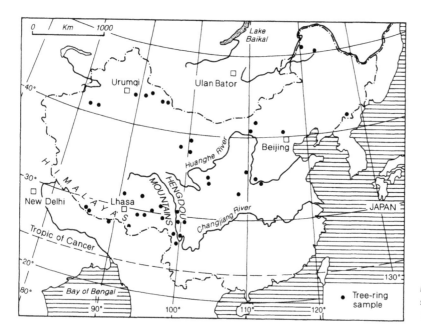

Figure 14.9 Distribution of tree-ring sample sites in China (*Source:* after Wu 1992)

14.5.1 Little Ice Age climate in China

China extends from Hai-nan Tao, less than 20°N, to 54°N at the border with Siberia, and over 60 degrees of longitude from the far northeast to the Tien Shan in the west (Figure 14.10). The present climate varies from cold temperate in the north to subtropical in the southeast, and montane periglacial on the Tibetan Plateau. Continentality is very pronounced: winters are cold, summers hot and annual temperature ranges wide. The seasonal precipitation and reversal of dominant wind direction reflect the influence of the Asian monsoon. About 85 per cent of the annual precipitation falls between April and October, when humid, maritime monsoon winds blow from the south. As far west as the Dunde Icecap over 80 per cent of the precipitation falls in the wet season between May and August (Thompson 1992). Winter winds from high pressure systems centred over Siberia frequently carry dust as far as the east coast.[24] Regional contrasts are so great that

the Central Weather Bureau identifies 9 climatic zones subdivided into 18 climatic areas and 36 regions plus a plateau-type climate, itself further subdivided (Jiacheng Zhang and Zhiguang Lin 1992).

The regional climatic differences are associated with varying distance from the sea and with great topographical contrasts between, for instance, the Tibetan Plateau and the great valleys of the Yellow and Yangtze Rivers. The Tibetan (Qinghai-Xizang) Plateau causes major blocking of zonal as well as meridional air flows. It has a determining effect on large-scale monsoon circulation, and is an important source of synoptic systems, large and small over China. A strong anticyclone centred over the plateau becomes a major heat source during summer.

The climate south of 25°30′N is subtropical to tropical, and snow is rare, yet even this region suffered from cold periods in the Little Ice Age.[25] As important as fluctuations of temperature have been major variations in precipitation, with swings from droughts to floods

---

24 Notably in the spring of 2002 at the end of an exceptionally cold winter in eastern Asia.
25 For instance the whole of southern Guangdong, with a mean annual temperature above 22 °C and a January mean above 13.3 °C, was covered by heavy snow on 15 and 16 January 1893, during a cold wave which affected all the coastal provinces of south China (Cai Fuxiang 1992).

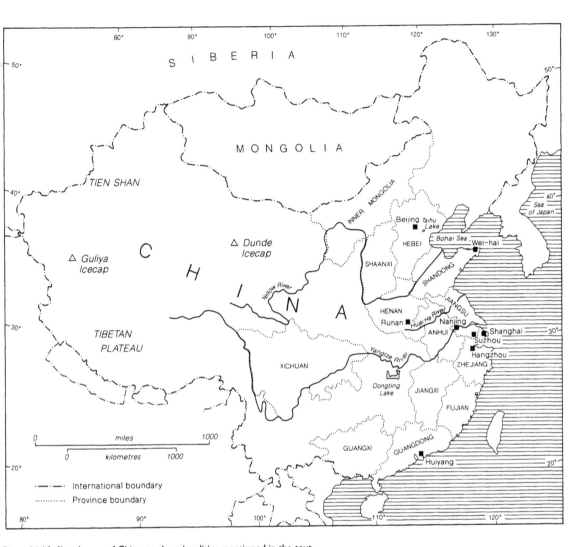

Figure 14.10 Sketch map of China, to show localities mentioned in the text

and back again (Wang Shaowu and Zhao Zong-ci 1981).

14.5.1.1 Climate in China before the Little Ice Age

Before the Little Ice Age, crop boundaries lay further north than during the last few centuries (Zhang De'er 1994). A tropical–subtropical perennial crop, *Boehmeria nivea*, which is known to have been grown in the vicinity of Huiyang and Runan (Figures 14.10 and 14.11), but does so no longer, requires a mean annual temperature of 15.5°–16.5 °C, and a mean January tem-

perature between 1.5 °C and 3 °C. Now, mean annual temperatures are 14.5 °C at Huiyang and 15.1 °C at Runan. It is concluded that mean annual temperatures were higher in the thirteenth century than in 1961–80, at Huiyang by almost 1 °C.

Other phenological evidence points in the same direction. In 1264 oranges were being grown in Tang and Deng counties, Xichuan, where this would now be impossible for climatic reasons. The northern boundary of citrus cultivation in the mid thirteenth century was further north than it had been in the Tang Dynasty (AD 618–709) and even further north than it is now.

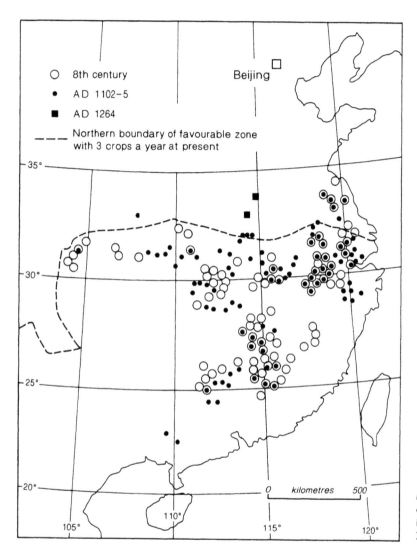

Figure 14.11 The location of cultivation of *Boehmeria nivea* in the eighth century, in AD 1102–5 and in 1264 (*Source:* after Zhang De'er 1994)

Central China evidently experienced the Medieval Warm Period (and possibly a preceding cool period), with the Little Ice Age beginning at some time after 1264.[26]

Ice core records from the Dunde icecap indicate that in Tibet between AD 1000 and 1180 somewhat cooler temperatures predominated, the cold being interrupted by short episodes of greater warmth. Then a warm period from AD 1200 to 1280 was followed by more prolonged cold (Figure 14.12). The timing of medieval warmth would seem to have differed in Xichuan and eastern Tibet. When ice core results from the Guliya icecap, high in the Kunlun of western Tibet, and from Dunde are compared, more striking regional

26 This conclusion conflicts with the view of Chu Kochen (1973) that records of snowfall from AD 1131 to 1264 in Hangzhou show this was a cold period. But the conversion of these documents from the lunar to the solar calendar has been revised and all the revised dates of spring snowfall have been shown to be significantly earlier in the year than those used by Chu. Therefore South Song time, around AD 1200, can no longer be considered one of the coldest periods in the last 5000 years in this part of China.

Table 14.2 Thirteenth- and twentieth-century temperatures in Guangdong and southern Henan provinces

	Mean annual temperature °C		Mean January temperature °C	
	13th century	20th century	13th century	20th century
Huiyang	15.5	14.5	1.5	0.9
Runan	15.5	15.1	>1.5	1.2

Source: after Zhang De'er 1994

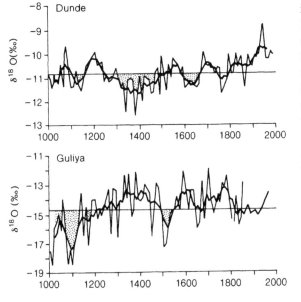

Figure 14.12 Warm and cold periods shown by the Dunde and Guliya ice cores. The shaded areas represent isotopically more negative, cooler periods, and the unshaded areas represent isotopically less negative, warmer periods, relative to the respective means of each record (*Source*: after Thompson 1992)

disparities are revealed; the Guliya core fails to register the warmth of the Medieval Warm Period and the cold of the Little Ice Age (Thompson *et al.* 1995b). The most negative $\partial^{18}O$ values, generally indicating low temperatures, occurred at the beginning of the millennium,

whereas values from 1300 to 1800 were higher suggesting warmer conditions. The reason why temperature conditions in the northeast of the Tibetan Plateau have differed from those in the far west has yet to be elucidated. Careful analysis of the characteristics of a selection of the other unstudied glaciers on the plateau may help to solve this problem.

14.5.1.2 Little Ice Age temperature reconstruction

Little Ice Age temperature series from the 1380s to the 1990s have been reconstructed for north[27] and east China[28] (Wang Shaowu and Wang Risheng 1990, Wang Shaowu 1991, 1992). Proxy data from tree rings and ice cores were used as well as documentary sources. Earlier studies had used records of cold events such as early frost, late last frost, heavy snow and freezing of lakes and rivers to construct severity indices. The bias towards cold events was now corrected by constructing seasonal anomaly series for spring, summer, autumn and winter (Figure 14.13) which differ somewhat from those published earlier, for instance by Chu Kochen (1973). The basis of the reconstruction was a 1982 study made by the Beijing Meteorological Centre in which temperatures observed since 1910 were translated into severity indices ranging from −2 to +2. Experiment showed that decadal mean temperature anomalies of −1.0 °C, −1.5 °C and −2.0 °C can be translated into severity indices of −0.5, −1.0 and −2.0 and vice versa. These relationships are found to be practically constant for all seasons and geographical locations (Wang Shaowu 1991: 755). Statistical analysis of modern data indicates that the frequency of cold events in a decade can be used to indicate the average temperature for that decade, and also that temperature correlates closely with the number of rainy days in summer. Severity indices for earlier centuries were constructed from the documentary data (Table 14.3).

The implications of the various indicators in Table 14.3 were established by reference to relationships during the period of instrumental observation. For instance, in the winter of 1976–7, when the seasonal temperature anomaly was −2 °C over most of China, a severity index of −2.0 was given. The Bohai Sea was

27   North China includes Beijing, Hebei, Henan, Shandong and Shaanxi.
28   East China includes Shanghai, Jiangsu, Jiangxi, Anhui and Zhejiang.

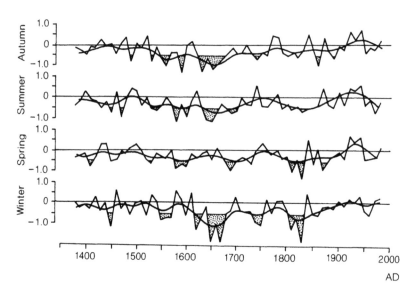

Figure 14.13 Seasonal decadal mean temperature anomalies for north China. The reference period was from the 1880s to the 1970s. The smooth curve shows the 50-year running mean. Anomalies greater than the average for the period from the 1380s to the 1870s are shaded (Source: after Wang Shaowu 1991)

Table 14.3 Severity indices for north and south China

Index	Spring	Summer	Autumn	Winter
North China				
0.5	snow, frost	prolonged rain, light snow	prolonged rain, snow, frost	heavy snow, glazing
−1.0	grain frosted, heavy snow, glazing	frost glazing, snow, floods	grain frosted, heavy snow, glazing	frozen ground, frozen wells, lakes and rivers
−2.0	frozen ground, bitter cold, frozen lakes, frozen rivers	severe frost, bitter cold	frozen lakes, frozen rivers, bitter cold	bitter cold, frozen sea, frozen rivers or great lakes
South China				
0.5	frost in April, heavy snow, high floods, prolonged rain	high floods, frost	prolonged rain in September, snow	heavy snow, severe glazing
−1.0	frost in May, grain damaged, snow in May, glazing	frost in August, rain for 20 days, light snow, glazing, ground frost	grain frosted, heavy snow, ground frost	heavy snow, ground frost for long periods, trees killed, bitter cold, frozen wells
−2.0	severe frost, grain damaged, bitter cold, frozen lakes, rivers or wells	frost in June or July, bitter cold like winter	snow lasting tens of days, bitter cold, travelling on frozen lakes or rivers	heavy snow not melted till spring, frozen lakes or rivers, frozen sea

Source: Wang Shaowu 1991

deeply ice-covered and the sea froze for more than 5 km along the coast near the port of Wei-hai in Shangdon Province (Figure 14.10). Taihu and Dongting lakes in central and south China froze for several days, and heavy snow and glazing frequently occurred. Spring temperatures were found to be 0.3–0.4 °C lower than the norm if the last frost was ten days later than usual, or 0.3– 0.4 °C higher if the last frost was ten days earlier. A reverse relationship applied to autumn. A severity index of 1.0 was judged proper for summers with prolonged rain because summers in Beijing with twenty or more days of rain have an average temperature anomaly of −1.3 °C. In this fashion, knowledge of the past decadal temperature was built up season by season.

Table 14.4 Decadal mean autumn temperature anomalies in Beijing

	Reconstructed	Observed
1880–9	–0.6	–0.51
1890–9	–0.1	–0.02
1900–9	–0.1	–0.07
1910–19	–0.5	–0.25
1920–9	1.5	0.90
1930–9	0.3	0.32
1940–9	0.4	0.44
1950–9	–0.1	–0.10
1960–9	–0.1	–0.11
1970–9	–0.3	–0.27

Source: Wang Shaowu 1991

Some instances of extreme cold of a type not experienced during the twentieth century appear in the literature. For example, in the winter of 1670–1 the Huai River froze for three months, and carriages were able to cross on the ice. Wells and springs froze and trees died. In such cases an index of –3.0 was given. Obversely, indexes of +1.5 were given for exceptionally warm periods, such as the winter of 1522–3 which was as warm as spring, and the summer of 1671, when drought from June to the end of August caused grass and trees to wither. In all, thirty-nine extremely cold

seasons were identified, none of them in the twentieth century, and thirty-two extremely warm seasons of which eleven were in the twentieth century.[29]

Comparison of the reconstructed and observed decadal anomalies for the period since meteorological measurements began gives results which are sufficiently satisfactory to confirm the basis of results for the historic period. The seasonal severity indices for the historic period were therefore used to derive temperature anomalies (Table 14.4).

Finally the seasonal decadal series were summed, to give annual temperature anomalies from the 1380s onwards. (Too little data are available before 1380 to go further back.) The decadal temperature series from east and north China are essentially similar in their general characteristics, but, as might be expected, not identical (Figure 14.14). Three cold periods were identified: in the mid fourteenth, mid sixteenth and early nineteenth centuries. Annual temperature in China has ranged through more than 1 °C from the coldest phase in the Little Ice Age to the warmest decades of the twentieth century.

Evidence for the first cold period comes only from the Dunde ice core (Figure 14.12). Wang Shaowu speculated that either it represents regional conditions only, or that the documentary data from north and east China are insufficient. Further study of early

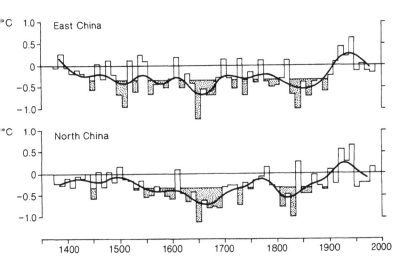

Figure 14.14 Decadal mean temperature anomalies in north and east China. Periods with temperature anomalies <–0.5 are shaded (Source: after Wang Shaowu 1992)

29 Both the Guliya and Dunde ice core records show warm conditions since 1940 (Thompson et al. 1995b).

phenological sources might well resolve this question. It may be that greater use could be made in China of detailed phenological data to add substance and calibrate reconstructions more closely, as has been done in Switzerland (see Chapter 6).

14.5.1.3 Regional variations in climate

Differences in the climatic sequence are to be expected within large regions. For example extreme cold affected the south of Guangxi Province (Figure 14.10) in 1522 and 1526 but not Guangdong and Fujian (Li Pingri 1992). Similarly, reconstruction of climate in the Yellow River basin since AD 1470 revealed a cold period lasting from the sixteenth to the nineteenth centuries, with the coldest years from the 1620s to the 1690s and from the 1830s to the 1860s (Wu Xiangding and Yin Xungang 1992). Differences in the exact timing of dry and wet years were found between the six regions of this enormous valley, though common to all of them were the dryness of the 1520s and 1630s, and the wetness of the mid seventeenth and late nineteenth centuries. Severe floods were less frequent than severe droughts. Climatic instability with more frequent extremes was found to have been particularly marked in the cold periods. (These conclusions do not conflict with the results of the decadal temperature index for north China shown in Figure 14.14.)

14.5.1.4 Droughts and floods

Spatial and temporal variations of precipitation are more complex than those of temperature. As drought and flood conditions are of major importance for agriculture, a great many records of their impact at particular places and times are to be found in historical documents. These have been used to construct Drought and Flood charts covering the period since 1470 for almost the whole of eastern China (Wang Shaowu and Zhao Zong-ci 1981). Data from 118 stations were classified according to severity, from very wet to very dry, and the results mapped (Figure 14.15). The findings proved to be consistent with reconstructions of eighteenth-century monthly precipitation at Nanjing, Suzhou and Hangzhou based independently on Clear and Rain day records (Wang and Zhang 1992). Six main distribution patterns were found:

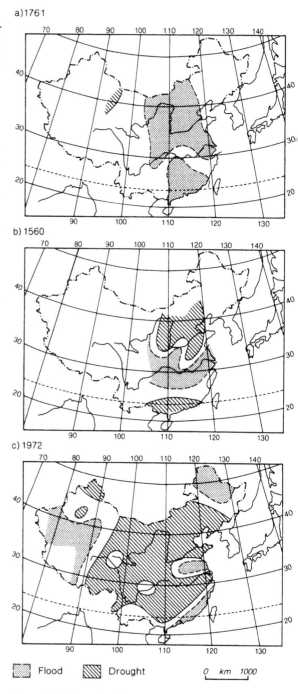

Figure 14.15 Examples of flood and drought distribution in China: (a) floods over almost all of China, (b) floods in the Yangtze River region with droughts to the north and south, (c) droughts over almost all of China (Source: after Wang Shaowu and Zhao Zong-ci 1981)

1 Floods all over China, but particularly in the Yangtze River region
2 Floods in the Yangtze River region with droughts
   to the north and south
3 Floods in the south and droughts in the north
4 Droughts in the Yangtze River region with floods
   to the north and south
5 Floods in the north and droughts in the south
6 Droughts almost all over China.

Power spectrum analysis of the Flood and Drought
series suggests connections between the zonal distribution of flood and drought over land and low
frequency variations of the intensity of the Southern
Oscillation Index, the presence or absence of high pressure over the Sea of Japan, and the frequency of occurrence of tropical cyclones over the southeast coast of
China. Precipitation and temperature fluctuations in
China are necessarily connected with large-scale climatic
conditions in the southern as well as the northern hemisphere (see Figure 14.12).

14.5.1.5 Variations in the strength and position of the monsoon

Climatic conditions in China are controlled by the
position and strength of the Asian monsoon. Fluctuations in the summer monsoon cause variations in the
timing and intensity of the rains, extremes involving
swings between devastating flooding and severe
drought. Winter temperatures are closely connected
with the strength of the monsoon winds blowing out
from central Asia. Sufficient reconstruction of associated parameters has now been accomplished to make it
possible to begin to trace variations over time in the
strength and position of the monsoon (Huang Jia-You
and Wang Shaowu 1985, Guo Qiyun 1992, Yoshino
1992).

14.5.1.5a Variations of the Western Pacific Subtropical High

The Subtropical High over the western Pacific is the
dominant synoptic system controlling summer rainfall
over China. In a 'normal' year the rainbelt is over south
China in the late spring; it occupies the Yangtze valley
in June and early July, and reaches its most northern
position in China and Inner Mongolia in July–August.

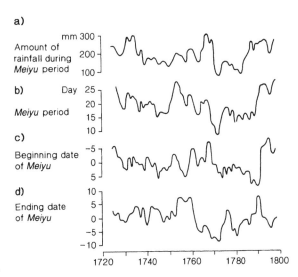

Figure 14.16 Variation of *Meiyu* rainfall between 1720 and 1800:
(a) amount of precipitation, (b) length of period, (c) beginning
date, (d) ending date (*Source:* after Yoshino 1992)

In the autumn it retreats south and persistent rain
occurs only in southwest China. The sequence is not a
gradual one; each June the general circulation over
eastern Asia changes suddenly. The upper Tibetan high
becomes stable, and the subtropical upper westerlies
and their associated jet streams suddenly jump northwards causing rapid disappearance of the westerly jet
stream over South Asia. As the Polar Front, which is
generally located at the northern periphery of the Subtropical High, moves northwards from south China the
Meiyu (plum or mildew) rains begin in the Yangtze
valley. The Subtropical High pushes further north in
the first half of July, ending the *Meiyu* in the Yangtze
and causing the start of the rainy season in north China
(Shi-yen Tao and Yi-hui Ding 1981).

Inter-annual variation in the position of the Polar
Front has been great even during the twentieth century. The average timing and duration of the *Meiyu*
rain, as well as its quantity, has varied during recent
centuries (Figure 14.16). The lower Yangtze region
was affected by the *Meiyu* rainbelt for about six days
longer in the interval 1723–1800 than in the modern
period, though drought occurred every five or six years
(cited Yoshino 1992). *Meiyu* precipitation seems to
correlate positively with snow cover over Tibet and to
be closely related to sea surface temperatures over the

eastern Pacific. In an El Niño year the *Meiyu* is generally deficient (Lau and Li 1984).

The long sequence of precipitation variations disclosed by the Flood and Drought records discussed earlier can be explained by shifts in the position of the Subtropical High over the western Pacific. The 500 mb level circulation over eastern Asia is closely correlated with surface temperature and precipitation values. The relationship is sufficiently close to allow statistical reconstruction of the circulation field over East Asia for the modern period on the basis of correlations between the surface indices and the known 500 mb height (Huang and Wang 1985). The reconstructed and observed data were considered to be close enough to justify reconstruction of the atmospheric circulation over China since 1471 from proxy data. It emerged that when the high spreads towards the west the eastern part of China suffers from greater frequency of drought and the central area from floods. When it expands northwards, floods occur in north China and droughts south of the Yangtze. The extent of the high's penetration westwards is the dominant factor controlling long-term variations of precipitation in lowland China.

Rainfall patterns compiled by Zhang and Crowley (1989), showed relatively dry seventeenth-century conditions in northern and western China which were thought to be explicable in terms of the summer monsoon being suppressed, perhaps by more extensive Eurasian snow cover.[30] Wang and Zhang (1992) found dry conditions in eastern China between AD 1500 and 1700 and again during the twentieth century to be accompanied by higher precipitation on the Tibetan Plateau, suggesting an anti-phase relationship. Low-frequency trends in accumulation indicated by the Dunde and Guliya ice cores, appear to be out of phase with the historical wetness/dryness index for eastern China since AD 1500 (Zhang De'er 1994, Thompson *et al.* 1995b).[31]

14.5.1.5b Variations of the winter monsoon

A significant correlation between winter temperature in Beijing and Shanghai and sea level pressure over Eurasia (Guo Qiyun 1992) underlines the close relationship between temperature and the strength of the winter monsoon in north and east China. Using a paleotemperature series for Beijing and Shanghai, Wang Shaowu (1991) reconstructed the winter monsoon circulation as it might have been from the 1380s to the 1980s. The winter monsoon appears to have been considerably stronger during several intervals than in the twentieth century, the most important of them being the 1490s–1510s, 1650s–1690s and 1830s–1890s. In the coldest decade of the Little Ice Age in this region, the 1650s (see Figure 14.14), atmospheric pressure in the centre of the anticyclone was postulated to have been 1040 hPa whereas in the 1460s, when winters were mild, it was about 1035 hPa (Figure 14.17).

14.5.1.6 A summary

China experienced warm conditions before the onset of the Little Ice Age (Zhang De'er 1994). Since the 1380s temperature has oscillated, with cold decades outnumbering warm from the late fourteenth to the late eighteenth century, except in the Kunlun. The inter-decadal variance of mean temperature was up to 20 per cent greater during the Little Ice Age than in the twentieth century (Wang Shaowu and Wang Rishong 1990, 1991, Wang Shaowu 1991). Three long cold periods occurred: in the mid fourteenth century, between the 1550s and 1690s, and in the nineteenth century. The earliest of the three was registered only in the Dunde icecap core on the Tibetan Plateau (Thompson 1992). Both later periods were characterized by winter cold; in the sixteenth/seventeenth-century episode low temperatures in spring and summer were marked; in the nineteenth century cold

---

30 The five worst years of consecutive drought in China as a whole, between 1637 and 1641, and severe famines in Japan from 1648 to 1653, coincided with high dust veil indices and sulphur-rich eruptions in Japan, the Philippines and Indonesia (Lamb 1970, Atwell 2001).

31 Net accumulation was reconstructed using visible dust layers, believed to be annual, and a simple flow model to estimate thinning. Thinning has to be taken into account because the original layer thickness is reduced over time as snow accumulates above and the ice at depth flows outwards. The Guliya core shows three extended wet periods (AD 1000–75, 1400–1775 and 1900–present), and two dry periods (1075–1375 and 1775–1900).

a) 1650s

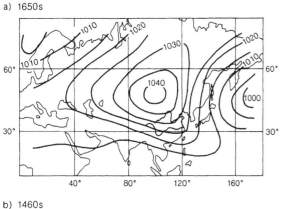

b) 1460s

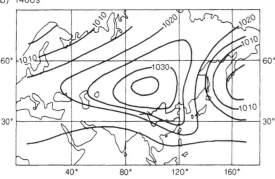

Figure 14.17 Reconstructed sea level pressure over China: (a) in the coldest decade, the 1650s, (b) in the mild decade, the 1460s (*Source:* after Guo Qiyun 1992)

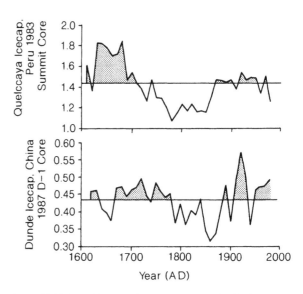

Figure 14.18 Decadal accumulation in metres in the Quelccaya summit core compared with the Dunde D-1 core for the period 1610–1980 (*Source:* after Thompson 1992)

autumns were important. The maximum mean temperature anomaly was −1.8 °C in winter and varied between −1.1 °C and −1.4 °C in other seasons. The coldest decade was around 1650, the second coldest was in the nineteenth century. Mean annual temperature has ranged through more than 1 °C, from the coldest phase in the Little Ice Age to the warmest decades of the twentieth century. No part of China was immune from the incidence of lower temperatures, frost and snow, the timing and extent of the phenomena varying in detail from region to region.

The distribution pattern of drought and flood has exhibited regional differences (Wang Shaowu and Zhao Zong-ci 1981). Clear and Rain records have provided much more precise information than records relating to temperature but are only available for some places (Wang and Zhang 1992). The varying location of the *Meiyu* rains over time is related to movements of the

Polar Front, which depend on shifts in the position and extent of the Subtropical High in the western Pacific (Huang Jia-You and Wang Shaowu 1985). Regional patterns of drought and flood are explicable in terms of pressure distribution. The winter monsoon was stronger during the Little Ice Age than in the twentieth century; periods when it was most pronounced correlate with those when greatest cold was experienced (Guo Qiyun 1992). Spectral analysis of the Flood and Drought series and of pressure indices indicates connections with low frequency variations of the Southern Oscillation Index, and sea surface temperatures over the western Pacific, especially in the region of the Kuroshio Current (Lau and Li 1984).

Climatic conditions in China depend upon the interplay between major elements of northern hemisphere circulation and are influenced by oceanic and atmospheric conditions in the southern hemisphere. The very strong similarity between the accumulation histories obtained from cores in the Dunde icecap in Tibet and the Quelccaya icecap far away in Peru (Figure 14.18) suggest long distance teleconnections between the two for low frequency events lasting several centuries, as well as for the high frequency El Niño–Southern Oscillation events (Thompson 1992).

14.6 A COMPARISON OF LITTLE ICE AGE CLIMATE IN EUROPE AND CHINA

Chinese glaciers show the same general pattern of advance and retreat as those in Europe and accumulated mass balance values for the recent past have similar trends right across Eurasia (Chapter 8). The main features of climate fluctuations over the last few centuries are also in agreement, though there are differences in detail, as Pfister *et al.* (1994c) have shown to be the case for the AD 1645–1715 Maunder Minimum.

Whereas in England, France and Switzerland the extreme cold of the 1690s and the subsequent warming are prominent features, the coldest period in China was around 1650, a time relatively mild in Switzerland and other parts of western Europe. It is quite understandable that the coldest periods during the Little Ice Age might not have been synchronous in western Europe and China. Should the Siberian high expand eastwards, China becomes colder but Europe may be unaffected or little affected. In the early eighteenth century, milder conditions in western Eurasia were accompanied by an increase in the frequency of southwesterly winds (see Figure 14.2). With higher temperatures to both east and west, it might well be presumed that the strength of the Siberian high diminished. This supposition is supported by a reduction in the number of dust events in China at that time, indicating a weakening of wind strength (Figure 14.19).

Precipitation in China is concentrated in summer and depends on the strength of the monsoon. When the summer monsoon is weak rainfall is depressed or absent, a relationship that was particularly marked during cool periods. While winters in both Switzerland and France were dry in the cold period of the late seventeenth century, summers were moderately wet. When warming began in the early eighteenth century, China got wetter in summer and western Europe drier.

Multi-proxy mapping began in Europe where many of the detailed records are precisely timed. Furthermore, Europe's climatic history is of special interest because deep water forms in the Atlantic to the northwest, so it helps to provide information relating to the thermohaline circulation of the oceans (Chapter 11, p. 568). The Chinese documentary data is exceptionally rich, but the time resolution is not so fine as in Europe.

a)

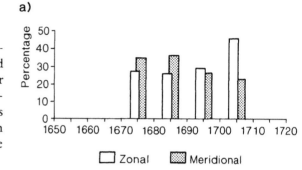

b)

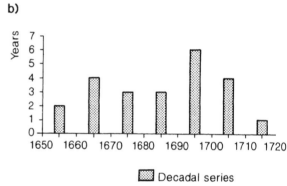

Figure 14.19 (a) Decadal frequencies of wind direction derived from direction of cloud movement in Paris, (b) decadal dust frequency in China, interpreted as a proxy for winter wind strength (*Source*: after Pfister *et al.* 1994c)

There is an obvious possibility that the type of analysis and multi-proxy synoptic mapping already initiated in Europe may be extended to China and Japan.

14.7 LITTLE ICE AGE CLIMATE IN AFRICA

Methods of tracing Little Ice Age climate in regions with fewer documentary records can be demonstrated from southern Africa. The current precipitation pattern in southern Africa is relatively simple: rain falls in summer, except in the southwest Cape where more than 70 per cent comes in winter, with a narrow belt along the southern coast liable to receive rain throughout the year (Figure 14.20). During the twentieth century a series of wet and dry spells alternated, each lasting about a decade; as they succeeded one another, meridional components of the circulation underwent reversal, though contrasts between east and west per-

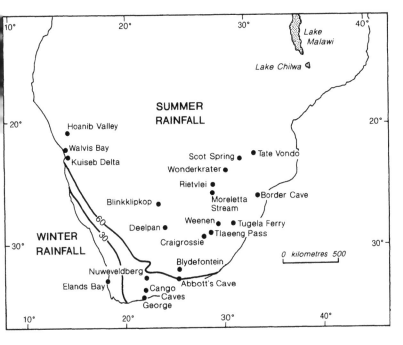

Figure 14.20 Sketch map to show present-day distribution of winter and summer precipitation regions in southern Africa. Isolines give the percentage of annual rainfall received in summer. The locations of sites of historic data mentioned in the text are marked (*Source:* after Cockcroft *et al.* 1987)

sisted (Tyson 1986, Cockcroft *et al.* 1987, Tyson and Lindesay 1992).[32]

The pattern of rainfall variability in southern and southeastern South Africa between 1820 and 1900 has been reconstructed from newspapers and other contemporary records such as missionary archives and bankers' documents (Vogel 1989) and a scatter of information has been retrieved for the eighteenth century (Nicholson 1981). Periodic occurrence of an inverse relationship between abundant rainfall in the eastern Cape and reduced rainfall in the southern Cape is a notable feature of the nineteenth-century reconstruction, as it is also of twentieth-century measurements.

Only the main features of the climate can be traced before the eighteenth century, but a multiplicity of data point to Africa south of 20°S having experienced the Little Ice Age, after a warmer period in medieval times.

A speleothem from one of the Cango Caves provides the longest record, with oxygen isotope analysis indicating that temperatures have fluctuated through more than 1 °C during the last thousand years (Figure 14.21) (Talma and Vogel 1991). The Medieval Warm Period peaked at about AD 900, after which temperatures dropped to reach the lowest levels of the millennium around AD 1300. The first phase of the Little Ice Age was interrupted by a warm interval about 1550, followed by eighteenth- and nineteenth-century cooling. A similar sequence of events is recorded by planktonic foraminifera in diatomaceous sediments on the continental shelf off Walvis Bay (Herbert 1987). Though the dating of the Walvis Bay core cannot be regarded as completely secure, the systematic variations displayed correlate reasonably with the speleothem evidence. Both register medieval warming followed by a cool period from about 1300 to 1850, broken by a warm episode

---

32   The degree to which the climate of the region responds to (i) variations of the atmospheric circulation on annual or shorter timescales, (ii) changes in synoptic circulation, (iii) the interaction of tropical and temperate systems, (iv) variations in mean poleward momentum flux, (v) changes in the meridional Hadley and Ferel Cells and (vi) modulation by the Walker Circulation is reviewed in detail by Tyson (1986).

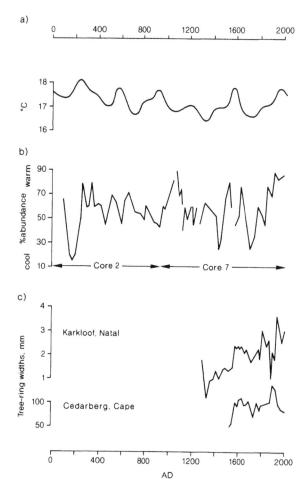

a pronounced warming, peaking about 1700. Pollen from hyrax middens at Blydefontein in the Cape, registers a long Little Ice Age, but does not show a warm phase within it (Scott and Bousman 1990). More detail is provided by tree-ring sequences from Karkloof and Cedarberg (Dunwiddie and LaMarche 1980), both of which show warming around 1600, and eighteenth and early nineteenth-century cooling, but also reveal regional differences.

No periglacial evidence of the Little Ice Age has been forthcoming, but geomorphological evidence has been interpreted in terms of variations in temperature and moisture. Buried soils and renewed incision after 1700 seem to reflect a sequence of climatic changes at Blydefontein (Bousman *et al.* 1988). The characteristics of peat deposits near George show that while in earlier medieval times both forests and wetlands were relatively extensive, in the thirteenth and fourteenth centuries both reacted adversely to drier conditions. That precipitation was greater during the early medieval period is shown by peat deposits dated to AD 1000–1100 in the Tlaeeng Pass area of the Lesotho Highlands. In the northern Namib desert a forest developed along the Hoanib River during the Medieval Warm Period but in the Little Ice Age, between 1640 and the nineteenth century, the Amspoort silts accumulated and buried the forest (Vogel and Rust 1987, 1990).

Tyson and Lindesay (1992) considered that a reasonably clear and coherent picture of changing climate was emerging from the available evidence, despite the imprecision of some of the dating, the varying nature of the proxy temperature records, and the problematic interpretation of the moisture records. They presented the following sequence of events:

Figure 14.21 Proxy temperature series based on: (a) oxygen isotope analyses of a cave speleothem at Cango Caves (from Talma and Vogel 1991), (b) foraminifera in diatomaceous sediment cores taken from shelf deposits off Walvis Bay (Johnson 1988), (c) tree-ring analyses from Karkloof, Natal (Hall 1976) and Cedarberg, Cape (Dunwiddie and LaMarche 1980) (*Source:* after Tyson and Lindesay 1992)

around 1600. A much less detailed record from oxygen isotope and aragonite/calcite ratios of *Patella granularis* and *P. gratina* shells from middens at Elands Bay, north of Cape Town, also shows the greatest Little Ice Age cooling between about 1300 and 1550, preceded by warming between 900 and 1300.

Inland, pollen data from Wonderkrater, Scot Spring, Tate Vondo and Rietvlei Dam (Figure 14.20) point to maximum cooling around 1400, preceded by medieval warming between 900 and 1300 and followed by

Year AD	
900–1300	Medieval Warm Period characterized by variable but warmer conditions, with the highest temperatures in the tenth and eleventh centuries.
1300–1850	Little Ice Age, characterized by considerable variability and instability, with the coldest period from 1300 to 1500, followed by sudden warming from 1500 to 1675 affecting all of southern Africa, and a later cool phase from about 1675 to 1850.

880 to the present A warming period following the Little Ice Age.

The alternating runs of dry and wet years observed during the instrumental period have been associated with changes in the relative positions of the subtropical high pressure cells over the subcontinent and the South Atlantic and Indian Oceans (Tyson 1980). Climatic changes in the late Quaternary and Holocene can be explained in the same sort of way. Wet periods are seen as forced by variations in the tropical easterlies, associated with northerly meridional circulation anomalies, while dry periods result from increases in mid latitude westerly disturbances and southerly meridional circulation anomalies (Figure 14.22).

The characteristics of the climate in the last millennium revealed by field data provide support for these ideas. The Little Ice Age was evidently drier as well as cooler, yet broken by phases when temperature was rather higher and rainfall greater. Rain comes to Namibia with the tropical easterlies; weakening of these winds brings drought. The growth of a forest along the Hoanib River in Namibia during medieval warmth, implies strong easterlies and plentiful moisture. This forest was buried by the Amspoort silts in the course of the subsequent cooler period, when the easterlies were presumably weaker and the climate drier.

In Kenya, the Naivasha diatom record points to a severe drought from AD 1000 to 1270, during the Medieval Warm Period (Verschuren *et al.* 2000). Lower levels in lakes Malawi and Chilwa both in the last cool phase from around AD 1700 to 1850 and during the great nineteenth-century drought which afflicted South Africa are significant. So is the accordance of the Walvis Bay foraminifera, the Karkloof tree-ring sequence and the Cango Cave isotope record, which all show slightly warmer conditions around 1800 at a time when documentary sources reported widespread rain and floods, with lakes at high levels (Nicholson 1981). Conversely the same three records show somewhat lower temperatures around 1840, when rainfall was less (Vogel 1989) (Figure 14.21).

In southern Africa during the Little Ice Age, the tropical easterlies seem to have weakened and droughts were consequently more frequent. Westerlies were stronger and the greater frequency of cold episodes associated with westerly disturbances were accompanied by increased precipitation over an expanded southwestern winter rainfall region, and decreased rainfall over the northeastern summer rainfall region (Figure 14.22). It seems probable that changes in the global thermohaline circulation as well as in the atmospheric circulation may have been involved. The Pacific–Indian Ocean–Atlantic warm water conveyor belt may have weakened, and upwelling of cold water along the west coast of southern Africa may have increased when North Atlantic deep water formation slackened (see Chapter 17). A possible correlation has been noted in southern Africa between extremely dry periods, lower temperature and the El Niño/Southern Oscillation (Vogel 1989). More field data are needed to confirm present ideas about the Little Ice Age in southern Africa, but Tyson's hypothesis seems to provide a satisfactory explanation for most of the existing evidence (Tyson and Lindesay 1992, Cockcroft *et al.* 1987).

The warmth in southern Africa between 1500 and 1675 overlaps with a longer period of warmth in the Antarctic Peninsula. Ice cores from Siple, at the base of the Peninsula, indicate that conditions were warmer than present from 1600 to 1830. Records from the South Pole and east Antarctica indicate cooler and dustier conditions (Mosley-Thompson *et al.* 1990). Intensification of the zonal westerlies leads to cooling at the Amundson-Scott Base and warming in the Peninsula (Rogers 1983); a prolonged period of intensified westerlies could explain this opposition. Further investigations in the Antarctic offer the prospect of placing the climatic sequence in a broader context.

14.8 THE GLOBAL PERSPECTIVE

Bradley and Jones (1992c) published an account of climate since AD 1500 which was based on data from twenty-nine regional studies. Only seven of these were from the southern hemisphere, none from Africa south of the Sahara, and all except one were based on tree rings or ice cores. Taken together these twenty-nine data sets demonstrated that the climate of the last thousand years was characterized by complex anomalies, each lasting no longer than a few decades. No globally synchronous cold period was revealed but there were seen to have been at least two widespread cool periods in the last thousand years, one beginning about 1275 and the other around 1510. Within the latter a few

WET

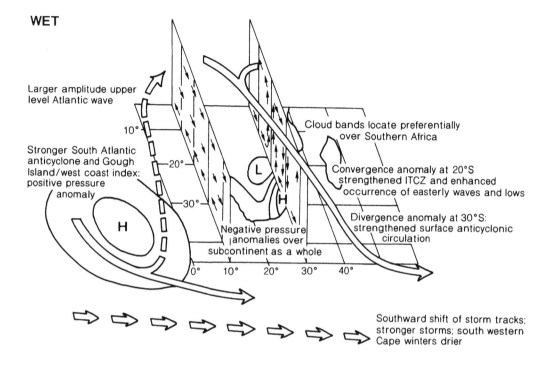

DRY

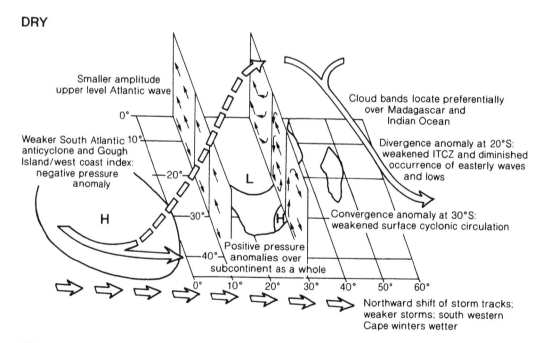

Figure 14.22 Tyson's (1986) models of anomalous meridional circulation over southern Africa during extended wet and dry spells. The relative positions of the upper tropospheric wave, preferred zones for cloud-band formation, the surface manifestation of the South Atlantic Anticyclone and location of westerly storm tracks are shown (*Source:* from Cockcroft *et al.* 1987)

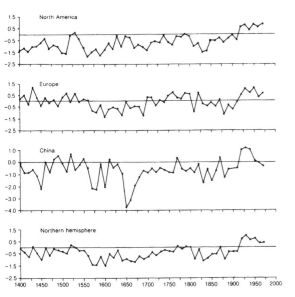

Figure 14.23 Composite temperature anomaly series for North America, Europe, China and the northern hemisphere. These figures were derived from normalized decadally averaged anomalies with reference to the period 1860–1959 (*Source:* from Bradley and Jones 1993)

short, cool episodes in the 1590s–1610s, the 1690s–1710s, the 1800s–1810s and the 1880s–1900s, were seen as synchronous on the hemispherical and global scale. Jones and Bradley concluded that the last five hundred years had not experienced an overall cold period, and that the coldest periods in one region were often not coincident with those in other regions.

The data series presented in Jones and Bradley (1992) vary in their resolution. Those covering the period from AD 1500 to the present and having a resolution of at least a decade were used to distinguish regions about which generalizations could be made (Figure 14.23) (Bradley and Jones 1993). The resultant composite northern hemisphere record indicated that there was a cool period corresponding with the Little Ice Age, but the cooling was less than 1 °C. The data were insufficient to produce a composite southern hemisphere record. Northern hemisphere temperatures fluctuated in such a way as to account for the glacial advances and retreats that have been identified, departures from the main trends being the result of regional differences in temperature and local variations in precipitation.

Bradley teamed up with Mann (Mann *et al.* 1995) and later with Hughes at Tree-Ring Research in Tucson (1998 and 1999) to produce graphs showing northern hemisphere temperatures over the last thousand years. The results played a prominent role in the sections dealing with the last millennium in 'Observed Climate Variability and Change' (chapter 2 of IPPC *Climate Change 2001*). The conclusion that global warming in the twentieth century as a whole, and above all in the 1990s, was greater than in any other century over the last thousand years conforms with the rapid retreat of non-surging glaciers all over the world, except in maritime Scandinavia and New Zealand where increased snowfall has been dominating glacier economies.

14.8.1 The instrumental record

The global rise in land surface temperatures between 1880 and 1998, as measured largely by instruments at weather stations, appears to have been 0.6 °C, equivalent to an increase in the equilibrium line altitude of glaciers worldwide of about 100 m. Most of the warming was concentrated in 1910–45 and 1976–98. The earlier warming was the greater one in northern hemisphere high latitudes. Then came a pause in the warming; for some reason not yet understood, global temperatures ceased to rise and even fell slightly between 1945 and 1976, though warming continued in much of the southern hemisphere and central Asia. Glacier termini responded to the temperature fluctuations with time lags varying from a few years in the case of small glaciers to a few decades in the case of medium-sized glaciers as diagrams in the earlier chapters indicate.

Around the Pacific and south of the Tropic of Cancer, the climate is strongly influenced by ENSO events. Increasing in strength and frequency after 1970, their effects on glaciers have varied regionally. In the Southern Alps of New Zealand and in parts of the southern Andes snowfall increased and temperatures fell at times of ENSO Low Index (El Niños), while in the northern Andes temperatures were higher during El Niños and snowfall less.

As global temperatures rose in the last two decades of the century, snow cover in the northern hemisphere contracted, the extent and thickness of Arctic sea ice diminished, and glaciers retreated even faster than they had in the first half of the century. In the Antarctic, ice shelves bordering the continent retreated throughout the twentieth century, but the sea ice cover slightly increased.

A feature of the change in temperature since 1950 has been a decrease in the diurnal temperature range in most parts of the world except northern Canada, eastern China and parts of Europe. Involving mainly an increase in night temperatures, it appears to be the result, at least in part, of increased cloudiness (IPPC 2001, figure 2.2).

The IPPC report *Climatic Change 2001* does not find that the evidence supports the existence of globally synchronous periods of cooling or warming associated with the 'Little Ice Age' and 'Medieval Warm Period'. Such a lack of synchroneity was also recognized in the first edition of *The Little Ice Age* (pp. 354–8). Nevertheless the report adds that 'reconstructed Northern Hemisphere temperatures do show a cooling during the fifteenth to nineteenth centuries and a relatively warm period during the eleventh to fourteenth centuries, though the latter period is still cooler than the late twentieth century'.

In *The Little Ice Age*, although the beginning of the main glacier expansion was seen as being a response to a sharp decline in summer temperatures about 1570, it was noted that 'in the European Alps the main advance was preceded by an advance of comparable magnitude in the decades around 1300' (p. 354). The positions of moraines formed in the Valais by thirteenth- and fourteenth-century advances were very close to those reached by glaciers in the middle of the nineteenth century. In the interval since the first edition appeared in 1988, radiocarbon dating of moraines associated with tree-ring studies has shown that the Little Ice Age was also under way by the thirteenth century or earlier in North America, New Zealand and parts of South America and Asia, as well as in the Swiss Alps (Grove 2001a, b, and earlier chapters here, especially pp. 159–60).

Large-scale estimates of decadal, annual or seasonal climatic variations must rely on sources that can resolve annual or seasonal climatic variations. Such proxy information includes width and density measurements from tree rings, layer thicknesses from sediment cores, isotope chemistry and accumulation from annually resolved ice cores and speleothems, isotopes from corals,

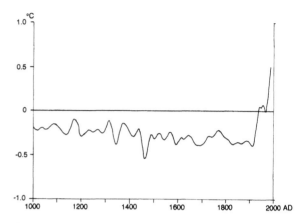

Figure 14.24 Millennial northern hemisphere temperature, from instrumental data 1900–99 and reconstruction from AD 1000–1900 (*Source:* from IPPC 2001)

and the sparse historical documentary evidence available over the globe during the past few centuries. 'Taken as a whole such proxy climate data can provide global scale sampling of climate variations several centuries into the past, with the potential to resolve large-scale patterns of climate change prior to the instrumental period, albeit with important limitations and uncertainties' (IPPC 2001: 130).

The millennial reconstruction of northern hemisphere temperature from AD 1000 to 1999 (Figure 14.24, taken from figure 2.20 in IPPC 2001, also corresponds with figure 3a in Mann *et al.* 1999) is a very striking diagram: it shows long-term cooling from 1000 to 1900 and then a sudden rise in temperature after 1900 as the age of industrial expansion gets into its stride. The cooling is tentatively attributed to astronomical forcing. 'The evidence', the IPPC report adds,

does not support globally-synchronous periods of anomalous cold or warmth over this timeframe, and the conventional terms of 'Little Ice Age' and 'Medieval Warm Period' appear to have limited utility in describing trends in hemispheric or global mean temperature changes in past centuries. With the more widespread proxy data and multi-proxy data and multi-proxy reconstructions of temperature change now available, the spatial and temporal character of these putative climatic epochs can be reassessed.

The Little Ice Age is seen in IPPC 2001 to have been most clearly expressed around the North Atlantic and especially in Eurasia, with cooling of about 1 °C

aused by more easterly winds associated with a negative North Atlantic Oscillation Index (higher pressure
over Iceland and lower pressure over the Azores). On
the other hand, the coolest part of the last millennium
in North America is seen to have been the nineteenth
century, with temperatures 1.5 °C colder than in the
late twentieth century. The evidence for temperature
changes in the southern hemisphere is described as quite
sparse, though 'glacier evidence from the Southern
Alps of New Zealand suggests cold conditions during
the mid-seventeenth and mid-nineteenth centuries'
(p. 135).

The Medieval Warm Period is also seen by the 2001
IPCC report as having been essentially a phenomenon
affecting regions around the North Atlantic, with northern hemisphere mean temperatures 'from the eleventh
to the fourteenth centuries about 0.2 °C warmer than
those from the fifteenth to nineteenth centuries, but
rather below mid twentieth century temperatures'.

The strongest cooling of the last millennium is somewhat surprisingly shown by Mann et al. (1999 and in
IPCC 2001) as having been in the second quarter of
the fifteenth century. In relatively few regions do we
find evidence of glaciers having been in advanced positions at this time. Mann's values for the pre-seventeenth
century are largely based on tree-ring records. The database shrinks considerably as we go further back in time
and reliance appears to have been placed on a relatively
small sample of records from around the northern
Pacific, the Arctic, coastal China and North America.

It has to be recognized that tree-ring records are
not readily interpreted to provide long-term, multicentennial temperature trends. As trees get older the
width of their annual rings diminishes and such growth
trends are removed by fitting a smooth mathematical
growth function. 'Thus, a 100-year-long tree-ring
series will not contain any climatic variance at periods
longer than 100 years if it is explicitly detrended by a
fitted growth curve' (Esper et al. 2002: 2250).

Esper et al. (2002) have interpreted tree-ring width
trends over the last 1200 years from fourteen sites, seven
of them around the Arctic Ocean and four in western
North America.[33] Their interpretation of the tree-ring
records gives very different long-term results from those

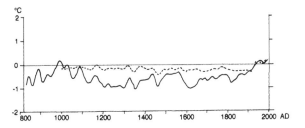

Figure 14.25 A comparison of northern hemisphere temperature
reconstructions from tree rings by Mann et al. (1999), dashed line,
and by Esper et al. (2002), solid line. The two are well related at
decadal to centennial timescales but disagree at multicentennial
timescales (Source: both taken from Esper et al. 2002)

obtained by Mann et al. (Figure 14.25). Over periods
of a few decades their temperature reconstructions are
not dissimilar, but the picture they present of temperature change over the last millennium is very different.
Esper and his colleagues have preserved possible growth
trends due to climate by dividing their data sets into
two groups, one with age trends with a 'linear' form
and the other more 'non-linear', and by using its own
mean biological growth curve for detrending each
group. The resultant tree-ring index curve and reconstructed temperature curve from AD 800 to 2000 are
very much closer to what one would expect from the
glacier evidence, much closer than the temperature
record reconstructed by Mann et al. (1999). In spite of
the smoothing, which conceals important short-term
variations, the Esper et al. curve depicts strong cooling
in the late thirteenth and late fourteenth centuries, a
cold seventeenth century, and early nineteenth-century
cooling. The Medieval Warm Period is evident and
appears as having been as warm as the late twentieth
century. Mann and Hughes (2002) criticize these
conclusions as depending on only six samples for the
period around AD 1000, and for not taking into account
tropical temperatures which are said to have varied less
than those in higher latitudes. Cook and Esper (2002)
replied that they have shown there were 'significant'
periods of above average growth (and inferred temperatures) during the Medieval Warm Period.

Tree rings, in spite of the above qualifications, are
generally assumed to be good indicators of summer

33 Two of the sites were used by Mann et al. (1999).

temperatures in the past. It is commonly concluded that they provide records of variations in annual mean temperature. However Jones *et al.* (2003) point out that for northern hemisphere land areas, from which most tree-ring records come, the winter season warmed by 1.05 °C between 1861 and 2000 while the summer season warmed by only 0.31 °C. In other words, the seasonal temperature range appears to have narrowed very markedly over the last 140 years. Furthermore, documentary sources show that the seasonal range has been greater than now in northwest Europe at least as far back as the fourteenth century. This suggests the possibility that the Little Ice Age was colder than tree rings have led us to suppose.

It may be concluded that although the climatic record for the northern hemisphere over the last 300 years is now known with some confidence, as Jones *et al.* (2001) emphasize, uncertainty increases for earlier centuries. They saw temperatures increasing from 1600 back to 1000, a view with which a good deal of glacial evidence conflicts (Grove 2001a, b). Such conflicts may be resolved as more information about past regional and seasonal variations in climate becomes available, and as a greater understanding is gained of the causal mechanisms for these variations. Glaciers and glacial moraines cannot provide annual climatic values but they can assist by providing a naturally smoothed record for regions where other data may be sparse.

Milton Keynes UK
Ingram Content Group UK Ltd.
UKHW051926141024
449569UK00027B/1378